AN INTRODUCTION TO
CELESTIAL MECHANICS

BY

FOREST RAY MOULTON

Second Revised Edition

D1248194

DOVER PUBLICATIONS, INC.
NEW YORK

Published in Canada by General Publishing Company, Ltd., 30 Lesmill Road, Don Mills, Toronto, Ontario.
Published in the United Kingdom by Constable and Company, Ltd.

This Dover edition, first published in 1970, is an unabridged and corrected republication of the second revised edition, 1914, of the work originally published in 1902 by The Macmillan Company.

The publisher gratefully acknowledges the cooperation of the University of Cincinnati Library, which kindly supplied a copy of this work for the purpose of reproduction.

International Standard Book Number: 0-486-64687-4
Library of Congress Catalog Card Number: 79-103400

Manufactured in the United States of America
Dover Publications, Inc.
31 East 2nd Street
Mineola, N.Y. 11501

PREFACE TO FIRST EDITION.

A N attempt has been made in this volume to give a somewhat satisfactory account of many parts of Celestial Mechanics rather than an exhaustive treatment of any special part. The aim has been to present the work so as to attain logical sequence, to make it progressively more difficult, and to give the various subjects the relative prominence which their scientific and educational importance deserves. In short, the aim has been to prepare such a book that one who has had the necessary mathematical training may obtain from it in a relatively short time and by the easiest steps a sufficiently broad and just view of the whole subject to enable him to stop with much of real value in his possession, or to pursue to the best advantage any particular portion he may choose.

In carrying out the plan of this work it has been necessary to give an introduction to the Problem of Three Bodies. This is not only one of the justly celebrated problems of Celestial Mechanics, but it has become of special interest in recent times through the researches of Hill, Poincaré, and Darwin. The theory of absolute perturbations is the central subject in mathematical Astronomy, and such a work as this would be inexcusably deficient if it did not give this theory a prominent place. A chapter has been devoted to geometrical considerations on perturbations. Although these methods are of almost no use in computing, yet they furnish in a simple manner a clear insight into the nature of the problem, and are of the highest value to beginners. The fundamental principles of the analytical methods have been given with considerable completeness, but many of the details in developing the formulas have been omitted in order that the size of the book might not defeat the object for which it has been prepared. The theory of orbits has not been given the unduly prominent position which it has occupied in this country, doubtless due to the influence of Watson's excellent treatise on this subject.

The method of treatment has been to state all problems in advance and, where the transformations are long, to give an outline of the steps which are to be made. The expression "order of small quantities" has not been used except when applied to power series in explicit parameters, thus giving the work all the definiteness and simplicity which are characteristic of operations with power series. This is exemplified particularly in the chapter on perturbations. Care has been taken to make note at all places where assumptions have been introduced or unjustified methods employed, for it is only by seeing where the points of possible weakness are that improvements can be made. The frequent references throughout the text and the bibliographies at the ends of the chapters, though by no means exhaustive, are sufficient to direct one in further reading to important sources of information.

This volume is the outgrowth of a course of lectures given annually by the author at the University of Chicago during the last six years. These lectures have been open to senior college students and to graduate students who have not had the equivalent of this work. They have been taken by students of Astronomy, by many making Mathematics their major work, and by some who, though specializing in quite distinct lines, have desired to get an idea of the processes by means of which astronomers interpret and predict celestial phenomena. Thus they have served to give many an idea of the methods of investigation and the results attained in Celestial Mechanics, and have prepared some for a detailed study extending into the various branches of modern investigations. The object of the work, the subjects covered, and the methods of treatment seem to have been amply justified by this experience.

Mr. A. C. Lunn, M.A., has read the entire manuscript with great care and a thorough insight into the subjects treated. His numerous corrections and suggestions have added greatly to the accuracy and the method of treatment in many places. Professor Ormond Stone has read the proofs of the first four chapters and the sixth. His experience as an investigator and as a teacher has made his criticisms and suggestions invaluable. Mr. W. O. Beal, M.A., has read the proofs of the whole book with great attention and he is responsible for many improvements. The author desires to express his sincerest thanks to all these gentlemen for the willingness and the effectiveness with which they have devoted so much of their time to this work.

F. R. MOULTON.

CHICAGO, July, 1902.

PREFACE TO SECOND EDITION.

THE necessity for a new edition of this work has given the opportunity of thoroughly revising it. The general plan which has been followed is the same as that of the first edition, because it was found that it satisfies a real need not only in this country, for whose students it was primarily written, but also in Europe. In spite of all temptations its elementary character has been preserved, and it has not been greatly enlarged. Very many improvements have been made, partly on the suggestion of numerous astronomers and mathematicians, and it is hoped that it will be found more worthy of the favor with which it has so far been received.

The most important single change is in the discussion of the methods of determining orbits. This subject logically follows the Problem of Two Bodies, and it is much more elementary in character than the Problem of Three Bodies and the Theory of Perturbations. For these reasons it was placed in chapter VI. The subject matter has also been very much changed. The methods of Laplace and Gauss, on which all other methods of general applicability are more or less directly based, are both given. The standard modes of presentation have not been followed because, however well they may be adapted to practice, they are not noted for mathematical clarity. Besides, there is no lack of excellent works giving details in the original forms and models of computation. The other changes and additions of importance are in the chapters on the Problem of Two Bodies, the Problem of Three Bodies, and in that on Geometrical Consideration of Perturbations.

It is a pleasure to make special acknowledgment of assistance to my colleague Professor W. D. MacMillan and to Mr. L. A. Hopkins who have read the entire proofs not only once but several times, and who have made important suggestions and have pointed out many defects that otherwise would have escaped notice. They are largely responsible for whatever excellence of form the book may possess.

F. R. MOULTON.

CHICAGO, January, 1914.

TABLE OF CONTENTS.

CHAPTER I.

FUNDAMENTAL PRINCIPLES AND DEFINITIONS.

CHAPTER II.

RECTILINEAR MOTION.

CHAPTER III.

CENTRAL FORCES.

CHAPTER IV.

THE POTENTIAL AND ATTRACTIONS OF BODIES.

CHAPTER V.

THE PROBLEM OF TWO BODIES.

CHAPTER VI.

THE DETERMINATION OF ORBITS.

CHAPTER VII.

THE GENERAL INTEGRALS OF THE PROBLEM OF n BODIES.

CHAPTER VIII.

THE PROBLEM OF THREE BODIES.

CHAPTER IX.

PERTURBATIONS—GEOMETRICAL CONSIDERATIONS.

CHAPTER X.

PERTURBATIONS—ANALYTICAL METHOD.

CHAPTER I.

FUNDAMENTAL PRINCIPLES AND DEFINITIONS.

1. Elements and Laws. The problems of every science are expressible in certain terms which will be designated as *elements*, and depend upon certain *principles* and *laws* for their solution. The elements arise from the very nature of the subject considered, and are expressed or implied in the formulation of the problems treated. The principles and laws are the relations which are known or are assumed to exist among the various elements. They are inductions from experiments, or deductions from previously accepted principles and laws, or simply agreements.

An explicit statement in the beginning of the type of problems which will be treated, and an enumeration of the elements which they involve, and of the principles and laws which relate to them, will lead to clearness of exposition. In order to obtain a complete understanding of the character of the conclusions which are reached, it would be necessary to make a philosophical discussion of the reality of the elements, and of the origin and character of the principles and laws. These questions cannot be entered into here because of the difficulty and complexity of metaphysical speculations. It is not to be understood that such investigations are not of value; they forever lead back, to simpler and more undeniable assumptions upon which to base all reasoning.

The method of procedure in this work will necessarily be to accept as true certain fundamental elements and laws without entering in detail into the questions of their reality or validity. It will be sufficient to consider whether they are definitions or have been inferred from experience, and to point out that they have been abundantly verified in their applications. They will be accepted with confidence, and their consequences will be derived, in the subjects treated, so far as the scope and limits of the work will allow.

1

2. Problems Treated. The motions of a material particle subject to a central force of any sort whatever will be briefly considered. It will be shown from the conclusions reached in this discussion, and from the observed motions of the planets and their satellites, that Newton's law of gravitation holds true in the solar system. The character of the motion of the binary stars shows that the probabilities are very great that it operates in them also, and that it may well be termed "the law of universal gravitation." This conclusion is confirmed by the spectroscope, which proves that the familiar chemical elements of our solar system exist in the stars also.

In particular, the motions of two free homogeneous spheres subject only to their mutual attractions and starting from arbitrary initial conditions will be investigated, and then their motions will be discussed when they are subject to disturbing influences of various sorts. The essential features of perturbations arising from the action of a third body will be developed, both from a geometrical and an analytical point of view. There are two somewhat different cases. One is that in which the motion of a satellite around a planet is perturbed by the sun; and the second is that in which the motion of one planet around the sun is perturbed by another planet.

Another class of problems which arises is the determination of the orbits of unknown bodies from the observations of their directions at different epochs, made from a body whose motion is known. That is, the theories of the orbits of comets and planetoids will be based upon observations of their apparent positions made from the earth. This incomplete outline of the questions to be treated is sufficient for the enumeration of the elements employed.

3. Enumeration of the Principal Elements. In the discussion of the problems considered in this work it will be necessary to employ the following elements:

(a) Real *numbers*, and complex numbers incidentally in the solution of certain problems.

(b) *Space* of three dimensions, possessing the same properties in every direction.

(c) *Time* of one dimension, which will be taken as the independent variable.

(d) *Mass*, having the ordinary properties of inertia, etc., which are postulated in elementary Physics.

(e) *Force*, with the content that the same term has in Physics.
Positive numbers arise in Arithmetic, and positive, negative,
and complex numbers, in Algebra. Space appears first as an
essential element in Geometry. Time appears first as an essential
element in Kinematics. Mass and force appear first and must be
considered as essential elements in physical problems. No definitions of these familiar elements are necessary here.

4. Enumeration of the Principles and Laws. In representing
the various physical magnitudes by numbers, certain agreements
must be made as to what shall be considered positive, and what
negative. The axioms of ordinary Geometry will be considered
as being true.

The fundamental principles upon which all work in Theoretical
Mechanics may be made to depend are Newton's three *Axioms*, or
Laws of Motion. The first two laws were known by Galileo and
Huyghens, although they were for the first time announced
together in all their completeness by Newton in the *Principia*,
in 1686. These laws are as follows:*

Law I. *Every body continues in its state of rest, or of uniform
motion in a straight line, unless it is compelled to change that state by
a force impressed upon it.*

Law II. *The rate of change of motion is proportional to the force
impressed, and takes place in the direction of the straight line in which
the force acts.*

Law III. *To every action there is an equal and opposite reaction;
or, the mutual actions of two bodies are always equal and oppositely
directed.*

5. Nature of the Laws of Motion. Newton calls the Laws of
Motion *Axioms*, and after giving each, makes a few remarks concerning its import. Later writers, among whom are Thomson and
Tait,† regard them as inferences from experience, but accept Newton's formulation of them as practically final, and adopt them
in the precise form in which they were given in the *Principia*. A
number of Continental writers, among whom is Dr. Ernest Mach,
have given profound thought to the fundamental principles of

---

* Other fundamental laws may be, and indeed have been, employed; but
they involve more difficult mathematical principles at the very start. They
are such as d'Alembert's principle, Hamilton's principle, and the systems of
Kirchhoff, Mach, Hertz, Boltzmann, etc.

† *Natural Philosophy*, vol. I., Art. 243.

Mechanics, and have concluded that they are not only *inductions* or simply *conventions*, but that Newton's statement of them is somewhat redundant, and lacks scientific directness and simplicity. There is no suggestion, however, that Newton's Laws of Motion are not in harmony with ordinary astronomical experience, or that they cannot be made the basis for Celestial Mechanics. But in some branches of Physics, particularly in Electricity and Light, certain phenomena are not fully consistent with the Newtonian principles, and they have recently led Einstein and others to the development of the so-called *Principle of Relativity*. The astronomical consequences of this modification of the principles of Mechanics are very slight unless the time under consideration is very long, and, whether they are true or not, they cannot be considered in an introduction to the subject.

6. Remarks on the First Law of Motion. In the first law the statement that a body subject to no forces moves with *uniform* motion, may be regarded as a definition of *time*. For, otherwise, it is implied that there exists some method of measuring time in which motion is not involved. Now it is a fact that in all the devices actually used for measuring time this part of the law is a fundamental assumption. For example, it is assumed that the earth rotates at a uniform rate because there is no force acting upon it which changes the rotation sensibly.*

The second part of the law, which affirms that the motion is in a *straight line* when the body is subject to no forces, may be taken as defining a straight line, if it is assumed that it is possible to determine when a body is subject to no forces; or, it may be taken as showing, together with the first part, whether or not forces are acting, if it is assumed that it is possible to give an independent definition of a straight line. Either alternative leads to troublesome difficulties when an attempt is made to employ strict and consistent definitions.

7. Remarks on the Second Law of Motion. In the second law the statement that the rate of change of motion is proportional to the force impressed, may be regarded as a definition of the relation between force and matter by means of which the magnitude of a force, or the amount of matter in a body can be measured, according as one or the other is supposed to be independently known. By rate of change of motion is meant the rate of change of velocity

* See memoir by R. S. Woodward, *Astronomical Journal*, vol. XXI. (1901).

multiplied by the mass of the body moved. This is usually called the rate of *change of momentum,* and the ideas of the second law may be expressed by saying, *the rate of change of momentum is proportional to the force impressed and takes place in the direction of the straight line in which the force acts.* Or, *the acceleration of motion of a body is directly proportional to the force to which it is subject, and inversely proportional to its mass, and takes place in the direction in which the force acts.*

It may appear at first thought that force can be measured without reference to velocity generated, and it is true in a sense. For example, the force with which gravity draws a body downward is frequently measured by the stretching of a coiled spring, or the intensity of magnetic action can be measured by the torsion of a fiber. But it will be noticed in all cases of this kind that the law of reaction of the machine has been determined in some other way. This may not have been directly by velocities generated, but it ultimately leads back to it. It is worthy of note in this connection that all the units of absolute force, as the *dyne,* contain explicitly in their definitions the idea of velocity generated.

In the statement of the second law it is implied that the effect of a force is exactly the same in whatever condition of rest or of motion the body may be, and to whatever other forces it may be subject. The change of motion of a body acted upon by a number of forces is the same at the end of an interval of time as if each force acted separately for the same time. Hence the implication in the second law is, *if any number of forces act simultaneously on a body, whether it is at rest or in motion, each force produces the same total change of momentum that it would produce if it alone acted on the body at rest.* It is apparent that this principle leads to great simplifications of mechanical problems, for in accordance with it the effects of the various forces can be considered separately.

Newton derived the *parallelogram of forces* from the second law of motion.* He reasoned that as forces are measured by the accelerations which they produce, the resultant of, say, two forces should be measured by the resultant of their accelerations. Since an acceleration has magnitude and direction it can be represented by a directed line, or *vector.* The resultant of two forces will then be represented by the diagonal of a parallelogram, of which two adjacent sides represent the two forces.

---

* *Principia,* Cor. i. to the Laws of Motion.

One of the most frequent applications of the parallelogram of forces is in the subject of Statics, which, in itself, does not involve the ideas of motion and time.   In it the idea of mass can also be entirely eliminated.   Newton's proof of the parallelogram of forces has been objected to on the ground that it requires the introduction of the fundamental conceptions of a much more complicated science than the one in which it is employed.   Among the demonstrations which avoid this objectionable feature is one due to Poisson,* which has for its fundamental assumption the axiom that the resultant of two equal forces applied at a point is in the line of the bisector of the angle which they make with each other.   Then the magnitude of the resultant is derived, and by simple processes the general law is obtained.

8. Remarks on the Third Law of Motion.   The first two of Newton's laws are sufficient for the determination of the motion of one body subject to any number of known forces; but another principle is needed when the investigation concerns the motion of a system of two or more bodies subject to their mutual interactions. The third law of motion expresses precisely this principle.   It is that if one body presses against another the second resists the action of the first with the same force.   And also, though it is not so easy to conceive of it, if one body acts upon another through any distance, the second reacts upon the first with an equal and oppositely directed force.

Suppose one can exert a given force at will; then, by the second law of motion, the relative masses of bodies can be measured since they are inversely proportional to the accelerations which equal forces generate in them.   When their relative masses have been found the third law can be tested by permitting the various bodies to act upon one another and measuring their relative accelerations.   Newton made several experiments to verify the law, such as measuring the rebounds from the impacts of elastic bodies, and observing the accelerations of magnets floating in basins of water.†
The chief difficulty in the experiments arises in eliminating forces external to the system under consideration, and evidently they cannot be completely removed.   Newton also concluded from a certain course of reasoning that to deny the third law would be to contradict the first.†

Mach points out that there is no accurate means of measuring

* *Traité de Mécanique*, vol. I., p. 45 *et seq.*
† *Principia*, Scholium to the Laws of Motion.

forces except by the accelerations they produce in masses, and therefore that effectively the reasoning in the preceding paragraph is in a circle. He objects also to Newton's definition that mass is proportional to the product of the volume and the density of a body. He prefers to rely upon experience for the fact that two bodies which act upon each other produce oppositely directed accelerations, and to *define* the relative values of the masses as inversely proportional to these accelerations. Experience proves further that if the relative masses of two bodies are determined by their interactions with a third, the ratio is the same whatever the third mass may be. In this way, when one body is taken as the unit of mass, the masses of all other bodies can be uniquely determined. These views have much to commend them.

In the Scholium appended to the Laws of Motion Newton made some remarks concerning an important feature of the third law. This was first stated in a manner in which it could actually be expressed in mathematical symbols by d'Alembert in 1742, and has ever since been known by his name.* It is essentially this: When a body is subject to an acceleration, it may be regarded as exerting a force which is equal and opposite to the force by which the acceleration is produced. This may be considered as being true whether the force arises from another body forming a system with the one under consideration, or has its source exterior to the system. In general, in a system of any number of bodies, the resultants of all the applied forces are equal and opposite to the reactions of the respective bodies. In other words, the *impressed* forces and the reactions, or the *expressed* forces, form systems which are in equilibrium for each body and for the whole system. This makes the whole science of Dynamics, in form, one of Statics, and formulates the conditions so that they are expressible in mathematical terms. This phrasing of the third law of motion has been made the starting point for the elegant and very general investigations of Lagrange in the subject of Dynamics.†

The primary purpose of fundamental principles in a science is to coördinate the various phenomena by stating in what respects their modes of occurring are common; the value of fundamental principles depends upon the completeness of the coördination of the phenomena, and upon the readiness with which they lead to the discovery of unknown facts; the characteristics of funda-

---

* See Appell's *Mécanique*, vol. II., chap. XXIII.
† *Collected Works*, vols. XI. and XII.

mental principles should be that they are self-consistent, that they are consistent with every observed phenomenon, and that they are simple and not redundant.

Newton's laws coördinate the phenomena of the mechanical sciences in a remarkable manner, while their value in making discoveries is witnessed by the brilliant achievements in the physical sciences in the last two centuries compared to the slow and uncertain advances of all the ancients. They have not been found to be mutually contradictory, and they are consistent with nearly all the phenomena which have been so far observed; they are conspicuous for their simplicity, but it has been claimed by some that they are in certain respects redundant. One naturally wonders whether they are primary and fundamental laws of nature, even as modified by the principle of relativity. In view of the past evolution of scientific and philosophical ideas one should be slow in affirming that any statement represents ultimate and absolute truth. The fact that several other sets of fundamental principles have been made the bases of systems of mechanics, points to the possibility that perhaps some time the Newtonian system, or the Newtonian system as modified by the principle of relativity, even though it may not be found to be in error, will be supplanted by a simpler one even in elementary books.

Definitions and General Equations.

9. Rectilinear Motion, Speed, Velocity. A particle is in *rectilinear motion* when it always lies in the same straight line, and when its distance from a point in that line varies with the time. It moves with *uniform speed* if it passes over equal distances in equal intervals of time, whatever their length. The speed is represented by a positive number, and is measured by the distance passed over in a unit of a time. The *velocity* of a particle is the directed speed with which it moves, and is positive or negative according to the direction of the motion. Hence in uniform motion the velocity is given by the equation

$$(1) \qquad\qquad v = \frac{s}{t}.$$

Since s may be positive or negative, v may be positive or negative, and the speed is the numerical value of v. The same value of v is obtained whatever interval of time is taken so long as the corresponding value of s is used.

The speed and velocity are *variable* when the particle does not describe equal distances in equal times; and it is necessary to define in this case what is meant by the speed and velocity at an instant. Suppose a particle passes over the distance Δs in the time Δt, and suppose the interval of time Δt approaches the limit zero in such a manner that it always contains the instant t. Suppose, further, that for every Δt the corresponding Δs is taken. Then the velocity at the instant t is defined as

$$(2) \qquad v = \lim_{\Delta t=0} \left(\frac{\Delta s}{\Delta t} \right) = \frac{ds}{dt},$$

and the speed is the numerical value of $\frac{ds}{dt}$.

Uniform and variable velocity may be defined analytically in the following manner. The distance s, counted from a fixed point, is considered as a function of the time, and may be written

$$s = \phi(t).$$

Then the velocity may be defined by the equation

$$v = \frac{ds}{dt} = \phi'(t),$$

where $\phi'(t)$ is the derivative of $\phi(t)$ with respect to t. The velocity is said to be constant, or uniform, if $\phi'(t)$ does not vary with t; or, in other words, if $\phi(t)$ involves t linearly in the form $\phi(t) = at + b$, where a and b are constants. It is said to be variable if the value of $\phi'(t)$ changes with t.

Some agreement must be made to denote the direction of motion. An arbitrary point on the line may be taken as the origin and the distances to the right counted as positive, and those to the left, negative. With this convention, if the value of s determining the position of the body increases algebraically with the time the velocity will be taken positive; if the value of s decreases as the time increases the velocity will be taken negative. Then, when v is given in magnitude and sign, the speed and direction of motion are determined.

10. Acceleration in Rectilinear Motion. Acceleration is the rate of change of velocity, and may be constant or variable. Since the case when it is variable includes that when it is constant, it will be sufficient to consider the former. The definition of acceleration at an instant t is similar to that for velocity, and is, if the acceleration is denoted by α,

$$(3) \qquad \alpha = \lim_{\Delta t = 0} \left(\frac{\Delta v}{\Delta t} \right) = \frac{dv}{dt}.$$

By means of (2) and (3) it follows that

$$(4) \qquad \alpha = \frac{d}{dt} \left(\frac{ds}{dt} \right) = \frac{d^2 s}{dt^2}.$$

There must be an agreement regarding the sign of the acceleration. When the velocity increases algebraically as the time increases, the acceleration will be taken positive; when the velocity decreases algebraically as the time increases, the acceleration will be taken negative.

11. Speed and Velocities in Curvilinear Motion. The speed with which a particle moves is the rate at which it describes a curve. If v represents the speed in this case, and s the arc of the curve, then

$$(5) \qquad v = \left| \frac{ds}{dt} \right|,$$

where $\left| \dfrac{ds}{dt} \right|$ represents the numerical value of $\dfrac{ds}{dt}$. As before, the velocity is the directed speed possessing the properties of vectors, and may be represented by a vector.* The vector can be resolved uniquely into three components parallel to any three coördinate axes; and conversely, the three components can be compounded uniquely into the vector. In other words, if the velocity is given, the components parallel to any coördinate axes are defined; and the components parallel to any non-coplanar coördinate axes define the velocity. It is generally simplest to use rectangular axes and to employ the components of velocity parallel to them. Let λ, μ, ν represent the angles between the line of motion and the x, y, and z-axes respectively. Then

$$(6) \qquad \cos \lambda = \frac{dx}{ds}, \quad \cos \mu = \frac{dy}{ds}, \quad \cos \nu = \frac{dz}{ds}.$$

Let v_x, v_y, v_z represent the components of velocity along the three axes. That is,

* Consult Appell's *Mécanique*, vol. i., p. 45 *et seq.*

$$(7) \quad \begin{cases} v_x = v \cos \lambda = \dfrac{ds}{dt}\dfrac{dx}{ds} = \dfrac{dx}{dt}, \\[2ex] v_y = v \cos \mu = \dfrac{ds}{dt}\dfrac{dy}{ds} = \dfrac{dy}{dt}, \\[2ex] v_z = v \cos \nu = \dfrac{ds}{dt}\dfrac{dz}{ds} = \dfrac{dz}{dt}. \end{cases}$$

From these equations it follows that

$$(8) \quad v = \sqrt{\left(\frac{dx}{dt}\right)^2 + \left(\frac{dy}{dt}\right)^2 + \left(\frac{dz}{dt}\right)^2}.$$

There must be an agreement as to a positive and a negative direction along each of the three coördinate axes.

12. Acceleration in Curvilinear Motion. As in the case of velocities, it is simplest to resolve the acceleration into component accelerations parallel to the coördinate axes. On constructing a notation corresponding to that used in Art. 11, the following equations result:

$$(9) \quad \alpha_x = \frac{d^2x}{dt^2}, \qquad \alpha_y = \frac{d^2y}{dt^2}, \qquad \alpha_z = \frac{d^2z}{dt^2}.$$

The numerical value of the whole acceleration is

$$(10) \quad \alpha = \sqrt{\left(\frac{d^2x}{dt^2}\right)^2 + \left(\frac{d^2y}{dt^2}\right)^2 + \left(\frac{d^2z}{dt^2}\right)^2}.$$

This is not, in general, equal to the component of acceleration along the curve; that is, to $\dfrac{d^2s}{dt^2}$. For, from (8) it follows that

$$v = \frac{ds}{dt} = \sqrt{\left(\frac{dx}{dt}\right)^2 + \left(\frac{dy}{dt}\right)^2 + \left(\frac{dz}{dt}\right)^2};$$

whence, by differentiation,

$$(11) \quad \begin{aligned} \frac{d^2s}{dt^2} &= \frac{\dfrac{dx}{dt}\dfrac{d^2x}{dt^2} + \dfrac{dy}{dt}\dfrac{d^2y}{dt^2} + \dfrac{dz}{dt}\dfrac{d^2z}{dt^2}}{\sqrt{\left(\dfrac{dx}{dt}\right)^2 + \left(\dfrac{dy}{dt}\right)^2 + \left(\dfrac{dz}{dt}\right)^2}} \\[3ex] &= \frac{dx}{ds}\frac{d^2x}{dt^2} + \frac{dy}{ds}\frac{d^2y}{dt^2} + \frac{dz}{ds}\frac{d^2z}{dt^2}. \end{aligned}$$

Thus, when the components of acceleration are known, the whole acceleration is given by (10), and the acceleration along the curve by (11). The fact that the two are different, in general, may cause some surprise at first thought. But the matter becomes clear if a body moving in a circle with constant speed is considered. The acceleration along the curve is zero because the speed is supposed not to change; but the acceleration is not zero because the body does not move in a straight line.

13. The Components of Velocity Along and Perpendicular to the Radius Vector. Suppose the path of the particle is in the xy-plane, and let the polar coördinates be r and θ. Then

$$(12) \qquad x = r \cos \theta, \quad y = r \sin \theta.$$

The components of velocity are therefore

$$(13) \qquad \begin{cases} \dfrac{dx}{dt} = v_x = -r \sin \theta \dfrac{d\theta}{dt} + \cos \theta \dfrac{dr}{dt}, \\[3mm] \dfrac{dy}{dt} = v_y = +r \cos \theta \dfrac{d\theta}{dt} + \sin \theta \dfrac{dr}{dt}. \end{cases}$$

Let QP be an arc of the curve described by the moving particle. When the particle is at P, it is moving in the direction PV, and the velocity may be represented by the vector PV. Let v_r and v_θ

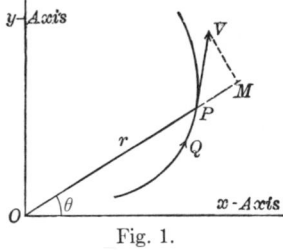

Fig. 1.

represent the components of velocity along and perpendicular to the radius vector. The resultant of the vectors v_r and v_θ is equal to the resultant of the vectors $\dfrac{dx}{dt}$ and $\dfrac{dy}{dt}$, that is, to PV. The sum of the projections of v_r and v_θ upon any line equals the sum of the projections of $\dfrac{dx}{dt}$ and $\dfrac{dy}{dt}$ upon the same line. Therefore, projecting v_r and v_θ upon the x and y-axes, it follows that

$$(14) \qquad \begin{cases} \dfrac{dx}{dt} = v_r \cos \theta - v_\theta \sin \theta, \\[3mm] \dfrac{dy}{dt} = v_r \sin \theta + v_\theta \cos \theta. \end{cases}$$

On comparing (13) and (14), the required components of velocity are found to be

$$(15) \qquad \begin{cases} v_r = \dfrac{dr}{dt}, \\[3mm] v_\theta = r\dfrac{d\theta}{dt}. \end{cases}$$

The square of the speed is

$$v_r{}^2 + v_\theta{}^2 = \left(\frac{dr}{dt}\right)^2 + r^2\left(\frac{d\theta}{dt}\right)^2.$$

The components of velocity, v_r and $v\theta$, can be found in terms of the components parallel to the x and y-axes by multiplying equations (14) by $\cos\theta$ and $\sin\theta$ respectively and adding, and then by $-\sin\theta$ and $\cos\theta$ and adding. The results are

$$(16) \qquad \begin{cases} v_r = +\cos\theta\,\dfrac{dx}{dt} + \sin\theta\,\dfrac{dy}{dt}, \\[3mm] v_\theta = -\sin\theta\,\dfrac{dx}{dt} + \cos\theta\,\dfrac{dy}{dt}. \end{cases}$$

14. The Components of Acceleration. The derivatives of equations (13) are

$$(17) \qquad \begin{cases} \alpha_x = \dfrac{d^2x}{dt^2} = \left[\dfrac{d^2r}{dt^2} - r\left(\dfrac{d\theta}{dt}\right)^2\right]\cos\theta - \left[r\dfrac{d^2\theta}{dt^2} + 2\dfrac{dr}{dt}\dfrac{d\theta}{dt}\right]\sin\theta, \\[4mm] \alpha_y = \dfrac{d^2y}{dt^2} = \left[r\dfrac{d^2\theta}{dt^2} + 2\dfrac{dr}{dt}\dfrac{d\theta}{dt}\right]\cos\theta + \left[\dfrac{d^2r}{dt^2} - r\left(\dfrac{d\theta}{dt}\right)^2\right]\sin\theta. \end{cases}$$

Let α_r and α_θ represent the components of acceleration along and perpendicular to the radius vector. As in Art. 13, it follows from the composition and resolution of vectors that

$$(18) \qquad \begin{cases} \alpha_x = \alpha_r\cos\theta - \alpha_\theta\sin\theta, \\[2mm] \alpha_y = \alpha_r\sin\theta + \alpha_\theta\cos\theta. \end{cases}$$

On comparing (17) and (18), it is found that

$$(19) \qquad \begin{cases} \alpha_r = \dfrac{d^2r}{dt^2} - r\left(\dfrac{d\theta}{dt}\right)^2, \\[4mm] \alpha_\theta = r\dfrac{d^2\theta}{dt^2} + 2\dfrac{dr}{dt}\dfrac{d\theta}{dt} = \dfrac{1}{r}\dfrac{d}{dt}\left(r^2\dfrac{d\theta}{dt}\right). \end{cases}$$

The components of acceleration along and perpendicular to the radius vector in terms of the components parallel to the x and y-axes are found from (17) to be

(20)
$$\begin{cases} \alpha_r = + \cos \theta \frac{d^2x}{dt^2} + \sin \theta \frac{d^2y}{dt^2}, \\ \alpha_\theta = - \sin \theta \frac{d^2x}{dt^2} + \cos \theta \frac{d^2y}{dt^2}. \end{cases}$$

By similar processes the components of velocity and acceleration parallel to any lines can be found.

15. Application to a Particle Moving in a Circle with Uniform Speed. Suppose the particle moves with uniform speed in a circle around the origin as center; it is required to determine the com-

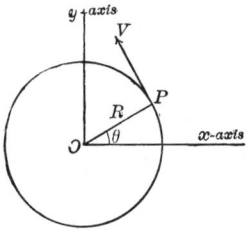

Fig. 2.

ponents of velocity and acceleration parallel to the x and y-axes, and parallel and perpendicular to the radius. Let R represent the radius of the circle; then

$$x = R \cos \theta, \qquad y = R \sin \theta.$$

Since the speed is uniform the angle θ is proportional to the time, or $\theta = ct$. The coördinates become

(21) $$x = R \cos (ct), \qquad y = R \sin (ct).$$

Since $\frac{d\theta}{dt} = c$ and $\frac{dR}{dt} = 0$, the components of velocity parallel to the x and y-axes are found from (13) to be

(22) $$v_x = - Rc \sin (ct), \qquad v_y = Rc \cos (ct).$$

From (15) it is found that

(23) $$v_r = 0, \qquad v_\theta = Rc.$$

The components of acceleration parallel to the x and y-axes, which are given by (17), are

$$(24) \qquad \begin{cases} \alpha_x = - Rc^2 \cos (ct), \\ \alpha_y = - Rc^2 \sin (ct). \end{cases}$$

And from (19) it is found that

$$(25) \qquad \alpha_r = - Rc^2, \qquad \alpha_\theta = 0.$$

It will be observed that, although the speed is uniform in this case, the velocity with respect to fixed axes is not constant, and the acceleration is not zero. If it is assumed that an exterior force is the only cause of the change of motion, or of acceleration of a particle, then it follows that a particle cannot move in a circle with uniform speed unless it is subject to some force. It follows also from (25) and the second law of motion that the force continually acts in a line which passes through the center of the circle.

16. The Areal Velocity. The rate at which the radius vector from a fixed point to the moving particle describes a surface is

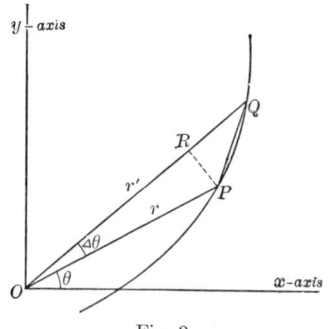

Fig. 3.

called the *areal velocity* with respect to the point. Suppose the particle moves in the xy-plane. Let ΔA represent the area of the triangle OPQ swept over by the radius vector in the interval of time Δt. Then

$$\Delta A = \frac{r' \cdot r}{2} \sin (\Delta\theta);$$

whence

$$(26) \qquad \frac{\Delta A}{\Delta t} = \frac{r' \cdot r}{2} \cdot \frac{\sin (\Delta\theta)}{\Delta\theta} \cdot \frac{\Delta\theta}{\Delta t}.$$

As the angle $\Delta\theta$ diminishes the ratio of the area of the triangle to that of the sector approaches unity as a limit. The limit of r' is r, and the limit of $\dfrac{\sin (\Delta\theta)}{\Delta\theta}$ is unity. Equation (26) gives, on passing to the limit $\Delta t = 0$ in both members,

(27)
$$\frac{dA}{dt} = \frac{1}{2} r^2 \frac{d\theta}{dt}$$

as the expression for the areal velocity. On changing to rectangular coördinates by the substitution

$$r = \sqrt{x^2 + y^2}, \qquad \tan \theta = \frac{y}{x},$$

equation (27) becomes

(28)
$$\frac{dA}{dt} = \frac{1}{2} \left(x \frac{dy}{dt} - y \frac{dx}{dt} \right).$$

If the motion is not in the xy-plane the projections of the areal velocity upon the three fundamental planes are used. They are respectively

(29)
$$\begin{cases} \dfrac{dA_{xy}}{dt} = \dfrac{1}{2} \left(x \dfrac{dy}{dt} - y \dfrac{dx}{dt} \right), \\[2mm] \dfrac{dA_{yz}}{dt} = \dfrac{1}{2} \left(y \dfrac{dz}{dt} - z \dfrac{dy}{dt} \right), \\[2mm] \dfrac{dA_{zx}}{dt} = \dfrac{1}{2} \left(z \dfrac{dx}{dt} - x \dfrac{dz}{dt} \right). \end{cases}$$

In certain mechanical problems the body considered moves so that the areal velocity is constant if the origin is properly chosen. In this case it is said that the body obeys the law of areas with respect to the origin. That is,

$$r^2 \frac{d\theta}{dt} = \text{constant.}$$

It follows from this equation and (19) that in this case

$$\alpha_\theta = 0;$$

that is, the acceleration perpendicular to the radius vector is zero.

17. Application to Motion in an Ellipse. Suppose a particle moves in an ellipse whose semi-axes are a and b in such a manner that it obeys the law of areas with respect to the center of the ellipse as origin; it is required to find the components of acceleration along and perpendicular to the radius vector. The equation of the ellipse may be written in the parametric form

(30) $$x = a \cos \phi, \qquad y = b \sin \phi;$$

for, if ϕ is eliminated, the ordinary equation

$$\frac{x^2}{a^2} + \frac{y^2}{b^2} = 1$$

is found. It follows from (30) that

(31) $$\frac{dx}{dt} = -a \sin \phi \frac{d\phi}{dt}, \qquad \frac{dy}{dt} = b \cos \phi \frac{d\phi}{dt}.$$

On substituting (30) and (31) in the expression for the law of areas,

$$x \frac{dy}{dt} - y \frac{dx}{dt} = c,$$

it is found that

$$\frac{d\phi}{dt} = \frac{c}{ab} = c_1.$$

The integral of this equation is

$$\phi = c_1 t + c_2;$$

and if $\phi = 0$ when $t = 0$, then $c_2 = 0$ and $\phi = c_1 t$.

On substituting the final expression for ϕ in (30), it is found that

$$\begin{cases} \dfrac{d^2x}{dt^2} = -c_1{}^2 a \cos \phi = -c_1{}^2 x, \\[2mm] \dfrac{d^2y}{dt^2} = -c_1{}^2 b \sin \phi = -c_1{}^2 y. \end{cases}$$

If these values of the derivatives are substituted in (20) the components of acceleration along and perpendicular to the radius vector are found to be

$$\begin{cases} \alpha_r = -c_1{}^2 r, \\ \alpha_\theta = 0. \end{cases}$$

I. PROBLEMS.

1. A particle moves with uniform speed along a helix traced on a circular cylinder whose radius is R; find the components of velocity and acceleration parallel to the x, y, and z-axes. The equations of the helix are

$$x = R \cos \omega, \quad y = R \sin \omega, \quad z = h\omega.$$

$$\textit{Ans.} \quad \begin{cases} v_x = -Rc\sin(ct), & v_y = +Rc\cos(ct), & v_z = hc; \\ \alpha_x = -Rc^2\cos(ct), & \alpha_y = -Rc^2\sin(ct), & \alpha_z = 0. \end{cases}$$

2. A particle moves in the ellipse whose parameter and eccentricity are p and e with uniform angular speed with respect to one of the foci as origin; find the components of velocity and acceleration along and perpendicular to the radius vector and parallel to the x and y-axes in terms of the radius vector and the time.

$$\textit{Ans.} \quad \begin{cases} v_r = \dfrac{ec}{p}\cdot r^2\sin(ct), \quad v_\theta = rc; \\[2mm] v_x = -cr\sin(ct) + \dfrac{ec}{2p}\cdot r^2\sin(2ct), \\[2mm] v_y = cr\cos(ct) + \dfrac{ec}{p}\cdot r^2\sin^2(ct); \\[2mm] \alpha_r = \dfrac{ec^2}{p}\cdot r^2\cos(ct) + \dfrac{2e^2c^2}{p^2}r^3\sin^2(ct) - c^2r, \\[2mm] \alpha_\theta = \dfrac{2ec^2}{p}\cdot r^2\sin(ct); \\[2mm] \alpha_x = -c^2r\cos(ct) + \dfrac{ec^2}{p}\cdot r^2 - \dfrac{3ec^2}{p}\cdot r^2\sin^2(ct) \\[2mm] \qquad\qquad\qquad\qquad\qquad + \dfrac{2e^2c^2}{p^2}\cdot r^3\sin^2(ct)\cos(ct), \\[2mm] \alpha_y = -c^2r\sin ct + \dfrac{3ec^2}{2p}\cdot r^2\sin(2ct) + \dfrac{2e^2c^2}{p^2}\cdot r^3\sin^3(ct). \end{cases}$$

3. A particle moves in an ellipse in such a manner that it obeys the law of areas with respect to one of the foci as an origin; it is required to find the components of velocity and acceleratio ι along and perpendicular to the radius vector and parallel to the axes in terms of the coördinates.

$$\textit{Ans.} \quad \begin{cases} v_r = \dfrac{eA}{p}\sin\theta, \quad v_\theta = \dfrac{A}{r}; \\[2mm] v_x = \dfrac{eA}{2p}\sin2\theta - \dfrac{A\sin\theta}{r}, \quad v_y = \dfrac{eA}{p}\sin^2\theta + \dfrac{A\cos\theta}{r}; \\[2mm] \alpha_r = -\dfrac{A^2}{p}\cdot\dfrac{1}{r^2}, \quad \alpha_\theta = 0; \\[2mm] \alpha_x = -\dfrac{A^2}{p}\cdot\dfrac{\cos\theta}{r^2}, \quad \alpha_y = -\dfrac{A^2}{p}\cdot\dfrac{\sin\theta}{r^2}. \end{cases}$$

4. The accelerations along the x and y-axes are the derivatives of the velocities along these axes; why are not the accelerations along and perpendicular to the radius vector given by the derivatives of the velocities in these respective directions? Find the accelerations along axes rotating with the angular velocity unity in terms of the accelerations with respect to fixed axes.

18. Center of Mass of n Equal Particles. The center of mass of a system of equal particles will be defined as that point whose distance from any plane is equal to the average distance of all of the particles from that plane. This must be true then for the three reference planes. Let (x_1, y_1, z_1), (x_2, y_2, z_2), etc., represent the rectangular coördinates of the various particles, and $\bar{x}$, $\bar{y}$, $\bar{z}$ the rectangular coördinates of their center of mass; then by the definition

$$(32) \quad \begin{cases} \bar{x} = \dfrac{x_1 + x_2 + \cdots + x_n}{n} = \dfrac{\sum\limits_{i=1}^{n} x_i}{n}, \\[3ex] \bar{y} = \dfrac{y_1 + y_2 + \cdots + y_n}{n} = \dfrac{\sum\limits_{i=1}^{n} y_i}{n}, \\[3ex] \bar{z} = \dfrac{z_1 + z_2 + \cdots + z_n}{n} = \dfrac{\sum\limits_{i=1}^{n} z_i}{n}. \end{cases}$$

Suppose the mass of each particle is m, and let M represent the mass of the whole system, or $M = nm$. On multiplying the numerators and denominators by m, equations (32) become

$$(33) \quad \begin{cases} \bar{x} = \dfrac{m \sum\limits_{i=1}^{n} x_i}{nm} = \dfrac{\sum\limits_{i=1}^{n} m x_i}{M}, \\[3ex] \bar{y} = \dfrac{m \sum\limits_{i=1}^{n} y_i}{nm} = \dfrac{\sum\limits_{i=1}^{n} m y_i}{M}, \\[3ex] \bar{z} = \dfrac{m \sum\limits_{i=1}^{n} z_i}{nm} = \dfrac{\sum\limits_{i=1}^{n} m z_i}{M}. \end{cases}$$

It remains to show that the distance from the point $(\bar{x}, \bar{y}, \bar{z})$ to any other plane is also the average distance of the particles from the plane. The equation of any other plane is

$$ax + by + cz + d = 0.$$

The distance of the point $(\bar{x}, \bar{y}, \bar{z})$ from this plane is

$$(34) \quad \bar{d} = \frac{a\bar{x} + b\bar{y} + c\bar{z} + d}{\sqrt{a^2 + b^2 + c^2}};$$

and similarly, the distance of the point (x_i, y_i, z_i) from the same plane is

$$(35) \qquad d_i = \frac{ax_i + by_i + cz_i + d}{\sqrt{a^2 + b^2 + c^2}}.$$

It follows from equations (32), (34), and (35) that

$$\bar{d} = \frac{a \sum_{i=1}^{n} x_i + b \sum_{i=1}^{n} y_i + c \sum_{i=1}^{n} z_i + nd}{n \sqrt{a^2 + b^2 + c^2}} = \frac{\sum_{i=1}^{n} d_i}{n}.$$

Therefore the point $(\bar{x}, \bar{y}, \bar{z})$ defined by (32) satisfies the definition of center of mass with respect to all planes.

19. Center of Mass of Unequal Particles. There are two cases, (a) that in which the masses are commensurable, and (b) that in which the masses are incommensurable.

(a) Select a unit m in terms of which all the n masses can be expressed integrally. Suppose the first mass is p_1m, the second p_2m, etc., and let $p_1m = m_1$, $p_2m = m_2$, etc. The system may be thought of as made up of $p_1 + p_2 + \cdots$ particles each of mass m, and consequently, by Art. 18,

$$(36) \qquad \begin{cases} \bar{x} = \dfrac{\sum_{i=1}^{n} mp_i x_i}{\sum_{i=1}^{n} mp_i} = \dfrac{\sum_{i=1}^{n} m_i x_i}{M}, \\[4ex] \bar{y} = \dfrac{\sum_{i=1}^{n} mp_i y_i}{\sum_{i=1}^{n} mp_i} = \dfrac{\sum_{i=1}^{n} m_i y_i}{M}, \\[4ex] \bar{z} = \dfrac{\sum_{i=1}^{n} mp_i z_i}{\sum_{i=1}^{n} mp_i} = \dfrac{\sum_{i=1}^{n} m_i z_i}{M} \end{cases}$$

(b) Select an arbitrary unit m smaller than any one of the n masses. They will be expressible in terms of it plus certain remainders. If the remainders are neglected equations (36) give the center of mass. Take as a new unit any submultiple of m and the remainders will remain the same, or be diminished, depending on their magnitudes. The submultiple of m can be taken so small that every remainder is smaller than any assigned quantity. Equations (36) continually hold where the m_i are the masses of the bodies minus the remainders. As the submultiples of m approach zero as a limit, the sum of the remainders approaches zero as a limit, and the expressions (36) approach as limits the expressions in which the m_i are the actual masses of the particles. Therefore in all cases equations (36) give a point which satisfies the definition of center of mass.

The fact that if the definition of center of mass is fulfilled for the three reference planes, it is also fulfilled for every other plane can easily be proved without recourse to the general formula for the distance from any point to any plane. It is to be observed that the yz-plane, for example, may be brought into any position whatever by a change of origin and a succession of rotations of the coördinate system around the various axes. It will be necessary to show, then, that equations (36) are not changed in form (1) by a change of origin, and (2) by a rotation around one of the axes.

(1) Transfer the origin along the x-axis through the distance a. The substitution which accomplishes the transfer is $x = x' + a$, and the first equation of (36) becomes

$$\bar{x}' + a = \frac{\sum_{i=1}^{n} m_i(x_i' + a)}{M} = \frac{\sum_{i=1}^{n} m_i x_i'}{M} + \frac{a \sum_{i=1}^{n} m_i}{M};$$

whence

$$\bar{x}' = \frac{\sum_{i=1}^{n} m_i x_i'}{M},$$

which has the same form as before.

(2) Rotate the x and y-axes around the z-axis through the angle θ. The substitution which accomplishes the rotation is

$$\begin{cases} x = x' \cos\theta - y' \sin\theta, \\ y = x' \sin\theta + y' \cos\theta. \end{cases}$$

The first two equations of (36) become by this transformation

$$\begin{cases} \bar{x}' \cos\theta - \bar{y}' \sin\theta = \cos\theta \dfrac{\sum_{i=1}^{n} m_i x_i'}{M} - \sin\theta \dfrac{\sum_{i=1}^{n} m_i y_i'}{M}, \\ \bar{x}' \sin\theta + \bar{y}' \cos\theta = \sin\theta \dfrac{\sum_{i=1}^{n} m_i x_i'}{M} + \cos\theta \dfrac{\sum_{i=1}^{n} m_i y_i'}{M}. \end{cases}$$

On solving these equations it is found that

$$\bar{x}' = \frac{\sum_{i=1}^{n} m_i x_i'}{M}, \qquad \bar{y}' = \frac{\sum_{i=1}^{n} m_i y_i'}{M}.$$

Therefore the point $(\bar{x},\ \bar{y},\ \bar{z})$ satisfies the definition of center of mass with respect to every plane.

20. The Center of Gravity. The members of a system of particles which are near together at the surface of the earth are subject to forces downward which are sensibly parallel and proportional to their respective masses. The *weight*, or *gravity*, of a particle will be defined as the intensity of the vertical force f,

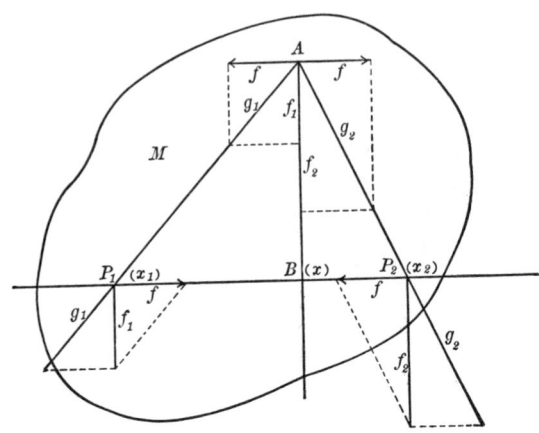

Fig. 4.

which is the product of the mass m of the particle and its acceleration g. The *center of gravity* of the system will be defined as the point such that, if the members of the system were rigidly connected and the sum of all the forces were applied at this point, then the effect on the motion of the system would be the same as that of the original forces for all orientations of the system.

It will now be shown that the center of gravity coincides with the center of mass. Consider two parallel forces f_1 and f_2 acting upon the rigid system M at the points P_1 and P_2. Resolve these two forces into the components f and g_1, and f and g_2 respectively. The components f, being equal and opposite, destroy each other. Then the components g_1 and g_2 may be regarded as acting at A. Resolve them again so that the oppositely directed components shall be equal and lie in a line parallel to P_1P_2; then the other components will lie in the same line AB, which is parallel to the direction of the original forces f_1 and f_2, and will be equal respectively to f_1 and f_2. Therefore the resultant of f_1 and f_2 is equal to $f_1 + f_2$ in magnitude and direction. It is found from similar triangles that

$$\frac{f_1}{f} = \frac{AB}{P_1B}, \qquad \frac{f_2}{f} = \frac{AB}{P_2B};$$

whence, by division,

$$\frac{f_1}{f_2} = \frac{P_2B}{P_1B} = \frac{x_2 - \bar{x}}{\bar{x} - x_1}.$$

The solution for $\bar{x}$ gives

$$\bar{x} = \frac{f_1 x_1 + f_2 x_2}{f_1 + f_2}.$$

If the resultant of these two forces be united with a third force f_3, the point where their sum may be applied with the same effects is found in a similar manner to be given by

$$\bar{x} = \frac{f_1 x_1 + f_2 x_2 + f_3 x_3}{f_1 + f_2 + f_3},$$

and so on for any number of forces. Similar equations are true for parallel forces acting in any other direction.

Suppose there are n particles m_i subject to n parallel forces f_i due to the attraction of the earth. The coördinates of their center of gravity with respect to the origin are given by

(37)

$$\begin{cases} \bar{x} = \dfrac{\sum\limits_{i=1}^{n} f_i x_i}{\sum\limits_{i=1}^{n} f_i} = \dfrac{\sum\limits_{i=1}^{n} g m_i x_i}{\sum\limits_{i=1}^{n} g m_i} = \dfrac{\sum\limits_{i=1}^{n} m_i x_i}{M}, \\[4ex] \bar{y} = \dfrac{\sum\limits_{i=1}^{n} m_i y_i}{M}, \\[4ex] \bar{z} = \dfrac{\sum\limits_{i=1}^{n} m_i z_i}{M}. \end{cases}$$

The center of gravity is thus seen to be coincident with the center of mass; nevertheless this would not in general be true if the system were not in such a position that the accelerations to which its various members are subject were both equal and parallel. Euler (1707–1783) proposed the designation of *center of inertia* for the center of mass.

21. Center of Mass of a Continuous Body. As the particles of a system become more and more numerous and nearer together it approaches as a limit a continuous body. In the case of the ordinary bodies of mechanics the particles are innumerable and indistinguishably close together; on this account such bodies are treated as continuous masses. For continuous masses, therefore, the limits of expressions (37), as m_i approaches zero, must be taken. At the limit m becomes dm and the sum becomes the definite integral. The equations which give the center of mass are therefore

$$(38) \quad \begin{cases} \bar{x} = \dfrac{\int x\,dm}{\int dm}, \\[2mm] \bar{y} = \dfrac{\int y\,dm}{\int dm}, \\[2mm] \bar{z} = \dfrac{\int z\,dm}{\int dm}, \end{cases}$$

where the integrals are to be extended throughout the whole body.

When the body is homogeneous the *density* is the quotient of any portion of the mass divided by its volume. When the body is not homogeneous the *mean density* is the quotient of the whole mass divided by the whole volume. The density at any point is the limit of the mean density of a volume including the point in question when this volume approaches zero as a limit. If the density is represented by σ, the element of mass is, when expressed in rectangular coördinates,

$$dm = \sigma\,dx\,dy\,dz.$$

Then equations (38) become

$$(39) \quad \begin{cases} \bar{x} = \dfrac{\int\int\int \sigma x\,dx\,dy\,dz}{\int\int\int \sigma\,dx\,dy\,dz}, \\[2mm] \bar{y} = \dfrac{\int\int\int \sigma y\,dx\,dy\,dz}{\int\int\int \sigma\,dx\,dy\,dz}, \\[2mm] \bar{z} = \dfrac{\int\int\int \sigma z\,dx\,dy\,dz}{\int\int\int \sigma\,dx\,dy\,dz}. \end{cases}$$

The limits of the integrals depend upon the shape of the body, and σ must be expressed as a function of the coördinates.

In certain problems the integrations are performed more simply if polar coördinates are employed. The element of mass when expressed in polar coördinates is

$$dm = \sigma \cdot \overline{ab} \cdot \overline{bc} \cdot \overline{cd}.$$

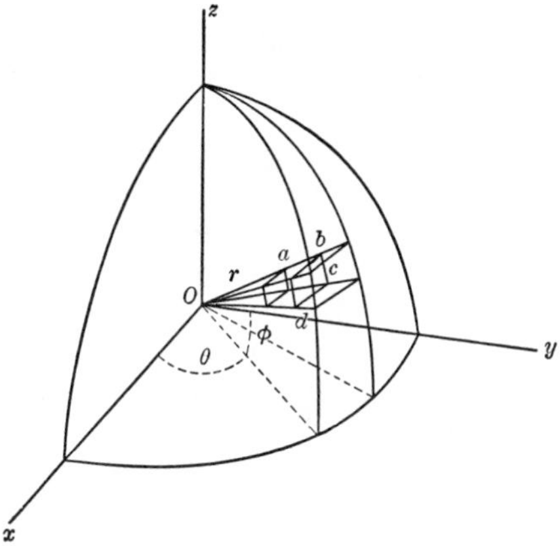

Fig. 5.

It is seen from the figure that

$$\begin{cases} \overline{ab} = dr, \\ \overline{bc} = r\,d\phi, \\ \overline{cd} = r\cos\phi\,d\theta. \end{cases}$$

Therefore

(40) $$dm = \sigma r^2 \cos\phi\,d\phi\,d\theta\,dr,$$

and

(41) $$\begin{cases} x = r\cos\phi\cos\theta, \\ y = r\cos\phi\sin\theta, \\ z = r\sin\phi. \end{cases}$$

Therefore equations (38) become

(42) $$\begin{cases} \bar{x} = \dfrac{\iiint \sigma r^3 \cos^2\phi\cos\theta\,d\phi\,d\theta\,dr}{\iiint \sigma r^2 \cos\phi\,d\phi\,d\theta\,dr}, \\[2mm] \bar{y} = \dfrac{\iiint \sigma r^3 \cos^2\phi\sin\theta\,d\phi\,d\theta\,dr}{\iiint \sigma r^2 \cos\phi\,d\phi\,d\theta\,dr}, \\[2mm] \bar{z} = \dfrac{\iiint \sigma r^3 \sin\phi\cos\phi\,d\phi\,d\theta\,dr}{\iiint \sigma r^2 \cos\phi\,d\phi\,d\theta\,dr}. \end{cases}$$

The density σ must be expressed as a function of the coördinates, and the limits must be so taken that the whole body is included. If the body is a line or a surface the equations admit of important simplifications.

22. Planes and Axes of Symmetry. If a homogeneous body is symmetrical with respect to any plane, the center of mass is in that plane, because each element of mass on one side of the plane can be paired with a corresponding element of mass on the other side, and the whole body can be divided up into such paired elements. This plane is called a *plane of symmetry*. If a homogeneous body is symmetrical with respect to two planes, the center of mass is in the line of their intersection. This line is called an *axis of symmetry*. If a homogeneous body is symmetrical with respect to three planes, intersecting in a point, the center of mass is at their point of intersection. From the consideration of the planes and axes of symmetry the centers of mass of many of the simple figures can be inferred without employing the methods of integration.

23. Application to a Non-Homogeneous Cube. Suppose the density varies directly as the square of the distance from one of the faces of the cube. Take the origin at one of the corners and let the yz-plane be the face of zero density. Then $\sigma = kx^2$, where k is the density at unit distance. Suppose the edge of the cube equals a; then equations (39) become

$$
\left\{
\begin{aligned}
\bar{x} &= \frac{k \displaystyle\int_0^a \int_0^a \int_0^a x^3 \, dx \, dy \, dz}{k \displaystyle\int_0^a \int_0^a \int_0^a x^2 \, dx \, dy \, dz}, \\[2ex]
\bar{y} &= \frac{k \displaystyle\int_0^a \int_0^a \int_0^a x^2 y \, dx \, dy \, dz}{k \displaystyle\int_0^a \int_0^a \int_0^a x^2 \, dx \, dy \, dz}, \\[2ex]
\bar{z} &= \frac{k \displaystyle\int_0^a \int_0^a \int_0^a x^2 z \, dx \, dy \, dz}{k \displaystyle\int_0^a \int_0^a \int_0^a x^2 \, dx \, dy \, dz}.
\end{aligned}
\right.
$$

These equations become, after integrating and substituting the limits,

$$
\bar{x} = \frac{3a}{4}, \qquad \bar{y} = \frac{a}{2}, \qquad \bar{z} = \frac{a}{2}.
$$

If polar coördinates were used in this problem the upper limits of the integrals would be much more complicated than they are with rectangular coördinates, and the integration would be correspondingly more difficult.

24. Application to the Octant of a Sphere. Suppose the sphere is homogeneous and that the density equals unity. It is preferable to use polar coördinates in this example, although it is by no means necessary. Either (39) or (42) can be used in any problem, and the choice should be determined by the form that the limits take in the two cases. It is desirable to have them all constants when they can be made so. If the origin is taken at the center of the sphere, and if the radius is a, equations (42) become

$$
\begin{cases}
\bar{x} = \dfrac{\displaystyle\int_0^{\frac{\pi}{2}} \int_0^{\frac{\pi}{2}} \int_0^a r^3 \cos^2\phi \cos\theta \, d\phi \, d\theta \, dr}{\displaystyle\int_0^{\frac{\pi}{2}} \int_0^{\frac{\pi}{2}} \int_0^a r^2 \cos\phi \, d\phi \, d\theta \, dr}, \\[4ex]
\bar{y} = \dfrac{\displaystyle\int_0^{\frac{\pi}{2}} \int_0^{\frac{\pi}{2}} \int_0^a r^3 \cos^2\phi \sin\theta \, d\phi \, d\theta \, dr}{\displaystyle\int_0^{\frac{\pi}{2}} \int_0^{\frac{\pi}{2}} \int_0^a r^2 \cos\phi \, d\phi \, d\theta \, dr}, \\[4ex]
\bar{z} = \dfrac{\displaystyle\int_0^{\frac{\pi}{2}} \int_0^{\frac{\pi}{2}} \int_0^a r^3 \sin\phi \cos\phi \, d\phi \, d\theta \, dr}{\displaystyle\int_0^{\frac{\pi}{2}} \int_0^{\frac{\pi}{2}} \int_0^a r^2 \cos\phi \, d\phi \, d\theta \, dr}.
\end{cases}
$$

Since the mass of a homogeneous sphere with radius a and density unity is $\frac{4}{3}\pi a^3$, each of the denominators of these expressions equals $\frac{1}{6}\pi a^3$. This can be verified at once by integration. On integrating the numerators with respect to ϕ and substituting the limits, the equations become

$$
\begin{cases}
\bar{x} = \dfrac{\dfrac{\pi}{4} \displaystyle\int_0^{\frac{\pi}{2}} \int_0^a r^3 \cos\theta \, d\theta \, dr}{\dfrac{\pi}{6} a^3}, \qquad
\bar{y} = \dfrac{\dfrac{\pi}{4} \displaystyle\int_0^{\frac{\pi}{2}} \int_0^a r^3 \sin\theta \, d\theta \, dr}{\dfrac{\pi}{6} a^3}, \\[5ex]
\bar{z} = \dfrac{\dfrac{1}{2} \displaystyle\int_0^{\frac{\pi}{2}} \int_0^a r^3 \, d\theta \, dr}{\dfrac{\pi}{6} a^3}.
\end{cases}
$$

On integrating with respect to θ, these equations give

$$\bar{x} = \frac{\dfrac{\pi}{4} \displaystyle\int_0^a r^3 dr}{\dfrac{\pi}{6}\, a^3}, \qquad \bar{y} = \frac{\dfrac{\pi}{4} \displaystyle\int_0^a r^3 dr}{\dfrac{\pi}{6}\, a^3}, \qquad \bar{z} = \frac{\dfrac{\pi}{4} \displaystyle\int_0^a r^3 dr}{\dfrac{\pi}{6}\, a^3};$$

and, finally, the integration with respect to r gives

$$\bar{x} = \bar{y} = \bar{z} = \tfrac{3}{8}a.$$

The octant of a sphere has three planes of symmetry, viz., the planes defined by the center of the sphere, the vertices of the bounding spherical triangle, and the centers of their respective opposite sides. Since these three planes intersect not only in a point but also in a line, they do not fully determine the center of mass.

As nearly all the masses occurring in astronomical problems are spheres or oblate spheroids with three planes of symmetry which intersect in a point but not in a line, the applications of the formulas just given are extremely simple, and no more examples need be solved.

II. PROBLEMS.

1. Find the center of mass of a fine straight wire of length R whose density varies directly as the nth power of the distance from one end.

$$Ans. \quad \frac{n+1}{n+2}\, R \ \text{ from the end of zero density.}$$

2. Find the coördinates of the center of mass of a fine wire of uniform density bent into a quadrant of a circle of radius R.

$$Ans. \quad \bar{x} = \bar{y} = \frac{2R}{\pi},$$

where the origin is at the center of the circle.

3. Find the coördinates of the center of mass of a thin plate of uniform density, having the form of a quadrant of an ellipse whose semi-axes are a and b.

$$Ans. \quad \begin{cases} \bar{x} = \dfrac{4a}{3\pi}, \\[2mm] \bar{y} = \dfrac{4b}{3\pi}. \end{cases}$$

4. Find the coördinates of the center of mass of a thin plate of uniform density, having the form of a complete loop of the lemniscate whose equation is $r^2 = a^2 \cos 2\theta$.

$$Ans. \quad \begin{cases} \bar{x} = \dfrac{\pi a}{2^{\frac{5}{2}}}, \\[2ex] \bar{y} = 0. \end{cases}$$

5. Find the coördinates of the center of mass of an octant of an ellipsoid of uniform density whose semi-axes are a, b, c.

$$Ans. \quad \begin{cases} \bar{x} = \dfrac{3a}{8}, \\[2ex] \bar{y} = \dfrac{3b}{8}, \\[2ex] \bar{z} = \dfrac{3c}{8}. \end{cases}$$

6. Find the coördinates of the center of mass of an octant of a sphere of radius R whose density varies directly as the nth power of the distance from the center.

$$Ans. \quad \bar{x} = \bar{y} = \bar{z} = \frac{n+3}{n+4} \cdot \frac{R}{2}.$$

7. Find the coördinates of the center of mass of a paraboloid of revolution cut off by a plane perpendicular to its axis.

$$Ans. \quad \begin{cases} \bar{x} = \tfrac{2}{3}h, \\ \bar{y} = \bar{z} = 0, \end{cases}$$

where h is the distance from the vertex of the paraboloid to the plane.

8. Find the coördinates of the center of mass of a right circular cone whose height is h and whose radius is R.

9. Find the coördinates of the center of mass of a double convex lens of homogeneous glass whose surfaces are spheres having the radii r_1 and $r_2 = 2r_1$ and whose thickness at the center is $\dfrac{r_1 + r_2}{4}$.

10. In a concave-convex lens the radius of curvature of the convex and concave surfaces are r_1 and $r_2 > r_1$. Determine the thickness and diameter of the lens so that the center of mass shall be in the concave surface.

HISTORICAL SKETCH FROM ANCIENT TIMES TO NEWTON.

25. The Two Divisions of the History. The history of the development of Celestial Mechanics is naturally divided into two distinct parts. The one is concerned with the progress of knowledge about the purely formal aspect of the universe, the natural divisions of time, the configurations of the constellations, and the determination of the paths and periods of the planets in their motions; the other treats of the efforts at, and the success in, attaining correct ideas regarding the physical aspects of natural phenomena, the fundamental properties of force, matter, space, and time, and, in particular, the way in which they are related. It is true that these two lines in the development of astrcnomical science have not always been kept distinct by those who have cultivated them; on the contrary, they have often been so intimately associated that the speculations in the latter have influenced unduly the conclusions in the former. While it is clear that the two kinds of investigation should be kept distinct in the mind of the investigator, it is equally clear that they should be constantly employed as checks upon each other. The object of the next two articles will be to trace, in as few words as possible, the development of these two lines of progress of the science of Celestial Mechanics from the times of the early Greek Philosophers to the time when Newton applied his transcendent genius to the analysis of the elements involved, and to their synthesis into one of the sublimest products of the human mind.

26. Formal Astronomy. The first division is concerned with phenomena altogether apart from their causes, and will be termed Formal Astronomy. The day, the month, and the year are such obvious natural divisions of time that they must have been noticed by the most primitive peoples. But the determination of the relations among these periods required something of the scientific spirit necessary for careful observations; yet, in the very dawn of Chaldean and Egyptian history they appear to have been known with a considerable degree of accuracy. The records left by these peoples of their earlier civilizations are so meager that little is known with certainty regarding their scientific achievements. The authentic history of Astronomy actually begins with the Greeks, who, deriving their first knowledge and inspiration from the Egyptians, pursued the subject with the enthusiasm and acuteness which was characteristic of the Greek race.

Thales (640–546 B.C.), of Miletus, went to Egypt for instruction, and on his return founded the Ionian School of Astronomy and Philosophy. Some idea of the advancement made by the Egyptians can be gathered from the fact that he taught the sphericity of the earth, the obliquity of the ecliptic, the causes of eclipses, and, according to Herodotus, predicted the eclipse of the sun of 585 B.C. According to Laertius he was the first to determine the length of the year. It is fair to assume that he borrowed

much of his information from Egypt, though the basis for pre-
dicting eclipses rests on the period of 6585 days, known as the
saros, discovered by the Chaldaeans. After the lapse of this
period eclipses of the sun and moon recur under almost identical
circumstances except that they are displaced about 120° westward
on the earth.

Anaximander (611–545 B.C.), a friend and probably a pupil of
Thales, constructed geographical maps, and is credited with
having invented the gnomon.

Pythagoras (about 569–470 B.C.) travelled widely in Egypt and
Chaldea, penetrating Asia even to the banks of the Ganges. On
his return he went to Sicily and founded a School of Astronomy
and Philosophy. He taught that the earth both rotates and
revolves, and that the comets as well as the planets move in orbits
around the sun. He is credited with being the first to maintain
that the same planet, Venus, is both evening and morning star at
different times.

Meton (about 465–385 B.C.) brought to the notice of the
learned men of Hellas the cycle of 19 years, nearly equalling 235
synodic months, which has since been known as the Metonic cycle.
After the lapse of this period the phases of the moon recur on the
same days of the year, and almost at the same time of day. The
still more accurate Callipic cycle consists of four Metonic cycles,
less one day.

Aristotle (384–322 B.C.) maintained the theory of the globular
form of the earth and supported it with many of the arguments
which are used at the present time.

Aristarchus (310–250 B.C.) wrote an important work entitled
Magnitudes and Distances. In it he calculated from the time at
which the earth is in quadrature as seen from the moon that the
latter is about one-nineteenth as distant from the earth as the sun.
The time in question is determined by observing when the moon
is at the first quarter. The practical difficulty of determining
exactly when the moon has any particular phase is the only thing
that prevents the method, which is theoretically sound, from
being entirely successful.

Eratosthenes (275–194 B.C.) made a catalogue of 475 of the
brightest stars, and is famous for having determined the size of
the earth from the measurement of the difference in latitude and
the distance apart of Syene, in Upper Egypt, and Alexandria.

Hipparchus (about 190–120 B.C.), a native of Bithynia, who

observed at Rhodes and possibly at Alexandria, was the greatest astronomer of antiquity. He added to zeal and skill as an observer the accomplishments of the mathematician. Following Euclid (about 330–275 B.C.) at Alexandria, he developed the important science of Spherical Trigonometry. He located places on the earth by their Longitudes and Latitudes, and the stars by their Right Ascensions and Declinations. He was led by the appearance of a temporary star to make a catalogue of 1080 fixed stars. He measured the length of the tropical year, the length of the month from eclipses, the motion of the moon's nodes and that of her apogee; he was the author of the first solar tables; he discovered the precession of the equinoxes, and made extensive observations of the planets. The works of Hipparchus are known only indirectly, his own writings having been lost at the time of the destruction of the great Alexandrian library by the Saracens under Omar, in 640 A.D.

Ptolemy (100–170 A.D.) carried forward the work of Hipparchus faithfully and left the *Almagest* as the monument of his labors. Fortunately it has come down to modern times intact and contains much information of great value. Ptolemy's greatest discovery is the evection of the moon's motion, which he detected by following the moon during the whole month, instead of confining his attention to certain phases as previous observers had done. He discovered refraction, but is particularly famous for the system of eccentrics and epicycles which he developed to explain the apparent motions of the planets.

A stationary period followed Ptolemy during which science was not cultivated by any people except the Arabs, who were imitators and commentators rather more than original investigators. In the Ninth Century the greatest Arabian astronomer, Albategnius (850–929), flourished, and a more accurate measurement of the arc of a meridian than had before that time been executed was carried out by him in the plain of Singiar, in Mesopotamia. In the Tenth Century Al-Sufi made a catalogue of stars based on his own observations. Another catalogue was made by the direction of Ulugh Beigh (1393–1449), at Samarkand, in 1433. At this time Arabian astronomy practically ceased to exist.

Astronomy began to revive in Europe toward the end of the Fifteenth Century in the labors of Peurbach (1423–1461), Waltherus (1430–1504), and Regiomontanus (1436–1476). It was given a great impetus by the celebrated German astronomer

Copernicus (1473–1543), and has been pursued with increasing zeal to the present time. Copernicus published his masterpiece, *De Revolutionibus Orbium Coelestium*, in 1543, in which he gave to the world the heliocentric theory of the solar system. The system of Copernicus was rejected by Tycho Brahe (1546–1601), who advanced a theory of his own, because he could not observe any parallax in the fixed stars. Tycho was of Norwegian birth, but did much of his astronomical work in Denmark under the patronage of King Frederick. After the death of Frederick he moved to Prague where he was supported the remainder of his life by a liberal pension from Rudolph II. He was an indefatigable and most painstaking observer, and made very important contributions to Astronomy. In his later years Tycho Brahe had Kepler (1571–1630) for his disciple and assistant, and it was by discussing the observations of Tycho Brahe that Kepler was enabled, in less than twenty years after the death of his preceptor, to announce the three laws of planetary motion. It was from these laws that Newton (1642–1727) derived the law of gravitation.

Galileo (1564–1642), an Italian astronomer, a contemporary of Kepler, and a man of greater genius and greater fame, first applied the telescope to celestial objects. He discovered four satellites revolving around Jupiter, the rings of Saturn, and spots on the sun. He, like Kepler, was an ardent supporter of the heliocentric theory.

27. Dynamical Astronomy. By Dynamical Astronomy will be meant the connecting of mechanical and physical causes with observed phenomena. Formal Astronomy is so ancient that it is not possible to go back to its origin; Dynamical Astronomy, on the other hand, did not begin until after the time of Aristotle, and then real advances were made at only very rare intervals.

Archimedes (287–212 B.C.), of Syracuse, is the author of the first sound ideas regarding mechanical laws. He stated correctly the principles of the lever and the meaning of the center of gravity of a body. The form and generality of his treatment were improved by Leonardo da Vinci (1452–1519) in his investigations of statical moments. The whole subject of Statics of a rigid body involves only the application of the proper mathematics to these principles.

It is a remarkable fact that no single important advance was made in the discovery of mechanical laws for nearly two thousand years after Archimedes, or until the time of Stevinus (1548–1620), who was the first, in 1586, to investigate the mechanics of the inclined plane, and of Galileo (1564–1642), who made the first important advance in Kinetics.   Thus, the mechanical principles involved in the motions of bodies were not discovered until relatively modern times.   The fundamental error in the speculations of most of the investigators was that they supposed that it required a continually acting force to keep a body in motion.   They thought it was natural for a body to have a position rather than a state of motion.   This is the opposite of the law of inertia (Newton's first law).   This law was discovered by Galileo quite incidentally in the study of the motion of bodies sliding down an inclined plane and out on a horizontal surface.   Galileo took as his fundamental principle that the change of velocity, or acceleration, is determined by the forces which act upon the body.   This contains nearly all of Newton's first two laws.   Galileo applied his principles with complete success to the discovery of the laws of falling bodies, and of the motion of projectiles.   The value of his discoveries is such that he is justly considered to be the founder of Dynamics. He was the first to employ the pendulum for the measurement of time.

Huyghens (1629–1695), a Dutch mathematician and scientist, published his *Horologium Oscillatorium* in 1675, containing the theory of the determination of the intensity of the earth's gravity from pendulum experiments, the theory of the center of oscillation, the theory of evolutes, and the theory of the cycloidal pendulum.

Newton (1642–1727) completed the formulation of the fundamental principles of Mechanics, and applied them with unparalleled success in the solution of mechanical and astronomical problems. He employed Geometry with such skill that his work has scarcely been added to by the use of his methods to the present day.

After Newton's time, mathematicians soon turned to the more general and powerful methods of analysis.   The subject of Analytical Mechanics was founded by Euler (1707–1783) in his work, *Mechanica sive Motus Scientia* (Petersburg, 1736); it was improved by Maclaurin (1698–1746) in his work, *A Complete System of Fluxions* (Edinburgh, 1742), and was highly perfected by Lagrange (1736–1813) in his *Mécanique Analytique* (Paris, 1788).   The *Mécanique Céleste* of Laplace (1749–1827) put Celestial Mechanics on a correspondingly high plane.

BIBLIOGRAPHY.

For the fundamental principles of Mechanics consult *Principien der Mechanik* (a history and exposition of the various systems of Mechanics from Archimedes to the present time), by Dr. E. Dühring; *Vorreden und Einleitungen der klassischen Werken der Mechanik*, edited by the Phil. Soc. of the Univ. of Vienna; *Die Principien der Mechanik*, by Heinrich Hertz, *Coll. Works*, vol. III; *The Science of Mechanics*, by E. Mach, translated by T. J. McCormack; *Principe der Mechanik*, by Boltzmann; *Newton's Laws of Motion*, by P. G. Tait; *Das Princip der Erhaltung der Energie*, by Planck; *Die geschichtliche Entwickelung des Bewegungsbegriffes*, by Lange.

For the theory of Relativity consult *Das Relativitätsprincip*, by M. Laue, and *The Theory of Relativity*, by R. D. Carmichael.

For velocity and acceleration and their resolution and composition consult the first parts of *Dynamics of a Particle*, by Tait and Steele; *Leçons de Cinématique*, by G. Koenigs; *Cinématique et Mécanismes*, by Poincaré; and the works on Dynamics (Mechanics) by Routh, Love, Budde, and Appell.

For the history of Celestial Mechanics and Astronomy consult *Histoire de l'Astronomie Ancienne* (old work), by Delambre; *Astronomische Beobachtungen der Alten* (old work), by L. Ideler; *Recherches sur l'Histoire de l'Astronomie Ancienne*, by Paul Tannery; *History of Astronomy*, by Grant; *Geschichte der Mathematik im Alterthum und Mittelalter*, by H. Hankel; *History of the Inductive Sciences* (2 vols.), by Whewell; *Geschichte der Mathematischen Wissenschaften* (2 vols.), by H. Süter; *Geschichte der Mathematik* (3 vols.), by M. Cantor; *A Short History of Mathematics*, by W. W. R. Ball; *A History of Mathematics*, by Florian Cajori; *Histoire des Sciences Mathématiques et Physiques* (12 vols.), by M. Maximilian Marie; *Geschichte der Astronomie*, by R. Wolf; *A History of Astronomy*, by Arthur Berry; *Histoire Abrégée de l'Astronomie*, by Ernest Lebon.

CHAPTER II.

RECTILINEAR MOTION.

28. A great part of the work in Celestial Mechanics consists of the solution of differential equations which, in most problems, are very complicated on account of the number of dependent variables involved. The ordinary Calculus is devoted, in a large part, to the treatment of equations in which there is but one independent variable and one dependent variable; and the step to simultaneous equations in several variables, requiring interpretation in connection with physical problems and mechanical principles, is one which is usually made not without some difficulty. The present chapter will be devoted to the formulation and to the solution of certain classes of problems in which the mathematical processes are closely related to those which are expounded in the mathematical text-books. It will form the bridge leading from the methods which are familiar in works on Calculus and elementary Differential Equations to those which are characteristic of mechanical and astronomical problems.

The examples chosen to illustrate the principles are taken largely from astronomical problems. They are of sufficient interest to justify their insertion, even though they were not needed as a preparation for the more complicated discussions which will follow. They embrace the theory of falling bodies, the velocity of escape, parabolic motion, and the meteoric and contraction theories of the sun's heat.

The Motion of Falling Particles.

29. The Differential Equation of Motion. Suppose the mass of the particle is m and let s represent the line in which it falls. Take the origin O at the surface of the earth and let the positive end of the line be directed upward. By the second law of motion the rate of change of momentum, or the product of the mass and the acceleration, is proportional to the force. Let k^2 represent the factor of proportionality, the numerical value of which will depend upon the units employed. Then, if f represents the force, the differential equation of motion is

$$(1) \qquad\qquad m\frac{d^2s}{dt^2} = -k^2f.$$

This is also the differential equation of motion for any case in which the resultant of all the forces is constantly in the same straight line and in which the body is not initially projected from that line. A more general treatment will therefore be given than would be required if f were simply the force arising from the earth's attraction for the particle m.

The force f will depend in general upon various things, such as the position of m, the time t, and the velocity v. This may be indicated by writing equation (1) in the form

$$(2) \qquad\qquad m\frac{d^2s}{dt^2} = -k^2f(s,\ t,\ v),$$

in which $f(s,\ t,\ v)$ simply means that the force may depend upon the quantities contained in the parenthesis. In order to solve this equation two integrations must be performed, and the character of the integrals will depend upon the manner in which f depends upon s, t, and v. It is necessary to discuss the different cases separately.

30. Case of Constant Force. This simplest case is nearly realized when particles fall through small distances near the earth's surface under the influence of gravity. If the second is taken as the unit of time and the foot as the unit of length then $k^2f = mg$, where g is the acceleration of gravity at the surface of the earth. Its numerical value, which varies somewhat with the latitude, is a little greater than 32. Then equation (1) becomes

$$(3) \qquad\qquad \frac{d^2s}{dt^2} = -g.$$

This equation gives after one integration

$$\frac{ds}{dt} = -gt + c_1,$$

where c_1 is the constant of integration. Let the velocity of the particle at the time $t = 0$ be $v = v_0$. Then the last equation becomes at $t = 0$

$$v_0 = c_1;$$

whence

$$\frac{ds}{dt} = -gt + v_0.$$

The integral of this equation is

$$s = -\frac{gt^2}{2} + v_0 t + c_2.$$

Suppose the particle is started at the distance s_0 from the origin at the time $t = 0$; then this equation gives

$$s_0 = c_2;$$

whence

(4) $$s = -\frac{gt^2}{2} + v_0 t + s_0.$$

When the initial conditions are given this equation determines the position of the particle at any time; or, it determines the time at which the body has any position by the solution of the quadratic equation in t.

If the acceleration were any other positive or negative constant than $-mg$, the equation for the space described would differ from (4) only in the coefficient of t^2.

It is also possible to obtain an important relation between the speed and the position of the particle. Multiply both members of equation (3) by $2\frac{ds}{dt}$. Then, since the derivative of $\left(\frac{ds}{dt}\right)^2$ is $2\frac{ds}{dt}\frac{d^2s}{dt^2}$, the integral of the equation is

$$\left(\frac{ds}{dt}\right)^2 = -2gs + c_3.$$

It follows from the initial conditions that

$$c_3 = v_0^2 + 2gs_0;$$

whence

(5) $$\left(\frac{ds}{dt}\right)^2 - \left(\frac{ds}{dt}\right)_0^2 = -2g(s - s_0).$$

31. Attractive Force Varying Directly as the Distance. Another simple case is that in which the force varies directly as the distance from the origin. Suppose it is always attractive toward the origin. This has been found by experiment to be very nearly the law of tension of an elastic string within certain limits of stretching. Then the velocity will decrease when the particle is on the positive side of the origin; therefore for these positions of the particle the acceleration must be taken with the negative sign, and the differential equation for positive values of s is

$$(6) \qquad m\frac{d^2s}{dt^2} = -k^2s.$$

For positions of the particle in the negative direction from the origin the velocity increases with the time, and therefore the acceleration is positive. The right member of equation (6) must be taken with such a sign that it will be positive. Since s is negative in the region under consideration the negative sign must be prefixed, and the equation remains as before. Equation (6) is, therefore, the general differential equation of motion for a body subject to an attractive force varying directly as the distance.

Multiply both members of equation (6) by $2\frac{ds}{dt}$ and integrate; the result is

$$m\left(\frac{ds}{dt}\right)^2 = -k^2s^2 + c_1.$$

If $s = s_0$ and $\frac{ds}{dt} = 0$, at the time $t = 0$, then this equation becomes

$$\left(\frac{ds}{dt}\right)^2 = \frac{k^2}{m}(s_0^2 - s^2),$$

which may be written, as is customary in separating the variables,

$$\frac{ds}{\sqrt{s_0^2 - s^2}} = \pm\frac{kdt}{\sqrt{m}}.$$

The integral of this equation is

$$\sin^{-1}\frac{s}{s_0} = \pm\frac{kt}{\sqrt{m}} + c_2.$$

It is found from the initial conditions that $c_2 = \frac{\pi}{2}$; whence

$$\sin^{-1}\frac{s}{s_0} = \pm\frac{kt}{\sqrt{m}} + \frac{\pi}{2}.$$

On taking the sine of both members, this equation becomes

$$(7) \qquad s = s_0 \sin\left(\pm\frac{kt}{\sqrt{m}} + \frac{\pi}{2}\right) = s_0 \cos\left(\frac{kt}{\sqrt{m}}\right).$$

From this equation it is seen that the motion is oscillatory and symmetrical with respect to the origin, with a period of $\frac{2\pi\sqrt{m}}{k}$.

For this reason it is called the equation for harmonic motion. Obviously $\frac{ds}{dt}$ vanishes at some time during the motion for all initial conditions, and there was no real restriction of the generality of the problem in supposing that it was zero at $t = 0$.

Equation (6) is in form the differential equation for many physical problems. When the initial conditions are assigned, it defines the motion of the simple pendulum, the oscillations of the tuning fork and most musical instruments, the vibrations of a radiating body, the small variations in the position of the earth's axis, etc. For this reason the method of finding its solution and the determination of the constants of integration should be thoroughly mastered.

III. PROBLEMS.

1. A particle is started with an initial velocity of 20 meters per sec. and is subject to an acceleration of 20 meters per sec. What will be its velocity at the end of 4 secs., and how far will it have moved?

$$Ans. \quad \begin{cases} v = 100 \text{ meters per sec.} \\ s = 240 \text{ meters.} \end{cases}$$

2. A particle starting with an initial velocity of 10 meters per sec. and moving with a constant acceleration describes 2050 meters in 5 secs. What is the acceleration?

Ans. $\alpha = 160$ meters per sec.

3. A particle is moving with an acceleration of 5 meters per sec. Through what space must it move in order that its velocity shall be increased from 10 meters per sec. to 20 meters per sec.?

Ans. 30 meters.

4. A particle starting with a positive initial velocity of 10 meters per sec. and moving under a positive acceleration of 20 meters per sec. described a space of 420 meters. What time was required?

Ans. $t = 6$ secs.

5. Show that, if a particle starting from rest moves subject to an attractive force varying directly as the distance, the time of falling from any point to the origin is independent of the distance of the particle.

6. Suppose a particle moves subject to an attractive force varying directly as the distance, and that the acceleration at a distance of 1 meter is 1 meter a sec. If the particle starts from rest how long will it take it to fall from a distance of 20 meters to 10 meters?

Ans. 1.0472 secs.

7. Suppose a particle moves subject to a force which is repulsive from the origin and which varies directly as the distance; show that if $v = 0$ and $s = s_0$ when $t = 0$, then

$$\log \left(\frac{s + \sqrt{s^2 - s_0{}^2}}{s_0} \right) = \frac{k}{\sqrt{m}} t;$$

whence, letting $\dfrac{k}{\sqrt{m}} = K$,

$$s = \frac{s_0}{2} \left(e^{Kt} + e^{-Kt} \right) = s_0 \cosh Kt.$$

Observe that equation (7) may be written in the similar form

$$s = \frac{s_0}{2} \left(e^{\sqrt{-1}Kt} + e^{-\sqrt{-1}Kt} \right) = s_0 \cos Kt.$$

8. The surface gravity of the sun is 28 times that of the earth. If a solar prominence 100,000 miles high was produced by an explosion, what must have been the velocity of the material when it left the surface of the sun?

Ans.   184 miles per sec.

32. Solution of Linear Equations by Exponentials.   The differential equation (6) and the corresponding one for a repulsive force are linear in s with constant coefficients. There is a general theory which shows that all linear equations having constant coefficients can be solved in terms of exponentials; or, in certain special cases, in terms of exponentials multiplied by powers of the independent variable t. This method has not only the advantage of generality, but is also very simple, and it will be illustrated by solving (5). Consider the two forms

(8)
$$\begin{cases} \dfrac{d^2 s}{dt^2} + k^2 s = 0, \\[2mm] \dfrac{d^2 s}{dt^2} - k^2 s = 0. \end{cases}$$

Assume $s = e^{\lambda t}$ and substitute in the differential equations; whence

$$\begin{cases} \lambda^2 e^{\lambda t} + k^2 e^{\lambda t} = 0, \\ \lambda^2 e^{\lambda t} - k^2 e^{\lambda t} = 0. \end{cases}$$

Since these equations must be satisfied by all values of t in order that $e^{\lambda t}$ shall be a solution, it follows that

(9)
$$\begin{cases} \lambda^2 + k^2 = 0, \\ \lambda^2 - k^2 = 0. \end{cases}$$

Let the roots of the first equation be λ_1 and λ_2; then the first equation of (8) is verified by both of the particular solutions $e^{\lambda_1 t}$ and $e^{\lambda_2 t}$. The general solution is the sum of these two particular solutions, each multiplied by an arbitrary constant. Precisely similar results hold for the second equation of (8). On putting in the value of the roots, the respective general solutions are

$$(10) \qquad \begin{cases} s = c_1 e^{\sqrt{-1}\,kt} + c_2 e^{-\sqrt{-1}\,kt}, \\ s = c_1' e^{kt} + c_2' e^{-kt}, \end{cases}$$

where c_1, c_2, c_1', and c_2' are the constants of integration.

Suppose $s = s_0$, and $\dfrac{ds}{dt} = s_0'$ when $t = 0$; therefore

$$\begin{cases} s_0 = c_1 + c_2, \\ s_0 = c_1' + c_2'. \end{cases}$$

The derivatives of (10) are

$$\begin{cases} \dfrac{ds}{dt} = c_1 \sqrt{-1}\, k e^{\sqrt{-1}\,kt} - c_2 \sqrt{-1}\, k e^{-\sqrt{-1}\,kt}, \\[2mm] \dfrac{ds}{dt} = c_1' k e^{kt} - c_2' k e^{-kt}. \end{cases}$$

On substituting $t = 0$ and $\dfrac{ds}{dt} = s_0'$, it follows that

$$\begin{cases} c_1 \sqrt{-1}\, k - c_2 \sqrt{-1}\, k = s_0', \\ c_1' k - c_2' k = s_0'. \end{cases}$$

Therefore

$$\begin{cases} c_1 = \dfrac{1}{2}\left(s_0 + \dfrac{s_0'}{k\sqrt{-1}} \right), \\[3mm] c_2 = \dfrac{1}{2}\left(s_0 - \dfrac{s_0'}{k\sqrt{-1}} \right), \\[3mm] c_1' = \dfrac{1}{2}\left(s_0 + \dfrac{s_0'}{k} \right), \\[3mm] c_2' = \dfrac{1}{2}\left(s_0 - \dfrac{s_0'}{k} \right). \end{cases}$$

Then the general solutions become

$$(11) \qquad \begin{cases} s = \dfrac{s_0}{2}\left(e^{\sqrt{-1}\,kt} + e^{-\sqrt{-1}\,kt} \right) + \dfrac{s_0'}{2k\sqrt{-1}}\left(e^{\sqrt{-1}\,kt} - e^{-\sqrt{-1}\,kt} \right), \\[3mm] s = \dfrac{s_0}{2}\left(e^{kt} + e^{-kt} \right) + \dfrac{s_0'}{2k}\left(e^{kt} - e^{-kt} \right). \end{cases}$$

Or these expressions can be written in the form

$$
\begin{cases}
s = s_0 \cos kt + \dfrac{s_0'}{k} \sin kt, \\[3mm]
s = s_0 \cosh kt + \dfrac{s_0'}{k} \sinh kt.
\end{cases}
$$

This method of treatment shows the close relation between the two problems much more clearly than the other methods of obtaining the solutions.

33. Attractive Force Varying Inversely as the Square of the Distance. For positions in the positive direction from the origin the velocity decreases algebraically as the time increases whether the motion is toward or from the origin; therefore in this region the acceleration is negative. Similarly, on the negative side of the origin the acceleration is positive. Since $\dfrac{k^2}{s^2}$ is always positive the right member has different signs in the two cases. For simplicity suppose the mass of the attracted particle is unity. Then the differential equation of motion for all positions of the particle in the positive direction from the origin is

$$
(12) \qquad \frac{d^2s}{dt^2} = -\frac{k^2}{s^2}.
$$

On multiplying both members of this equation by $2\dfrac{ds}{dt}$ and integrating, it is found that

$$
(13) \qquad \left(\frac{ds}{dt}\right)^2 = \frac{2k^2}{s} + c_1.
$$

Suppose $v = v_0$ and $s = s_0$ when $t = 0$; then

$$
c_1 = v_0{}^2 - \frac{2k^2}{s_0}.
$$

On substituting this expression for c_1 in (13), it is found that

$$
\frac{ds}{dt} = \pm \sqrt{\frac{2k^2}{s} + v_0{}^2 - \frac{2k^2}{s_0}}.
$$

If $v_0{}^2 - \dfrac{2k^2}{s_0} < 0$ there will be some finite distance s_1 at which $\dfrac{ds}{dt}$ will vanish; if the direction of motion of the particle is such that it reaches that point it will turn there and move in the opposite direction. If $v_0{}^2 - \dfrac{2k^2}{s_0} = 0$, $\dfrac{ds}{dt}$ will vanish at $s = \infty$; and if the particle moves out from the origin toward infinity its distance will become indefinitely great as the velocity approaches zero. If $v_0^2 - \dfrac{2k^2}{s_0} > 0$, $\dfrac{ds}{dt}$ never vanishes, and if the particle moves out from the origin toward infinity its distance will become indefinitely great and its velocity will not approach zero.

Suppose $v_0^2 - \dfrac{2k^2}{s_0} < 0$ and that $\dfrac{ds}{dt} = 0$ when $s = s_1$. Then equation (13) gives

(14)
$$\frac{ds}{dt} = \pm \sqrt{\frac{2}{s_1}} k \sqrt{\frac{s_1 - s}{s}}.$$

The positive or negative sign is to be taken according as the particle is receding from, or approaching toward, the origin.

This equation can be written in the form

$$\frac{s\,ds}{\sqrt{s_1 s - s^2}} = \pm \sqrt{\frac{2}{s_1}} k\,dt,$$

and the integral is therefore

$$- \sqrt{s_1 s - s^2} + \frac{s_1}{2} \sin^{-1}\left(\frac{2s - s_1}{s_1} \right) = \pm \sqrt{\frac{2}{s_1}} kt + c_2.$$

Since $s = s_0$ when $t = 0$, it follows that

$$c_2 = - \sqrt{s_1 s_0 - s_0^2} + \frac{s_1}{2} \sin^{-1}\left(\frac{2s_0 - s_1}{s_1} \right);$$

whence

(15)
$$\frac{s_1}{2}\left[\sin^{-1}\left(\frac{2s - s_1}{s_1} \right) - \sin^{-1}\left(\frac{2s_0 - s_1}{s_1} \right) \right]$$
$$+ \sqrt{s_1 s_0 - s_0^2} - \sqrt{s_1 s - s^2} = \pm \sqrt{\frac{2}{s_1}} kt.$$

This equation determines the time at which the particle has any position at the right of the origin whose distance from it is less than s_1. For values of s greater than s_1, and for all negative values of s, the second term becomes imaginary. That means that the equation does not hold for these values of the variables; this was indeed certain because the differential equations (13) and (14) were valid only for

$$0 < s \leqq s_1.$$

Suppose the particle is approaching the origin; then the negative sign must be used in the right member of (15). The time at which the particle was at rest is obtained by putting $s = s_1$ in (15), and is

$$T_1 = -\frac{1}{k}\sqrt{\frac{s_1}{2}}\sqrt{s_1 s_0 - s_0^2} - \frac{1}{k}\left(\frac{s_1}{2}\right)^{\frac{3}{2}}\left[+\frac{\pi}{2} - \sin^{-1}\left(\frac{2s_0 - s_1}{s_1}\right)\right].$$

The time required for the particle to fall from s_0 to the origin is obtained by putting $s = 0$ in (15), and is

$$T_2 = -\frac{1}{k}\sqrt{\frac{s_1}{2}}\sqrt{s_1 s_0 - s_0^2} - \frac{1}{k}\left(\frac{s_1}{2}\right)^{\frac{3}{2}}\left[-\frac{\pi}{2} - \sin^{-1}\left(\frac{2s_0 - s_1}{s_1}\right)\right].$$

The time required for the particle to fall from rest at $s = s_1$ to the origin is

(16) $$T = T_2 - T_1 = \frac{\pi}{k}\left(\frac{s_1}{2}\right)^{\frac{3}{2}}.$$

34. The Height of Projection. When the constant c_1 has been determined by the initial conditions, equation (13) becomes

$$\left(\frac{ds}{dt}\right)^2 - v_0^2 = v^2 - v_0^2 = 2k^2\left(\frac{1}{s} - \frac{1}{s_0}\right).$$

It follows from this equation that the speed depends only on the distance of the particle from the center of force and not on the direction of its motion. The greatest distance to which the particle recedes from the origin, or the height of projection from the origin, is obtained by putting $v = 0$, which gives

$$\frac{1}{s_1} = \frac{1}{s_0} - \frac{v_0^2}{2k^2}.$$

But if the height of projection is measured from the point of projection s_0, as would be natural in considering the projection of a body away from the surface of the earth, the formula becomes

$$s_2 = s_1 - s_0 = \frac{v_0^2 s_0^2}{2k^2 - v_0^2 s_0}.$$

35. The Velocity from Infinity. When the particle starts from s_0 with zero velocity, equation (13) becomes

(17) $$\left(\frac{ds}{dt}\right)^2 = 2k^2\left(\frac{1}{s} - \frac{1}{s_0}\right).$$

If the particle falls from an infinite distance, s_0 is infinite and the equation reduces to

(18) $$\left(\frac{ds}{dt}\right)^2 = \frac{2k^2}{s}.$$

From the investigations of Art. 34 it follows that, if the particle is projected from any point s in the positive direction with the velocity defined by (18), it will recede to infinity. The law of attraction in deriving (18) is Newton's law of gravitation; therefore this equation can be used for the computation of the velocity which a particle starting from infinity would acquire in falling to the surfaces of the various planets, satellites, and the sun. Then, if the particle were projected from the surfaces of the various members of the solar system with these respective velocities, it would recede to an infinite distance if it were not acted on by other forces. But if its velocity were only enough to take it away from a satellite or a planet, it would still be subject to the attraction of the remaining members of the solar system, chief of which is of course the sun, and it would not in general recede to infinity and be entirely lost to the system.

36. Application to the Escape of Atmospheres. The kinetic theory of gases is that the volumes and pressures of gases are maintained by the mutual impacts of the individual molecules, which are, on the average, in a state of very rapid motion. The rarity of the earth's atmosphere and the fact that the pressure is about fifteen pounds to the square inch, serve to give some idea of the high speed with which the molecules move and of the great frequency of their impacts. The different molecules do not all move with the same speed in any given gas under fixed conditions; but the number of those which have a rate of motion different from the mean of the squares becomes less very rapidly as the differences increase. Theoretically, in all gases the range of the values of the velocities is from zero to infinity, although the extreme cases occur at infinitely rare intervals compared to the others. Under constant pressure the velocities are directly proportional to the square root of the absolute temperature, and inversely proportional to the square root of the molecular weight.

Since in all gases all velocities exist, some of the molecules of the gaseous envelopes of the heavenly bodies must be moving with velocities greater than the *velocity from infinity*, as defined in Art. 35. If the molecules are near the upper limits of the atmosphere, and start away from the body to which they belong, they may miss collisions with other molecules and escape never to return*. Since the kinetic theory of gases is supported by very strong observational evidence, and since if it is true it is probable that some molecules move with velocities greater than the velocity

---

* This theory is due to Johnstone Stoney, *Trans. Royal Dublin Soc.*, 1892.

from infinity, it is probable that the atmospheres of all celestial bodies are being depleted by this process; but in most cases it is an excessively slow one, and is compensated, to some extent at least, by the accretion of meteoric matter and atmospheric molecules from other bodies. In the upper regions of the gaseous envelopes, from which alone the molecules escape, the temperatures are low, at least for planetary bodies like the earth, and high velocities are of rare occurrence. If the mean square velocity were as great as, or exceeded, the velocity from infinity the depletion would be a relatively rapid process. In any case the elements and compounds with low molecular weights would be lost most rapidly, and thus certain ones might freely escape and others be largely retained.

The manner in which the velocity from infinity with respect to the surface of an attracting sphere varies with its mass and radius will now be investigated. The mass of a body is proportional to the product of its density and cube of its radius. Then k^2, which is the attraction at unit distance, varies directly as the mass, and therefore directly as the product of the density and the cube of the radius. Hence it follows from (18) that the velocity from infinity at the surface of a body varies as the product of its radius and the square root of its density. The densities and the radii of the members of the solar system are usually expressed in terms of the density and the radius of the earth; hence the velocity from infinity can be easily computed for each of them after it has been determined for the earth.

Let R represent the radius of the earth, and g the acceleration of gravitation at its surface. Then it follows that

(19) $$g = \frac{k^2}{R^2}.$$

On substituting the value of k^2 determined from this equation into (18), the expression for the square of the velocity becomes

$$\left(\frac{ds}{dt}\right)^2 = \frac{2gR^2}{s}.$$

Let $V = \dfrac{ds}{dt}$ when $s = R$; whence

$$V^2 = 2gR.$$

Let a second be taken as the unit of time, and a meter as the unit

of length. Then $R = 6,371,000$, and $g = 9.8094$*. On substituting in the last equation and carrying out the computation, it
is found that $V = 11,180$ meters, or about 6.95 miles, per sec.
On taking the values of the relative densities and radii of the
planets as given in Chapter XI of Moulton's Introduction to
Astronomy, the following results are found:

Body	Velocity of Escape		
Earth	11,180 meters, or	6.95	miles, per sec.
Moon	2,396 "	" 1.49	" " "
Sun	618,200 "	" 384.30	" " "
Mercury	3,196 "	" 1.99	" " "
Venus	10,475 "	" 6.51	" " "
Mars	5,180 "	" 3.22	" " "
Jupiter	61,120 "	" 38.04	" " "
Saturn	37,850 "	" 23.53	" " "
Uranus	23,160 "	" 14.40	" " "
Neptune	20,830 "	" 12.95	" " "

The velocity from infinity decreases as the distance from the
center of a planet increases, and the necessary velocity of projection in order that a molecule may escape decreases correspondingly; and the centrifugal acceleration of the planet's rotation
adds to the velocities of certain molecules.

The question arises whether, under the conditions prevailing
at the surfaces of the various bodies enumerated, the average
molecular velocities of the atmospheric elements do not equal or
surpass the corresponding velocity from infinity.

It is possible to find experimentally the pressure exerted by a
gas having a given density and temperature upon a unit surface,
from which the mean square velocity can be computed. It is
shown in the kinetic theory of gases that the square root of the
mean square velocity of hydrogen molecules at the temperature
$0°$ Centigrade under atmospheric pressure is about 1,700 meters per
second. Under the same conditions the velocities of oxygen and
nitrogen molecules are about one-fourth as great.

The surface temperatures of the inferior planets are certainly
much greater than zero degrees Centigrade in the parts where
they receive the rays of the sun most directly, even if all the heat
which may ever have been received from their interiors is neglected.
It seems probable from the geological evidences of igneous action

* *Annuaire du Bureau des Long.* g is given for the lat. of Paris, $48° 50'$.

upon the earth that in the remote past they were at a much higher temperature, and the superior planets have not yet cooled down to the solid state. There is evidence that the most refractory substances have been in a molten state at some time, which implies that they have had a temperature of 3000 or 4000 degrees Centigrade. Therefore the square root of the mean square velocity may have been much greater than the approximate mile a second for hydrogen given above, and it probably continued much greater for a long period of time. On comparing these results with the table of velocities from infinity, it is seen that the moon and inferior planets, according to this theory, could not possibly have retained free hydrogen and other elements of very small molecular weights, such as helium, in their envelopes; in the case of the moon, Mercury, and Mars, the escape of heavier molecules as nitrogen and oxygen must have been frequent. This is especially probable if the heated atmospheres extended out to great distances. The superior planets, and especially the sun, could have retained all of the familiar terrestrial elements, and from this theory it should be expected that these bodies would be surrounded with extensive gaseous envelopes.

The observed facts are that the moon has no appreciable atmosphere whatever; Mercury an extremely rare one, if any at all; Mars, one much less dense than that of the earth; Venus, one perhaps somewhat more dense than that of the earth; on the other hand the superior planets are surrounded by extensive gaseous envelopes.

37. The Force Proportional to the Velocity. When a particle moves in a resisting medium the forces to which it is subject depend upon its velocity. Experiments have shown that in the earth's atmosphere when the velocity is less than 3 meters per second the resistance varies nearly as the first power of the velocity; for velocities from 3 to 300 meters per second it varies nearly as the second power of the velocity; and for velocities about 400 meters per second, nearly as the third power of the velocity.

(a) Consider first the case where the resistance varies as the first power of the velocity, and suppose the motion is on the earth's surface in a horizontal direction with no force acting except that arising from the resistance. Then the differential equation of motion is

(20) $$\frac{d^2s}{dt^2} + k^2 \frac{ds}{dt} = 0,$$

where k^2 is a positive constant which depends upon the units employed, the nature of the body, and the character of the resisting medium. Equation (20) is linear in the dependent variable s, and the general method of solving linear equations can be applied.

Assume the particular solution

$$s = e^{\lambda t}.$$

Substitute in (20) and divide by $e^{\lambda t}$; then

$$\lambda^2 + k^2\lambda = 0.$$

The roots of this equation are

$$\begin{cases} \lambda_1 = 0, \\ \lambda_2 = -k^2, \end{cases}$$

and the general solution is

$$(21) \quad \begin{cases} s = c_1 + c_2 e^{-k^2 t}, \\ \dfrac{ds}{dt} = -c_2 k^2 e^{-k^2 t}. \end{cases}$$

Suppose $\dfrac{ds}{dt} = v_0$ and $s = s_0$ when $t = 0$. Then the constants c_1 and c_2 can be determined in terms of v_0 and s_0, and the solution becomes

$$(22) \qquad s = s_0 + \frac{v_0}{k^2} - \frac{v_0}{k^2} e^{-k^2 t}.$$

(b) Consider the case where the resistance varies as the first power of the velocity and suppose the motion is in the vertical line. Take the positive end of the axis upward. When the motion is upward the velocity is positive and the resistance diminishes the velocity. Therefore when the motion is upward the resistance produces a negative acceleration, and the differential equation of motion is

$$(23) \qquad \frac{d^2 s}{dt^2} + k^2 \frac{ds}{dt} = -g.$$

When the motion is downward the resistance algebraically increases the velocity; therefore in this case the resistance produces a positive acceleration. But since the velocity is opposite in sign in the two cases, equation (23) holds whether the particle is ascending or descending.

Equation (23) is linear, but not homogeneous, and it can easily be solved by the method known as the *Variation of Parameters*.

This method is so important in astronomical problems that it will be introduced in the present simple connection, though it is not at all necessary in order to obtain the solution of (23). In order to apply the method consider first the equation

$$(24) \qquad \frac{d^2 s}{dt^2} + k^2 \frac{ds}{dt} = 0,$$

which is obtained from (23) simply by omitting the term which does not involve s. The general solution of this equation is the first of (21). The method of the variation of parameters, or constants, consists in so determining c_1 and c_2 as functions of t that the differential equation shall be satisfied when the right member is included. This imposes only one condition upon the two quantities c_1 and c_2, and another can therefore be added.

If the coefficients c_1 and c_2 are regarded as functions of t, it is found on differentiating the first of (21) that

$$\frac{ds}{dt} = - k^2 c_2 e^{-k^2 t} + \frac{dc_1}{dt} + e^{-k^2 t} \frac{dc_2}{dt}.$$

As the supplementary condition on c_1 and c_2 these quantities will be made subject to the relation

$$(25) \qquad \frac{dc_1}{dt} + e^{-k^2 t} \frac{dc_2}{dt} = 0,$$

which simplifies the expression for $\frac{ds}{dt}$. Then it is found that

$$(26) \qquad \frac{d^2 s}{dt^2} = k^4 c_2\, e^{-k^2 t} - k^2 e^{-k^2 t} \frac{dc_2}{dt}$$

and equation (23) gives

$$(27) \qquad k^2 \frac{dc_1}{dt} = - g.$$

It follows from this equation and (25) that

$$(28) \qquad \begin{cases} \dfrac{dc_1}{dt} = - \dfrac{g}{k^2}, \qquad \dfrac{dc_2}{dt} = \dfrac{g}{k^2} e^{k^2 t}, \\[2ex] c_1 = - \dfrac{g}{k^2} t + c_1{}', \quad c_2 = \dfrac{g}{k^4} e^{k^2 t} + c_2{}', \end{cases}$$

where $c_1{}'$ and $c_2{}'$ are new constants of integration. When these values of c_1 and c_2 are substituted in (21), it is found that

$$(29) \qquad s = c_1{}' + c_2{}' e^{-k^2 t} - \frac{g}{k^2} t + \frac{g}{k^4}.$$

Since c_1' is arbitrary it can of course be supposed to include the constant $\dfrac{g}{k^4}$.

The expression (29) is the general solution of (23) because it contains two arbitrary constants, c_1' and c_2', and when substituted in (23) satisfies it identically in t. It will be observed that the part of the solution depending on c_1' and c_2' has the same form as the solution of (20). It is clear that the general solution could have been found by the same method if the right member of (23) had been a known function of t, instead of the constant $-g$.

The velocity of the particle is found from (29) to be given by the equation

$$(30) \qquad \frac{ds}{dt} = -k^2 c_2' e^{-k^2 t} - \frac{g}{k^2}.$$

Suppose $s = s_0$, $\dfrac{ds}{dt} = v_0$ at $t = 0$. On putting these values in equations (29) and (30), it is found that

$$\begin{cases} s_0 = c_1' + c_2' + \dfrac{g}{k^4}, \\[2mm] v_0 = -k^2 c_2' - \dfrac{g}{k^2}; \end{cases}$$

whence

$$\begin{cases} c_1' = s_0 + \dfrac{v_0}{k^2}, \\[2mm] c_2' = -\dfrac{v_0}{k^2} - \dfrac{g}{k^4}. \end{cases}$$

Consequently, when the constants are determined by the initial conditions, the general solution (29) becomes

$$(31) \qquad \begin{cases} s = s_0 + \dfrac{v_0}{k^2} + \dfrac{g}{k^4} - \dfrac{g}{k^2} t - \left(\dfrac{v_0}{k^2} + \dfrac{g}{k^4} \right) e^{-k^2 t}, \\[3mm] \dfrac{ds}{dt} = -\dfrac{g}{k^2} + \left(v_0 + \dfrac{g}{k^2} \right) e^{-k^2 t}. \end{cases}$$

The particle reaches its highest point when $\dfrac{ds}{dt}$ is zero. Let T represent the time it reaches this point and $S - s_0$ the height of this point; then it is found from equations (31) that

$$\begin{cases} e^{k^2 T} = 1 + \dfrac{k^2 v_0}{g}, \\[3mm] S - s_0 = \dfrac{v_0}{k^2} - \dfrac{g}{k^4} \log \left(1 + \dfrac{k^2 v_0}{g} \right). \end{cases}$$

38. The Force Proportional to the Square of the Velocity. At the velocity of a strong wind, or of a body falling any considerable distance, or of a ball thrown, the resistance varies very nearly as the square of the velocity. An investigation will now be made of the character of the motion of a particle when projected *upward* against gravity, and subject to a resistance from the atmosphere varying as the square of the velocity. For simplicity in writing, the acceleration due to resistance at unit velocity will be taken as $k^2 g$. Then the differential equation of motion for a unit particle is

$$(32) \qquad \frac{d^2 s}{dt^2} = - g - k^2 g \left(\frac{ds}{dt} \right)^2.$$

This equation may be written in the form

$$\frac{\dfrac{d}{dt}\left(k \dfrac{ds}{dt} \right)}{1 + k^2 \left(\dfrac{ds}{dt} \right)^2} = - kg,$$

of which the integral is

$$(33) \qquad \tan^{-1} \left(k \frac{ds}{dt} \right) = - kgt + c_1.$$

If $\dfrac{ds}{dt} = v_0$ and $s_0 = 0$ when $t = 0$, then

$$c_1 = \tan^{-1} (kv_0).$$

On substituting in (33) and taking the tangent of both members, it is found that

$$(34) \qquad \frac{ds}{dt} = \frac{1}{k} \frac{v_0 k - \tan (kgt)}{1 + v_0 k \tan (kgt)}.$$

This equation expresses the velocity in terms of the time. On multiplying both numerator and denominator of the right member of (34) by $\cos (kgt)$, the numerator becomes the derivative of the denominator with respect to the time. Then integrating, the final solution becomes

$$(35) \qquad s = \frac{1}{k^2 g} \log \left[v_0 k \sin (kgt) + \cos (kgt) \right] + c_2.$$

It follows from the initial conditions that $c_2 = 0$. This equation expresses the distance passed over in terms of the time.

The equations can be so treated that the velocity will be expressed in terms of the distance. Equation (32) can be written

$$\frac{\frac{d}{dt}\left\{k^2\left(\frac{ds}{dt}\right)^2\right\}}{1 + k^2\left(\frac{ds}{dt}\right)^2} = -2gk^2\frac{ds}{dt},$$

of which the integral is

$$\log\left\{1 + k^2\left(\frac{ds}{dt}\right)^2\right\} = -2gk^2s + c_1'.$$

From the initial conditions it follows that

$$c_1' = \log(1 + k^2v_0^2).$$

Therefore

(36) $$\left(\frac{ds}{dt}\right)^2 = \frac{1}{k^2}(1 + k^2v_0^2)e^{-2gk^2s} - \frac{1}{k^2}.$$

The maximum height, which is reached when the velocity becomes zero, is found from (36) to be

$$S = \frac{1}{2gk^2}\log(1 + k^2v_0^2).$$

The time of reaching the highest point, which is found by putting $\frac{ds}{dt}$ equal to zero in (34), is given by

$$T = \frac{1}{kg}\tan^{-1}(v_0k).$$

When the particle falls the resistance acts in the opposite direction and the sign of the last term in (32) is changed. This may be accomplished by writing $k\sqrt{-1}$ instead of k, and if this change is made throughout the solution the results will be valid. Of course the results should be written in the exponential form, instead of the trigonometric as they were in (34) and (35), in order to avoid the appearance of imaginary expressions. If the initial velocity is zero, $v_0 = 0$ and the equations corresponding to (34), (35), and (36) are repectively

(37) $$\begin{cases} \dfrac{ds}{dt} = -\dfrac{1}{k}\dfrac{e^{kgt} - e^{-kgt}}{e^{kgt} + e^{-kgt}}, \\[2mm] e^{-gk^2s} = \dfrac{e^{kgt} + e^{-kgt}}{2}, \\[2mm] \left(\dfrac{ds}{dt}\right)^2 = \dfrac{1}{k^2}(1 - e^{2gk^2s}). \end{cases}$$

IV. PROBLEMS.

1. Show that

$$\frac{d^2s}{dt^2} = -k^2 \frac{s}{\sqrt{s^6}},$$

where the positive square root of s^6 is always taken, holds for the problem of Art. 33 whichever side of the origin the particle may be. Integrate this equation.

2. Let $s = s' - \frac{g}{k^2}t$ in equation (23); integrate directly and show that the result is the same as that found by the variation of parameters.

3. Find equations (37) by direct integration of the differential equations.

4. Suppose a particle starts from rest and moves subject to a repulsive force varying inversely as the square of the distance; find the velocity and time elapsed in terms of the space described.

$$\text{Ans.} \quad \begin{cases} v^2 = 2k^2 \left(\frac{1}{s_0} - \frac{1}{s} \right), \\[2ex] k\sqrt{\frac{2}{s_0}}\, t = \sqrt{s^2 - s_0 s} + \frac{s_0}{2} \log \left(\dfrac{\sqrt{s^2 - s_0 s} + s - \dfrac{s_0}{2}}{\dfrac{s_0}{2}} \right). \end{cases}$$

5. What is the velocity from infinity with respect to the sun at the earth's distance from the sun?

Ans. 42,220 meters, or 26.2 miles, per sec.

6. Suppose a particle moves subject to an attractive force varying directly as the distance, and to a resistance which is proportional to the speed; solve the differential equation by the general method for linear equations.

Ans. Let k^2 be the factor of proportionality for the velocity and l^2 for the distance. Then the solutions are

$$s = c_1 e^{\lambda_1 t} + c_2 e^{\lambda_2 t},$$

where

$$\begin{cases} \lambda_1 = \dfrac{-k^2 + \sqrt{k^4 - 4l^2}}{2}, \\[2ex] \lambda_2 = \dfrac{-k^2 - \sqrt{k^4 - 4l^2}}{2}. \end{cases}$$

Discuss more in detail the form of the solution and its physical meaning when (a) $k^4 - 4l^2 < 0$, (b) $k^4 - 4l^2 = 0$, (c) $k^4 - 4l^2 > 0$.

7. Suppose that in addition to the forces of problem 6 there is a force $\mu e^{\nu t}$; derive the solution by the method of the variation of parameters and discuss the motion of the particle.

8. Develop the method of the variation of parameters for a linear differential equation of the third order.

9. If $k^2 = 0$ equation (23) becomes that which defines the motion of a freely falling body. Show that the limit of the solution (32) as k^2 approaches zero is

$$s = s_0 + v_0 t - \tfrac{1}{2}g t^2.$$

39. Parabolic Motion. There is a class of problems involving for their solution mathematical processes which are similar to those employed thus far in this chapter, although the motion is not in a straight line. On account of the similarity in the analysis a short discussion of these problems will be inserted here.

Suppose the particle is subject to a constant acceleration downward; the problem is to find the character of the curve described when the particle is projected in any manner. The orbit will be in a plane which will be taken as the xy-plane. Let the y-axis be vertical with the positive end directed upward. Then the differential equations of motion are

(38)
$$\begin{cases} \dfrac{d^2x}{dt^2} = 0, \\[2mm] \dfrac{d^2y}{dt^2} = -g. \end{cases}$$

Since these equations are independent of each other, they can be integrated separately, and give

$$\begin{cases} x = a_1 t + a_2, \\[2mm] y = -\dfrac{gt^2}{2} + b_1 t + b_2. \end{cases}$$

Suppose $x = y = 0$, $\dfrac{dx}{dt} = v_0 \cos \alpha$, $\dfrac{dy}{dt} = v_0 \sin \alpha$ when $t = 0$,

where α is the angle between the line of initial projection and the plane of the horizon, and v_0 is the speed of the projection. Then

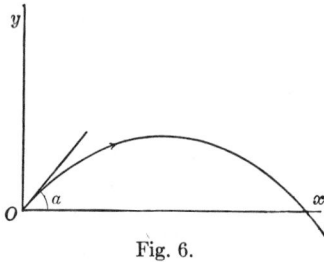

Fig. 6.

the constants of integration are found to be

$$a_1 = v_0 \cos \alpha, \qquad a_2 = 0,$$
$$b_1 = v_0 \sin \alpha, \qquad b_2 = 0;$$

and therefore

$$(39) \quad \begin{cases} x = v_0 \cos \alpha \cdot t, \\ y = -\dfrac{gt^2}{2} + v_0 \sin \alpha \cdot t. \end{cases}$$

The equation of the curve described, which is found by eliminating t between these two equations, is

$$(40) \quad y = x \tan \alpha - \frac{g \sec^2 \alpha}{2v_0{}^2} x^2.$$

This is the equation of a parabola whose axis is vertical with its vertex upward. It can be written in the form

$$\left(x - \frac{v_0{}^2}{g} \sin \alpha \cos \alpha \right)^2 = - \frac{2v_0{}^2}{g \sec^2 \alpha} \left(y - \frac{v_0{}^2 \sin^2 \alpha}{2g} \right).$$

The equation of a parabola with its vertex at the origin has the form

$$x^2 = - 2py,$$

where $2p$ is the parameter. On comparing this equation with the equation of the curve described by the particle, the coördinates of the vertex, or highest point, are seen to be

$$\begin{cases} \bar{x} = \dfrac{v_0{}^2}{g} \sin \alpha \cos \alpha, \\ \bar{y} = \dfrac{v_0{}^2}{2g} \sin^2 \alpha. \end{cases}$$

The distance of the directrix from the vertex is one-fourth of the parameter; therefore the equation of the directrix is

$$y = \bar{y} + \frac{p}{2} = \frac{v_0{}^2 \sin^2 \alpha}{2g} + \frac{v_0{}^2 \cos^2 \alpha}{2g} = \frac{v_0{}^2}{2g}.$$

The square of the velocity is found to be

$$v^2 = \left(\frac{dx}{dt} \right)^2 + \left(\frac{dy}{dt} \right)^2 = v_0{}^2 - 2gy.$$

To find the place where the particle will strike the horizontal plane put $y = 0$ in (40). The solutions for x are $x = 0$ and

$$x = \frac{2v_0{}^2}{g} \sin \alpha \cos \alpha = \frac{v_0{}^2}{g} \sin 2\alpha.$$

From this it follows that the range is greatest for a given initial velocity if $\alpha = 45°$. From (39) the horizontal velocity is seen to be $v_0 \cos \alpha$; hence the time of flight is $\dfrac{2v_0}{g} \sin \alpha$. Therefore, if the other initial conditions are kept fixed, the whole time of flight varies directly as the sine of the angle of elevation.

The angle of elevation to attain a given range is found by solving

$$x = a = \frac{v_0^2}{g} \sin 2\alpha$$

for α. If $a > \dfrac{v_0^2}{g}$ there is no solution. If $a < \dfrac{v_0^2}{g}$ there are two solutions differing from the value for maximum range ($\alpha = 45°$) by equal amounts.

If the variation in gravity at different heights above the earth's surface, the curvature of the earth, and the resistance of the air are neglected, the investigation above applies to projectiles near the earth's surface. For bodies of great density the results given by this theory are tolerably accurate for short ranges. When the acceleration is taken toward the center of the earth, and gravity is supposed to vary inversely as the square of the distance, the path described by the particle is an ellipse with the center of the earth as one of the foci. This will be proved in the next chapter.

V. PROBLEMS.

1. Prove that, if the accelerations parallel to the x and y-axes are any constant quantities, the path described by the particle is a parabola for general initial conditions.

2. Find the direction of the major axis, obtained in problem 1, in terms of the constant components of acceleration.

3. Under the assumptions of Art. 39 find the range on a line making an angle β with the x-axis.

4. Show that the direction of projection for the greatest range on a given line passing through the point of projection is in a line bisecting the angle between the given line and the y-axis.

5. Show that the locus of the highest points of the parabolas as α takes all values is an ellipse whose major axis is $\dfrac{v_0^2}{g}$, and minor axis, $\dfrac{v_0^2}{2g}$.

6. Prove that the velocity at any point equals that which the particle would have at the point if it fell from the directrix of the parabola.

THE HEAT OF THE SUN.

40. Work and Energy. When a force moves a particle against any resistance it is said to do *work*. The amount of the work is proportional to the product of the resistance and the distance through which the particle is moved. In the case of a free particle the resistance comes entirely from the inertia of the mass; if there is friction this is also resistance.

Energy is the power of doing work. If a given amount of work is done upon a particle free to move, the particle acquires a motion that will enable it to do exactly the same amount of work. The energy of motion is called *kinetic energy*. If the particle is retarded by friction part of the original work expended will be used in overcoming the friction, and the particle will be capable of doing only as much work as has been done in giving it motion. Until about 1850 it was generally supposed that work done in overcoming friction is partly, or perhaps entirely, lost. In other words, it was believed that the total amount of energy in an isolated system might continually decrease. It was observed, however, that friction generates heat, sound, light, and electricity, depending upon the circumstances, and that these manifestations of energy are of the same nature as the original, but in a different form. It was then proved that these modified forms of energy are in every case quantitatively equivalent to the waste of that originally considered. The brilliant experiments of Joule and others, made in the middle of the nineteenth century, have established with great certainty the fact that the total amount of energy remains the same whatever changes it may undergo. This principle, known as the *conservation of energy*, when stated as holding throughout the universe, is one of the most far-reaching generalizations that has been made in the natural sciences in a hundred years.*

41. Computation of Work. The amount of work done by a Newtonian force in moving a free particle any distance will now be computed. Let m equal the mass of the particle moved, and k^2 a constant depending upon the mass of the attracting body and the units employed. Then

$$(41) \qquad\qquad m\frac{d^2s}{dt^2} = -\frac{k^2m}{s^2}.$$

* Herbert Spencer regards the principle as being axiomatic, and states his views in regard to it in his *First Principles*, part II., chap. VI.

The right member is the force to which the particle is subject. By Newton's third law it is numerically equal to the reaction, or the resistance due to inertia. Then the work done in moving the particle through the element of distance ds is

$$m \frac{d^2s}{dt^2} ds = - \frac{k^2m}{s^2} ds = dW.$$

The work done in moving the particle through the interval from s_0 to s_1 is found by taking the definite integral of this expression between the limits s_0 and s_1. On performing the integrations and substituting the limits, it is found that

$$\frac{m}{2} \left(\frac{ds_1}{dt} \right)^2 - \frac{m}{2} \left(\frac{ds_0}{dt} \right)^2 = k^2m \left(\frac{1}{s_1} - \frac{1}{s_0} \right).$$

Suppose the initial velocity is zero; then the kinetic energy equals the work W done upon the particle, and

$$(42) \qquad W = \frac{m}{2} \left(\frac{ds_1}{dt} \right)^2 = k^2m \left(\frac{1}{s_1} - \frac{1}{s_0} \right).$$

By hypothesis, the particle has no kinetic energy on the start, and therefore the power of doing work equals the product of one half the mass and the square of the velocity. If the particle falls from infinity, s_0 is infinite, and the formula for the kinetic energy becomes

$$(43) \qquad \frac{m}{2} \left(\frac{ds_1}{dt} \right)^2 = \frac{k^2m}{s_1}.$$

If the particle is stopped by striking a body when it reaches the point s_1, its kinetic energy is changed into some other form of energy such as heat. It has been found by experiment that a body weighing one kilogram falling 425 meters* in the vicinity of the earth's surface, under the influence of the earth's attraction, generates enough heat to raise the temperature of one kilogram of water one degree Centigrade. This quantity of heat is called the *calory*.† The amount of heat generated is proportional to the product of the square of the velocity and the mass of the moving particle. Then, letting Q represent the number of calories, it follows that

$$(44) \qquad\qquad Q = Cmv^2.$$

---

* Joule found 423.5; Rowland 427.8. For results of experiments and references see Preston's *Theory of Heat*, p. 594.

† One-thousandth of this unit, defined in using the gram instead of the kilogram, is also called a calory.

Let m be expressed in kilograms and v in meters per second. In order to determine the constant C, take Q and m each equal to unity; then the velocity is that acquired by the body falling through 425 meters. It was shown in Art. 30 that, if the body falls from rest,

$$\begin{cases} s = -\tfrac{1}{2}gt^2, \\ v = -gt. \end{cases}$$

On eliminating t between these equations, it is found that

$$v^2 = 2gs.$$

In the units employed $g = 9.8094$, and since $s = 425$ and $v^2 = 8338$, it follows from (44) that

$$C = \frac{1}{8338}.$$

Then the general formula (44) becomes

(45) $$Q = \frac{mv^2}{8338},$$

where Q is expressed in calories if the kilogram, meter, and second are taken as the units of mass, distance, and time.

42. The Temperature of Meteors. The increase of temperature of a body, when the proper units are employed, is equal to the number of calories of heat acquired divided by the product of the mass and the specific heat of the substance. Suppose a meteor whose specific heat is unity (in fact it would probably be much less than unity) should strike the earth with any given velocity; it is required to compute its increase of temperature if it took up all the heat generated. The specific heat has been taken so that the increase of temperature is numerically equal to the number of calories generated per unit mass. Meteors usually strike the earth with a velocity of about 25 miles, or 40,233 meters, per second. On substituting 40,233 for v and unity for m in (45), it is found that $T = Q = 194{,}134$, the number of calories generated per unit mass, or the number of degrees through which the temperature of the meteor would be raised. As a matter of fact a large part of the heat would be imparted to the atmosphere; but the quantity of heat generated is so enormous that it could not be expected that any but the largest meteors would last long enough to reach the earth's surface.

A meteor falling into the sun from an infinite distance would strike its surface, as has been seen in Art. 36, with a velocity of about 384 miles per second. The heat generated would be therefore $(\frac{384}{25})^2$, or 236, times as great as that produced in striking the earth. Thus it follows that a kilogram would generate, in falling into the sun, 45,815,624 calories.

43. The Meteoric Theory of the Sun's Heat. When it is remembered what an enormous number of meteors (estimated by H. A. Newton* as being as many as 8,000,000 daily) strike the earth, it is easily conceivable that enough strike the sun to maintain its temperature. Indeed, this has been advanced as a theory to account for the replenishment of the vast amount of heat which the sun radiates. There can be no question of its qualitative correctness, and it only remains to examine it quantitatively.

Let it be assumed that the sun radiates heat equally in every direction, and that meteors fall upon it in equal numbers from every direction. Under this assumption, the amount of heat radiated by any portion of the surface will equal that generated by the impact of meteors upon that portion. The amount of heat received by the earth will be to the whole amount radiated from the sun as the surface which the earth subtends as seen from the sun is to the surface of the whole sphere whose radius is the distance from the earth to the sun. The portion of the sun's surface which is within the cone whose base is the earth and vertex the center of the sun, is to the whole surface of the sun as the surface subtended by the earth is to the surface of the whole sphere whose radius is the distance to the sun. Therefore, the earth receives as much heat as is radiated by, and consequently generated upon, the surface cut out by this cone. But the earth would intercept precisely as many meteors as would fall upon this small area, and would, therefore, receive heat from the impact of a certain number of meteors upon itself, and as much heat from the sun as would be generated by the impact of an equal number upon the sun.

The heat derived by the earth from the two sources would be as the squares of the velocities with which the meteors strike the earth and sun. It was seen in Art. 42 that this number is $\frac{1}{236}$. Therefore, if this meteoric hypothesis of the maintenance of the sun's heat is correct, the earth should receive $\frac{1}{236}$ as much heat from the impact of meteors as from the sun. This is certainly

* *Mem. Nat. Acad. of Sci.*, vol. i.

millions of times more heat than the earth receives from meteors, and consequently the theory that the sun's heat is maintained by the impact of meteors is not tenable.

44. Helmholtz's Contraction Theory. The amount of work done upon a particle is proportional to the product of the resistance overcome by the distance moved. There is nothing whatever said about how long the motion shall take, and if the work is converted into heat the total amount is the same whether the particle falls the entire distance at once, or covers the same distance by a succession of any number of shorter falls. When a body contracts it is equivalent to a succession of very short movements of all its particles in straight lines toward the center, and it is evident that, knowing the law of density, the amount of heat which will be generated can be computed.

In 1854 Helmholtz applied this idea to the computation of the heat of the sun in an attempt to explain its source of supply. He made the supposition that the sun contracts in such a manner that it always remains homogeneous. While this assumption is certainly incorrect, nevertheless the results obtained are of great value and give a good idea of what doubtless actually takes place under contraction. The mathematical part of the theory is given in the *Philosophical Magazine* for 1856, p. 516.

Consider a homogeneous gaseous sphere whose radius is R_0 and density σ. Let M_0 represent its mass. Let dM represent an element of mass taken anywhere in the interior or at the surface of the sphere. Let R be the distance of dM from the center of the sphere, and let M represent the mass of the sphere whose radius is R. The element of mass in polar coördinates is (Art. 21)

$$(46) \qquad dM = \sigma R^2 \cos \phi \, d\phi \, d\theta \, dR.$$

The element is subject to the attraction of the whole sphere within it. As will be shown in Chapter IV, the attraction of the spherical shell outside of it balances in opposite directions so that it need not be considered in discussing the forces acting upon dM. Every element in the infinitesimal shell whose radius is R is attracted toward the center by a force equal to that acting on dM; therefore the whole shell can be treated at once. Let dM_s represent the mass of the elementary shell whose radius is R. It is found by integrating (46) with respect to θ and ϕ. Thus

$$(47) \quad dM_s = \sigma R^2 dR \int_0^{2\pi} \left\{ \int_{-\frac{\pi}{2}}^{\frac{\pi}{2}} \cos \phi \, d\phi \right\} d\theta = 4\pi\sigma R^2 \, dR.$$

The force to which dM_s is subject is $-\dfrac{k^2 M \, dM_s}{R^2}$. The element of work done in moving dM_s through the element of distance dR is

$$dW_s = - dM_s \frac{k^2 M}{R^2} \, dR.$$

The work done in moving the shell from the distance CR to R is the integral of this expression between the limits CR and R, or

$$W_s = - dM_s k^2 M \int_{CR}^{R} \frac{dR}{R^2} = \frac{dM_s k^2 M}{R} \left(\frac{C-1}{C} \right).$$

But $M = \frac{4}{3}\pi\sigma R^3$; hence, substituting the value of dM_s from (47) and representing the work done on the elementary shell by $W_s = dW$, it follows that

$$dW = \frac{16}{3} \pi^2 \sigma^2 k^2 \left(\frac{C-1}{C} \right) R^4 \, dR.$$

The integral of this expression from 0 to R_0 gives the total amount of work done in the contraction of the homogeneous sphere from radius CR_0 to R_0. That is,

$$W = \frac{16}{3} \pi^2 \sigma^2 k^2 \left(\frac{C-1}{C} \right) \int_0^{R_0} R^4 \, dR = \frac{16}{15} \pi^2 \sigma^2 k^2 \left(\frac{C-1}{C} \right) R_0^5,$$

which may be written

(48) $$W = \tfrac{3}{5} k^2 \left(\frac{C-1}{C} \right) \frac{M_0^2}{R_0}.$$

If C equals infinity, then

(49) $$W = \tfrac{3}{5} k^2 \frac{M_0^2}{R_0}.$$

If the second is taken as the unit of time, the kilogram as the unit of mass, and the meter as the unit of distance, and if k^2 is computed from the value of g for the earth, then, after dividing W by $\frac{8338}{2}$, the result will be numerically equal to the amount of heat in calories that will be generated if the work is all transformed into this kind of energy. The temperature to which the body will be raised, which is this quantity divided by the product of the mass and the specific heat, is

(50) $$T = \frac{Q}{\eta M_0} = \frac{2W}{8338 \eta M_0},$$

where η is the specific heat of the substance. Or, substituting (48) in (50), it is found that

(51)
$$T = \frac{3k^2}{5\eta} \cdot \frac{C-1}{C} \cdot \frac{M_0}{R_0} \cdot \frac{2}{8338}.$$

By definition, k^2 is the attraction of the unit of mass at unit distance; therefore, if m is the mass of the earth and r its radius, it follows that

$$g = \frac{k^2 m}{r^2}.$$

On solving for k^2 and substituting in (51), the expression for T becomes

(52)
$$T = \frac{3(C-1)}{5\eta C} \cdot \frac{r^2}{R_0} \cdot \frac{M_0}{m} \cdot \frac{2g}{8338}.$$

If the body contracted from infinity ($C = \infty$), the amount of heat generated would be sufficient to raise its temperature T degrees Centigrade, where T is given by the equation

(53)
$$T = \frac{3}{5} \cdot \frac{1}{\eta} \cdot \frac{r^2}{R_0} \cdot \frac{M_0}{m} \cdot \frac{2g}{8338}.$$

Suppose the specific heat is taken as unity, which is that of water.* The value of g is 9.8094 and

$$\begin{cases} \dfrac{r^2}{R_0} = \dfrac{116,356}{2} = 58,178 \\ \dfrac{M_0}{m} = 332,000. \end{cases}$$

On substituting these numbers in (53) and reducing, it is found that

$$T = 27,268,000° \text{ Centigrade.}$$

Therefore, *the sun contracting from infinity in such a way as to always remain homogeneous would generate enough heat to raise the temperature of an equal mass of water more than twenty-seven millions of degrees Centigrade.*

If it is supposed that the sun has contracted only from Neptune's orbit equation (52) can be used, which will give a value of T about $\frac{1}{6600}$ less. In any case it is not intended to imply that it did ever contract from such great dimensions in the particular manner assumed; the results given are nevertheless significant and throw much light on the question of evolution of the solar system from a vastly extended nebula. If the contraction of the

---

* No other ordinary terrestrial substance has a specific heat so great as unity except hydrogen gas, whose specific heat is 3.409. But the lighter gases of the solar atmosphere may also have high values.

sun were the only source of its energy, this discussion would give a rather definite idea as to the upper limit of the age of the earth. But the limit is so small that it is not compatible with the conclusions reached by several lines of reasoning from geological evidence, and it is utterly at variance with the age of certain uranium ores computed from the percentage of lead which they contain. The recent discovery of enormous sub-atomic energies which become manifest in the disintegration of radium and several other substances prove the existence of sources of energy not heretofore considered, and suggest that the sun's heat may be supplied partly, if not largely, from these sources. It is certainly unsafe at present to put any limits on the age of the sun.

The experiments of Abbott have shown that, under the assumption that the sun radiates heat equally in every direction, the amount of heat emitted yearly would raise the temperature of a mass of water equal to that of the sun 1.44 degrees Centigrade. In order to find how great a shrinkage in the present radius would be required to generate enough heat to maintain the present radiation 10,000 years, substitute 14,400 for T in (52) and solve for C. On carrying out the computation, it is found that

$$C = 1.000528.$$

Therefore, *the sun would generate enough heat in shrinking about one four-thousandth of its present diameter to maintain its present radiation 10,000 years.*

The sun's mean apparent diameter is 1924″, so a contraction of its diameter of .000528 would make an apparent change of only 1.″0, a quantity far too small to be observed on such an object by the methods now in use. On reducing the shrinkage to other units, it is found that a contraction of the sun's radius of 36.8 meters annually would account for all the heat that is being radiated at present.

VI. PROBLEMS.

1. According to the recent work of Abbott, of the Smithsonian Institution, a square meter exposed perpendicularly to the sun's rays at the earth's distance would receive 19.5 calories per minute. The average amount received per square meter on the earth's surface is to this quantity as the area of a circle is to the surface of a sphere of the same radius, or 1 to 4. The earth's surface receives, therefore, on the average 5 calories per square meter per minute. How many kilograms of meteoric matter would have to strike the earth with a velocity of 25 miles (40,233 meters) per sec. to generate $\frac{1}{216}$ this amount of heat?

        Ans. .000,000,1∪94 kilograms.

2. How many pounds would have to fall per day on every square mile on the average? Tons on the whole earth?

$$Ans. \quad \begin{cases} 824 \text{ pounds/day/square mile} \\ 81,200,000 \text{ tons/day} \end{cases}$$

3. Find the amount of work done in the contraction of any fraction of the radius of a sphere when the law of density is $\sigma = \dfrac{l}{R^2}$.

$Ans.$ $W = 16\pi^2 k^2 l^2 \left(\dfrac{C-1}{C} \right) \cdot R = k^2 \left(\dfrac{C-1}{C} \right) \cdot \dfrac{M_0^2}{R_0}$, or $\frac{5}{3}$ of the work done when the sphere is homogeneous.

4. Laplace assumed that the resistance of a fluid against compression is directly proportional to its density, and on the basis of this assumption he found that the law of density of a spherical body would be

$$\sigma = \frac{G \sin \left(\mu \dfrac{R}{a} \right)}{\dfrac{R}{a}},$$

where G and μ are constants depending on the material of which the body is composed, and where a is the radius of the sphere. This law of density is in harmony, when applied to the earth, with a number of phenomena, such as the precession of the equinoxes. Find the amount of heat generated by contraction from infinite dimensions to radius R_0 of a body having the Laplacian law of density.

5. Find how much the heat generated by the contraction of the earth from the density of meteorites, 3.5, to the present density of 5.6 would raise the temperature of the whole earth, assuming that the specific heat is 0.2.

$Ans.$ $T = 6523.5$ degrees Kelvin

HISTORICAL SKETCH AND BIBLIOGRAPHY.

The laws of falling bodies under constant acceleration were investigated by Galileo and Stevinus, and for many cases of variable acceleration by Newton. Such problems are comparatively simple when treated by the analytical processes which have come into use since the time of Newton. Parabolic motion was discussed by Galileo and Newton.

The kinetic theory of gases seems to have been first suggested by J. Bernouilli about the middle of the 18th century, but it was first developed mathematically by Clausius. Maxwell, Boltzmann, and O. E. Meyer have made important contributions to the subject, and more recently Burbury, Jeans, and Hilbert. Some of the principal books on the subject are: Risteen's *Molecules and the Molecular Theory* (descriptive work); L. Boltzmann's *Gastheorie;* H. W. Watson's *Kinetic Theory of Gases;* O. E. Meyer's *Die Kinetische Theorie der Gase;* S. H. Burbury's *Kinetic Theory of Gases;* J. H. Jean's *Kinetic Theory of Gases.*

The meteoric theory of the sun's heat was first suggested by R. Mayer. The contraction theory was first announced in a public lecture by Helmholtz at Königsberg Feb. 7, 1854, and was published later in *Phil. Mag.* 1856. An important paper by J. Homer Lane appeared in the *Am. Jour. of Sci.* July, 1870. The amount of heat generated depends upon the law of density of the gaseous sphere. Investigations covering this point are 16 papers by Ritter in *Wiedemann's Annalen*, vol. v., 1878, to vol. xx., 1883; by G. W. Hill, *Annals of Math.*, vol. iv., 1888; and by G. H. Darwin, *Phil. Trans.*, 1888. The original papers must be read for an exposition of the subject of the heat of the sun. Sub-atomic energies are discussed in E. Rutherford's *Radioactive Substances and their Radiations.*

For evidences of the great age of the earth consult Chamberlin and Salisbury's *Geology*, vol. ii., and vol. iii., p. 413 et seq.; for a general discussion of the age of the earth see Arthur Holmes' *The Age of the Earth.*

CHAPTER III.

45. Central Force. This chapter will be devoted to the discussion of the motion of a material particle when subject to an attractive or repelling force whose line of action always passes through a fixed point. This fixed point is called the *center of force.* It is not implied that the force emanates from the center or that there is but one force, but simply that the resultant of all the forces acting on the particle always passes through this point. The force may be directed toward the point or from it, or part of the time toward and part of the time from it. It may be zero at any time, but if the particle passes through a point where the force to which it is subject becomes infinite, a special investigation, which cannot be taken up here, is required to follow it farther. Since attractive forces are of most frequent occurrence in astronomical and physical problems, the formulas developed will be for this case; a change of sign of the coefficient of intensity of the force for unit distance will make the formulas valid for the case of repulsion.

The origin of coördinates will be taken at the center of force, and the line from the origin to the moving particle is called the *radius vector.* The path described by the particle is called the *orbit.* The orbits of this chapter are plane curves. The planes are defined by the position of the center of force and the line of initial projection. The xy-plane will be taken as the plane of the orbit.

46. The Law of Areas. The first problem will be to derive the general properties of motion which hold for all central forces. The first of these, which is of great importance, is the *law of areas,* and constitutes the first Proposition of Newton's *Principia.* It is, *if a particle is subject to a central force, the areas which are swept over by the radius vector are proportional to the intervals of time in which they are described.* The following is Newton's demonstration of it.

Let O be the center of force, and let the particle be projected from A in the direction of B with the velocity AB. Then, by the first law of motion, it would pass to C' in the first two units of

time if there were no external forces acting upon it.   But suppose
that when it arrives at B an instantaneous force acts upon it in
the direction of the origin with such intensity that it would move

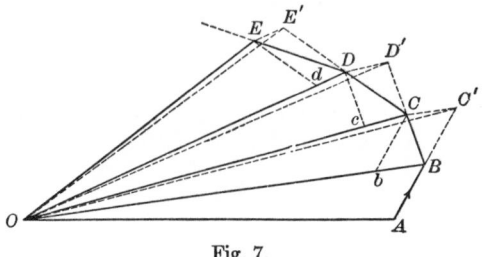

Fig. 7.

to b in a unit of time if it had no previous motion.   Then, by the
second law of motion, it will move along the diagonal of the
parallelogram $BbCC'$ to C.   If no other force were applied it
would move with uniform velocity to D' in the next unit of time.
But suppose that when it arrives at C another instantaneous force
acts upon it in the direction of the origin with such intensity
that it would move to c in a unit of time if it had no previous
motion.   Then, as before, it will move along the diagonal of the
parallelogram and arrive at D at the end of the unit of time.   This
process can be repeated indefinitely.

The following equalities among the areas of the triangles involved hold, since they have sequentially equal bases and altitudes:

$$OAB = OBC' = OBC = OCD' = OCD = \text{etc.}$$

Therefore, it follows that $OAB = OBC = OCD = ODE$, etc.
That is, the areas of the triangles swept over in the succeeding
units of time are equal; and, therefore, the sums of the areas of the
triangles described in any intervals of time are proportional to
the intervals.

The reasoning is true without any changes however small the
intervals of time may be.   Let the path be considered for some
fixed finite period of time.   Let the intervals into which it is divided
be taken shorter and shorter; the impulses will become closer and
closer together.   Suppose the ratio of the magnitudes of the impulses
to the values of the intervals between them remains finite; then the
broken line will become more and more nearly a smooth curve.
Suppose the intervals of time approach zero as a limit; the succession of impulses will approach a continuous force as a limit, and

the broken line will approach a smooth curve as a limit.   The areas swept over by the radius vector in any finite intervals of time are proportional to these intervals during the whole limiting process. Therefore, the proportionality of areas holds at the limit, and the theorem is true for a continuous central force.

It will be observed that it is not necessary that the central force shall vary continuously.   It may be attractive and instantaneously change to repulsion, or become zero, and the law will still hold; but it is necessary to exclude the case where it becomes infinite unless a special investigation is made.

The linear velocity varies inversely as the perpendicular from the origin upon the tangent to the curve at the point of the moving particle; for, the area described in a unit of time is equal to the product of the velocity and the perpendicular upon the tangent. Since the area described in a unit of time is always the same, it follows that the linear velocity of the particle varies inversely as the perpendicular from the origin to the tangent of its orbit.

47. Analytical Demonstration of the Law of Areas.   Although the language of Geometry was employed in the demonstration of Art. 46, yet the essential elements of the methods of the Differential and Integral Calculus were used.   Thus, in passing to the limit, the process was essentially that of expressing the problem in differential equations; and, in insisting upon comparing intervals of finite size when the units of measurement were indefinitely decreased, the process of integration was really employed.   It will be found that in the treatment of all problems involving variable forces and motions the methods are in essence those of the Calculus, even though the demonstrations be couched in geometrical language.   It is perhaps easier to follow the reasoning in geometrical form when one encounters it for the first time; but the processes are all special and involve fundamental difficulties which are often troublesome.   On the other hand, the development of the Calculus is of the precise form to adapt it to the treatment of these problems, and after its principles have been once mastered, the application of it is characterized by comparative simplicity and great generality.   A few problems will be treated by both methods to show their essential sameness, and to illustrate the advantages of analysis.

Let f represent the acceleration to which the particle is subject. By hypothesis, the line of force always passes through a fixed point, which will be taken as the origin of coördinates.

Let O be the center of force, and P any position of the particle whose rectangular coördinates are x and y, and whose polar coördinates are r and θ. Then the components of acceleration

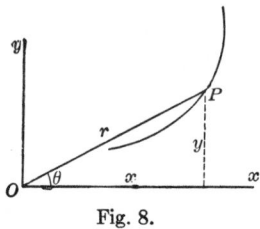

Fig. 8.

along the x and y-axes are respectively $\mp f \cos \theta$ and $\mp f \sin \theta$, and the differential equations of motion are

$$(1) \quad \begin{cases} \dfrac{d^2x}{dt^2} = \mp f \cos \theta = \mp f \dfrac{x}{r}, \\[2mm] \dfrac{d^2y}{dt^2} = \mp f \sin \theta = \mp f \dfrac{y}{r}. \end{cases}$$

The negative sign must be taken before the right members of these equations if the force is attractive, and the positive if it is repulsive.

Multiply the first equation of (1) by $-y$ and the second one by $+x$ and add. The result is

$$x \frac{d^2y}{dt^2} - y \frac{d^2x}{dt^2} = 0.$$

On integrating this expression by parts, it is found that

$$(2) \quad x \frac{dy}{dt} - y \frac{dx}{dt} = h,$$

where h is the constant of integration.

The integrals of differential equations generally lead to important theorems even though the whole problem has not been solved, and in what follows they will be discussed as they are obtained.

On referring to Art. 16, it is seen that (2) may be written

$$x \frac{dy}{dt} - y \frac{dx}{dt} = r^2 \frac{d\theta}{dt} = 2 \frac{dA}{dt} = h,$$

where A is the area swept over by the radius vector. The integral of this equation is

$$A = \tfrac{1}{2}ht + c,$$

which shows that the area is directly proportional to the time. This is the theorem which was to be proved.

48. Converse of the Theorem of Areas.   By hypothesis

$$A = c_1 t + c_2.$$

On taking the derivative with respect to t, it is found that

$$\frac{dA}{dt} = c_1.$$

This equation becomes in polar coördinates

$$r^2 \frac{d\theta}{dt} = 2c_1,$$

and in rectangular coördinates

$$x \frac{dy}{dt} - y \frac{dx}{dt} = 2c_1.$$

The derivative of this expression with respect to t is

$$x \frac{d^2y}{dt^2} - y \frac{d^2x}{dt^2} = 0;$$

or

$$\frac{d^2x}{dt^2} : \frac{d^2y}{dt^2} = x : y.$$

That is, the components of acceleration are proportional to the coördinates; therefore, if the law of areas is true with respect to a point, the resultant of the accelerations passes through that point.

Or, since $r^2 \dfrac{d\theta}{dt} = 2c_1$, it follows that $\dfrac{d}{dt}\left(r^2 \dfrac{d\theta}{dt} \right) = 0$.   Hence, by (19), Art. 14, the acceleration perpendicular to the radius vector is zero; that is, the acceleration is in the line passing through the origin.

49. The Laws of Angular and Linear Velocity.   From the expression for the law of areas in polar coördinates, it follows that

(3) $$\frac{d\theta}{dt} = \frac{h}{r^2};$$

therefore, *the angular velocity is inversely proportional to the square of the radius vector.*

The linear velocity is

$$\frac{ds}{dt} = \frac{ds}{d\theta}\frac{d\theta}{dt} = \frac{ds}{d\theta}\frac{h}{r^2}.$$

Let p represent the perpendicular from the origin upon the tangent; then it is known from Differential Calculus that

$$\frac{ds}{d\theta} = \frac{r^2}{p}.$$

Hence the expression for the linear velocity becomes

$$(4) \qquad \frac{ds}{dt} = \frac{h}{p};$$

therefore, *the linear velocity is inversely proportional to the perpendicular from the origin upon the tangent.*

SIMULTANEOUS DIFFERENTIAL EQUATIONS.

50. The Order of a System of Simultaneous Differential Equations. One integral, equation (2), of the differential equations (1) has been found which the motion of the particle must fulfill. The question is how many more integrals must be found in order to have the complete solution of the problem.

The number of integrals which must be found to completely solve a system of differential equations is called the *order* of the system. Thus, the equation

$$(5) \qquad \frac{d^n x}{dt^n} = c$$

is of the nth order, because it must be integrated n times to be reduced to an integral form. Similarly, the general equation

$$(6) \qquad f_n \frac{d^n x}{dt^n} + f_{n-1} \frac{d^{n-1} x}{dt^{n-1}} + \cdots + f_1 \frac{dx}{dt} + f_0 = 0,$$

where $f_n, \cdots\cdots, f_0$ are functions of x and t, must be integrated n times in order to express x as a function of t, and is of the nth order.

An equation of the nth order can be reduced to an equivalent system of n simultaneous equations each of the first order. Thus, to reduce (6) to a simultaneous system, let

$$x_1 = \frac{dx}{dt}, \qquad x_2 = \frac{dx_1}{dt}, \qquad \cdots\cdots, \qquad x_{n-1} = \frac{dx_{n-2}}{dt};$$

whence

$$(7) \quad \begin{cases} \dfrac{dx}{dt} = x_1, \\[2mm] \dfrac{dx_1}{dt} = x_2, \\[2mm] \dfrac{dx_2}{dt} = x_3, \\[2mm] \cdot \quad \cdot \quad \cdot \quad \cdot \quad \cdot \quad \cdot \\[1mm] \dfrac{dx_{n-1}}{dt} = -\dfrac{f_{n-1}}{f_n} x_{n-1} - \cdots - \dfrac{f_1}{f_n} x_1 - \dfrac{f_0}{f_n}. \end{cases}$$

Therefore, these n simultaneous equations, each of the first order, constitute a system of the nth order. An equation, or a system of equations, reduced to the form (7) is said to be reduced to the *normal form*, and the system is called a *normal system*.

Two simultaneous equations of order m and n can be reduced to a normal system of order $m + n$. Consider the equations

$$(8) \quad \begin{cases} f_m \dfrac{d^m x}{dt^m} + \cdots + f_1 \dfrac{dx}{dt} + f_0 = 0, \\[3mm] \phi_n \dfrac{d^n y}{dt^n} + \cdots + \phi_1 \dfrac{dy}{dt} + \phi_0 = 0, \end{cases}$$

where the f_i and the ϕ_i are functions of x, y, and t. By a substitution similar to that used in reducing (6), it follows that they are equivalent to the normal system

$$(9) \quad \begin{cases} \dfrac{dx}{dt} = x_1, \\[1mm] \cdot \quad \cdot \quad \cdot \quad \cdot \quad \cdot \quad \cdot \\[1mm] \dfrac{dx_{m-1}}{dt} = -\dfrac{f_{m-1}}{f_m} x_{m-1} - \cdots - \dfrac{f_1}{f_m} x_1 - \dfrac{f_0}{f_m}, \\[3mm] \dfrac{dy}{dt} = y_1, \\[1mm] \cdot \quad \cdot \quad \cdot \quad \cdot \quad \cdot \quad \cdot \\[1mm] \dfrac{dy_{n-1}}{dt} = -\dfrac{\phi_{n-1}}{\phi_n} y_{n-1} - \cdots - \dfrac{\phi_1}{\phi_n} y_1 - \dfrac{\phi_0}{\phi_n}, \end{cases}$$

which is of the order $m + n$. Evidently a similar reduction is possible when each equation contains derivatives with respect to both of the variables, either separately or as products.

Conversely, *a normal system of order n can in general be transformed into a single equation of order n with one dependent variable.* To fix the ideas, consider the system of the second order

(10)
$$\begin{cases} \dfrac{dx}{dt} = f(x,\ y,\ t), \\[2mm] \dfrac{dy}{dt} = \phi(x,\ y,\ t). \end{cases}$$

In addition to these two equations form the derivative of one of them, for example the first, with respect to t. The result is

(11)
$$\frac{d^2x}{dt^2} = \frac{\partial f}{\partial x}\frac{dx}{dt} + \frac{\partial f}{\partial y}\frac{dy}{dt} + \frac{\partial f}{\partial t}.$$

If y and $\dfrac{dy}{dt}$ are eliminated between (10) and (11) the result will be an equation of the form

$$\frac{d^2x}{dt^2} + F\left(x,\ \frac{dx}{dt}\right) = 0,$$

where F is a function of both x and $\dfrac{dx}{dt}$. Of course, f and ϕ of equations (10) may have such properties that the elimination of y and $\dfrac{dy}{dt}$ is very difficult.

If the normal system were of the third order in the dependent variables x, y, and z, the first and second derivatives of the first equation would be taken, and the first derivative of the second and third equations. These four new equations with the original three make seven from which y, z, $\dfrac{dy}{dt}$, $\dfrac{dz}{dt}$, $\dfrac{d^2y}{dt^2}$, and $\dfrac{d^2z}{dt^2}$ can in general be eliminated, giving an equation of the third order in x alone. This process can be extended to a system of any order.

The differential equations (1) can be reduced by the substitution $x' = \dfrac{dx}{dt}$, $y' = \dfrac{dy}{dt}$ to the normal system

(12)
$$\begin{cases} \dfrac{dx}{dt} = x', & \dfrac{dx'}{dt} = \mp f\dfrac{x}{r}, \\[2mm] \dfrac{dy}{dt} = y', & \dfrac{dy'}{dt} = \mp f\dfrac{y}{r}, \end{cases}$$

which is of the fourth order. Therefore four integrals must be found in order to have the complete solution of the problem. The components of velocity, x' and y', play rôles similar to the coördinates in (12), and, for brevity, they will be spoken of frequently in the future as being coördinates.

51. Reduction of Order. When an integral of a system of differential equations has been found two methods can be followed in completing the solution. The remaining integrals can be found from the original differential equations as though none was already known; or, by means of the known integral, the order of the system of differential equations can be reduced by one. That the order of the system can be reduced by means of the known integrals will be shown in the general case. Consider the system of differential equations

$$(13) \quad \begin{cases} \dfrac{dx_1}{dt} = f_1(x_1, \cdots, x_n, t), \\[2mm] \dfrac{dx_2}{dt} = f_2(x_1, \cdots, x_n, t), \\[1mm] \cdot \quad \cdot \quad \cdot \quad \cdot \quad \cdot \quad \cdot \quad \cdot \\[1mm] \dfrac{dx_n}{dt} = f_n(x_1, \cdots, x_n, t). \end{cases}$$

Suppose an integral

$$F(x_1, x_2, \cdots, x_n, t) = \text{constant} = c,$$

has been found. Suppose this equation is solved for x_n in terms of $x_1, \cdots, x_{n-1}, c$, and t. The result may be written

$$x_n = \psi(x_1, x_2, \cdots, x_{n-1}, c, t).$$

Substitute this expression for x_n in the first $n - 1$ equations of (13); they become

$$(14) \quad \begin{cases} \dfrac{dx_1}{dt} = \phi_1(x_1, \cdots, x_{n-1}, c, t), \\[2mm] \dfrac{dx_2}{dt} = \phi_2(x_1, \cdots, x_{n-1}, c, t), \\[1mm] \cdot \quad \cdot \quad \cdot \quad \cdot \quad \cdot \quad \cdot \quad \cdot \quad \cdot \\[1mm] \dfrac{dx_{n-1}}{dt} = \phi_{n-1}(x_1, \cdots, x_{n-1}, c, t). \end{cases}$$

This is a simultaneous system of order $n - 1$, and is independent of the variable x_n.

It is apparent from these theorems and remarks that the order of a simultaneous system of differential equations is equal to the sum of the orders of the individual equations; that the equations can be written in several ways, e. g., as one equation of the nth order, or n equations of the first order; and that the integrals may all be derived from the original system, or that the order may be reduced after each integral is found. In mechanical and physical problems the intuitions are important in suggesting methods of treatment, so it is generally advantageous to use such variables that their geometrical and physical meanings shall be easily perceived. For this reason, it is generally simpler not to reduce the order of the problem after each integral is found.

VII. PROBLEMS.

1. Prove the converse of the law of areas by the geometrical method, and show that the steps agree with the analysis of Art. 48.

2. Prove the law of angular velocity by the geometrical method.

3. Why cannot equations (1) be integrated separately?

4. Derive the law of areas directly from equation (2) without passing to polar coördinates.

5. Show in detail that a normal system of order four can be reduced to a single equation of order four, and the converse.

6. Reduce the system of equations (12) to one of the third order by means of the integral of areas.

52. The Vis Viva Integral. Suppose the acceleration is toward the origin; then the negative sign must be taken before the right members of equations (1). Multiply the first of (1) by $2 \frac{dx}{dt}$, the second by $2 \frac{dy}{dt}$, and add. The result is

$$2\frac{d^2x}{dt^2}\frac{dx}{dt} + 2\frac{d^2y}{dt^2}\frac{dy}{dt} = -\frac{2f}{r}\left(x\frac{dx}{dt} + y\frac{dy}{dt}\right).$$

It follows from $r^2 = x^2 + y^2$ that

$$x\frac{dx}{dt} + y\frac{dy}{dt} = r\frac{dr}{dt};$$

therefore

$$2\frac{d^2x}{dt^2}\frac{dx}{dt} + 2\frac{d^2y}{dt^2}\frac{dy}{dt} = -2f\frac{dr}{dt}.$$

Suppose f depends upon r alone, as it does in most astronomical and physical problems. Then $f = \phi(r)$, and

$$2\frac{d^2x}{dt^2}\frac{dx}{dt} + 2\frac{d^2y}{dt^2}\frac{dy}{dt} = -2\phi(r)\frac{dr}{dt}.$$

The integral of this equation is

$$(15) \qquad \left(\frac{dx}{dt}\right)^2 + \left(\frac{dy}{dt}\right)^2 = v^2 = -2\int\phi(r)dr + c.$$

When the form of the function $\phi(r)$ is given the integral on the right can be found. Suppose the integral is $\Phi(r)$; then

$$(16) \qquad v^2 = - 2\Phi(r) + c.$$

If $\Phi(r)$ is a single-valued function of r, as it is in physical problems, it follows from (16) that, if the central force is a function of the distance alone, the speed is the same at all points equally distant from the origin. Its magnitude at any point depends upon the initial distance and speed, and not upon the path described. Since the force of gravitation varies inversely as the square of the distance between the attracting bodies, it follows that a body, for example a comet, has the same speed at a given distance from the sun whether it is approaching or receding.

EXAMPLES WHERE f IS A FUNCTION OF THE COÖRDINATES ALONE.

53. Force Varying Directly as the Distance. In order to find integrals of equations (1) other than that of areas, the value of f in terms of the coördinates must be known. In the case in which the intensity of the force varies directly as the distance the integration becomes particularly simple. Let k^2 represent the acceleration at unit distance. Then $f = k^2 r$, and, in case the force is attractive, equations (1) become

$$(17) \qquad \begin{cases} \dfrac{d^2x}{dt^2} = - k^2 x, \\[2mm] \dfrac{d^2y}{dt^2} = - k^2 y. \end{cases}$$

An important property of these equations is that each is independent of the other, as the first one contains the dependent variable x alone and the second one y alone. It is observed, moreover, that they are linear and the solution can be found by the method given in Art. 32. If $x = x_0$, $\dfrac{dx}{dt} = x_0'$, $y = y_0$, $\dfrac{dy}{dt} = y_0'$ at $t = 0$, then the solutions expressed in the trigonometrical form are

$$(18) \qquad \begin{cases} x = + x_0 \cos kt + \dfrac{x_0'}{k} \sin kt, \\[2mm] \dfrac{dx}{dt} = - kx_0 \sin kt + x_0' \cos kt, \\[2mm] y = + y_0 \cos kt + \dfrac{y_0'}{k} \sin kt, \\[2mm] \dfrac{dy}{dt} = - ky_0 \sin kt + y_0' \cos kt. \end{cases}$$

The equation of the orbit is obtained by eliminating t between the first and third equations of (18). On multiplying by the appropriate factors and adding, it is found that

(19) $\qquad \begin{cases} (x_0 y_0' - y_0 x_0') \sin kt = k(x_0 y - y_0 x), \\ (x_0 y_0' - y_0 x_0') \cos kt = y_0' x - x_0' y. \end{cases}$

The result of squaring and adding these equations is

(20) $\quad (k^2 y_0^2 + y_0'^2)x^2 + (k^2 x_0^2 + x_0'^2)y^2 - 2(k^2 x_0 y_0 + x_0' y_0')xy$
$$= (x_0 y_0' - y_0 x_0')^2.$$

This is the equation of an ellipse with the origin at the center unless $x_0 y_0' - y_0 x_0' = 0$, when the orbit degenerates to two straight lines which must be coincident; for, then

$$\frac{x_0}{x_0'} = \frac{y_0}{y_0'} = \text{constant} = c;$$

from which

$$x_0 = c x_0', \qquad y_0 = c y_0'.$$

In this case equation (20) becomes

(21) $\qquad (k^2 c^2 + 1)(y_0' x - x_0' y)^2 = 0,$

and the motion is rectilinear and oscillatory. In every case both the coördinates and the components of velocity are periodic with the period $\dfrac{2\pi}{k}$, whatever the initial conditions may be.

54. Differential Equation of the Orbit. The curve described by the moving particle, independently of the manner in which it may move along this curve, is of much interest. A general method of finding the orbit is to integrate the differential equations and then to eliminate the time. This is often a complicated process, and the question arises whether the time cannot be eliminated before the integration is performed, so that the integration will lead directly to the orbit. It will be shown that this is the case when the force does not depend upon the time.

The differential equations of motion are [Art. 47]

(22) $\qquad \begin{cases} \dfrac{d^2 x}{dt^2} = -f\dfrac{x}{r}, \\ \dfrac{d^2 y}{dt^2} = -f\dfrac{y}{r}. \end{cases}$

Since f does not involve the time t enters only in the derivatives.

But a second differential quotient cannot be separated as though it were an ordinary fraction; therefore, the order of the derivatives must be reduced before the direct elimination of t can be made. In order to do this most conveniently polar coördinates will be employed. Equations (22) become in these variables

(23)
$$\begin{cases} \dfrac{d^2r}{dt^2} - r\left(\dfrac{d\theta}{dt}\right)^2 = -f, \\ r\dfrac{d^2\theta}{dt^2} + 2\dfrac{dr}{dt}\dfrac{d\theta}{dt} = 0. \end{cases}$$

The integral of the second of these equations is

$$r^2\frac{d\theta}{dt} = h.$$

On eliminating $\dfrac{d\theta}{dt}$ from the first of (23) by means of this equation, the result is found to be

(24)
$$\frac{d^2r}{dt^2} = \frac{h^2}{r^3} - f.$$

Now let $r = \dfrac{1}{u}$; therefore

$$\begin{cases} \dfrac{dr}{dt} = -\dfrac{1}{u^2}\dfrac{du}{dt} = -\dfrac{1}{u^2}\dfrac{du}{d\theta}\dfrac{d\theta}{dt} = -h\dfrac{du}{d\theta}, \\ \dfrac{d^2r}{dt^2} = -h\dfrac{d}{dt}\left(\dfrac{du}{d\theta}\right) = -h\dfrac{d^2u}{d\theta^2}\dfrac{d\theta}{dt} = -h^2u^2\dfrac{d^2u}{d\theta^2}. \end{cases}$$

When this value of the second derivative of r is equated to the one given in (24), it is found that

(25)
$$f = h^2u^2\left(u + \frac{d^2u}{d\theta^2}\right).$$

This differential equation is of the second order, but one integral has been used in determining it; therefore the problem of finding the path of the body is of the third order. The complete problem was of the fourth order; the fourth integral expresses the relation between the coördinates and the time, or defines the position of the particle in its orbit.

Since the integral of (25) expresses u, and therefore r, in terms of θ, the equation

$$r^2\frac{d\theta}{dt} = h,$$

when integrated, gives the relation between θ and t.

Conversely, equation (25) can be used to find the law of central force which will cause a particle to describe a given curve. It is only necessary to write the equation of the curve in polar coördinates and to compute the right member of (25). This is generally a simpler process than the reverse one of finding the orbit when the law of force is given.

55. Newton's Law of Gravitation. In the early part of the seventeenth century Kepler announced three laws of planetary motion, which he had derived from a most laborious discussion of a long series of observations of the planets, especially of Mars. They are the following:

LAW I.   *The radius vector of each planet with respect to the sun as the origin sweeps over equal areas in equal times.*

LAW II.   *The orbit of each planet is an ellipse with the sun at one of its foci.*

LAW III.   *The squares of the periods of the planets are to each other as the cubes of the major semi-axes of their respective orbits.*

It was on these laws that Newton based his demonstration that the planets move subject to forces directed toward the sun, and varying inversely as the squares of their distances from the sun. The Newtonian law will be derived here by employing the analytical method instead of the geometrical methods of the *Principia*.*

From the converse of the theorem of areas and Kepler's first law, it follows that the planets move subject to central forces directed toward the sun. The curves described are given by the second law, and equation (25) can, therefore, be used to find the expression for the acceleration in terms of the coördinates. Let a represent the major semi-axis of the ellipse, and e its eccentricity; then its equation in polar coördinates with origin at a focus is

$$r = \frac{a(1 - e^2)}{1 + e \cos \theta}.$$

Therefore

$$u + \frac{d^2u}{d\theta^2} = \frac{1}{a(1 - e^2)}.$$

On substituting this expression in (25), it is found that the equation for the acceleration is

$$f = \frac{h^2}{a(1 - e^2)} \cdot \frac{1}{r^2} = \frac{k^2}{r^2}.$$

* Book I., Proposition XI.

Therefore, the acceleration to which any planet is subject varies inversely as the square of its distance from the sun.

If the distance r is eliminated by the polar equation of the conic the expression for f has the form

$$f = k_1^2 (1 + e \cos \theta)^2,$$

which depends only upon the direction of the attracted body and not upon its distance. Now for points on the ellipse the two expressions for f give the same value, but elsewhere they give different values. It is clear that many other laws of force, all agreeing in giving the same numerical values of f for points on the ellipse, can be obtained by making other uses of the equation of the conic to eliminate r. For example, since it follows from the polar equation of the ellipse for points on its circumference that

$$\frac{(1 + e \cos \theta)r}{a(1 - e^2)} = 1,$$

one such law is

$$f = \frac{k^2(1 + e \cos \theta)^3 r}{a(1 - e^2)},$$

and this value of f, which depends both upon the direction and distance of the attracted body, differs from both of the preceding for points not on the ellipse. All of these laws are equally consistent with the motion of the planet in question as expressed by Kepler's laws. But the laws of Kepler hold for each of the eight planets and the twenty-six known satellites of the solar system, besides for more than seven hundred small planets which have so far been discovered. It is natural to impose the condition, if possible, that the force shall vary according to the same law for each body. Since the eccentricities and longitudes of the perihelia of their orbits are all different, the law of force is the same for all these bodies only when it has the form

$$f = \frac{k^2}{r^2}.$$

Another reason for adopting this expression for f is that in case of all the others the attraction would depend upon the direction of the attracted body, and this seems improbable. This conclusion is further supported by the fact that the forces to which comets are subject when they move through the entire system of planets vary according to this law. And finally, as will be shown in Art. 89, the accelerations to which the various planets are subject vary from one to another according to this law.

From the consideration of Kepler's laws, the gravity at the earth's surface, and the motion of the moon around the earth, Newton was led to the enunciation of the Law of Universal Gravitation, which is, *every two particles of matter in the universe attract each other with a force which acts in the line joining them, and whose intensity varies as the product of their masses and inversely as the squares of their distance apart.*

It will be observed that the law of gravitation involves considerably more than can be derived from Kepler's laws of planetary motion; and it was by a master stroke of genius that Newton grasped it in its immense generality, and stated it so exactly that it has stood without change for more than 200 years. When contemplated in its entirety it is one of the grandest conceptions in the physical sciences.

56. Examples of Finding the Law of Force. (*a*) If a particle describes a circle passing through the origin, the law of force (depending on the distance alone) under which it moves is a very simple expression. Let *a* represent the radius; then the polar equation of the circle is

$$r = 2a \cos \theta, \qquad u = \frac{1}{2a \cos \theta}.$$

Therefore

$$\frac{d^2u}{d\theta^2} + u = 8a^2u^3.$$

On substituting this expression in (25), it is found that

$$f = \frac{8a^2h^2}{r^5} = \frac{k^2}{r^5}.$$

(*b*) Suppose the particle describes an ellipse with the origin at the center. The polar equation of an ellipse with the center as origin is

$$r^2 = \frac{b^2}{1 - e^2 \cos^2 \theta}.$$

From this it follows that

$$\begin{cases} bu = \sqrt{1 - e^2 \cos^2 \theta}, \\ b\dfrac{d^2u}{d\theta^2} = \dfrac{e^2 \cos^2 \theta - e^2 \sin^2 \theta}{\sqrt{1 - e^2 \cos^2 \theta}} - \dfrac{e^4 \sin^2 \theta \cos^2 \theta}{(1 - e^2 \cos^2 \theta)^{\frac{3}{2}}}, \\ u + \dfrac{d^2u}{d\theta^2} = \dfrac{1 - e^2}{b^4} \cdot \dfrac{1}{u^3}. \end{cases}$$

On substituting in (25), the expression for f is found to be

$$f = \frac{h^2(1 - e^2)}{b^4} \cdot r = k^2r.$$

THE UNIVERSALITY OF NEWTON'S LAW.

57. Double Star Orbits. The law of gravitation is proved from Kepler's laws and certain assumptions as to its uniqueness to hold in the solar system; the question whether it is actually *universal* naturally presents itself. The fixed stars are so remote that it is not possible to observe planets revolving around them, if indeed they have such attendants. The only observations thus far obtained which throw any light upon the subject are those of the motions of the double stars.

Double star astronomy started about 1780 with the search for close stars by Sir William Herschel for the purpose of determining parallax by the differential method. A few years were sufficient to show him, to his great surprise, that in some cases the two components of a pair were revolving around each other, and that, therefore, they were physically connected as well as being apparently in the same part of the sky. The discovery and measurement of these systems has been pursued with increasing interest and zeal by astronomers. Burnham's great catalogue of double stars contains about 13,000 of these objects. The relative motions are so slow in most cases that only a few have yet completed one revolution, or enough of one revolution so that the shapes of their orbits are known with certainty. There are now about thirty pairs whose observed angular motions have been sufficiently great to prove, within the errors of the observations, that they move in ellipses with respect to each other in such a manner that the law of areas is fulfilled. In no case is the primary at the focus, or at the center, of the relative ellipse described by the companion, but it occupies some other place within the ellipse, the position varying greatly in different systems.

From the observations and the converse of the law of areas it follows that the resultant of the forces acting upon one star of a pair is always directed toward the other. The law of variation of the intensity of the force depends upon the position in the ellipse which the center of force occupies. It must not be overlooked at this point that the orbits of the stars are not observed directly, but that it is their projections upon the planes tangent to the celestial sphere at their respective places which are seen. The effect of this sort of projection is to change the true ellipse into a different apparent eilipse whose major axis has a different direction, and one that is differently situated with respect to the central star; indeed, it might happen that if one of the stars was really in the focus of the true ellipse described by the other, the projection would be such as to make it lie on the minor axis of the apparent ellipse.

Astronomers have assumed that the orbits are plane curves and that the apparent departure of the central star from the focus of the ellipse described by the companion is due to projection, and have then computed the angle of the line of nodes and the inclination. No inconsistencies are introduced in this way, but the

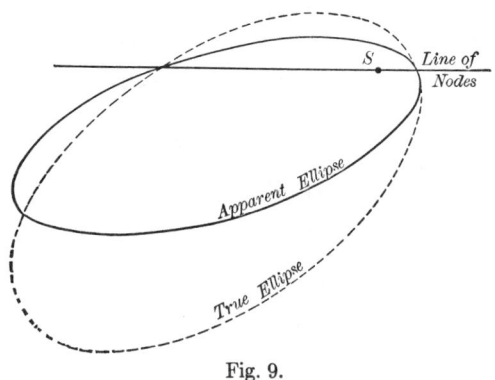

Fig. 9.

possibility remains that the assumptions are not true. The question of what the law of force must be if it is not Newton's law of gravitation will now be investigated.

58. Law of Force in Binary Stars. If the force varied directly as the distance the primary star would be at the center of the ellipse described by the secondary (Art. 53). No projection would change this relative position, and since such a condition has never been observed, it is inferred that the force does not vary directly as the distance.

The condition will now be imposed that the curve shall be a conic with general position for the origin, and the expression for the central force will be found. The equation of the general conic is

(26) $ax^2 + 2bxy + cy^2 + 2dx + 2fy = g.$

On transforming to polar coördinates and putting $r = \dfrac{1}{u}$, this equation gives

(27) $\quad u = A \sin \theta + B \cos \theta \pm \sqrt{C \sin 2\theta + D \cos 2\theta + H}$,

where

$$
\begin{cases}
A = \dfrac{f}{g}, \qquad B = \dfrac{d}{g}, \qquad C = \dfrac{fd + bg}{g^2}, \\[2mm]
D = \dfrac{d^2 + ag - f^2 - cg}{2g^2}, \qquad H = \dfrac{d^2 + ag + f^2 + cg}{2g^2}.
\end{cases}
$$

On differentiating (27) twice, it is found that

(28)
$$
\frac{d^2u}{d\theta^2} = - A \sin \theta - B \cos \theta
$$
$$
\pm \frac{-C^2 - D^2 - (C \sin 2\theta + D \cos 2\theta)^2 - 2H(C \sin 2\theta + D \cos 2\theta)}{(C \sin 2\theta + D \cos 2\theta + H)^{\frac{3}{2}}}.
$$

On substituting (27) and (28) in (25), it follows that

(29) $\qquad f = \pm \dfrac{h^2}{r^2} \dfrac{(H^2 - C^2 - D^2)}{(C \sin 2\theta + D \cos 2\theta + H)^{\frac{3}{2}}}.$

This becomes as a consequence of (27)

(30) $\qquad f = \pm \dfrac{h^2}{r^2} \dfrac{(H^2 - C^2 - D^2)}{\left(\dfrac{1}{r} - A \sin \theta - B \cos \theta \right)^3}.$

There are also infinitely many other laws, all giving the same values of f for points on the ellipse in question, which are obtained by multiplying these expressions by any functions of u and θ which are unity on the ellipse in virtue of equation (27).

It does not seem reasonable to suppose that the attraction of two stars for each other depends upon their orientation in space. Equation (29) becomes independent of θ if $C = D = 0$, and (30), if $A = B = 0$. The first gives

$$
f = \pm \frac{\text{constant}}{r^2},
$$

and the second,

$$
f = \pm \text{constant} \cdot r.
$$

The first is Newton's law, and the second is excluded by the fact that no primary star has been found in the center of the orbit described by the companion. It is clear that θ can be eliminated from (29) and (30) by means of (27) without imposing the conditions that $A = B = C = D = 0$. But Griffin has shown[*] that for all such laws, except the Newtonian, the force either vanishes when the distance between the bodies vanishes, or becomes imaginary for certain values of r. Clearly both of these laws are improbable from a physical point of view. Hence it is extremely probable that the law of gravitation holds throughout the stellar systems; and this conclusion is supported by the fact that the spectroscope shows the stars are composed of familiar terrestrial elements.

59. Geometrical Interpretation of the Second Law. The expression for the central force given in (30) may be put in a very simple and interesting form. Let $g^3h^2(H^2 - C^2 - D^2) = N$, and transform $\dfrac{1}{r} - A \sin \theta - B \cos \theta$ into rectangular coördinates and the original constants; then (30) becomes

$$(31) \qquad f = \frac{\mp Nr}{(dx + fy - g)^3}.$$

The equation of the polar of the point (x', y') with respect to the general conic (26) is[†]

$$ax_1x' + b(x_1y' + y_1x') + cy_1y' + d(x_1 + x') + f(y_1 + y') - g = 0,$$

where x_1 and y_1 are the running variables. When (x', y') is the origin this equation becomes

$$(32) \qquad dx_1 + fy_1 - g = 0,$$

and has the same form as the denominator of (31). The values of x and y in (31) are such that they satisfy the equation of the conic, while x_1 and y_1 of (32) satisfy the equation of the polar line. They are, therefore, in general numerically different from x and y. But the distance from any point (x, y) of the conic to the polar line with respect to the origin is given by the equation

$$p = \frac{dx + fy - g}{\sqrt{d^2 + f^2}},$$

where x and y are the coördinates of points on the conic. Let

$$N' = \frac{N}{(d^2 + f^2)^{\frac{3}{2}}};$$

then (31) becomes

[*] *American Journal of Mathematics*, vol. 31 (1909), pp. 62–85.
[†] Salmon's *Conic Sections*, Art. 89.

$$(33) \qquad\qquad f = \mp \frac{N'r}{p^3}.$$

Therefore, *if a particle moving subject to a central force describes any conic, the intensity of the force varies directly as the distance of the particle from the origin, and inversely as the cube of its distance from the polar of the origin with respect to the conic.*

60. Examples of Conic Section Motion. (*a*) When the orbit is a central conic with the origin at the center, the polar line recedes to infinity, and $\frac{N'}{p^3}$ must be regarded as a constant. Then the force varies directly as the distance, as was shown in Art. 56 (*b*).

(*b*) When the origin is at one of the foci of the conic the polar line is the directrix, and $p = \frac{r}{e}$, where e is the eccentricity. Then (33) becomes

$$f = \mp \frac{N'e^3}{r^2}.$$

This is Newton's law which was derived from the same conditions in Art. 55.

VIII. PROBLEMS.

1. Find the vis viva integral when $f = \frac{c}{r^2}$, $f = cr$, $f = \frac{c}{r^n}$.

2. Suppose that in Art. 53 the particle is projected orthogonally from the x-axis; find the equations corresponding to (19) and (20). Suppose still further that $k = 1$, $x_0 = 1$; find the initial velocity such that the eccentricity of the ellipse may be 1/2.

$$Ans. \quad \begin{cases} v_0 = \tfrac{1}{2}\sqrt{3}, \quad \text{or} \\ v_0 = \tfrac{2}{3}\sqrt{3}. \end{cases}$$

3. Find the central force as a function of the distance under which a particle may describe the spiral $r = \frac{1}{c\theta}$; the spiral $r = e^\theta$.

$$Ans. \quad f = \frac{h^2}{r^3}, \qquad f = \frac{2h^2}{r^3}.$$

4. Find the central force as a function of the distance under which a particle may describe the lemniscate $r^2 = a^2 \cos 2\theta$.

$$Ans. \quad f = \frac{3h^2a^4}{r^7}.$$

5. Find the central force as a function of the distance under which a particle may describe the cardioid $r = a(1 + \cos \theta)$.

$$Ans. \quad f = \frac{3ah^2}{r^4}.$$

6. Suppose the particle describes an ellipse with the origin in its interior, at a distance n from the x-axis and m from the y-axis. (a) Show that two of the laws of force are

$$
\begin{cases}
f = \dfrac{h^2}{r^2} \dfrac{(ac)^{\frac{1}{2}}}{[2mn \sin\theta \cos\theta + (a - c - n^2 + m^2)\cos^2\theta + c - m^2]^{\frac{3}{2}}}, \\
f = \dfrac{h^2 a^2 c^2 r}{[ac - am^2 - cn^2 + cny + amx]^3},
\end{cases}
$$

where a and c have the same meaning as in (26), and where the polar axis is parallel to the major axis of the ellipse. (b) If the origin is between the center and the focus show that the force at unit distance is a maximum for $\theta = 0$ and is a minimum for $\theta = \dfrac{\pi}{2}$; that if the origin is between a focus and the nearest apse the maximum is for $\theta = \dfrac{\pi}{2}$ and the minimum for $\theta = 0$; and that if the origin is on the minor axis the maximum is for $\theta = 0$, and the minimum for $\theta = \dfrac{\pi}{2}$.

7. Interpret equation (29) geometrically.

Hint. $C \sin 2\theta + D \cos 2\theta + H = \dfrac{(dx + fy)^2 + g(ax^2 + cy^2 + 2bxy)}{g^2 r^2}$.

The numerator of this expression set equal to zero is the equation of the tangents (real or imaginary) from the origin to the conic. (Salmon's *Conic Sections*, Art. 92.)

8. Find expressions for the central force when the orbit is an ellipse with the origin at an end of the major and minor axes respectively. Show that they reduce to $\dfrac{k^2}{r^5}$ when the ellipse becomes a circle.

$$
Ans. \quad
\begin{cases}
f = \dfrac{h^2 \sqrt{c}}{ar^2} \cdot \dfrac{1}{\cos^3\theta}, \\
f = \dfrac{h^2 \sqrt{a}}{cr^2} \cdot \dfrac{1}{\sin^3\theta}.
\end{cases}
$$

DETERMINATION OF THE ORBIT FROM THE LAW OF FORCE.

61. Force Varying as the Distance.   The problem of finding the orbit when the law of force is given is generally more difficult than the converse, since it involves the integration of (25).   The method of integration varies with the different laws of force, and the character of the integrals depends upon the initial conditions. The process will be illustrated first in the case in which the force varies as the distance, a problem treated by another method in Art. 53.

If $f = k^2 r$, equation (25) becomes

$$\frac{k^2}{u} = h^2 u^2 \left[u + \frac{d^2 u}{d\theta^2} \right],$$

or

$$\frac{d^2 u}{d\theta^2} = \frac{k^2}{h^2} \frac{1}{u^3} - u.$$

The first integral of this equation is

$$\left(\frac{du}{d\theta} \right)^2 = -\frac{k^2}{h^2} \frac{1}{u^2} - u^2 + c_1;$$

whence

(34) $$d\theta = \frac{\pm\, u\, du}{\left[\dfrac{c_1{}^2}{4} - \dfrac{k^2}{h^2} - \left(\dfrac{c_1}{2} - u^2 \right)^2 \right]^{\frac{1}{2}}}.$$

Let

$$\frac{c_1}{2} - u^2 = z, \qquad \frac{c_1{}^2}{4} - \frac{k^2}{h^2} = A^2.$$

The constant A^2 must be positive in order that $\dfrac{du}{d\theta}$ may be real, as it is if the particle is started with real initial conditions.

If the upper sign is used, equation (34) becomes

(35) $$2d\theta = \frac{-\,dz}{\sqrt{A^2 - z^2}}.$$

It is easily verified that the same equation (36) would be reached, when the initial conditions are substituted, if the lower sign were used. The integral of (35) is

$$\cos^{-1}\frac{z}{A} = 2(\theta + c_2),$$

or

$$z = A \cos 2(\theta + c_2).$$

On going back to the variable r, this equation becomes

(36) $$r^2 = \frac{2}{c_1 - 2A \cos 2(\theta + c_2)}.$$

This is the polar equation of an ellipse with the origin at the center. Hence, a particle moving subject to an attractive force varying directly as the distance describes an ellipse with the origin at the center. The only exceptions are when the particle passes through the origin, and when it describes a circle. In the first $h = 0$, and equation (25) ceases to be valid; in the second, c_1 has such a value that it satisfies the equation

$$\left(\frac{du}{d\theta}\right)_0^2 = -\frac{k^2}{h^2} \cdot \frac{1}{u_0{}^2} - u_0{}^2 + c_1 = 0,$$

and the equation of the orbit becomes $u = u_0$. In this case equation (34) fails.

62. Force Varying Inversely as the Square of the Distance. Suppose a particle moves under the influence of a central attraction the intensity of which varies inversely as the square of the distance; it is required to determine its orbit when it is projected in any manner. Equation (25) is in this case

(37) $$\frac{d^2u}{d\theta^2} = \frac{k^2}{h^2} - u.$$

This equation can be written in the form

$$\frac{d^2u}{d\theta^2} + u = \frac{k^2}{h^2}.$$

This is a linear non-homogeneous differential equation and can be integrated by the method of variation of parameters, which was explained in Art. 37. When its right member is neglected the general solution is

$$u = c_1 \cos \theta + c_2 \sin \theta.$$

It is clear that if $\dfrac{k^2}{h^2}$ is added to this value of u the differential equation will be identically satisfied. Consequently the general solution of (37), which is the same as that found by the variation of parameters, is

$$u = \frac{k^2}{h^2} + c_1 \cos \theta + c_2 \sin \theta.$$

On taking the reciprocal of this equation, it is found that

$$r = \frac{1}{\dfrac{k^2}{h^2} + c_1 \cos \theta + c_2 \sin \theta}.$$

Now let $c_1 = A \cos \theta_0$, $c_2 = A \sin \theta_0$, where A and θ_0 are constants. It is clear that A can always be taken positive and equal to $\sqrt{c_1{}^2 + c_2{}^2}$ and a real θ_0 can be determined so that these equations will be satisfied whatever real values c_1 and c_2 may have. Then the equation for the orbit becomes

(38) $$r = \frac{1}{\dfrac{k^2}{h^2} + A \cos (\theta - \theta_0)}.$$

This is the polar equation of a conic with the origin at one of the foci.

From this investigation and that of Art. 55 it follows that if the orbit is a conic section with the origin at one of the foci, and the force depends on the distance alone, then the body moves subject to a central force varying inversely as the square of the distance; and conversely, if the force varies inversely as the square of the distance, then the body will describe a conic section with the origin at one of the foci.

Let p represent the parameter of the conic and e its eccentricity. Then, comparing (38) with the ordinary polar equation of the conic, $r = \dfrac{p}{1 + e \cos \phi}$, it is found that

(39)
$$\begin{cases} p = \dfrac{h^2}{k^2}, \\[2mm] e = \dfrac{h^2}{k^2} A, \end{cases}$$

and θ_0 is the angle between the polar axis and the end of the major axis directed to the farthest apse. The constants h^2 and A are determined by the initial conditions, and they in turn define p and e by (39). If $e < 1$, the conic is an ellipse; if $e = 1$, the conic is a parabola; if $e > 1$, the conic is a hyperbola; and if $e = 0$, the conic is a circle.

63. Force Varying Inversely as the Fifth Power of the Distance.

In this case $f = \dfrac{k^2}{r^5}$, and (25) becomes

(40)
$$k^2 u^5 = h^2 u^2 \left(u + \frac{d^2 u}{d\theta^2} \right).$$

On solving for $\dfrac{d^2 u}{d\theta^2}$ and integrating, it is found that

(41)
$$\left(\frac{du}{d\theta} \right)^2 = \frac{1}{2} \frac{k^2}{h^2} u^4 - u^2 + c_1.$$

Therefore

(42)
$$d\theta = \frac{du}{\sqrt{c_1 + \dfrac{1}{2} \dfrac{k^2}{h^2} u^4 - u^2}}.$$

The right member of this equation cannot in general be integrated in terms of the elementary functions, but it can be transformed into an elliptic integral of the first kind. Then u, and consequently r, is expressible in terms of θ by elliptic functions, and the orbits in general either wind into the origin or pass out to infinity, their particular character depending upon the initial conditions.

There are certain special cases which are integrable in terms of elementary functions.

(a) If such a constant value of u is taken that it fulfills (41) when its right member is set equal to zero, then r is a constant and the orbit is a circle with the origin at the center. It is easily seen that a similar special case will occur for a central force varying as any power of the distance.

(b) Another special case is that in which the initial conditions are such that $c_1 \neq 0$ and the right member of (41) is a perfect square. That is, $c_1 = \dfrac{h^2}{2k^2}$. Then equation (41) becomes

$$\left(\frac{du}{d\theta}\right)^2 = \left(\frac{1}{\sqrt{2}}\frac{k}{h}u^2 - \frac{1}{\sqrt{2}}\frac{h}{k}\right)^2 = \frac{1}{2}\left(A^2u^2 - \frac{1}{A^2}\right)^2.$$

The integral of this equation is

$$\log\frac{1 + A^2u}{1 - A^2u} = \sqrt{2}(\pm\,\theta - c_2);$$

whence

(43) $$r = -\,A^2\frac{\left[1 + e^{\sqrt{2}(\pm\theta-c_2)}\right]}{\left[1 - e^{\sqrt{2}(\pm\theta-c_2)}\right]} = +\,A^2\coth\frac{\sqrt{2}}{2}(\pm\,\theta - c_2),$$

where $\coth\dfrac{\sqrt{2}}{2}(\pm\,\theta - c_2)$ is the hyperbolic cotangent of

$$\tfrac{1}{2}\sqrt{2}(\pm\,\theta - c_2).$$

(c) If the initial conditions are such that $c_1 = 0$, equation (41) gives

$$\pm\,d\theta = \frac{du}{u\sqrt{\dfrac{1}{2}\dfrac{k^2}{h^2}u^2 - 1}},$$

the integral of which is

$$\pm\,\theta = \cos^{-1}\left(\frac{\sqrt{2}h}{ku}\right) + c_2.$$

On taking the cosines of both members and solving for r, the polar equation of the orbit is found to be

(44) $$r = \frac{k}{\sqrt{2}h}\cos(c_2 \mp \theta),$$

which is the equation of a circle with the origin on the circumference.

(d) If none of these conditions is fulfilled the right member of (41) is a biquadratic, and equation (42) can be written in the form

$$(45) \qquad \pm\, d\theta = \frac{C\,du}{\sqrt{\pm\, (1 \pm \alpha^2 u^2)(1 \pm \beta^2 u^2)}},$$

where C, α^2, and β^2 are constants which depend upon the coefficients of (41) in a simple manner. Equation (45) leads to an elliptic integral which expresses θ in terms of u. On taking the inverse functions and the reciprocals, r is expressed as an elliptic function of θ. The curves are spirals of which the circle through the origin, and the one around the origin as center, are limiting cases.

If the curve is a circle through the origin the force varies inversely as the fifth power of the distance (Art. 56); but if the force varies inversely as the fifth power of the distance, the orbits which the particle will describe are curves of which the circle is a particular limiting case. On the other hand, if the orbit is a conic with the origin at the center or at one of the foci, the force varies directly as the distance, or inversely as the square of the distance; and conversely, if force varies directly as the distance, or inversely as the square of the distance, the orbits are *always* conics with the origin at the center, or at one of the foci respectively [Arts. 53, 55, 56 (b)]. A complete investigation is necessary for every law to show this converse relationship.

IX. PROBLEMS.

1. Discuss the motion of the particle by the general method for linear equations when the force varies inversely as the cube of the distance. Trace the curves in the various special cases.

2. Express C, α^2, and β^2 of equation (45) in terms of the initial conditions. For original projections at right angles to the radius vector investigate all the possible cases, reducing the integrals to the normal form, and expressing r as elliptic functions of θ. Draw the curves in each case.

3. Suppose the law of force is that given in (29); i. e.

$$f = \frac{M}{r^2(C \sin 2\theta + D \cos 2\theta + H)^{\frac{3}{2}}} = \frac{M}{r^2[\phi(\theta)]^{\frac{3}{2}}}.$$

Integrate the differential equation of the orbit, (25), by the method of variation of parameters, and show that the general solution has the form

$$\frac{1}{r} = c_1 \cos\theta + c_2 \sin\theta + \sqrt{\phi(\theta)},$$

where c_1 and c_2 are constants of integration. Show that the curve is a conic.

4. When the force is $f = \frac{\mu}{r^2} + \frac{\nu}{r^3}$ show that, if $\nu < h^2$, the general equation of the orbit described has the form

$$r = \frac{a}{1 - e \cos (k\gamma)} \, ,$$

where a, e, and k are the constants depending upon the initial conditions and μ and ν. Observe that this may be regarded as being a conic section whose major axis revolves around the focus with the mean angular velocity

$$n = (1 - k) \frac{2\pi}{T} \, ,$$

where T is the period of revolution.

5. In the case of a central force the motion along the radius vector is defined by the equation

$$\frac{d^2 r}{dt^2} = -f + \frac{h^2}{r^3} \, .$$

Discuss the integration of this equation when

$$f = \frac{k^2}{r^5} \, .$$

6. Suppose the law of force is that given by (30); i. e.,

$$f = \frac{N}{r^2 \left(\dfrac{1}{r} - A \sin \theta - B \cos \theta \right)^3} \, .$$

Substitute in (25) and derive the general equation of the orbit described

Hint. Let $u = v + A \sin \theta + B \cos \theta$; then (25) becomes

$$\frac{d^2 v}{d\theta^2} + v = \frac{N h^{-2}}{v^3} \, .$$

Ans. $\dfrac{1}{r} = A \sin \theta + B \cos \theta + \sqrt{c_1 \cos^2 \theta + c_2 \sin 2\theta + c_3 \sin^2 \theta}$,

which is the equation of a conic section.

7. Suppose the law of force is

$$f = \frac{c_1 + c_2 \cos 2\theta}{r^2} \, ,$$

show that, for all initial projections, the orbit is an algebraic curve of the fourth degree unless $c_2 = 0$, when it reduces to a conic.

HISTORICAL SKETCH AND BIBLIOGRAPHY.

The subject of central forces was first discussed by Newton. In Sections II. and III. of the First Book of the *Principia* he gave a splendid geometrical treatment of the subject, and arrived at some very general theorems. These portions of the *Principia* especially deserve careful study.

All the simpler cases were worked out in the eighteenth century by analytical methods. A few examples are given in detail in Legendre's *Traité des Fonctions Elliptiques*. An exposition of principles and a list of examples are given in nearly every work on analytical mechanics; among the best of these treatments are the Fifth Chapter in Tait and Steele's *Dynamics of a Particle*, and the Tenth Chapter, vol. I., of Appell's *Mécanique Rationelle*. Stader's memoir, vol. XLVI., *Journal für Mathematik*, treats the subject in great detail. The special problem where the force varies inversely as the fifth power of the distance has been given a complete and elegant treatment by MacMillan in *The American Journal of Mathematics*, vol. XXX, pp. 282–306.

The problem of finding the general expression for the possible laws of force operating in the binary star systems was proposed by M. Bertrand in vol. LXXXIV. of the *Comptes Rendus*, and was immediately solved by MM. Darboux and Halphen, and published in the same volume. The treatment given above in the text is similar to that given by M. Darboux, which has also been reproduced in a note at the end of the *Mécanique* of M. Despeyrous. The method of M. Halphen is given in Tisserand's *Mé anique Céleste*, vol. I., p. 36, and in Appell's *Mécanique Rationelle*, vol. I., p. 372. It seems to have been generally overlooked that Newton had reated the same problem in the *Principia*, Book I., Scholium to Proposition XVII. This was reproduced and shown to be equivalent to the work of MM. Darboux and Halphen by Professo Glaisher in the *Monthly Notices f R.A.S.*, vol. XXXIX.

M. Bertrand has shown (*Comptes Rendus*, vol. LXXVII.) that the only laws of central force under the action of which a particle will describe a conic section for all initial conditions are $f = \pm \dfrac{k^2}{r^2}$ and $f = \pm k^2 r$. M. Koenigs has proved (*Bulletin de la Société Mathématique*, vol. XVII.) that the only laws of central force depending upon the distance alone, for which the curves described are algebraic for all initial conditions are $f = \pm \dfrac{k^2}{r^2}$ and $f = \pm k^2 r$.

Griffin has shown (*American Journal of Mathematics*, vol. XXXI.) that the only law, where the force is a function of the distance alone, where it does not vanish at the center of force, and where it is real throughout the plane, giving an elliptical orbit is the Newtonian law.

CHAPTER IV.

THE POTENTIAL AND ATTRACTIONS OF BODIES.

64. THE previous chapters have been concerned with problems in which the law of force was given, or with the discovery of the law of force when the orbits were given. All the investigations were made as though the masses were mere points instead of being of finite size. When forces exist between every two particles of all the masses involved, bodies of finite size cannot be assumed to attract one another according to the same laws. Hence it is necessary to take up the problem of determining the way in which finite bodies of different shapes attract one another.

It follows from Kepler's laws and the principles of central forces that, if the planets are regarded as being of infinitesimal dimensions compared to their distances from the sun, they move under the influence of forces which are directed toward the center of the sun and which vary inversely as the squares of their distances from it. This suggests the idea that the law of inverse squares may account for the motions still more exactly if the bodies are regarded as being of finite size, with every particle attracting every other particle in the system. The appropriate investigation shows that this is true.

This chapter will be devoted to an exposition of general methods of finding the attractions of bodies of any shape on unit particles in any position, exterior or interior, when the forces vary inversely as the squares of the distances. The astronomical applications will be to the attractions of spheres and oblate spheroids, to the variations in the surface gravity of the planets, and to the perturbations of the motions of the satellites due to the oblateness of the planets.

65. Solid Angles. If a straight line constantly passing through a fixed point is moved until it retakes its original position, it generates a conical surface of two sheets whose vertices are at the given point. The area which one end of this double cone cuts out of the surface of the unit sphere whose center is at the given point is called the *solid angle* of the cone; or, the area cut out of any concentric sphere divided by the square of its radius measures the solid angle.

Since the area of a spherical surface equals the product of 4π and the square of its radius, it follows that the sum of all the solid angles about a point is 4π. The sum of the solid angles of one-half of all the double cones which can be constructed about a point without intersecting one another is 2π.

The volume contained within an infinitesimal cone whose solid angle is ω and between two spherical surfaces whose centers are at the vertex of the cone, approaches as a limit, as the surfaces approach each other, the product of the solid angle, the square of the distance of the spherical surfaces from the vertex, and the distance between them. If the centers of the spherical surfaces are at a point not in the axis of the cone, the volume approaches as a limit the product of the solid angle, the square of the distance

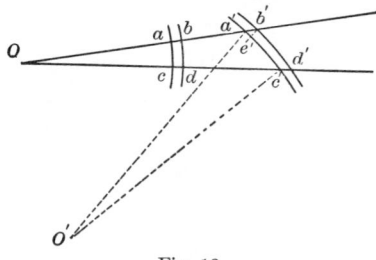

Fig. 10.

from the vertex, the distance between the spherical surfaces, and the reciprocal of the cosine of the angle between the axis of the cone and the radius from the center of the sphere; or, it is the product of the solid angle, the square of the distance from the vertex, and the intercept on the cone between the spherical surfaces. Thus, the volume of $abdc$, Fig. 10, is $V = \omega \overline{aO}^2 \cdot ab$. The volume of $a'b'd'c'$ is

$$V' = \frac{\omega \overline{a'O}^2 \cdot b'e'}{\cos (Oa'O')} = \omega \overline{a'O}^2 \cdot a'b'.$$

Sometimes it will be convenient to use one of these expressions and sometimes the other.

66. The Attraction of a Thin Homogeneous Spherical Shell upon a Particle in its Interior. The attractions of spheres and other simple figures were treated by Newton in the *Principia*,

Book I., Section 12. The following demonstration is essentially as given by him.

Consider the spherical shell contained between the infinitely near spherical surfaces S and S', and let P be a particle of unit mass situated within it. Construct an infinitesimal cone whose

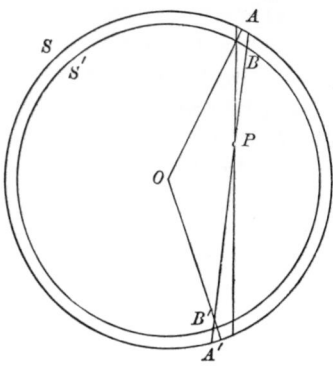

Fig. 11.

solid angle is ω with its vertex at P. Let σ be the density of the shell. Then the mass of the element of the shell at A is $m = \sigma \overline{AB} \, \omega \, \overline{AP}^2$; likewise the mass of the element of A' is $m' = \sigma \overline{A'B'} \omega \overline{A'P}^2$. The attractions of m and m' upon P are respectively

$$\alpha = \frac{k^2 m}{\overline{AP}^2}, \qquad \alpha' = \frac{k^2 m'}{\overline{A'P}^2}.$$

Since $\overline{A'B'} = \overline{AB}$, $\alpha = k^2 \overline{AB} \omega \sigma = \alpha'$. This holds for every infinitesimal solid angle with vertex at P; *therefore a thin homogeneous spherical shell attracts particles within it equally in opposite directions.*

This holds for any number of thin spherical shells and, therefore, for shells of finite thickness.

67. The Attraction of a Thin Homogeneous Ellipsoidal Shell upon a Particle in its Interior. The theorem of this article is given in the *Principia*, Book I., Prop. XCI., Cor. 3.

Let a *homoeoid* be defined as a thin shell contained between two similar surfaces similarly placed. Thus, an *elliptic homoeoid* is a thin shell contained between two similar ellipsoidal surfaces similarly placed.

Consider the attraction of the elliptic homoeoid whose surfaces are the similar ellipsoids E and E' upon the interior unit particle P. Construct an infinitesimal cone whose solid angle is ω with vertex at P. The masses of the two infinitesimal elements at A and A' are respectively $m = \sigma \overline{AB} \omega \overline{AP}^2$ and $m' = \sigma \overline{A'B'} \omega \overline{A'P}^2$. The attractions are $\alpha = \dfrac{k^2 m}{\overline{AP}^2}$ and $\alpha' = \dfrac{k^2 m'}{\overline{A'P}^2}$. Construct a diameter $\overline{CC'}$ parallel to $\overline{AA'}$ in the elliptical section of a plane through the cone and the center of the ellipsoids, and draw its conjugate $\overline{DD'}$. They are conjugate diameters in both elliptical sections, E and E'; therefore $\overline{DD'}$ bisects every chord parallel to $\overline{CC'}$, and hence $\overline{AB} = \overline{A'B'}$. The attractions of the elements at A and A' upon

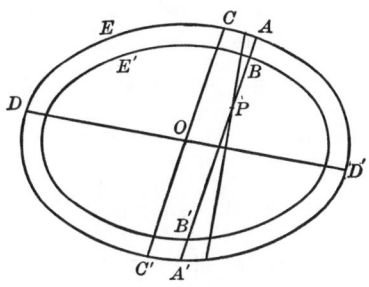

Fig. 12.

P are therefore equal. This holds for every infinitesimal solid angle whose vertex is at P; therefore *the attractions of a thin elliptic homoeoid upon an interior particle are equal in opposite directions.*

This holds for any number of thin shells and, therefore, for shells of finite thickness.

68. The Attraction of a Thin Homogeneous Spherical Shell upon an Exterior Particle. Newton's Method. Let $AHKB$ and $ahkb$ be two equal thin spherical shells with centers at O and o

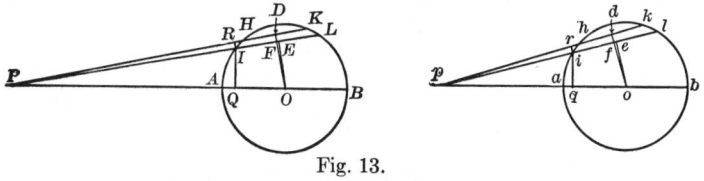

Fig. 13.

respectively. Let two unit particles be placed at P and p, unequal distances from the centers of the shells. Draw any secants from p cutting off the arcs il and hk, and let the angle kpl approach zero as a limit. Draw from P the secants PL and PK, cutting off the arcs IL and HK equal respectively to il and hk. Draw oe perpendicular to pl, od perpendicular to pk, iq perpendicular to pb, and ir perpendicular to pk. Draw the corresponding lines in the other figure.

Rotate the figures around the diameters PB and pb, and call the masses of the circular rings generated by HI and hi, M and m respectively; then

(1) $$HI \times IQ : hi \times iq = M : m.$$

The attractions of unit masses situated at I and i are respectively proportional to the inverse squares of PI and pi. The components of these attractions in the directions PO and po are the respective attractions multiplied by $\dfrac{PQ}{PI}$ and $\dfrac{pq}{pi}$ respectively. If A' and a' represent the components of attraction toward O and o, then

(2) $$A' : a' = \frac{1}{\overline{PI}^2}\frac{PQ}{PI} : \frac{1}{\overline{pi}^2}\frac{pq}{pi}.$$

Now consider the attractions of the rings upon P and p. Their resultants are in the directions of O and o respectively because of the symmetry of the figures with respect to the lines PO and po, and they are respectively M and m times those of the unit particles. Let A and a represent the attractions of M and m; then

(3) $$A : a = \frac{M}{\overline{PI}^2}\frac{PQ}{PI} : \frac{m}{\overline{pi}^2}\frac{pq}{pi} = \frac{HI \times IQ}{\overline{PI}^2}\frac{PF}{PO} : \frac{hi \times iq}{\overline{pi}^2}\frac{pf}{po}.$$

In order to reduce the right member of (3) consider the similar triangles PIR and PFD and the corresponding triangles in the other figure. At the limit as the angles KPL and kpl approach zero, $DF : df = 1$ because the secants IL and HK respectively equal il and hk. Therefore

$$PI : PF = RI : DF,$$
$$pf : pi = DF(= df) : ri,$$

and the product of these proportions is

(4) $$PI \times pf : PF \times pi = RI : ri = HI : hi.$$

From the similar triangles PIQ and POE, it follows that

$$PI : PO = IQ : OE,$$

and similarly

$$po : pi = OE(= oe) : iq.$$

The product of these two proportions is

(5) $PI \times po : PO \times pi = IQ : iq.$

The product of (4) and (5) is

$$\overline{PI}^2 \times pf \times po : \overline{pi}^2 \times PF \times PO = HI \times IQ : hi \times iq.$$

Consequently equation (3) becomes

(6) $A : a = \overline{po}^2 : \overline{PO}^2.$

Therefore, the circular rings attract the exterior particles toward the centers of the shells with forces which are inversely proportional to the squares of the respective distances of the particles from these centers. In a similar manner the same can be proved for the rings KL and kl.

Now let the lines PK and pk vary from coincidence with the diameters PB and pb to tangency with the spherical shells. The results are true at every position separately, and hence for all at once. Therefore, *the resultants of the attractions of thin spherical shells upon exterior particles are directed toward their centers, and the intensities of the forces vary inversely as the squares of the distances of the particles from the centers.*

If the body is a homogeneous sphere, or is made up of homogeneous spherical layers, the theorem holds for each layer separately, and consequently for all of them combined.

69. Comments upon Newton's Method. While the demonstration above is given in the language of Geometry, it really depends upon the principles which are fundamental in the Calculus. Letting the angle kpl approach zero as a limit is equivalent to taking a differential element; the rotation around the diameters is equivalent to an integration with respect to one of the polar angles; the variation of the line pk from coincidence with the diameter to tangency with the shell is equivalent to an integration with respect to the other polar angle; and the summation of the infinitely thin shells to form a solid sphere is equivalent to an integration with respect to the radius.

Since Newton's method gives only the ratios of the attraction of equal spherical shells at different distances, it does not give the manner in which the attraction depends upon the masses of the finite bodies. This is of scarcely less importance than a knowledge of the manner in which it varies with the distance.

In order to find the manner in which the attraction depends upon the mass of the attracting body, take two equally dense spherical shells, S_1 and S_2, internally tangent to the cone C. Let $PO_1 = a_1$, $PO_2 = a_2$, and M_1 and M_2 be the masses of S_1 and S_2 respectively. The two shells attract the particle P equally; for, any solid angle which includes part of one shell also includes a similar part of the other. The masses of these included parts are as the squares of

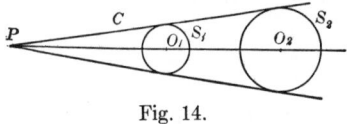

Fig. 14.

their distances, and their attractions are inversely as the squares of their distances, whence the equality of their attractions upon P. Let A represent the common attraction; then remove S_1 so that its center is also at O_2. Let A' represent the intensity of the attraction of S_1 in the new position; then, by the theorem of Art. 68,

$$\frac{A'}{A} = \frac{a_1^2}{a_2^2} = \frac{M_1}{M_2}.$$

Therefore, *the two shells attract a particle at the same distance with forces directly proportional to their masses.* From this and the previous theorem, it follows that *a particle exterior to a sphere which is homogeneous in concentric layers is attracted toward its center with a force which is directly proportional to the mass of the sphere and inversely as the square of the distance from its center; or, as though the mass of the sphere were all at its center.*

Since the heavenly bodies are nearly homogeneous in concentric spherical layers they can be regarded as material points in the discussion of their mutual interactions except when they are relatively near each other as in the case of the planets and their respective satellites.

70. The Attraction of a Thin Homogeneous Spherical Shell upon an Exterior Particle. Thomson and Tait's Method. Let O be the center of the spherical shell whose radius is a and whose thickness is Δa, P the position of the attracted particle and PO a line from the attracted particle to the center cutting the spherical surface in C. Take the point A so that $PO : OC = OC : OA$, and construct the infinitesimal cone whose solid angle is ω with its vertex at A. Let σ be the density of the shell. Then the elements of mass at B and B' are respectively

$$m = \sigma\omega\overline{AB}^2\,\frac{\Delta a}{\cos(OBA)}, \qquad m' = \sigma\omega\overline{AB'}^2\,\frac{\Delta a}{\cos(OB'A)}.$$

The attractions of the two masses upon P are respectively

(7)
$$\begin{cases} \alpha = k^2\sigma\omega\,\dfrac{\overline{AB}^2}{\overline{BP}^2}\cdot\dfrac{\Delta a}{\cos(OBA)}, \\[3mm] \alpha' = k^2\sigma\omega\,\dfrac{\overline{AB'}^2}{\overline{B'P}^2}\cdot\dfrac{\Delta a}{\cos(OB'A)} \end{cases}$$

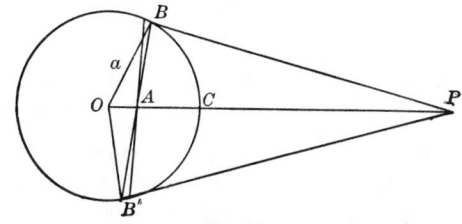

Fig. 15.

From the construction of A it follows that

$$PO : OB = OB : OA.$$

Hence the triangles POB and BOA, having a common angle included between proportional sides, are similar.   Therefore

$$\frac{AB}{BP} = \frac{OB}{OP} = \frac{a}{OP}.$$

Similarly

$$\frac{AB'}{B'P} = \frac{a}{OP}.$$

The angle OBA equals the angle $OB'A$. Then equations (7) become

(8)
$$\begin{cases} \alpha = k^2\sigma\omega\,\dfrac{a^2}{\overline{OP}^2}\cdot\dfrac{\Delta a}{\cos(OBA)}, \\[3mm] \alpha' = k^2\sigma\omega\,\dfrac{a^2}{\overline{OP}^2}\cdot\dfrac{\Delta a}{\cos(OBA)} = \alpha. \end{cases}$$

The angles BPO and $B'PO$ are respectively equal to OBA and $OB'A$; therefore they are equal to each other. The resultant of the two equal attractions α and α' is in the line bisecting the angle between them, or in the direction of O, and is given in magnitude by the equation

$$\Delta R = \alpha \cos (BPO) + \alpha' \cos (B'PO) = 2\alpha \cos (OBA).$$

This becomes, as a consequence of (8),

$$\Delta R = 2k^2\sigma\omega \frac{a^2\Delta a}{\overline{OP}^2}.$$

This equation is true for every solid angle with vertex at A, and consequently for their sum. Therefore the attraction of the whole spherical shell upon the exterior particle is, on summing with respect to ω,

$$R = 4\pi k^2\sigma \frac{a^2\Delta a}{\overline{OP}^2} = \frac{k^2 M}{\overline{OP}^2};$$

or, the attraction varies directly as the mass of the shell and inversely as the square of the distance of the particle from its center.

71. The Attraction upon a Particle in a Homogeneous Spherical Shell. In Arts. 66–69 the attractions of a thin homogeneous spherical shell upon an interior and an exterior particle, respectively, have been discussed; the problem is now completed by treating the case where the attracted particle is a part of the shell itself.

Let O be the center of the spherical shell of thickness Δa, and P the position of the attracted particle. Construct a cone whose solid angle is ω with its vertex at P. Let σ be the den-

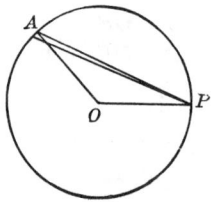

Fig. 16.

sity of the shell; then the mass of the section cut out at A by the cone is $\sigma\omega\overline{AP}^2 \dfrac{\Delta a}{\cos (OAP)}$. The attraction of the element along AP is $\alpha = k^2\sigma\omega \dfrac{\overline{AP}^2}{\overline{AP}^2} \dfrac{\Delta a}{\cos (OAP)}$. The resultant attraction of the shell is in the direction PO since the mass is symmetrically situated with respect to this line. The component in the direction PO is

$$\Delta R = \alpha \cos (APO) = \alpha \cos (OAP) = k^2\sigma\omega\Delta a.$$

The attraction of the whole shell is

$$R = 2k^2\sigma\pi\Delta a = \frac{k^2M}{2a^2}.$$

It follows from this equation and the results obtained in Arts. 66 and 69 that the attraction on an interior particle infinitely near the shell is zero, on a particle in the shell, $\frac{k^2M}{2a^2}$, and on an exterior particle infinitely near the shell, $\frac{k^2M}{a^2}$.* The discontinuity in the attraction is due to the fact that the mass of any finite area of the shell is assumed to be finite although it is supposed to be infinitely thin. There is no such discontinuity at the surface of a solid sphere because an infinitely thin shell taken from it has only an infinitesimal mass.

X. PROBLEMS.

1. Suppose any two similar bodies are similarly placed in perspective. Show that a particle at their center of perspectivity is attracted inversely as their linear dimensions if they are thin rods of equal density; equally, if they are thin shells of equal density; and directly as their linear dimensions if they are solids of equal density. Consider a nebula which is apparently as large as the sun. Suppose its distance is one million times that of the sun and that its density is one millionth that of the sun. Compare its attraction for the earth with that of the sun.

2. Prove that the attractions of two homogeneous spheres of equal density for particles upon their surfaces are to each other as their radii.

3. Prove that the attraction of a homogeneous sphere upon a particle in its interior varies directly as the distance of the particle from the center.

4. Prove that all the frustums of equal height of any homogeneous cone attract a particle at its vertex equally.

5. Find the law of density such that the attraction of a sphere for a particle upon its surface shall be independent of the size of the sphere.

6. Prove that the attraction of a uniform thin rod, bent in the form of an arc of a circle, upon a particle at the center of the circle is the same as that which the mass of a similar rod equal to the chord joining the extremities would exert if it were concentrated at the middle point of the arc.

7. Prove that the attraction of a thin uniform straight rod on an exterior particle is the same in magnitude and direction as that of a circular arc of the same density, with its center at the particle and subtending the same angle as the rod, and which is tangent to the rod.

* See note on the attraction of spherical shells, Lagrange, *Collected Works*, vol. VII., p. 591.

8. Prove that if straight uniform rods form a polygon all of whose sides are tangent to a circle, a particle at the center of the circle is attracted equally in opposite directions by the rods.

9. Prove that two spheres, homogeneous in concentric spherical layers, attract each other as though their masses were all at their respective centers.

72. The General Equations for the Components of Attraction and for the Potential when the Attracted Particle is not a Part of the Attracting Mass. The geometrical methods of the preceding articles are special, being efficient only in the particular cases to which they are applied; the analytical methods which follow are characterized by their uniformity and generality, and illustrate again the advantages of processes of this nature.

Consider the attraction of the finite mass M whose density is σ upon the unit particle P, which is not a part of it. That is, P is exterior to M or within some cavity in it. Let the coördinates of P be x, y, z. Let the coördinates of any element of mass dm

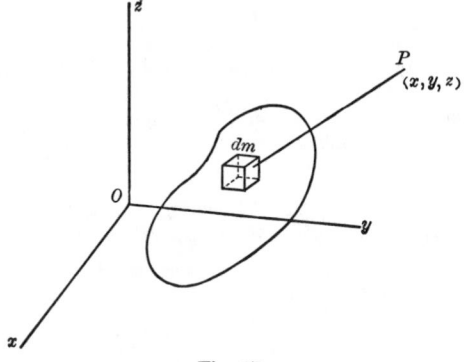

Fig. 17.

be ξ, η, ζ, and the distance from dm to P be ρ. Then the components of attraction parallel to the coördinate axes are respectively

$$(9) \begin{cases} X = -k^2 \int_{(M)} \dfrac{dm}{\rho^2} \cdot \dfrac{(\xi - x)}{\rho} = -k^2 \int_{(M)} \dfrac{(\xi - x)}{\rho^3} \, dm, \\[2ex] Y = -k^2 \int_{(M)} \dfrac{(\eta - y)}{\rho^3} \, dm, \\[2ex] Z = -k^2 \int_{(M)} \dfrac{(\zeta - z)}{\rho^3} \, dm, \end{cases}$$

where

$$dm = \sigma \, d\xi \, d\eta \, d\zeta,$$
$$\rho^2 = (x - \xi)^2 + (y - \eta)^2 + (z - \zeta)^2,$$
$$\sigma = f(\xi, \eta, \zeta).$$

The integral sign $\int_{(M)}$ signifies that the integral must be extended over the whole mass M. Then, if σ is a finite continuous function of the coördinates, as will always be the case in what follows, X, Y, and Z are finite definite quantities. In practice dm is expressed in terms of σ and the ordinary rectangular or polar coördinates, and X, Y, and Z are found by triple integrations.

The three integrals (9) can be made to depend upon a single integral in a very simple manner. Let

$$(10) \qquad V = \int_{(M)} \frac{dm}{\rho}.$$

V is called the *potential function*, the term having been introduced by Green in 1828. It is a function of x, y, and z and will be spoken of as the potential of M upon P at the point (x, y, z).

Since P is not a part of the mass M, ρ does not vanish in the region of integration. The limits of the integral are independent of the position of the attracted particle; therefore the function under the integral sign can be differentiated with respect to x, y, z which are treated as constants in computing the definite integrals. The partial derivatives of V with respect to x, y, and z are

$$\begin{cases} \dfrac{\partial V}{\partial x} = - \int_{(M)} \dfrac{(x - \xi)}{\rho^3} \, dm, \\[2mm] \dfrac{\partial V}{\partial y} = - \int_{(M)} \dfrac{(y - \eta)}{\rho^3} \, dm, \\[2mm] \dfrac{\partial V}{\partial z} = - \int_{(M)} \dfrac{(z - \zeta)}{\rho} \, dm. \end{cases}$$

On comparing these equations with (9), it is found that

$$(11) \qquad \begin{cases} X = k^2 \dfrac{\partial V}{\partial x}, \\[2mm] Y = k^2 \dfrac{\partial V}{\partial y}, \\[2mm] Z = k^2 \dfrac{\partial V}{\partial z}. \end{cases}$$

Therefore, in the case in which P is *not* a part of M, the solution of the problem of finding the components of attraction depends upon the computation of the single function V.

73. Case where the Attracted Particle is a Part of the Attracting Mass. It will now be proved that the components of attraction and the potential have finite, definite values when the particle *is* a part of the attracting mass, and that equations (11) also hold in this case.

In order to show first that X, Y, Z, and V have finite, determinate values in this case, let dm and its position be expressed in polar coördinates with the origin at the attracted particle P. The equations expressing the rectangular coördinates in terms of the polar with the origin at P are

$$\begin{cases} \xi - x = \rho \cos \varphi \cos \theta, \\ \eta - y = \rho \cos \varphi \sin \theta, \\ \zeta - z = \rho \sin \varphi, \\ dm = \sigma \rho^2 \cos \varphi \, d\varphi \, d\theta \, d\rho. \end{cases}$$

Then the expressions for the components of attraction and the potential become

$$\begin{cases} X = - k^2 \iiint \sigma \cos^2 \varphi \cos \theta \, d\varphi \, d\theta \, d\rho, \\ Y = - k^2 \iiint \sigma \cos^2 \varphi \sin \theta \, d\varphi \, d\theta \, d\rho, \\ Z = - k^2 \iiint \sigma \sin \varphi \cos \varphi \, d\varphi \, d\theta \, d\rho, \\ V = + \iiint \sigma \rho \cos \varphi \, d\varphi \, d\theta \, d\rho, \end{cases}$$

where the limits are to be so determined that the integration shall be extended throughout the whole body M. The integrands are all finite for all points in M, and therefore the integrals have finite, determinate values.

The simplest method of proving that equations (11) hold when P is in the attracting mass M is to start from the definition of the derivative of V with respect to x. By definition

$$\frac{\partial V}{\partial x} = \lim_{\Delta x = 0} \frac{V' - V}{\Delta x},$$

where V' is the potential at the point P' whose coördinates are $(x + \Delta x, y, z)$. Construct a small sphere of radius ϵ enclosing both P and P'. Let the mass contained within the sphere ϵ be represented by M_1 and that outside of it by M_2. Let the corresponding parts of the components of attraction and the potential be distinguished by the subscripts 1 and 2.   Then

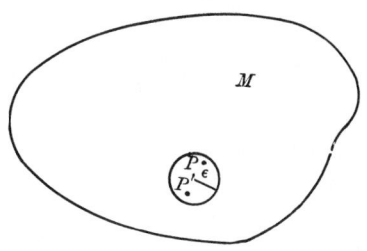

Fig. 18.

$$(12) \qquad X = X_1 + X_2, \quad \cdots, \quad V = V_1 + V_2,$$

because all of these quantities are uniquely defined.   Moreover, it follows from Art. 72 that

$$X_2 = k^2 \frac{\partial V_2}{\partial x}, \qquad Y_2 = k^2 \frac{\partial V_2}{\partial y}, \qquad Z_2 = k^2 \frac{\partial V_2}{\partial z}.$$

Now consider the derivative of V with respect to x.   It becomes

$$(13) \qquad \begin{aligned} \frac{\partial V}{\partial x} &= \lim_{\Delta x = 0} \frac{V_1' - V_1}{\Delta x} + \lim_{\Delta x = 0} \frac{V_2' - V_2}{\Delta x} \\ &= \lim_{\Delta x = 0} \frac{V_1' - V_1}{\Delta x} + \frac{1}{k^2} X_2. \end{aligned}$$

Let the distance from P to dm be ρ, and from P' to dm be ρ'. Then

$$\frac{V_1' - V_1}{\Delta x} = \int_{(M_1)} \left(\frac{1}{\rho'} - \frac{1}{\rho} \right) \frac{dm}{\Delta x}.$$

It follows from the triangle $P\, dm\, P'$ that $|\Delta x| \geqq |\rho' - \rho|$, where the vertical lines on a quantity indicate that its numerical value is taken.   Hence it follows that

$$\left| \left(\frac{1}{\rho'} - \frac{1}{\rho} \right) \frac{1}{\Delta x} \right| \leqq \frac{1}{\rho \rho'} \leqq \frac{1}{2} \left(\frac{1}{\rho^2} + \frac{1}{\rho'^2} \right).$$

Therefore

$$\left| \frac{V_1' - V_1}{\Delta x} \right| \leqq \frac{1}{2} \int_{(M_1)} \frac{dm}{\rho^2} + \frac{1}{2} \int_{(M_1)} \frac{dm}{\rho'^2}.$$

When dm is expressed in polar coördinates this inequality becomes

$$\left| \frac{V_1' - V_1}{\Delta x} \right| \leqq \frac{1}{2} \int_{-\frac{\pi}{2}}^{\frac{\pi}{2}} \int_0^{2\pi} \int_0^{\rho} \sigma \cos \varphi \, d\varphi \, d\theta \, d\rho$$

$$+ \frac{1}{2} \int_{-\frac{\pi}{2}}^{\frac{\pi}{2}} \int_0^{2\pi} \int_0^{\rho'} \sigma \cos \varphi' d\varphi' d\theta' d\rho'.$$

Let σ_0 be the maximum value of σ in ϵ. The result of integrating with respect to ρ and ρ' is

$$\left| \frac{V_1' - V_1}{\Delta x} \right| \leqq \frac{\sigma_0}{2} \int_{-\frac{\pi}{2}}^{\frac{\pi}{2}} \int_0^{2\pi} \rho \cos \varphi \, d\varphi \, d\theta + \frac{\sigma_0}{2} \int_{-\frac{\pi}{2}}^{\frac{\pi}{2}} \int_0^{2\pi} \rho' \cos \varphi' d\varphi' d\theta'.$$

Since P and P' are in the sphere ϵ the distances ρ and ρ' cannot exceed 2ϵ. Then

$$\left| \frac{V_1' - V_1}{\Delta x} \right| < \sigma_0 \epsilon \int_{-\frac{\pi}{2}}^{\frac{\pi}{2}} \int_0^{2\pi} \cos \varphi \, d\varphi \, d\theta$$

$$+ \sigma_0 \epsilon \int_{-\frac{\pi}{2}}^{\frac{\pi}{2}} \int_0^{2\pi} \cos \varphi' d\varphi' d\theta' = 8\pi\sigma_0 \epsilon,$$

and

$$\lim_{\Delta x = 0} \left| \frac{V_1' - V_1}{\Delta x} \right| < 8\pi\sigma_0 \epsilon.$$

It follows from this inequality and (13) that

$$k^2 \frac{\partial V}{\partial x} - 8k^2 \pi \sigma_0 \epsilon < X_2 < k^2 \frac{\partial V}{\partial x} + 8k^2 \pi \sigma_0 \epsilon.$$

Now pass to the limit $\epsilon = 0$. The limit of X_1, for $\epsilon = 0$, is easily proved to be zero by using polar coördinates. Hence it follows from (12) that

$$\lim_{\epsilon = 0} X_2 = X,$$

and consequently, from the last inequalities,

$$X = k^2 \frac{\partial V}{\partial x}.$$

The corresponding relations for derivatives with respect to y and z are proved similarly, and therefore equations (11) hold whether or not P is a part of M.

74. Level Surfaces. The equation $V = c$, where c takes constant values, defines what are called *level surfaces* or *equipotential surfaces.*

Any displacement δx, δy, δz, of the particle from the point (x_0, y_0, z_0) in a level surface must fulfill the equation

$$\frac{\partial V}{\partial x} \delta x + \frac{\partial V}{\partial y} \delta y + \frac{\partial V}{\partial z} \delta z = 0,$$

which is the condition that the points (x_0, y_0, z_0) and $(x_0 + \delta x, y_0 + \delta y, z_0 + \delta z)$ shall both be in the same level surface. This equation becomes as a consequence of (11)

(14) $$X\delta x + Y\delta y + Z\delta z = 0.$$

The direction cosines of the resultant attraction to which the particle is subject are proportional to X, Y, Z, and the direction cosines of the line of the displacement are proportional to δx, δy, δz. Since the sum of the products of these direction cosines in corresponding pairs is zero, it follows that the *resultant attraction is perpendicular to the level surfaces.* Consequently, if the particle starts from rest it will begin to move perpendicularly to the level surface through its initial position; but after it has acquired an appreciable velocity it will not in general move perpendicularly to the level surfaces because the motion depends not only upon the forces, which have been shown to be orthogonal to the level surfaces, but also upon the velocity.

75. The Potential and Attraction of a Thin Homogeneous Circular Disc upon a Particle in its Axis. Take the origin at the center of the disc whose radius is R. Let the coördinates of P be $x, 0, 0$. Then

$$V = \int \frac{dm}{\rho} = \sigma \int_0^R \int_0^{2\pi} \frac{r\, dr\, d\theta}{\sqrt{x^2 + r^2}}.$$

Upon integrating, it is found that

(15) $$\begin{cases} V = 2\pi\sigma[\sqrt{x^2 + R^2} - \sqrt{x^2}], \\ X = k^2 \dfrac{\partial V}{\partial x} = 2\pi k^2 \sigma \left[\dfrac{x}{\sqrt{x^2 + R^2}} - \dfrac{x}{\sqrt{x^2}} \right]. \end{cases}$$

If x is kept constant and R is made to approach infinity as a limit, the attraction becomes

(16) $$X = \mp 2\pi k^2 \sigma,$$

according as the particle is on the positive or negative side of the yz-plane. The right member of this equation does not depend upon x; therefore a thin circular disc of infinite extent attracts a particle above it with a force which is independent of its altitude. Any

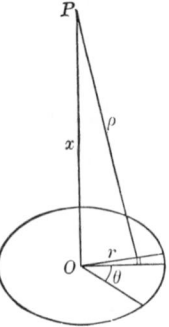

Fig. 19.

number of superposed discs would act jointly in the same manner. Hence, if the earth were a plane of infinite extent, as the ancients commonly supposed, bodies would gravitate toward it with constant forces at all altitudes, and the laws of falling bodies derived under the hypothesis of constant acceleration would be rigorously true.

76. The Potential and Attraction of a Thin Homogeneous Spherical Shell upon an Interior or an Exterior Particle. Let

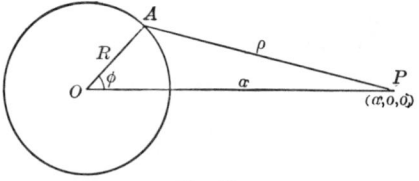

Fig. 20.

ϕ represent the angle between OP and the radius, and θ* the angle between the fundamental plane and the plane OAP. Then

$$(17) \qquad V = \int \frac{dm}{\rho} = \sigma \int_0^\pi \int_0^{2\pi} \frac{R^2 \sin \phi \, d\phi \, d\theta}{\rho}.$$

* It must be noticed that the ϕ and θ here are not the ordinary polar angles used elsewhere.

One of the three variables ϕ, θ, ρ must be expressed in terms of the remaining two. From the figure it is seen that

$$\rho^2 = x^2 + R^2 - 2xR \cos \phi;$$

whence

(18) $$\rho \, d\rho = xR \sin \phi \, d\phi.$$

Then (17) becomes, if P is exterior,

(19) $$V_E = \frac{R\sigma}{x} \int_{x-R}^{x+R} \int_0^{2\pi} d\rho \, d\theta;$$

and if P is interior,

(20) $$V_I = \frac{R\sigma}{x} \int_{R-x}^{R+x} \int_0^{2\pi} d\rho \, d\theta.$$

The integrals of these equations are respectively

(21)
$$\begin{cases} V_E = \dfrac{4\pi\sigma R^2}{x} = \dfrac{M}{x}, \\[2mm] V_I = 4\pi\sigma R = \dfrac{M}{R}. \end{cases}$$

The x-components of attraction are respectively

(22)
$$\begin{cases} X_E = k^2 \dfrac{\partial V_E}{\partial x} = -\dfrac{k^2 M}{x^2}, \\[2mm] X_I = k^2 \dfrac{\partial V_I}{\partial x} = 0, \end{cases}$$

which agree with the results obtained in Arts. 66 and 70.

The attraction of a solid homogeneous sphere also can be found at once. Considering the shell as an element of the sphere, the M of (22) is given by the equation

$$M = 4\pi\sigma r^2 dr.$$

Let $\overline{X}$ represent the attraction of the whole sphere $\overline{M}$; then

$$\overline{X} = -\frac{4k^2\pi\sigma}{x^2} \int_0^a r^2 dr = -\frac{4k^2}{3} \frac{\pi\sigma a^3}{x^2} = -k^2 \frac{\overline{M}}{x^2}.$$

Consider the mutual attraction of two spheres. In accordance with the results which have just been obtained, each one attracts every particle of the other as it would if its mass were all at its center. Hence the two spheres attract each other as they would if their masses were all at their respective centers.

77. Second Method of Computing the Attraction of a Homogeneous Sphere. A very simple method will now be given of finding the attraction of a solid homogeneous sphere upon an exterior particle when it is known for interior particles. It is a trivial matter in this case and is introduced only because the corresponding device in the much more difficult case of the attractions of ellipsoids is of the greatest value, and constitutes Ivory's celebrated method.

Let it be required to find the attraction of the sphere S upon the exterior particle P', supposing it is known how to find the attraction upon interior particles. Construct the concentric

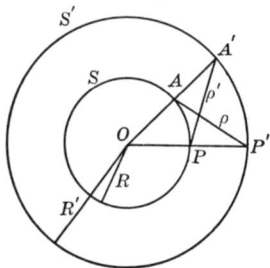

Fig. 21.

sphere S' through P' and suppose it has the same density as S. A one-to-one correspondence between the points on the surfaces of the two spheres is established by the relations

$$(23) \qquad x = \frac{R}{R'}x', \qquad y = \frac{R}{R'}y', \qquad z = \frac{R}{R'}z'.$$

The corresponding points are in lines passing through the common center of the spheres, and P corresponds to P'. Let X and X' represent the attractions of S' and S upon P and P' respectively. They are given by the equations

$$(24) \quad \begin{cases} X = -k^2 \displaystyle\int_{(S')} \frac{R-x'}{\rho'^3} dm' = -k^2\sigma \iiint \frac{R-x'}{\rho'^3} dx'dy'dz', \\[3mm] X' = -k^2 \displaystyle\int_{(S)} \frac{R'-x}{\rho^3} dm = -k^2\sigma \iiint \frac{R'-x}{\rho^3} dx\,dy\,dz. \end{cases}$$

But it follows from the definition of ρ and ρ' that

$$(25) \quad \begin{cases} k^2\sigma \displaystyle\iiint \frac{R-x'}{\rho'^3} dx'dy'dz' = +k^2\sigma \iiint \frac{\partial\left(\dfrac{1}{\rho'}\right)}{\partial x'} dx'dy'dz' \\[4mm] \qquad\qquad = k^2\sigma \displaystyle\iint \left(\frac{1}{\rho_2'} - \frac{1}{\rho_1'}\right) dy'dz', \\[4mm] k^2\sigma \displaystyle\iiint \frac{R'-x}{\rho^3} dx\,dy\,dz = k^2\sigma \iint \left(\frac{1}{\rho_2} - \frac{1}{\rho_1}\right) dy\,dz, \end{cases}$$

where ρ_2 and ρ_1 are the extreme values of ρ obtained by integrating with respect to x. That is, the first integration gives the attraction of an elementary column extending through the sphere parallel to the x-axis, and ρ_1 and ρ_2 are the distances from the attracted particle P' to the ends of this column. In completing the integration the sum of all of these elementary columns is taken. The corresponding statements with respect to the first equation of (25) are true.

Suppose the integrals (25) are computed in such a manner that corresponding columns of the two spheres are always taken at the same time. Consider any two pairs of corresponding elements, as those at A and A'. For these $\rho = \rho'$, and this relation holds throughout the integration as arranged above. Hence it follows from equations (24) and (25) that

$$X' = - k^2\sigma \iint \left(\frac{1}{\rho_2} - \frac{1}{\rho_1} \right) dy\, dz = - k^2\sigma \iint \left(\frac{1}{\rho_2{}'} - \frac{1}{\rho_1{}'} \right) dy\, dz.$$

But, from (23),

$$dy = \frac{R}{R'} dy', \qquad dz = \frac{R}{R'} dz';$$

therefore

$$X' = - k^2\sigma \frac{R^2}{R'^2} \iint \left(\frac{1}{\rho_2{}'} - \frac{1}{\rho_1{}'} \right) dy'dz' = \frac{R^2}{R'^2} X.$$

Let M represent the mass of the sphere S, and M' that of S'. The attraction of S' upon the interior particle P is given by

$$X = - \frac{k^2 M}{R^2};$$

therefore it follows from the relation $R'^2 X' = R^2 X$ that the attraction of S upon the exterior particle P' is

$$X' = - \frac{k^2 M}{R'^2},$$

agreeing with results previously obtained (Arts. 69, 70).

XI. PROBLEMS.

1. Prove by the limiting process that the potential and components of attraction have finite, determinate values, and that equations (11) hold when the particle is on the surface of the attracting mass.

2. Find the expression for the potential function for a particle exterior to the attracting body when the force varies inversely as the nth power of the distance.

$$Ans. \quad V = \frac{1}{n-1} \int_{(M)} \frac{dm}{\rho^{n-1}}.$$

3. Find by the limiting process for what values of n the potential in the last problem is finite and determinate when the particle is a part of the attracting mass.

4. Show that the level surfaces for a straight homogeneous rod are prolate spheroids whose foci are the extremities of the rod.

5. Find the components of attraction of a uniform hemisphere, whose radius is R, upon a particle on its edge: (a) in the direction of the center of its base; (b) perpendicular to this direction in the plane of the base; (c) perpendicular to these two directions.

$$Ans. \quad (a) \ X = \tfrac{2}{3}\pi\sigma k^2 R; \quad (b) \ Y = 0; \quad (c) \ Z = \tfrac{1}{4}\sigma k^2 R.$$

6. Find the deviation of the plumb-line due to a hemispherical hill of radius r and density σ_1. Let R represent the radius of the earth, assumed to be spherical, and σ_2 its mean density.

$Ans.$ If λ is the angle of deviation,

$$\tan \lambda = \frac{\tfrac{2}{3}\pi\sigma_1 r}{\tfrac{4}{3}\pi\sigma_2 R - \tfrac{4}{3}\sigma_1 r} = \frac{\tfrac{1}{2}\pi\sigma_1 r}{\pi\sigma_2 R - \sigma_1 r},$$

or

$$\tan \lambda = \frac{1}{2}\frac{\sigma_1}{\sigma_2}\frac{r}{R} \text{ approximately.}$$

7. Prove that if the attraction varies directly as the distance, a body of any shape attracts a particle as though its whole mass were concentrated at its center of mass.

78. The Potential and Attraction of a Solid Homogeneous Oblate Spheroid upon a Distant Unit Particle. The planets are very nearly oblate spheroids, and they are so nearly homogeneous that the results obtained in this article will represent the actual facts with sufficient approximation for most astronomical applications.

Suppose the attracted particle is remote compared to the dimensions of the attracting spheroid. Take the origin of co-

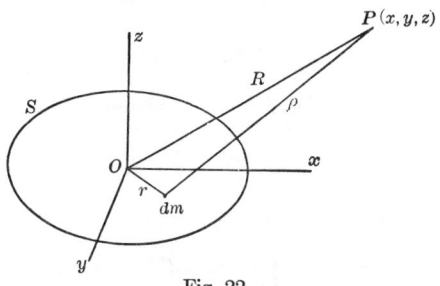

Fig. 22.

ordinates at the center of the spheroid with the z-axis coinciding with the axis of revolution. Let R represent the distance from O to P, and r the distance from O to the element of mass. Then

(26)
$$\begin{cases} V = \int_{(S)} \dfrac{dm}{\rho}, \\[2mm] \rho = \sqrt{(x - \xi)^2 + (y - \eta)^2 + (z - \zeta)^2}, \\[2mm] R = \sqrt{x^2 + y^2 + z^2}, \\[2mm] r = \sqrt{\xi^2 + \eta^2 + \zeta^2}. \end{cases}$$

It follows from these equations that

$$\frac{1}{\rho} = \frac{1}{\sqrt{R^2 + r^2 - 2(x\xi + y\eta + z\zeta)}} = \frac{1}{R\sqrt{1 + \dfrac{r^2 - 2(x\xi + y\eta + z\zeta)}{R^2}}}.$$

Let $\dfrac{\xi}{R}$, $\dfrac{\eta}{R}$, and $\dfrac{\zeta}{R}$ be taken as small quantities of the first order; then, on expanding the expression for ρ^{-1} by the binomial theorem, it is found that, up to small quantities of the third order,

$$\frac{1}{\rho} = \frac{1}{R}\left\{ 1 + \frac{x\xi + y\eta + z\zeta}{R^2} - \frac{r^2}{2R^2} \right.$$
$$\left. + \frac{3}{2} \frac{(x^2\xi^2 + y^2\eta^2 + z^2\zeta^2 + 2xy\xi\eta + 2yz\eta\zeta + 2zx\zeta\xi)}{R^4} + \cdots \right\}.$$

Therefore

$$(27)\begin{cases} V = \dfrac{1}{R}\int dm + \dfrac{x}{R^3}\int \xi\, dm + \dfrac{y}{R^3}\int \eta\, dm + \dfrac{z}{R^3}\int \zeta\, dm \\[2mm] \qquad - \dfrac{1}{2R^3}\int r^2 dm + \dfrac{3}{2}\dfrac{x^2}{R^5}\int \xi^2 dm + \dfrac{3}{2}\dfrac{y^2}{R^5}\int \eta^2 dm \\[2mm] \qquad + \dfrac{3}{2}\dfrac{z^2}{R^5}\int \zeta^2 dm + \dfrac{3xy}{R^5}\int \xi\eta\, dm + \dfrac{3yz}{R^5}\int \eta\zeta\, dm \\[2mm] \qquad + \dfrac{3zx}{R^5}\int \zeta\xi\, dm + \cdots. \end{cases}$$

Let M represent the mass of the spheroid; then

$$\int dm = M,$$

and, since the origin is at the center of gravity,

$$\int \xi\, dm = 0, \qquad \int \eta\, dm = 0, \qquad \int \zeta\, dm = 0.$$

Let σ represent the density; then

$$\begin{cases} dm = \sigma r^2 \cos\phi\, d\phi\, d\theta\, dr, \\ \xi = r\cos\phi\cos\theta, \\ \eta = r\cos\phi\sin\theta, \\ \zeta = r\sin\phi, \end{cases}$$

and (27) becomes

$$V = \dfrac{M}{R} - \dfrac{\sigma}{2R^3}\iiint r^4\cos\phi\, d\phi\, d\theta\, dr + \dfrac{3x^2\sigma}{2R^5}\iiint r^4\cos^3\phi\cos^2\theta\, d\phi\, d\theta\, dr$$

$$+ \dfrac{3y^2\sigma}{2R^5}\iiint r^4\cos^3\phi\sin^2\theta\, d\phi\, d\theta\, dr + \dfrac{3z^2\sigma}{2R^5}\iiint r^4\sin^2\phi\cos\phi\, d\phi\, d\theta\, dr$$

$$+ \dfrac{3xy\sigma}{R^5}\iiint r^4\cos^3\phi\sin\theta\cos\theta\, d\phi\, d\theta\, dr$$

$$+ \dfrac{3yz\sigma}{R^5}\iiint r^4\sin\phi\cos^2\phi\sin\theta\, d\phi\, d\theta\, dr$$

$$+ \dfrac{3zx\sigma}{R^5}\iiint r^4\sin\phi\cos^2\phi\cos\theta\, d\phi\, d\theta\, dr + \cdots,$$

where the limits of integration are: for r, 0 and r; for ϕ, $-\dfrac{\pi}{2}$ and $\dfrac{\pi}{2}$; and for θ, 0 and 2π. Since r and ϕ are independent of θ, the integration can be performed with respect to θ first, giving

$$(28)\begin{cases} V = \dfrac{M}{R} - \dfrac{\pi\sigma}{R^3}\int_{-\frac{\pi}{2}}^{\frac{\pi}{2}}\int_0^r r^4\cos\phi\,d\phi\,dr \\[2ex] \qquad + \dfrac{3}{2}\dfrac{\pi x^2\sigma}{R^5}\int_{-\frac{\pi}{2}}^{\frac{\pi}{2}}\int_0^r r^4\cos^3\phi\,d\phi\,dr \\[2ex] \qquad + \dfrac{3}{2}\dfrac{\pi y^2\sigma}{R^5}\int_{-\frac{\pi}{2}}^{\frac{\pi}{2}}\int_0^r r^4\cos^3\phi\,d\phi\,dr \\[2ex] \qquad + \dfrac{3\pi z^2\sigma}{R^5}\int_{-\frac{\pi}{2}}^{\frac{\pi}{2}}\int_0^r r^4\sin^2\phi\cos\phi\,d\phi\,dr + \cdots, \end{cases}$$

the last three integrals being zero.

The next integration must be made with respect to r, as this variable depends upon ϕ. Let the major and minor semi-axes of a meridian section of the spheroid be a and b respectively, and let e be the eccentricity. Then

$$r^2 = \frac{b^2}{1 - e^2\cos^2\phi}.$$

Upon integrating (28) with respect to r and expanding in powers of e, it is found that, up to terms of the second order inclusive,

$$V = \frac{M}{R} - \frac{\pi\sigma b^5}{5R^3}\int_{-\frac{\pi}{2}}^{\frac{\pi}{2}}(1 + \tfrac{5}{2}e^2\cos^2\phi + \cdots)\cos\phi\,d\phi$$

$$+ \frac{3}{10}\frac{\pi\sigma b^5}{R^5}(x^2 + y^2)\int_{-\frac{\pi}{2}}^{\frac{\pi}{2}}(1 + \tfrac{5}{2}e^2\cos^2\phi + \cdots)\cos^3\phi\,d\phi$$

$$+ \frac{3}{5}\frac{\pi\sigma b^5 z^2}{R^5}\int_{-\frac{\pi}{2}}^{\frac{\pi}{2}}(1 + \tfrac{5}{2}e^2\cos^2\phi + \cdots)\sin^2\phi\cos\phi\,d\phi$$

$$+ \cdot \quad \cdot \quad \cdot \quad \cdot \quad \cdot \quad \cdot \quad \cdot \quad \cdot \quad \cdot \quad \cdot \quad \cdot \quad \cdot \quad \cdot \quad \cdot.$$

On integrating with respect to ϕ and arranging in powers of e, the expression for V becomes

$$V = \frac{M}{R} + \frac{2}{15}\frac{\pi\sigma b^5}{R^5}(x^2 + y^2 - 2z^2)e^2 + \cdots.$$

But

$$M = \tfrac{4}{3}\pi\sigma a^2 b,$$

$$b^2 = a^2(1 - e^2);$$

therefore

(29) $\qquad V = \dfrac{M}{R} \left[1 + \dfrac{b^2}{10} \dfrac{(x^2 + y^2 - 2z^2)}{R^4} e^2 + \cdots \right].$

The components of attraction are found from equations (11) and (29) to be

(30) $\begin{cases} X = -\dfrac{k^2 M x}{R^3} \left[1 + \dfrac{3}{10} b^2 \dfrac{(x^2 + y^2 - 4z^2)}{R^4} e^2 + \cdots \right], \\[2mm] Y = -\dfrac{k^2 M y}{R^3} \left[1 + \dfrac{3}{10} b^2 \dfrac{(x^2 + y^2 - 4z^2)}{R^4} e^2 + \cdots \right], \\[2mm] Z = -\dfrac{k^2 M z}{R^3} \left[1 + \dfrac{3}{10} b^2 \dfrac{3(x^2 + y^2) - 2z^2}{R^4} e^2 + \cdots \right]. \end{cases}$

If the spheroid should become a sphere of the same mass, the expressions for the components of attraction would reduce to the first terms of the right members of equations (30). If the attracted particle is in the plane of the equator of the attracting spheroid, $z = 0$; and if it is in the polar line, $x = y = 0$. Hence it follows from (30) that *the attraction of an oblate spheroid upon a particle at a given distance from the center in the plane of its equator is greater than that of a sphere of equal mass; and in the polar line, less than that of a sphere of equal mass.* As the particle recedes from the attracting body the attraction approaches that of a sphere of equal mass. Therefore, *as the particle recedes in the plane of the equator the attraction decreases more rapidly than the square of the distance increases; and as it approaches, the attraction increases more rapidly than the square of the distance decreases.* The opposite results are true when the particle is in the polar line.

79. The Potential and Attraction of a Solid Homogeneous Ellipsoid upon a Unit Particle in its Interior. Let the equation of the surface of the ellipsoid be

(31) $\qquad\qquad \dfrac{\xi^2}{a^2} + \dfrac{\eta^2}{b^2} + \dfrac{\zeta^2}{c^2} - 1 = 0,$

and let the attracted particle be situated at the interior point (x, y, z). Take this point for the origin of the polar coördinates ρ, θ, and ϕ. On taking the fundamental planes of this system parallel to those of the first system, these variables are related to the rectangular coördinates by the equations

$$(32) \quad \begin{cases} \xi = x + \rho \cos \phi \cos \theta, \\ \eta = y + \rho \cos \phi \sin \theta, \\ \zeta = z + \rho \sin \phi. \end{cases}$$

The potential of the ellipsoid upon the unit particle P is

$$V = \int_{(M)} \frac{dm}{\rho} = \sigma \int_{-\frac{\pi}{2}}^{\frac{\pi}{2}} \int_0^{2\pi} \int_0^{\rho_1} \rho \cos \phi \, d\phi \, d\theta \, d\rho.$$

Since the value of ρ depends upon the polar angles the integration must be made first with respect to this variable. The integration gives

$$(33) \qquad V = \frac{\sigma}{2} \int_{-\frac{\pi}{2}}^{\frac{\pi}{2}} \int_0^{2\pi} \rho_1^2 \cos \phi \, d\phi \, d\theta.$$

To express ρ_1 in terms of the polar angles substitute (32) in (31); whence it is found that

$$(34) \qquad A\rho_1^2 + 2B\rho_1 + C = 0,$$

where

$$(35) \quad \begin{cases} A = \dfrac{\cos^2 \phi \cos^2 \theta}{a^2} + \dfrac{\cos^2 \phi \sin^2 \theta}{b^2} + \dfrac{\sin^2 \phi}{c^2}, \\[2mm] B = \dfrac{x \cos \phi \cos \theta}{a^2} + \dfrac{y \cos \phi \sin \theta}{b^2} + \dfrac{z \sin \phi}{c^2}, \\[2mm] C = \dfrac{x^2}{a^2} + \dfrac{y^2}{b^2} + \dfrac{z^2}{c^2} - 1. \end{cases}$$

From (34) it is found that

$$\rho_1 = \frac{-B \pm \sqrt{B^2 - AC}}{A}.$$

The only ρ_1 having a meaning in this problem is positive; A is essentially positive, and C is negative because (x, y, z) is within the ellipsoidal surface. Therefore the positive sign must be taken before the radical. On substituting this value of ρ_1 in (33), it is found that

$$(36) \quad V = \frac{\sigma}{2} \int_{-\frac{\pi}{2}}^{\frac{\pi}{2}} \int_0^{2\pi} \frac{(2B^2 - 2B\sqrt{B^2 - AC} - AC)}{A^2} \cos \phi \, d\phi \, d\theta.$$

Consider the integral

$$V_1 = \int_{-\frac{\pi}{2}}^{\frac{\pi}{2}} \int_0^{2\pi} \frac{B\sqrt{B^2 - AC}}{A^2} \cos \phi \, d\phi \, d\theta.$$

It follows from the expression for B that the differential elements corresponding to $\theta = \theta_0$, $\phi = \phi_0$ and to $\theta = \pi + \theta_0$, $\phi = -\phi_0$ are equal in numerical value but opposite in sign. Since all the elements entering in the integral can be paired in this way, it follows that $V_1 = 0$, after which (36) becomes

(37)
$$
\begin{cases}
V = \dfrac{\sigma}{2} \displaystyle\int_{-\frac{\pi}{2}}^{\frac{\pi}{2}} \int_{0}^{2\pi} \left\{ \dfrac{\cos^2 \phi \cos^2 \theta}{a^2} \left(\dfrac{2x^2}{a^2} - C \right) + \dfrac{\cos^2 \phi \sin^2 \theta}{b^2} \right. \\[2ex]
\qquad\qquad \times \left. \left(\dfrac{2y^2}{b^2} - C \right) + \dfrac{\sin^2 \phi}{c^2} \left(\dfrac{2z^2}{c^2} - C \right) \right\} \dfrac{\cos \phi \, d\phi \, d\theta}{A^2} \\[2ex]
+ 2\sigma \displaystyle\int_{-\frac{\pi}{2}}^{\frac{\pi}{2}} \int_{0}^{2\pi} \left\{ \dfrac{xy \cos^2 \phi \sin \theta \cos \theta}{a^2 b^2} + \dfrac{yz \sin \phi \cos \phi \sin \theta}{b^2 c^2} \right. \\[2ex]
\qquad\qquad + \left. \dfrac{zx \sin \phi \cos \phi \cos \theta}{c^2 a^2} \right\} \dfrac{\cos \phi \, d\phi \, d\theta}{A^2}.
\end{cases}
$$

By comparing the elements properly paired, it is seen that the second integral is zero.

Let

(38) $W = \dfrac{\sigma}{2} \displaystyle\int_{-\frac{\pi}{2}}^{\frac{\pi}{2}} \int_{0}^{2\pi} \dfrac{\cos \phi \, d\phi \, d\theta}{\dfrac{\cos^2 \phi \cos^2 \theta}{a^2} + \dfrac{\cos^2 \phi \sin^2 \theta}{b^2} + \dfrac{\sin^2 \phi}{c^2}}$;

then (37) can be written in the form

(39) $V = -CW + \dfrac{x^2}{a} \dfrac{\partial W}{\partial a} + \dfrac{y^2}{b} \dfrac{\partial W}{\partial b} + \dfrac{z^2}{c} \dfrac{\partial W}{\partial c}.$

For a given ellipsoid W is a constant, and the equation of the level surfaces has the form

$$C_1 x^2 + C_2 y^2 + C_3 z^2 = \text{constant},$$

which is the equation of concentric similar ellipsoids, whose axes are proportional to $C_1^{-\frac{1}{2}}$, $C_2^{-\frac{1}{2}}$, and $C_3^{-\frac{1}{2}}$.

In order to reduce W to an integrable form, let

(40)
$$
\begin{cases}
M = \dfrac{\cos^2 \phi}{a^2} + \dfrac{\sin^2 \phi}{c^2}, \\[2ex]
N = \dfrac{\cos^2 \phi}{b^2} + \dfrac{\sin^2 \phi}{c^2} ;
\end{cases}
$$

then (38) becomes

$$W = \frac{\sigma}{2} \int_{-\frac{\pi}{2}}^{\frac{\pi}{2}} \int_{0}^{2\pi} \frac{\cos \phi \, d\phi \, d\theta}{M \cos^2 \theta + N \sin^2 \theta}$$

$$= 4\sigma \int_{0}^{\frac{\pi}{2}} \int_{0}^{\frac{\pi}{2}} \frac{\cos \phi \, d\phi \, d\theta}{M \cos^2 \theta + N \sin^2 \theta} \cdot$$

M and N are independent of θ; hence, on integrating with respect to this variable, it is found that*

(41)
$$W = 2\pi\sigma \int_{0}^{\frac{\pi}{2}} \frac{\cos \phi \, d\phi}{\sqrt{MN}}$$

$$= 2\pi\sigma abc^2 \int_{0}^{\frac{\pi}{2}} \frac{\cos \phi \, d\phi}{\sqrt{(a^2 \sin^2 \phi + c^2 \cos^2 \phi)(b^2 \sin^2 \phi + c^2 \cos^2 \phi)}} \cdot$$

To return to the symmetry in a, b, and c which existed in (38), Jacobi introduced the transformation

$$\sin \phi = \frac{c}{\sqrt{c^2 + s}} ;$$

whence

$$W = \pi\sigma abc \int_{0}^{\infty} \frac{ds}{\sqrt{(a^2 + s)(b^2 + s)(c^2 + s)}} \cdot$$

On forming the derivatives with respect to a, b, and c, and substituting in (39), it follows that

(42)
$$V = \pi\sigma abc \int_{0}^{\infty} \left(1 - \frac{x^2}{a^2 + s} - \frac{y^2}{b^2 + s} - \frac{z^2}{c^2 + s} \right)$$
$$\times \frac{ds}{\sqrt{(a^2 + s)(b^2 + s)(c^2 + s)}} \cdot$$

The components of attraction are

(43)
$$\begin{cases} X = k^2 \dfrac{\partial V}{\partial x} = - \displaystyle\int_{0}^{\infty} \frac{2\pi\sigma abcx k^2 \, ds}{(a^2 + s) \sqrt{(a^2 + s)(b^2 + s)(c^2 + s)}} , \\[2mm] Y = k^2 \dfrac{\partial V}{\partial y} = - \displaystyle\int_{0}^{\infty} \frac{2\pi\sigma abcy k^2 \, ds}{(b^2 + s) \sqrt{(a^2 + s)(b^2 + s)(c^2 + s)}} , \\[2mm] Z = k^2 \dfrac{\partial V}{\partial z} = - \displaystyle\int_{0}^{\infty} \frac{2\pi\sigma abcz k^2 \, ds}{(c^2 + s) \sqrt{(a^2 + s)(b^2 + s)(c^2 + s)}} . \end{cases}$$

Equation (41) is homogeneous of the second degree in a, b,

* Letting $\tan \theta = x$, the integral reduces to one of the standard forms.

and c; and therefore $\dfrac{\partial V}{\partial x}$, $\dfrac{\partial V}{\partial y}$, $\dfrac{\partial V}{\partial z}$, computed from (39), are homogeneous of degree zero in the same quantities. It follows, therefore, that if a, b, and c are increased by any factor ν the components of attraction X, Y, and Z, will not be changed; or, *the elliptic homoeoid contained between the ellipsoidal surfaces whose axes are a, b, c and νa, νb, and νc attracts the interior particle P equally in opposite directions.* (Compare Art. 67.)

The component of attraction, X, is independent of y and z and involves x to the first degree; therefore *the x-component of attraction is proportional to the x-coördinate of the particle and is constant everywhere within the ellipsoid in the plane $\xi = x$.* Similar results are true for the two other coördinates.

Suppose the notation has been chosen so that $a > b > c$. Then (41) can be put in the normal form for an elliptic integral of the first kind by the substitution

$$\sin \varphi = \frac{c}{\sqrt{a^2 - c^2}} \, \frac{u}{\sqrt{1 - u^2}} \, ,$$

$$\kappa^2 = \frac{a^2 - b^2}{a^2 - c^2} < 1,$$

which gives

(44) $$W = \frac{2\pi\sigma abc}{\sqrt{a^2 - c^2}} \int_0^{\frac{\sqrt{a^2-c^2}}{a}} \frac{du}{\sqrt{(1 - u^2)(1 - \kappa^2 u^2)}}.$$

This integral can be readily computed, when κ^2 is small, by expanding the integrand as a power series in κ^2 and integrating term by term.

XII. PROBLEMS.

1. Discuss the level surfaces given by equation (29).

2. Set up the expressions for the components of attraction instead of that for the potential as in Art. 79. Determine what parts of the integrals vanish, integrate with respect to θ, and show that the results are

$$\begin{cases} X = -4\pi\sigma bcxk^2 \displaystyle\int_0^{\frac{\pi}{2}} \frac{\sin^2 \phi \, \cos \phi \, d\phi}{\sqrt{(b^2 \sin^2 \phi + a^2 \cos^2 \phi)(c^2 \sin^2 \phi + a^2 \cos^2 \phi)}}, \\[3ex] Y = -4\pi\sigma cayk^2 \displaystyle\int_0^{\frac{\pi}{2}} \frac{\sin^2 \phi \, \cos \phi \, d\phi}{\sqrt{(c^2 \sin^2 \phi + b^2 \cos^2 \phi)(a^2 \sin^2 \phi + b^2 \cos^2 \phi)}}, \\[3ex] Z = -4\pi\sigma abzk^2 \displaystyle\int_0^{\frac{\pi}{2}} \frac{\sin^2 \phi \, \cos \phi \, d\phi}{\sqrt{(a^2 \sin^2 \phi + c^2 \cos^2 \phi)(b^2 \sin^2 \phi + c^2 \cos^2 \phi)}}. \end{cases}$$

Hint. Derive the results for Z, and since it is immaterial in what order the axes are chosen, derive the others by a permutation of the letters a, b, c.

3. Transform the equations of problem 2 by

$$\sin \phi = \frac{a}{\sqrt{a^2 + s}}, \qquad \sin \phi = \frac{b}{\sqrt{b^2 + s}}, \qquad \sin \phi = \frac{c}{\sqrt{c^2 + s}},$$

respectively, and show that equations (43) result.

4. Show that the potential of an ellipsoid upon a particle at its center is

$$V_0 = \pi \sigma abc \int_0^\infty \frac{ds}{\sqrt{(a^2 + s)(b^2 + s)(c^2 + s)}} = W.$$

5. From the value of V_0 and equations (43) derive the value of the potential (42).

6. Transform the equations of problem 2 so that they take the form

$$\int \frac{u^2 du}{\sqrt{(1 - u^2)(1 - \kappa^2 u^2)}}.$$

7. Integrate equations (28) without expanding the expression for r^2 as a power series in e^2.

80. The Attraction of a Solid Homogeneous Ellipsoid upon an Exterior Particle. Ivory's Method. The integrals become so complicated in the case of an exterior particle that the components of attraction have not been found by direct integration except in series. They are computed indirectly by expressing them in

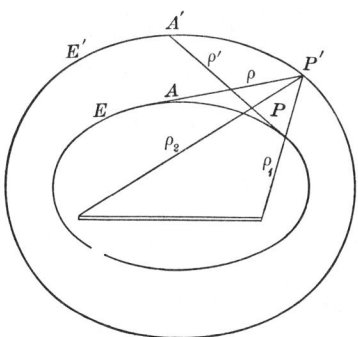

Fig. 23.

terms of the components of attraction of a related ellipsoid upon particles in its interior. This artifice constitutes Ivory's method.*

Let it be required to find the attraction of the ellipsoid E upon the exterior particle P' at the point (x', y', z'). Let the semi-

* *Philosophical Transactions,* 1809.

axes of E be a, b, and c. Construct through P' an ellipsoid E', confocal with E, with the semi-axes a', b', c', and suppose it has the same density as E. The axes of the two ellipsoids are related by the equations

$$(45) \qquad \begin{cases} a' = \sqrt{a^2 + \kappa}, \\ b' = \sqrt{b^2 + \kappa}, \\ c' = \sqrt{c^2 + \kappa}, \end{cases}$$

where κ is defined by the equation

$$(46) \qquad \frac{x'^2}{a^2 + \kappa} + \frac{y'^2}{b^2 + \kappa} + \frac{z'^2}{c^2 + \kappa} - 1 = 0.$$

The only value of κ admissible in this problem is real and positive. Equation (46) is a cubic in κ and has one positive and two negative roots; for, the left member considered as a function of κ is negative

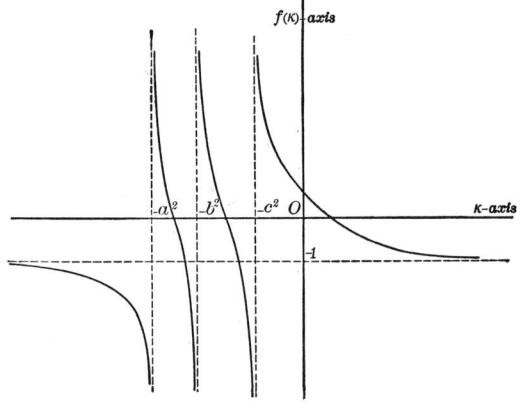

Fig. 24.

when $\kappa = + \infty$; positive, when $\kappa = 0$ (because (x', y', z') is exterior to the ellipsoid E); positive, when $\kappa = -c^2 + \epsilon$ (where ϵ is a very small positive quantity); negative, when $\kappa = -c^2 - \epsilon$; positive, when $\kappa = -b^2 + \epsilon$; negative, when $\kappa = -b^2 - \epsilon$; positive, when $\kappa = -a^2 + \epsilon$; negative, when $\kappa = -a^2 - \epsilon$; and negative when $\kappa = -\infty$. The graph of the function is given in Fig. 24. When the positive root is taken, a', b', and c' are determined uniquely.

A one-to-one correspondence between the points upon the two

ellipsoids will now be established by the equations (compare Art. 77)

$$(47) \qquad \xi' = \frac{a'}{a}\,\xi, \qquad \eta' = \frac{b'}{b}\,\eta, \qquad \zeta' = \frac{c'}{c}\,\zeta.$$

Let P be the point corresponding to P'. It will be shown that the attraction of E upon P' is related in a very simple manner to that of E' upon P.

Let X, Y, and Z represent the components of attraction of E' upon the interior particle P at the point (x, y, z). They can be computed by the methods of Art. 79, and will be supposed known. Let X', Y', and Z' be the components of attraction of E upon P', which are required. The expressions for the x-components are

$$(48) \begin{cases} X = -k^2\sigma \iiint \dfrac{x-\xi'}{\rho'^3}d\xi'd\eta'd\zeta' = k^2\sigma \iiint \dfrac{\partial\left(\dfrac{1}{\rho'}\right)}{\partial\xi'}d\xi'd\eta'd\zeta', \\[4ex] X' = -k^2\sigma \iiint \dfrac{x'-\xi}{\rho^3}d\xi\,d\eta\,d\zeta = k^2\sigma \iiint \dfrac{\partial\left(\dfrac{1}{\rho}\right)}{\partial\xi}d\xi\,d\eta\,d\zeta. \end{cases}$$

On performing the integration with respect to ξ, it is found that

$$(49) \begin{cases} X = k^2\sigma \iint \left(\dfrac{1}{\rho_2'} - \dfrac{1}{\rho_1'}\right)d\eta'd\zeta', \\[3ex] X' = k^2\sigma \iint \left(\dfrac{1}{\rho_2} - \dfrac{1}{\rho_1}\right)d\eta\,d\zeta, \end{cases}$$

where ρ_2 and ρ_1 are the distances from P' to the ends of the elementary column obtained by integrating with respect to ξ. The solution is completed by integrating over the whole surface of E. The first equation of (49) is interpreted similarly.

Now X' will be related to X in a simple manner by the aid of the following lemma:

If P and A are any two points on the surface of E, and if P' and A' are the respective corresponding points on the surface of E', then the distances $\overline{PA'}$ and $\overline{P'A}$ are equal.

Let $\overline{PA'} = \rho'$ and $\overline{AP'} = \rho$. Then $\rho = \rho'$. For, let the coördinates of P and A be respectively ξ_1, η_1, ζ_1 and ξ_2, η_2, ζ_2; and of P' and A', $\xi_1', \eta_1', \zeta_1'$ and $\xi_2', \eta_2', \zeta_2'$. Then

$$\begin{cases} \rho'^2 = (\xi_1 - \xi_2')^2 + (\eta_1 - \eta_2')^2 + (\zeta_1 - \zeta_2')^2, \\ \rho^2 = (\xi_2 - \xi_1')^2 + (\eta_2 - \eta_1')^2 + (\zeta_2 - \zeta_1')^2. \end{cases}$$

On making use of equations (45) and (47), it is found that

$$\rho'^2 - \rho^2 = \kappa\left(\frac{\xi_2^2}{a^2} + \frac{\eta_2^2}{b^2} + \frac{\zeta_2^2}{c^2}\right) - \kappa\left(\frac{\xi_1^2}{a^2} + \frac{\eta_1^2}{b^2} + \frac{\zeta_1^2}{c^2}\right).$$

Since P and A are on the surface of the ellipsoid whose semi-axes are a, b, and c, each parenthesis equals unity. Therefore $\rho'^2 - \rho^2 = 0$, or $\rho = \rho'$.

Suppose the integrals (49) are computed so that the elements at *corresponding points* of the two surfaces are always taken simultaneously. Then $\rho_1 = \rho_1'$ and $\rho_2 = \rho_2'$ throughout the integration. Moreover, it follows from (47) that $d\eta = \frac{b}{b'} d\eta'$ and $d\zeta = \frac{c}{c'} d\zeta'$. Therefore

$$(50) \quad \begin{cases} X = k^2\sigma \displaystyle\iint \left(\frac{1}{\rho_2'} - \frac{1}{\rho_1'}\right) d\eta'd\zeta', \\[2mm] X' = k^2\sigma \dfrac{bc}{b'c'} \displaystyle\iint \left(\frac{1}{\rho_2'} - \frac{1}{\rho'}\right) d\eta'd\zeta' = \dfrac{bc}{b'c'} X; \end{cases}$$

and similarly

$$(51) \quad \begin{cases} Y' = \dfrac{ca}{c'a'} Y, \\[3mm] Z' = \dfrac{ab}{a'b'} Z. \end{cases}$$

The letters a, b, c, and s of equations (43) should be given accents to agree with the notations of this article; and, since P and P' are corresponding points, $x = \frac{a}{a'} x'$, $y = \frac{b}{b'} y'$, $z = \frac{c}{c'} z'$. After making these changes in equations (43) and substituting them in (50) and (51), it is found that

$$\begin{cases} X' = -2\pi\sigma abck^2x' \displaystyle\int_0^\infty \frac{ds'}{(a'^2+s')\sqrt{(a'^2+s')(b'^2+s')(c'^2+s')}}, \\[4mm] Y' = -2\pi\sigma abck^2y' \displaystyle\int_0^\infty \frac{ds'}{(b'^2+s')\sqrt{(a'^2+s')(b'^2+s')(c'^2+s')}}, \\[4mm] Z' = -2\pi\sigma abck^2z' \displaystyle\int_0^\infty \frac{ds'}{(c'^2+s')\sqrt{(a'^2+s')(b'^2+s')(c'^2+s')}}. \end{cases}$$

It follows from equations (45) that

$$a'^2 = a^2 + \kappa, \qquad b'^2 = b^2 + \kappa, \qquad c'^2 = c^2 + \kappa;$$

hence, on letting $s = s' + \kappa$, it follows that

$$(52) \begin{cases} X' = -2\pi\sigma abck^2 x' \displaystyle\int_\kappa^\infty \frac{ds}{(a^2+s)\,\sqrt{(a^2+s)(b^2+s)(c^2+s)}}, \\[2ex] Y' = -2\pi\sigma abck^2 y' \displaystyle\int_\kappa^\infty \frac{ds}{(b^2+s)\,\sqrt{(a^2+s)(b^2+s)(c^2+s)}}, \\[2ex] Z' = -2\pi\sigma abck^2 z' \displaystyle\int_\kappa^\infty \frac{ds}{(c^2+s)\,\sqrt{(a^2+s)(b^2+s)(c^2+s)}}. \end{cases}$$

It follows from equations (40) and (41) that the components of attraction for interior particles are homogeneous of degree zero in a, b, and c, and that they are proportional to the respective coördinates of the attracted particle. Let X, as above, represent the attraction of the ellipsoid E', whose semi-axes are a', b', c', upon the interior particle at (x, y, z); let X'' represent the attraction of E' upon an interior particle at (x'', y'', z''), which will be supposed to be related to (x, y, z) by equations of the same form as (47). Then it follows that

$$\frac{X''}{X} = \frac{x''}{x}, \qquad \frac{Y''}{Y} = \frac{y''}{y}, \qquad \frac{Z''}{Z} = \frac{z''}{z}.$$

Let the point (x'', y'', z''), always corresponding to (x, y, z), approach the surface of E' as a limit. Then at the limit

$$\frac{X''}{X} = \frac{a'}{a}, \qquad \frac{Y''}{Y} = \frac{b'}{b}, \qquad \frac{Z''}{Z} = \frac{c'}{c}.$$

On combining these equations with (50) and (51), it is found that

$$(53) \qquad \frac{X''}{X'} = \frac{Y''}{Y'} = \frac{Z''}{Z'} = \frac{a'b'c'}{abc} = \frac{M'}{M}.$$

That is, *the attraction of a solid ellipsoid upon an exterior particle is to the attraction of a confocal ellipsoid passing through the particle, as the mass of the first ellipsoid is to that of the second ellipsoid.*

Consider another ellipsoid confocal with the one passing through the particle and interior to it; by the same reasoning the ratios of the components of attraction of these two ellipsoids are as their masses. Let X''', Y''', Z''' be the components of attraction of the new ellipsoid whose semi-axes are a''', b''', c'''. Then

$$\frac{X''}{X'''} = \frac{Y''}{Y'''} = \frac{Z''}{Z'''} = \frac{a'b'c'}{a'''b'''c'''} = \frac{M'}{M'''}.$$

On combining this proportion with (53), it is found that

$$\frac{X'}{X'''} = \frac{Y'}{Y'''} = \frac{Z'}{Z'''} = \frac{M}{M'''}.$$

Therefore, *two confocal ellipsoids attract particles which are exterior to both of them in the same direction and with forces which are proportional to their masses.* This theorem was found by Maclaurin and Lagrange for ellipsoids of revolution, and was extended by Laplace to the general case where the three axes are unequal. It is established most easily, however, by Ivory's method as above, and it is frequently called Ivory's theorem.

The right members of equations (52) can be transformed to forms which are more convenient for computation by putting, in the first, $\dfrac{a}{\sqrt{a^2 + s}} = u$; in the second, $\dfrac{b}{\sqrt{b^2 + s}} = u$; and in the third, $\dfrac{c}{\sqrt{b^2 + s}} = u$. The results of the substitutions are

$$(54)\begin{cases} X' = -4\pi\sigma bck^2 x' \displaystyle\int_0^{\frac{a}{\sqrt{a^2+\kappa}}} \frac{u^2 du}{\sqrt{[a^2 - (a^2-b^2)u^2][a^2 - (a^2-c^2)u^2]}}, \\[3ex] Y' = -4\pi\sigma cak^2 y' \displaystyle\int_0^{\frac{b}{\sqrt{b^2+\kappa}}} \frac{u^2 du}{\sqrt{[b^2 - (b^2-c^2)u^2][b^2 - (b^2-a^2)u^2]}}, \\[3ex] Z' = -4\pi\sigma abk^2 z' \displaystyle\int_0^{\frac{c}{\sqrt{c^2+\kappa}}} \frac{u^2 du}{\sqrt{[c^2 - (c^2-a^2)u^2][c^2 - (c^2-b^2)u^2]}}. \end{cases}$$

When the attracted particle is in the interior of the ellipsoid the forms of the integrals are the same except that the upper limits are unity.

81. The Attraction of Spheroids. The components of attraction will be obtained from (54), which hold for exterior particles. Suppose the attracting body is an oblate spheroid in which $a = b > c$ and let e represent the eccentricity of a meridian section. Then

$$c^2 = a^2(1 - e^2),$$

and equations (54) become

$$(55)\begin{cases} \dfrac{X'}{x'} = \dfrac{Y'}{y'} = -4\pi\sigma k^2 \sqrt{1 - e^2} \displaystyle\int_0^{\frac{a}{\sqrt{a^2+\kappa}}} \frac{u^2 du}{\sqrt{1 - e^2 u^2}}, \\[3ex] \dfrac{Z'}{z'} = -4\pi\sigma k^2 \displaystyle\int_0^{\frac{c}{\sqrt{c^2+\kappa}}} \frac{u^2 du}{1 - e^2 + e^2 u^2}. \end{cases}$$

The integrals of these equations are

$$(56) \begin{cases} \dfrac{X'}{x'} = \dfrac{Y'}{y'} = -2\pi\sigma k^2 \dfrac{\sqrt{1-e^2}}{e^3} \left[\dfrac{-ae}{\sqrt{a^2+\kappa}} \sqrt{1 - \dfrac{a^2 e^2}{a^2+\kappa}} \right. \\ \qquad\qquad\qquad\qquad\qquad\qquad \left. + \sin^{-1}\left(\dfrac{ae}{\sqrt{a^2+\kappa}} \right) \right], \\[2em] \dfrac{Z'}{z'} = -4\pi\sigma \dfrac{k^2}{e^3} \left[\dfrac{ce}{\sqrt{c^2+\kappa}} - \sqrt{1-e^2} \right. \\ \qquad\qquad\qquad\qquad\qquad \left. \times \tan^{-1}\left(\dfrac{ce}{\sqrt{(1-e^2)(c^2+\kappa)}} \right) \right]. \end{cases}$$

The components of attraction for interior particles are obtained from equations (56) by putting $\kappa = 0$.

Now suppose the attracting body is a prolate spheroid and that $a = b < c$. Then $a^2 = b^2 = c^2(1 - \epsilon^2)$, and equations (54) become

$$(57) \begin{cases} \dfrac{X'}{y'} = \dfrac{Y'}{x'} = -4\pi\sigma k^2 \displaystyle\int_0^{\frac{a}{\sqrt{a^2+\kappa}}} \dfrac{u^2 du}{\sqrt{1 - \epsilon^2 + \epsilon^2 u^2}}, \\[2em] \dfrac{Z'}{z'} = -4\pi\sigma k^2 (1-\epsilon^2) \displaystyle\int_0^{\frac{c}{\sqrt{c^2+\kappa}}} \dfrac{u^2 du}{1 - \epsilon^2 u^2}. \end{cases}$$

The integrals of these equations are

$$(58) \begin{cases} \dfrac{X'}{x'} = \dfrac{Y'}{y'} = -\dfrac{2\pi\sigma k^2}{\epsilon^3} \left[\dfrac{a\epsilon}{\sqrt{a^2+\kappa}} \sqrt{1 - \epsilon^2 + \dfrac{a^2\epsilon^2}{a^2+\kappa}} \right. \\ \qquad\qquad - (1-\epsilon^2) \log\left(\dfrac{a\epsilon}{\sqrt{(1-\epsilon^2)(a^2+\kappa)}} \right. \\ \qquad\qquad\qquad\qquad \left.\left. + \sqrt{1 + \dfrac{a^2\epsilon^2}{(1-\epsilon^2)(a^2+\kappa)}} \right) \right], \\[2em] \dfrac{Z'}{z'} = -2\pi\sigma k^2 \dfrac{(1-\epsilon^2)}{\epsilon^3} \left(\dfrac{-2c\epsilon}{\sqrt{c^2+\kappa}} + \log \dfrac{1 + \dfrac{c\epsilon}{\sqrt{c^2+\kappa}}}{1 - \dfrac{c\epsilon}{\sqrt{c^2+\kappa}}} \right). \end{cases}$$

When the particle is interior to the spheroid the equations for the components of attraction are the same except that $\kappa = 0$.

82. The Attraction at the Surfaces of Spheroids. The components of attraction for an interior particle, which are obtained in the case of an oblate spheroid from (56) by putting $\kappa = 0$, are, omitting the accents,

$$(59) \begin{cases} \dfrac{X}{x} = \dfrac{Y}{y} = -2\pi\sigma k^2 \dfrac{\sqrt{1-e^2}}{e^3}\left[-e\sqrt{1-e^2} + \sin^{-1}e \right], \\[2em] \dfrac{Z}{z} = -4\pi\sigma\dfrac{k^2}{e^3}\left[e - \sqrt{1-e^2}\tan^{-1}\left(\dfrac{e}{\sqrt{1-e^2}}\right) \right]. \end{cases}$$

The limits of these expressions as the attracted particle approaches the surface of the spheroid are the components of attraction for a particle at the surface. As the attracted particle passes outward through the surface, κ, in equations (56), starts with the value zero and increases continuously in such a manner that it always fulfills equation (46). Therefore equations (59), having no discontinuity as the attracted particle reaches the surface, hold when x, y, z fulfill the equation of the ellipsoid.

When e is small, as in the case of the planets, equations (59) are convenient when expanded as power series in e. On substituting the expansions

$$\begin{cases} \sqrt{1-e^2} = 1 - \dfrac{e^2}{2} - \dfrac{e^4}{8} - \cdots, \\[1.5em] \sin^{-1}e = e + \dfrac{e^3}{6} + \dfrac{3e^5}{40} + \cdots, \\[1.5em] \sqrt{1-e^2}\tan^{-1}\left(\dfrac{e}{\sqrt{1-e^2}}\right) = e - \dfrac{e^3}{3} - \dfrac{2e^5}{15} + \cdots, \end{cases}$$

in equations (59), it is found that

$$(60) \begin{cases} \dfrac{X}{x} = \dfrac{Y}{y} = -\tfrac{4}{3}\pi\sigma k^2(1 - \tfrac{1}{5}e^2 + \cdots), \\[1.5em] \dfrac{Z}{z} = -\tfrac{4}{3}\pi\sigma k^2(1 + \tfrac{2}{5}e^2 + \cdots). \end{cases}$$

The mass of the spheroid is

$$M = \tfrac{4}{3}\pi\sigma a^2 c = \tfrac{4}{3}\pi\sigma a^3\sqrt{1-e^2}.$$

The radius of a sphere having equal mass is defined by the equation

$$M = \tfrac{4}{3}\pi\sigma R^3 = \tfrac{4}{3}\pi\sigma a^3\sqrt{1-e^2};$$

whence

$$R = a(1-e^2)^{\frac{1}{3}}.$$

The attraction of this sphere for a particle upon its surface is given by the equation

$$(61) \qquad F = -\dfrac{k^2 M}{R^2} = -\tfrac{4}{3}\pi\sigma k^2 a(1-e^2)^{\frac{1}{3}}.$$

When the attracted particle is at the equator of the spheroid $\sqrt{x^2 + y^2} = a$; hence the ratio of the attraction of the spheroid for a particle at its equator to that of an equal sphere for a particle upon its surface is

$$\frac{\sqrt{X^2 + Y^2}}{F} = \frac{(1 - \frac{1}{5}e^2 \cdots)}{(1 - e^2)^1} = 1 - \frac{e^2}{30} + \cdots.$$

This is less than unity when e is small; therefore the attraction of the spheroid for a particle on its surface at its equator is less than that of a sphere having equal mass and volume for a particle on its surface.

When the attracted particle is at the pole of the spheroid $z = c = a\sqrt{1 - e^2}$; hence in this case

$$\frac{Z}{F} = \sqrt{1 - e^2}\frac{(1 + \frac{2}{5}e^2 \cdots)}{(1 - e^2)^{\frac{1}{3}}} = 1 + \frac{e^2}{15} + \cdots.$$

This is greater than unity when e is small; therefore the attraction of the spheroid for a particle on its surface at its pole is greater than that of a sphere having equal mass and volume for a particle on its surface.

There is some place between the equator and pole at which the attractions are just equal. The latitude of this place will now be found. The coördinates of the particle must fulfill the equation of the spheroid; therefore

$$(62) \qquad f(x, y, z) \equiv \frac{x^2 + y^2}{a^2} + \frac{z^2}{c^2} - 1 = 0.$$

The direction cosines of the normal to the surface at the point (x, y, z) are

$$\frac{\dfrac{\partial f}{\partial x}}{\sqrt{\left(\dfrac{\partial f}{\partial x}\right)^2 + \left(\dfrac{\partial f}{\partial y}\right)^2 + \left(\dfrac{\partial f}{\partial z}\right)^2}}, \qquad \frac{\dfrac{\partial f}{\partial y}}{\sqrt{\left(\dfrac{\partial f}{\partial x}\right)^2 + \left(\dfrac{\partial f}{\partial y}\right)^2 + \left(\dfrac{\partial f}{\partial z}\right)^2}},$$

$$\frac{\dfrac{\partial f}{\partial z}}{\sqrt{\left(\dfrac{\partial f}{\partial x}\right)^2 + \left(\dfrac{\partial f}{\partial y}\right)^2 + \left(\dfrac{\partial f}{\partial z}\right)^2}}.$$

The last is the cosine of the angle between the normal at the point (x, y, z) and the z-axis, and is, therefore, the sine of the geographical latitude, which will be represented by ϕ. Hence, it follows from (62) that

$$(63) \qquad \sin \phi = \frac{\dfrac{\partial f}{\partial z}}{\sqrt{\left(\dfrac{\partial f}{\partial x}\right)^2 + \left(\dfrac{\partial f}{\partial y}\right)^2 + \left(\dfrac{\partial f}{\partial z}\right)^2}}$$

$$= \frac{z}{\sqrt{(x^2 + y^2)(1 - e^2)^2 + z^2}}.$$

From (62) and (63) it is found that

$$(64) \quad \begin{cases} x^2 + y^2 = \dfrac{a^2 \cos^2 \phi}{1 - e^2 \sin^2 \phi} = a^2 \cos^2 \phi \{1 + e^2 \sin^2 \phi + \cdots\}, \\[3mm] z^2 = \dfrac{a^2(1 - e^2)^2 \sin^2 \phi}{1 - e^2 \sin^2 \phi} \\[3mm] \qquad = a^2 \sin^2 \phi \{1 - e^2(1 + \cos^2 \phi) + \cdots\}. \end{cases}$$

Let G represent the whole attraction of the spheroid; then it is found from (60) and (64) that

$$G = -\sqrt{X^2 + Y^2 + Z^2}$$

$$= -\tfrac{4}{3}\pi\sigma k^2 \sqrt{(1 - \tfrac{1}{5}e^2 \cdots)^2(x^2 + y^2) + (1 + \tfrac{2}{5}e^2 \cdots)^2 z^2}$$

$$= -\tfrac{4}{3}\pi\sigma k^2 a \left\{1 - \frac{e^2}{10}(1 + \cos^2 \phi) + \cdots\right\}.$$

The ratio of this expression to that for the attraction of a sphere of equal mass and volume, given by (61), is

$$(65) \qquad \frac{G}{F} = \frac{1 - \dfrac{e^2}{10}(1 + \cos^2 \phi) \cdots}{(1 - e^2)^{\frac{1}{3}}} = 1 - \frac{e^2(1 - 3 \sin^2 \phi)}{30} \cdots.$$

This becomes equal to unity up to terms of the fourth order in e when $3 \sin^2 \phi = 1$, from which it is found that

$$\phi = 35° \ 15' \ 52''.$$

Let r represent the radius of the spheroid; then

$$r^2 = \frac{a^2(1 - e^2)}{1 - e^2 \cos^2 \psi},$$

where ψ is the angle between the radius and the plane of the equator. Since this angle differs from ϕ only by terms of the second and higher orders in e, it follows that, with the degree of approximation employed,

$$r^2 = \frac{a^2(1 - e^2)}{1 - e^2 \cos^2 \phi} = a^2(1 - e^2)(1 + e^2 \cos^2 \phi + \cdots).$$

When $\phi = 35° \; 15' \; 52''$,

$$r^2 = a^2 \left(1 - \frac{e^2}{3} + \cdots \right).$$

The radius of a sphere of equal volume has been found to be given by the equation

$$R^2 = a^2(1 - e^2)^{\frac{1}{3}} = a^2 \left(1 - \frac{e^2}{3} + \cdots \right),$$

which is seen to be equal to the radius of the spheroid up to terms of the second order inclusive in the eccentricity. Therefore, in the case of an oblate spheroid of small ecentricity, the intensity of the attraction is sensibly the same for a particle on its surface in latitude $35° \; 15' \; 52''$ as that of a sphere having equal mass and volume for a particle on its surface; or, because of the equality of R and r, a spheroid of small eccentricity attracts a particle on its surface in latitude $35° \; 15' \; 52''$ with sensibly the same force it would exert if its mass were all at its center.

XIII. PROBLEMS.

1. Show that Ivory's method can be applied when the attraction varies as any power of the distance.

2. Show why Ivory's method cannot be used to find the potential of a solid ellipsoid upon an exterior particle when it is known for an interior particle.

3. Find the potential of a thin ellipsoidal shell contained between two similar ellipsoids upon an interior particle. *Hint.* It has been proved (Art. 79) that the resultant attraction is zero at all interior points; therefore the potential is constant and it is sufficient to find it for the center. Let the semi-axes of the two surfaces be a, b, c and $(1 + \mu)a$, $(1 + \mu)b$, $(1 + \mu)c$; then the distance between the two surfaces measured along the radius from the center will be $\mu\rho$. Therefore

$$V = \sigma\mu \int_{-\frac{\pi}{2}}^{\frac{\pi}{2}} \int_0^{2\pi} \frac{\rho^3 \cos \phi \, d\phi \, d\theta}{\rho}$$

$$= \sigma\mu \int_{-\frac{\pi}{2}}^{\frac{\pi}{2}} \int_0^{2\pi} \frac{\cos \phi \, d\phi \, d\theta}{\dfrac{\cos^2 \phi \cos^2 \theta}{a^2} + \dfrac{\cos^2 \phi \sin^2 \theta}{b^2} + \dfrac{\sin^2 \phi}{c^2}}$$

$$= 2\pi\sigma\mu abc \int_0^{\infty} \frac{ds}{\sqrt{(a^2 + s)(b^2 + s)(c^2 + s)}}.$$

4. Show that in the case of two thin confocal shells similar elements of mass at points which correspond according to the definition (47) are proportional to the products of the three axes of the respective ellipsoids. Then show, using problem 3 and Ivory's method, that the potential of an ellipsoidal shell upon an exterior particle is

$$V' = 2\pi\sigma\mu abc \int_0^\infty \frac{ds'}{\sqrt{(a'^2 + s')(b'^2 + s')(c'^2 + s')}}$$

$$= 2\pi\sigma\mu abc \int_\kappa^\infty \frac{ds}{\sqrt{(a^2 + s)(b^2 + s)(c^2 + s)}}.$$

5. Prove that the level surfaces of thin homogeneous ellipsoids are confocal ellipsoids. What are the lines of force which are orthogonal to these surfaces?

6. Discuss the form of level surfaces when they are entirely exterior to homogeneous solid ellipsoids.

HISTORICAL SKETCH AND BIBLIOGRAPHY.

The attractions of bodies were first investigated by Newton. His results are given in the *Principia*, Book I., Secs. XII. and XIII., and are derived by synthetic processes similar to those used in the first part of this chapter. The problem of the attraction of ellipsoids has been the subject of many memoirs, and the case in which they are homogeneous was completely solved early in the nineteenth century. Among the important papers are those by Stirling, 1735, *Phil. Trans.*; by Euler, 1738, *Petersburg;* by Lagrange, 1773 and 1775, *Coll. Works*, vol. III., p. 619; by Laplace, 1782, *Méc. Cél.*, vol. II.; by Ivory, 1809–1828, *Phil. Trans.*; by Legendre, 1811, *Mém. de l'Inst. de France*, vol. XI.; by Gauss, *Coll. Works*, vol. V.; by Rodriguez, 1816, *Corres. sur l'École Poly.*, vol. III.; by Poisson, 1829, *Conn. des Temps;* by Green, 1835, *Math. Papers*, vol. VIII.; Chasles, 1837–1846, *Jour. l'École Poly.* and *Mém. des Savants Étrangers*, vol. IX.; MacCullagh, 1847, *Dublin Proc.*, vol. III.; Lejeune-Dirichlet, *Journal de Liouville*, vol. IV., and *Crelle*, vol. XXXII.

The earlier papers were devoted for the most part to the attractions of homogeneous ellipsoids of revolution upon particles in particular positions, as on the axis. Lagrange gave the general solution for the attractions of general homogeneous ellipsoids upon interior particles. This was extended by Ivory and Maclaurin (with Laplace's generalizations) to exterior particles. Ivory's theorem has been extended in a most interesting manner by Darboux in Note XVI. to the second volume of the *Mécanique* of Despeyrous. Chasles gave a synthetic proof of the theorems regarding the attractions of homogeneous ellipsoids in *Mémoires des Savants Étrangers*, vol. IX., and Lejeune-Dirichlet embraced in a most elegant manner in one discussion the case of both interior and exterior points by using a discontinuous factor (*Liouville's Journal*, vol. IV.).

Laplace proved that the potential for an exterior particle fulfills the partial differential equation

$$\frac{\partial^2 V}{\partial x^2} + \frac{\partial^2 V}{\partial y^2} + \frac{\partial^2 V}{\partial z^2} = 0,$$

and determined V by the condition that it must be a function satisfying this equation. This is a process of great generality, and is relatively simple except in the trivial cases. This has been made the starting-point of most of the investigations of the latter part of the last century, especially where the attracting bodies are not homogeneous. In a paper on Electricity and Magnetism, in 1828, Green introduced the term *potential function* for V, and discussed many of its mathematical properties. Green's memoir remained nearly unknown until about 1846, and in the meantime many of his theorems had been rediscovered by Chasles, Gauss, Sturm, and Thomson. One of Green's theorems has found an extremely useful application, when the independent variables are two in number, in the Theory of Functions.

Poisson showed that the potential function for an interior particle fulfills the partial differential equation

$$\frac{\partial^2 V}{\partial x^2} + \frac{\partial^2 V}{\partial y^2} + \frac{\partial^2 V}{\partial z^2} = -4\pi\sigma.$$

Among the books treating the subject of attractions and potential may be mentioned Thomson and Tait's *Natural Philosophy*, part II., Neumann's *Potential*, Poincaré's *Potential*, Routh's *Analytical Statics*, vol. II., and Tisserand's *Mécanique Céleste*, vol. II. The last-mentioned develops most fully the astronomical applications and should be used in further reading.

The attractions of spheroids and ellipsoids has been fundamental in the discussions of possible figures of equilibrium of rotating fluids. The reason is, of course, that the conditions for equilibrium involve the components of attraction. Maclaurin proved in 1742 that for slow rotation an oblate spheroid, whose eccentricity is a function of the rate of rotation and the density of the fluid, is a figure of equilibrium. There are, indeed, two such figures; for slow rotation one is nearly spherical and the other is very much flattened. For faster rotation the figures are more nearly of the same shape; for a certain greater rate of rotation they are identical; and for still faster rotation no spheroid is a figure of equilibrium. In 1834 Jacobi proved that when the rate of rotation is not too great there is an ellipsoid of three unequal axes which is a figure of equilibrium, which for a certain rate of rotation coincides with the more nearly spherical of the Maclaurin spheroids. For this work Tisserand's *Mecanique Céleste*, vol. II., should be consulted. In a very important memoir (*Acta Mathematicæ*, vol. VII.) Poincaré proved that there are infinitely many other figures of equilibrium which, for certain values of the rate of rotation, coincide with the corresponding Jacobian ellipsoid, as it, for a certain rate of rotation, coincides with the Maclaurin spheroid. The least elongated of these figures is larger at one end than it is at the other, and was called the *apioid*, that is, the pear-shaped figure. Later computations by Sir George Darwin (*Philosophical Transactions*, vol. 198) have shown it is so elongated that it might well be called a cucumber-shaped figure.

CHAPTER V.

THE PROBLEM OF TWO BODIES.

83. Equations of Motion. It will be assumed in this chapter that the two bodies are spheres and homogeneous in concentric layers. Then, in accordance with the results obtained in Art. 69, they will attract each other with a force which is proportional to the product of their masses and which varies inversely as the square of the distance between their centers.

Let m_1 and m_2 represent the masses of the two bodies, and $m_1 + m_2 = M$. Choose an arbitrary system of rectangular axes in space and let the coördinates of m_1 and m_2 referred to it be respectively (ξ_1, η_1, ζ_1) and (ξ_2, η_2, ζ_2). Let the distance between m_1 and m_2 be denoted by r; then it follows from the laws of motion and the law of gravitation that the differential equations which the coördinates of the bodies satisfy are

(1)
$$\begin{cases} m_1 \dfrac{d^2\xi_1}{dt^2} = -k^2 m_1 m_2 \dfrac{(\xi_1 - \xi_2)}{r^3}, \\[2mm] m_1 \dfrac{d^2\eta_1}{dt^2} = -k^2 m_1 m_2 \dfrac{(\eta_1 - \eta_2)}{r^3}, \\[2mm] m_1 \dfrac{d^2\zeta_1}{dt^2} = -k^2 m_1 m_2 \dfrac{(\zeta_1 - \zeta_2)}{r^3}, \\[2mm] m_2 \dfrac{d^2\xi_2}{dt^2} = -k^2 m_2 m_1 \dfrac{(\xi_2 - \xi_1)}{r^3}, \\[2mm] m_2 \dfrac{d^2\eta_2}{dt^2} = -k^2 m_2 m_1 \dfrac{(\eta_2 - \eta_1)}{r^3}, \\[2mm] m_2 \dfrac{d^2\zeta_2}{dt^2} = -k^2 m_2 m_1 \dfrac{(\zeta_2 - \zeta_1)}{r^3}. \end{cases}$$

In order to solve these six simultaneous equations of the second order twelve integrals must be found. They will introduce twelve arbitrary constants of integration which can be determined in any particular case by the three initial coördinates and the three components of the initial velocity of each of the bodies.

84. The Motion of the Center of Mass. On adding the first and fourth, the second and fifth, and the third and sixth equations of the system (1), it is found that

$$
\begin{cases}
m_1 \dfrac{d^2\xi_1}{dt^2} + m_2 \dfrac{d^2\xi_2}{dt^2} = 0, \\[2mm]
m_1 \dfrac{d^2\eta_1}{dt^2} + m_2 \dfrac{d^2\eta_2}{dt^2} = 0, \\[2mm]
m_1 \dfrac{d^2\zeta_1}{dt^2} + m_2 \dfrac{d^2\zeta_2}{dt^2} = 0.
\end{cases}
$$

These equations are immediately integrable, and give

$$
(2) \quad
\begin{cases}
m_1 \dfrac{d\xi_1}{dt} + m_2 \dfrac{d\xi_2}{dt} = \alpha_1, \\[2mm]
m_1 \dfrac{d\eta_1}{dt} + m_2 \dfrac{d\eta_2}{dt} = \beta_1, \\[2mm]
m_1 \dfrac{d\zeta_1}{dt} + m_2 \dfrac{d\zeta_2}{dt} = \gamma_1.
\end{cases}
$$

On integrating again they become

$$
(3) \quad
\begin{cases}
m_1\xi_1 + m_2\xi_2 = \alpha_1 t + \alpha_2, \\
m_1\eta_1 + m_2\eta_2 = \beta_1 t + \beta_2, \\
m_1\zeta_1 + m_2\zeta_2 = \gamma_1 t + \gamma_2.
\end{cases}
$$

Thus, six of the twelve integrals are found, the arbitrary constants of integration being α_1, α_2, β_1, β_2, γ_1, γ_2.

Let $\bar{\xi}$, $\bar{\eta}$, and $\bar{\zeta}$ be the coördinates of the center of mass of the system; then it follows from Art. 19 and equations (3) that

$$
(4) \quad
\begin{cases}
M\bar{\xi} = m_1\xi_1 + m_2\xi_2 = \alpha_1 t + \alpha_2, \\
M\bar{\eta} = m_1\eta_1 + m_2\eta_2 = \beta_1 t + \beta_2, \\
M\bar{\zeta} = m_1\zeta_1 + m_2\zeta_2 = \gamma_1 t + \gamma_2.
\end{cases}
$$

From these equations it follows that the coördinates increase directly as the time, and, therefore, that the center of mass moves with uniform velocity. Or, taking their derivatives, squaring, and adding, it is found that

$$
M^2 \left\{ \left(\frac{d\bar{\xi}}{dt} \right)^2 + \left(\frac{d\bar{\eta}}{dt} \right)^2 + \left(\frac{d\bar{\zeta}}{dt} \right)^2 \right\} = \alpha_1{}^2 + \beta_1{}^2 + \gamma_1{}^2;
$$

whence

$$\overline{V} = \frac{\sqrt{\alpha_1{}^2 + \beta_1{}^2 + \gamma_1{}^2}}{M};$$

where $\overline{V}$ represents the speed with which the center of mass moves. The speed is therefore constant.

On eliminating t from (4), it is found that

$$\frac{M\overline{\xi} - \alpha_2}{\alpha_1} = \frac{M\overline{\eta} - \beta_2}{\beta_1} = \frac{M\overline{\zeta} - \gamma_2}{\gamma_1}.$$

The coördinates of the center of mass fulfill these relations which are the symmetrical equations of a straight line in space; therefore, *the center of mass moves in a straight line with constant speed.*

85. The Equations for Relative Motion. Take a new system of axes parallel to the old, but with the origin at the center of mass of the two bodies. Let the coördinates of m_1 and m_2 referred to this new system be x_1, y_1, z_1 and x_2, y_2, z_2 respectively. They are related to the old coördinates by the equations

$$(5) \quad \begin{cases} x_1 = \xi_1 - \overline{\xi}, & x_2 = \xi_2 - \overline{\xi}, \\ y_1 = \eta_1 - \overline{\eta}, & y_2 = \eta_2 - \overline{\eta}, \\ z_1 = \zeta_1 - \overline{\zeta}; & z_2 = \zeta_2 - \overline{\zeta}. \end{cases}$$

On substituting in (1), the differential equations of motion in the new variables are found to be

$$(6) \quad \begin{cases} m_1 \dfrac{d^2 x_1}{dt^2} = -k^2 m_1 m_2 \dfrac{(x_1 - x_2)}{r^3}, \\[2mm] m_1 \dfrac{d^2 y_1}{dt^2} = -k^2 m_1 m_2 \dfrac{(y_1 - y_2)}{r^3}, \\[2mm] m_1 \dfrac{d^2 z_1}{dt^2} = -k^2 m_1 m_2 \dfrac{(z_1 - z_2)}{r^3}, \\[2mm] m_2 \dfrac{d^2 x_2}{dt^2} = -k^2 m_2 m_1 \dfrac{(x_2 - x_1)}{r^3}, \\[2mm] m_2 \dfrac{d^2 y_2}{dt^2} = -k^2 m_2 m_1 \dfrac{(y_2 - y_1)}{r^3}, \\[2mm] m_2 \dfrac{d^2 z_2}{dt^2} = -k^2 m_2 m_1 \dfrac{(z_2 - z_1)}{r^3}, \end{cases}$$

which are of the same form as the equations for absolute motion.

The coördinates of the center of mass are given by equations (4); therefore if x_1, y_1, $\cdots\cdots$, z_2 were known, and if the constants

α_1, α_2, β_1, β_2, γ_1, and γ_2 were known, the absolute positions in space could be found. But, since there is no way of determining these constants, the problem of relative motion, as expressed in (6), is all that can be solved.

Since the new origin is at the center of mass, the coördinates are related by the equations

(7)
$$\begin{cases} m_1x_1 + m_2x_2 = 0, \\ m_1y_1 + m_2y_2 = 0, \\ m_1z_1 + m_2z_2 = 0. \end{cases}$$

Therefore, when the coördinates of one body with respect to the center of mass of the two are known the coördinates of the second body are given by equations (7).

Equations (7) can be used to eliminate x_2, y_2, and z_2 from the first three equations of (6), and x_1, y_1, and z_1 from the last three. The results of the elimination are

(8)
$$\begin{cases} \dfrac{d^2x_1}{dt^2} = - k^2M\dfrac{x_1}{r^3}, \\[2mm] \dfrac{d^2y_1}{dt^2} = - k^2M\dfrac{y_1}{r^3}, \\[2mm] \dfrac{d^2z_1}{dt^2} = - k^2M\dfrac{z_1}{r^3}; \\[2mm] \dfrac{d^2x_2}{dt^2} = - k^2M\dfrac{x_2}{r^3}, \\[2mm] \dfrac{d^2y_2}{dt^2} = - k^2M\dfrac{y_2}{r^3}, \\[2mm] \dfrac{d^2z_2}{dt^2} = - k^2M\dfrac{z_2}{r^3}. \end{cases}$$

In the first three equations the r which appears in the right member must be expressed in terms of x_1, y_1, and z_1; and in the second three it must be expressed in terms of x_2, y_2, and z_2. It follows from equations (7) that

$$\frac{M}{m_2}\sqrt{x_1^2 + y_1^2 + z_1^2} = \frac{M}{m_2}r_1 = \frac{M}{m_1}r_2 = r.$$

The equations in x_1, y_1, z_1 are now independent of those in x_2, y_2, z_2, and conversely. But what is really desired in practice is the motion of one body with respect to the other. Let x, y, and z represent the coördinates of m_2 with respect to m_1, then

$$x = x_2 - x_1, \quad y = y_2 - y_1, \quad z = z_2 - z_1.$$

Hence if the first, second, and third equations of (8) are subtracted from the fourth, fifth, and sixth equations respectively, the results are, as a consequence of these relations,

$$(9) \quad \begin{cases} \dfrac{d^2x}{dt^2} = -k^2 M \dfrac{x}{r^3}, \\[2mm] \dfrac{d^2y}{dt^2} = -k^2 M \dfrac{y}{r^3}, \\[2mm] \dfrac{d^2z}{dt^2} = -k^2 M \dfrac{z}{r^3}. \end{cases}$$

The problem is now of the sixth order, having been reduced from the twelfth by means of the six integrals (2) and (3). The six new constants of integration which will be introduced in integrating equations (9) will be determined by the three initial coördinates, and the three projections of the initial velocity of m_1 with respect to m_2.

86. The Integrals of Areas. Multiply the first equation of (9) by $-y$, and the second by $+x$, and add; the result is

$$\begin{cases} x\dfrac{d^2y}{dt^2} - y\dfrac{d^2x}{dt^2} = 0, \text{ and similarly,} \\[2mm] y\dfrac{d^2z}{dt^2} - z\dfrac{d^2y}{dt^2} = 0, \\[2mm] z\dfrac{d^2x}{dt^2} - x\dfrac{d^2z}{dt^2} = 0. \end{cases}$$

The integrals of these equations are

$$(10) \quad \begin{cases} x\dfrac{dy}{dt} - y\dfrac{dx}{dt} = a_1, \\[2mm] y\dfrac{dz}{dt} - z\dfrac{dy}{dt} = a_2, \\[2mm] z\dfrac{dx}{dt} - x\dfrac{dz}{dt} = a_3. \end{cases}$$

It follows from Art. 16 that a_1, a_2, a_3 are the projections of twice the areal velocity upon the xy, yz, and zx-planes respectively.

Upon multiplying equations (10) by z, x, and y respectively, and adding, it is found that

(11) $a_1 z + a_2 x + a_3 y = 0.$

This is the equation of a plane passing through the origin, and it follows from its derivation that the coördinates of m_1 always fulfill it; therefore, *the motion of one body with respect to the other is in a plane which passes through the center of the other.*

The constants a_1, a_2, and a_3 determine the position of the plane of the orbit with respect to the axes of reference. In polar coördinates equation (11) becomes

(12) $a_1 \sin \varphi + a_2 \cos \varphi \cos \theta + a_3 \cos \varphi \sin \theta = 0.$

The xy-plane and the plane of the orbit intersect in a line L (Fig. 25). Suppose OL is that half line which passes through

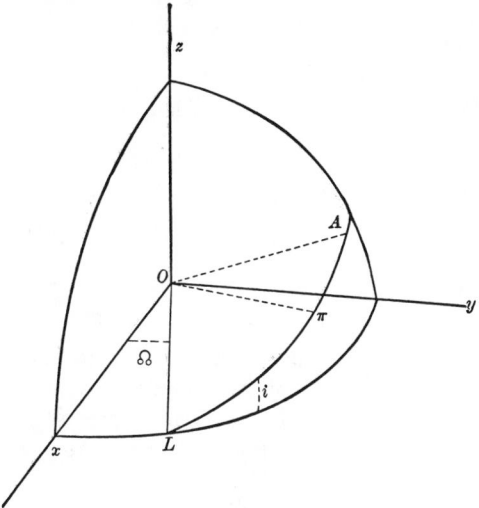

Fig. 25.

the point at which the body m_1 goes from the negative to the positive side of the xy-plane. Let Ω represent the angle between the positive end of the x-axis and the line OL counted in the positive direction from Ox. This angle may have any value from $0°$ to $360°$. Let i represent the inclination between the two planes counted in the direction of positive rotation around OL. The angle i may have any value from $0°$ to $180°$. It is less or greater than $90°$ according as a_1 is positive or negative. Then, when $\varphi = 0$ the value of θ is $\mathcal{B}$ or $\mathcal{B} + \pi$. When $\theta = \mathcal{B} + \frac{1}{2}\pi$ the value of φ is i 'or $\pi - i$ according as i is less than or greater than 90°. In these cases equation (12) becomes respectively

(13)
$$
\begin{cases}
a_2 \cos \mathcal{B} + a_3 \sin \mathcal{B} = 0, \\
a_1 \sin i \mp a_2 \cos i \sin \mathcal{B} \pm a_3 \cos i \cos \mathcal{B} = 0,
\end{cases}
$$

where the signs of the second equation are the upper if i is less than 90°, and the lower if it is greater than 90°.

Since the projections of the areal velocity upon the three fundamental planes are constants (viz., $\frac{1}{2}a_1$, $\frac{1}{2}a_2$, and $\frac{1}{2}a_3$), the areal velocity in the plane of the orbit is also constant. Let this constant be represented by $\frac{1}{2}c_1$; then

(14)
$$
c_1 = \sqrt{a_1{}^2 + a_2{}^2 + a_3{}^2},
$$

where the positive value of the square root is taken. On solving (13) and (14) for a_1, a_2, and a_3, it is found that

(15)
$$
\begin{cases}
a_1 = + c_1 \cos i, \\
a_2 = \pm c_1 \sin i \sin \mathcal{B}, \\
a_3 = \mp c_1 \sin i \cos \mathcal{B},
\end{cases}
$$

where the upper or lower signs are to be taken in the last two equations according as i is less than or greater than 90°; that is, according as a_1 is positive or negative. With this understanding equations (15) uniquely determine i and $\mathcal{B}$, which uniquely determine the position of the plane of the orbit.

87. Problem in the Plane. Since the orbit lies in a known plane, the coördinate axes may be chosen so that the x and y-axes lie in this plane. If the coördinates are represented by x and y as before, the differential equations of motion are

(16)
$$
\begin{cases}
\dfrac{d^2x}{dt^2} = - k^2 M \dfrac{x}{r^3}, \\
\dfrac{d^2y}{dt^2} = - k^2 M \dfrac{y}{r^3}.
\end{cases}
$$

The problem is now of the fourth order instead of the sixth as it was in (9), having been reduced by means of the integrals (10). It will be observed that, since the position of the plane is defined by the two elements $\mathcal{B}$ and i, or by the ratios of a_1, a_2, and a_3 in (11), only two of the arbitrary constants were involved in the reduction. This problem might be solved by deriving the differential equation of the orbit as in Art. 54 and integrating as in Art. 62, the last integral being derived from the integral of areas; but, it is preferable to obtain the results directly by the method which is usually employed in Celestial Mechanics.

Equations (16) give

$$x \frac{d^2y}{dt^2} - y \frac{d^2x}{dt^2} = 0.$$

The integral of this equation is

$$x \frac{dy}{dt} - y \frac{dx}{dt} = c_1,$$

which becomes in polar coördinates

$$(17) \qquad r^2 \frac{d\theta}{dt} = c_1.$$

Let A represent the area swept over by the radius vector r; then

$$2 \frac{dA}{dt} = r^2 \frac{d\theta}{dt} = c_1;$$

whence

$$(18) \qquad 2A = c_1 t + c_2,$$

from which it follows that the areas swept over by the radius vector are proportional to the times in which they are described.

On multiplying (16) by $2 \frac{dx}{dt}$ and $2 \frac{dy}{dt}$ respectively, and adding, the result is

$$2 \frac{d^2x}{dt^2} \frac{dx}{dt} + 2 \frac{d^2y}{dt^2} \frac{dy}{dt} = -2 \frac{k^2M}{r^3} \left(x \frac{dx}{dt} + y \frac{dy}{dt} \right) = -\frac{2k^2M}{r^2} \frac{dr}{dt}.$$

The integral of this equation is

$$(19) \qquad \left(\frac{dx}{dt} \right)^2 + \left(\frac{dy}{dt} \right)^2 = \frac{2k^2M}{r} + c_3.$$

This equation, which involves only the square of the velocity and the distance, is known as the *vis viva* integral (Art. 52). On transforming the left member to polar coördinates, this equation becomes

$$\left(\frac{dr}{dt} \right)^2 + r^2 \left(\frac{d\theta}{dt} \right)^2 = \frac{2k^2M}{r} + c_3.$$

But

$$\frac{dr}{dt} = \frac{dr}{d\theta} \frac{d\theta}{dt};$$

therefore

$$\left(\frac{d\theta}{dt}\right)^2 \left\{ \left(\frac{dr}{d\theta}\right)^2 + r^2 \right\} = \frac{2k^2 M}{r} + c_3.$$

On eliminating $\dfrac{d\theta}{dt}$ by means of (17), this equation gives

$$d\theta = \frac{c_1\, dr}{r \ \sqrt{-\, c_1{}^2 + 2k^2 Mr + c_3 r^2}},$$

which may be written in the form

$$(20) \qquad d\theta = \frac{-\, d\left(\dfrac{c_1}{r}\right)}{\sqrt{c_3 + \dfrac{k^4 M^2}{c_1{}^2} - \left(\dfrac{k^2 M}{c_1} - \dfrac{c_1}{r}\right)^2}}.$$

Let B^2 and u be defined by

$$\begin{cases} c_3 + \dfrac{k^4 M^2}{c_1{}^2} = B^2, \\[2mm] \dfrac{k^2 M}{c_1} - \dfrac{c_1}{r} = -\, u, \end{cases}$$

in which B^2 must be positive for a real orbit; then (20) becomes

$$d\theta = \frac{-\, du}{\sqrt{B^2 - u^2}}.$$

The integral of this equation is

$$\theta = \cos^{-1}\frac{u}{B} + c_4.$$

On changing from u, B, and c_4 to r and the original constants, it is found that

$$(21) \qquad r = \frac{c_1}{\dfrac{k^2 M}{c_1} + \sqrt{c_3 + \dfrac{k^4 M^2}{c_1{}^2}} \cos\left(\theta - c_4\right)},$$

which is the polar equation of a conic section with the origin at one of its foci.

88. The Elements in Terms of the Constants of Integration. The node and inclination are expressed in terms of the constants of integration by (15).

The ordinary equation of a conic section with the origin at the right-hand focus is

$$r = \frac{p}{1 + e \cos\left(\theta - \omega\right)},$$

where p is the semi-parameter, and ω is the angle between the polar axis and the major axis of the conic. On comparing this equation with (21), it is found that

(22)
$$
\begin{cases}
p = \dfrac{c_1^2}{k^2 M}, \\[2mm]
e^2 = 1 + \dfrac{c_1^2 c_3}{k^4 M^2}, \\[2mm]
\omega = c_4 - \pi; \\[2mm]
c_1 = k\sqrt{Mp}, \\[2mm]
c_3 = -\dfrac{k^2(1-e^2)}{p} M.
\end{cases}
$$

When $e^2 < 1$, the orbit is an ellipse and $p = a(1 - e^2)$, where a is the major semi-axis; when $e^2 = 1$, the orbit is a parabola and $p = 2q$, where q is the distance from the origin to the vertex of the parabola; and when $e^2 > 1$, the orbit is an hyperbola and $p = a(e^2 - 1)$.

Let A_0 represent the area described at the time the body passes perihelion;[*] then the time of perihelion passage is found from equation (18) to be

$$
(23) \qquad\qquad T = \frac{2A_0 - c_2}{c_1}.
$$

This completes the determination of the elements in terms of the constants of integration. They are defined in terms of the initial coördinates and components of velocity by the equations where they first occur, viz., (10), (17), (18), (19), and (21).

89. Properties of the Motion. Suppose the orbit is an ellipse. Then, when the values of the constants of integration given in (22) are substituted in (17) and (19), these equations become

(24)
$$
\begin{cases}
r^2 \dfrac{d\theta}{dt} = k\sqrt{Ma(1 - e^2)}, \\[3mm]
\left(\dfrac{dx}{dt}\right)^2 + \left(\dfrac{dy}{dt}\right)^2 = V^2 = k^2 M\left(\dfrac{2}{r} - \dfrac{1}{a}\right),
\end{cases}
$$

where V is the speed in the orbit at the distance r from the origin.

When the orbit is a circle, $r = a$ and

---

[*] Unless m_2 is specified to be some body other than the sun the nearest apse will be called the perihelion point.

$$V_c{}^2 = \frac{k^2 M}{r}.$$

When the orbit is a parabola, $a = \infty$ and

$$V_p{}^2 = \frac{2k^2 M}{r}.$$

Therefore, at a given distance from the origin the ratio of the speed in a parabolic orbit to that in a circular orbit is

(25) $$V_p : V_c = \sqrt{2} : 1.$$

Thus, in the motion of comets around the sun they cross the planets' orbits with velocities about 1.414 times those with which the respective planets move.

The speed that a body will acquire in falling from the distance s to the distance r toward the center of force $k^2 M$ is given by (see Art. 35)

$$V^2 = 2k^2 M \left(\frac{1}{r} - \frac{1}{s} \right).$$

If s is determined by the condition that this shall equal the speed in the orbit, it is found, after equating the right member of this

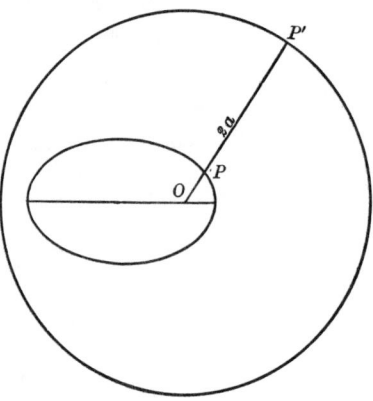

Fig. 26.

expression to the right member of the second of (24), that $s = 2a$ and

(26) $$V^2 = 2k^2 M \left(\frac{1}{r} - \frac{1}{2a} \right).$$

Therefore, *the speed of a body moving in an ellipse is at every*

point equal to that which it would acquire in falling from the circum-
ference of a circle, with center at the origin and radius equal to the
major axis of the conic, to the ellipse.

The speed at P in the ellipse is equal to that which would be acquired in falling from P' to P.

Equation (26) gives an interesting conclusion about the possible motion of m_1 on the basis of this equation alone, and without making any use of the detailed properties of motion in a conic section. Since the left member is necessarily positive (or zero) r can take only such values that the right member shall be positive (or zero). Consequently $r \leq 2a$ in all the motion whatever it may be. This result is trivial in this simple case in which all the circumstances of motion are fully known, but the corresponding discussion in the *Problem of Three Bodies* (Chap. VIII.) gives valuable information which has not been otherwise obtained.

Consider the second equation of (24) and suppose the body m_1 is projected from a point which is distant r from the body m_2. It follows at once that *the major axis of the conic depends upon the initial distance from the origin and the initial speed, but not upon the direction of projection.* If $V^2 < \dfrac{2k^2 M}{r} = U^2$, which is the velocity the body m_1 would acquire in falling from infinity, a is positive and the orbit is an ellipse; if $V^2 = U^2$, a is infinite and the orbit is a parabola; if $V^2 > U^2$, a is negative and the orbit is an hyperbola.

Let t_1 and t_2 be two epochs, and A_1 and A_2 the corresponding values of the area described by the radius vector. Then equation (18) gives

$$2(A_2 - A_1) = (t_2 - t_1)c_1.$$

Suppose $t_2 - t_1 = P$, the period of revolution; then $2(A_2 - A_1)$ equals twice the area of the ellipse, which equals $2\pi ab$. The expression for the period, found by substituting the value of c_1 given in (22) and solving, is

$$(27) \qquad\qquad P = \frac{2\pi a^{\frac{3}{2}}}{k\sqrt{M}}.$$

From this equation it follows that *the period is independent of every element except the major axis;* or, because of (26), the period depends only upon the initial distance from the origin and the initial speed, and not upon the direction of projection. The major semi-axis will be called the *mean distance*, although it must be understood that it is *not* the *average* distance when the time is used as the independent variable. (See Probs. 4 and 5, p. 154.)

The three orbits drawn in Fig. 27 have the same length of major axis and are consequently described in the same time. The speed of projection from A is the same in each case, the differences in the shapes and positions resulting from the different directions of projection.

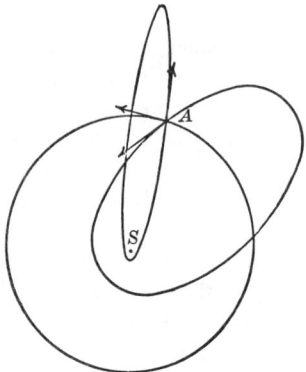

Fig. 27.

If the two systems m_1, m_2, and m_2, m_3 are considered, and the ratio of their periods is taken, it is found that

$$\frac{P^2_{1,\,2}}{P^2_{3,\,2}} = \frac{a^3_{1,\,2}}{a^3_{3,\,2}} \cdot \frac{M_{3,\,2}}{M_{1,\,2}}.$$

If the two systems are composed of the sun and two planets respectively, then $M_{1,\,2}$ and $M_{3,\,2}$ are very nearly equal because the masses of the planets are exceedingly small compared to that of the sun. Therefore, this equation becomes very nearly

$$\frac{P^2_{1,\,2}}{P^2_{3,\,2}} = \frac{a^3_{1,\,2}}{a^3_{3,\,2}};$$

or, *the squares of the periodic times of the planets are proportional to the cubes of their mean distances.* This is Kepler's third law.

It is to be observed that, in taking the ratios of the periods, it was assumed that k has the same value for the different planets; that is, that the sun's acceleration of the two planets would be the same at unit distance. On the other hand, it follows from the last equation, which Kepler established directly by observations, that k has the same value for the various planets. This means that the force of gravitation between the sun and the several planets is proportional to their respective masses, as measured by their inertias. This result is not self-evident for the force of gravitation conceivably might depend upon the chemical constitution or physical condition of a body, just as chemical affinity, magnetism and all other known forces depend upon one or both of these things. In fact, it is remarkable that gravitation is proportional to inertia and independent of everything else.

90. Selection of Units and the Determination of the Constant k. When the units of time, mass, and distance are chosen k can be determined from (27). It is evident that they can all be taken arbitrarily, but it will be convenient to employ those units in which astronomical problems are most frequently treated. The mean solar day will be taken as the unit of time; the mass of the sun will be taken as the unit of mass; and the major semi-axis of the earth's orbit will be taken as the unit of distance. When these units are employed the k determined by them is called the Gaussian constant, having been defined in this way by Gauss in the *Theoria Motus*, Art. 1.

Let m_2 represent the mass of the sun and m_1 that of the earth and moon together; then it has been found from observation that in these units

$$(28) \qquad \begin{cases} m_1 = \dfrac{m_2}{354710} = \dfrac{1}{354710}, \\ P = 365.2563835. \end{cases}$$

On substituting these numbers in (27), it is found that

$$(29) \qquad \begin{cases} k = \dfrac{2\pi}{P\sqrt{1 + m_1}} = 0.01720209895, \\ \log k = 8.2355814414 - 10. \end{cases}$$

Since m_1 is very small $k = \dfrac{2\pi}{P}$ nearly, and is, therefore, nearly the mean daily motion of the earth in its orbit, or about $\frac{1}{60}$. The mean daily motion of a planet whose mass is m_i is $\dfrac{2\pi}{P_i}$, and is usually designated by n_i. This is found from (27) to be

$$(30) \qquad n_i = \dfrac{k\sqrt{1 + m_i}}{a_i^{\frac{3}{2}}}.$$

The period of the earth's revolution around the sun and its mean distance were not known with perfect exactness at the time of Gauss, nor are they yet, and it is clear that the value of k varies with the different determinations of these quantities. If astronomers held strictly to the definitions of the units given above it would be necessary to recompute those tables which depend upon k every time an improvement in the values of the constants is made. These inconveniences are avoided by keeping the numerical value of k that which Gauss determined, and choosing the unit of distance so that (27) will always be fulfilled.

If the mean distance between two bodies is taken as the unit of distance and the sum of their masses as the unit of mass, and if the unit of time is taken so that k equals unity, then the units form what is called a *canonical system*. Since $M = 1$ and $k^2 = 1$ in this system, and from (30) $n = 1$, the equations become somewhat simplified and are advantageous in purely theoretical investigations.

XIV. PROBLEMS.

1. Find the differential equations for the problem of the relative motion of two bodies in polar coördinates.

$Ans.$
$$\frac{d^2r}{dt^2} = r\left(\frac{d\theta}{dt}\right)^2 - \frac{k^2M}{r^2}, \qquad \frac{d}{dt}\left(r^2\frac{d\theta}{dt}\right) = 0.$$

2. Integrate the equations of problem 1 and interpret the constants of integration.

3. The earth moves in its orbit, which may be assumed to be circular, with a speed of 18.5 miles per second. Suppose the meteors approach the sun in parabolas; between what limits will be their relative speed when they strike into the earth's atmosphere?

$Ans.$ 7.66 to 44.66 miles per second. (The Nov. 14 meteors meet the earth and have a relative speed near the upper limit; the Nov. 27 meteors overtake the earth and have a relative speed near the lower limit.)

4. Find the average length of the radius vector of an ellipse in terms of a and e, taking the time as the independent variable.

$Ans.$
$$\text{Average } r = \frac{\int r\,dt}{\int dt} = a\left(1 + \frac{e^2}{2}\right).$$

5. Find the average length of the radius vector of an ellipse, taking the angle as the independent variable.

$Ans.$
$$\text{Average } r = \frac{\int r\,d\theta}{\int d\theta} = \frac{2\pi a\sqrt{1 - e^2}}{2\pi} = b.$$

6. Prove that the amount of heat received from the sun by the planets per unit area is on the average proportional to the reciprocals of the products of the major and minor axes of their orbits. For a fixed major axis how does the total amount of heat received in a revolution depend upon the eccentricity of the orbit?

7. If particles are projected from a given point with a given velocity but in different directions, find the locus of (a) perihelion points; (b) aphelion points; (c) centers of ellipses; (d) ends of minor axes.

8. If particles are projected from a given point in a given direction but with different speeds, find the loci of the same points as in problem 7, and express the coördinates of these points in terms of the initial values of the coördinates and the components of velocity.

9. Suppose a comet moving in a parabolic orbit with perihelion distance q collides with and combines with an equal mass which is at rest before the collision. Find the eccentricity and the perihelion distance of the orbit of the combined mass.

10. Suppose the mass of Jupiter is $1/1047$ when expressed in terms of the mass of the sun, and that its mean distance from the sun is 483,300,000 miles (the mean distance from the earth to the sun is 92,900,000 miles). Find Jupiter's period of revolution around the sun, and the size of the orbit which the sun describes with respect to the center of gravity of itself and Jupiter.

91. Position in Parabolic Orbits. Having found the curves in which the bodies move, it remains to find their positions in their orbits at any given epoch. The case of the parabolic orbit being the simplest will be considered first, and it will be supposed, to fix the ideas, that the motion is that of a comet with respect to the sun. Since the masses of the comets are negligible, $M = 1$ and equation (17) becomes

$$(31) \qquad r^2 \frac{d\theta}{dt} = k \sqrt{p} = k \sqrt{2q}.$$

When the polar angle in the orbit is counted from the vertex of the parabola it is denoted by v, and is called the *true anomaly*. Then

$$\begin{cases} \dfrac{d\theta}{dt} = \dfrac{dv}{dt}, \\[2mm] r = \dfrac{p}{1 + \cos v} = q \sec^2 \dfrac{v}{2}. \end{cases}$$

Hence, equation (31) gives

$$\frac{\sqrt{2}k}{q^{\frac{3}{2}}} \, dt = \sec^4 \frac{v}{2} \, dv = \left(\sec^2 \frac{v}{2} + \sec^2 \frac{v}{2} \tan^2 \frac{v}{2} \right) dv.$$

The integral of this expression is

$$(32) \qquad \tan\frac{v}{2} + \frac{1}{3}\tan^3\frac{v}{2} = \frac{k(t-T)}{\sqrt{2}q^{\frac{3}{2}}},$$

where T is the time of perihelion passage. This is a cubic equation in $\tan\frac{v}{2}$. On taking the right member to the left side it is seen that for $t - T > 0$, the function is negative when $v = 0$, and that it increases continually with v until it equals infinity for $v = 180°$. Therefore, there is but one real solution of (32) for $\tan\frac{v}{2}$, and it is positive. For $t - T < 0$ it is seen in a similar manner that there is one real negative solution.

Equation (32) may be written

$$25\tan^3\frac{v}{2} + 75\tan\frac{v}{2} = \frac{75k}{\sqrt{2}}\frac{(t-T)}{q^{\frac{3}{2}}} = K\frac{(t-T)}{q^{\frac{3}{2}}}.$$

Tables have been constructed giving the value of the right member of this equation for different values of v. From these tables v can be found by interpolation when $t - T$ is given; or, conversely, $t - T$ can be found when v is given. These tables are known as Barker's, and are VI. in Watson's *Theoretical Astronomy*, and IV. in Oppolzer's *Bahnbestimmung*.*

In order to find the direct solution of the cubic equation let

$$\tan\frac{v}{2} = 2\cot 2w = \cot w - \tan w;$$

whence

$$\tan^3\frac{v}{2} = -3\tan\frac{v}{2} + \cot^3 w - \tan^3 w.$$

This substitution reduces (32) to

$$\cot^3 w - \tan^3 w = \frac{3k(t-T)}{\sqrt{2}q^{\frac{3}{2}}}.$$

Let

$$\cot w = \sqrt[3]{\cot\frac{s}{2}};$$

whence

$$\cot s = \frac{3k(t-T)}{2^{\frac{3}{2}}q^{\frac{3}{2}}}.$$

Therefore the formulas for the computation of $\tan\frac{v}{2}$ are, in the

* In Oppolzer's *Bahnbestimmung* the factor 75 is not introduced.

order of their application,

$$
(33)
\begin{cases}
\cot\ s = \dfrac{3k(t - T)}{(2q)^{\frac{3}{2}}}, \\[2mm]
\cot w = \sqrt[3]{\cot \dfrac{s}{2}}, \\[2mm]
\tan \dfrac{v}{2} = 2 \cot 2w.
\end{cases}
$$

After v has been found r is determined by the polar equation of the parabola, $r = \dfrac{p}{1 + \cos v} = q \sec^2 \dfrac{v}{2}$.

92. Equation involving Two Radii and their Chord. Euler's Equation. Consider the positions of the comet at the instants t_1 and t_2. Let the corresponding radii be r_1 and r_2, and the chord joining their extremities s. Let the corresponding true anomalies be v_1 and v_2. Then it follows that

$$
\begin{cases}
\dfrac{k(t_1 - T)}{\sqrt{2}q^{\frac{3}{2}}} = \tan \dfrac{v_1}{2} + \dfrac{1}{3}\tan^3 \dfrac{v_1}{2}, \\[2mm]
\dfrac{k(t_2 - T)}{\sqrt{2}q^{\frac{3}{2}}} = \tan \dfrac{v_2}{2} + \dfrac{1}{3}\tan^3 \dfrac{v_2}{2}.
\end{cases}
$$

The difference of these equations is

$$
\frac{k(t_2 - t_1)}{\sqrt{2}q^{\frac{3}{2}}} = \tan \frac{v_2}{2} - \tan \frac{v_1}{2} + \frac{1}{3}\left(\tan^3\frac{v_2}{2} - \tan^3 \frac{v_1}{2} \right);
$$

or,

$$
(34)\quad \frac{3k(t_2 - t_1)}{\sqrt{2}q^{\frac{3}{2}}} = \left(\tan \frac{v_2}{2} - \tan \frac{v_1}{2} \right) \left[3 \left(1 + \tan \frac{v_1}{2} \tan \frac{v_2}{2} \right) \right.
$$

$$
\left. + \left(\tan \frac{v_2}{2} - \tan \frac{v_1}{2} \right)^2 \right].
$$

The equation for the chord is

$$
s^2 = r_1{}^2 + r_2{}^2 - 2r_1 r_2 \cos (v_2 - v_1)
$$

$$
= (r_1 + r_2)^2 - 4 r_1 r_2 \cos^2 \left(\frac{v_2 - v_1}{2} \right).
$$

From this equation it is found that

$$
(35)\quad 2\sqrt{r_1 r_2}\ \cos \left(\frac{v_2 - v_1}{2} \right) = \pm \sqrt{(r_1 + r_2 + s)(r_1 + r_2 - s)}.
$$

The $+$ sign is to be taken before the radical if $v_2 - v_1 < \pi$, and the $-$ sign if $v_2 - v_1 > \pi$.

It follows from the polar equation of the parabola that

$$r_1 = q \sec^2 \frac{v_1}{2}, \qquad r_2 = q \sec^2 \frac{v_2}{2}.$$

These expressions for r_1 and r_2, substituted in (35), give

$$(36) \quad 1 + \tan \frac{v_1}{2} \tan \frac{v_2}{2} = \pm \frac{\sqrt{(r_1 + r_2 + s)(r_1 + r_2 - s)}}{2q}.$$

It also follows from the expressions for r_1 and r_2 that

$$r_1 + r_2 = q \left(2 + \tan^2 \frac{v_1}{2} + \tan^2 \frac{v_2}{2} \right).$$

The last two equations give

$$\frac{(r_1 + r_2 + s) + (r_1 + r_2 - s) \mp 2\sqrt{(r_1 + r_2 + s)(r_1 + r_2 - s)}}{2q}$$
$$= \left(\tan \frac{v_2}{2} - \tan \frac{v_1}{2} \right)^2;$$

whence

$$(37) \quad \frac{\sqrt{r_1 + r_2 + s} \mp \sqrt{r_1 + r_2 - s}}{\sqrt{2q}} = \tan \frac{v_2}{2} - \tan \frac{v_1}{2}.$$

Equation (34) becomes, as a consequence of (36) and (37),

$$(38) \qquad 6k(t_2 - t_1) = (r_1 + r_2 + s)^{\frac{3}{2}} \mp (r_1 + r_2 - s)^{\frac{3}{2}}.$$

This equation is remarkable in that it does not involve q. It was discovered by Euler and bears his name. It is of the first importance in some methods of determining the elements of a parabolic orbit from geocentric observations.

There is a corresponding equation, due to Lambert, for elliptic orbits. The right member is developed as a power series in $1/a$, the first term constituting the right member of Euler's equation.

93. Position in Elliptic Orbits. The integral of areas and the vis viva integral are respectively

$$\begin{cases} r^2 \dfrac{dv}{dt} = k \sqrt{(1 + m)a(1 - e^2)}, \\ \left(\dfrac{dr}{dt} \right)^2 + r^2 \left(\dfrac{dv}{dt} \right)^2 = k^2(1 + m) \left(\dfrac{2}{r} - \dfrac{1}{a} \right). \end{cases}$$

The result of eliminating $\dfrac{dv}{dt}$ from the second of these equations by means of the first is

$$(39) \quad \left(\frac{dr}{dt}\right)^2 + \frac{k^2(1+m)a(1-e^2)}{r^2} = k^2(1+m)\left(\frac{2}{r} - \frac{1}{a}\right).$$

Let n represent the mean angular motion of the body in its orbit; then

$$n = \frac{2\pi}{P} = \frac{k\sqrt{1+m}}{a^{\frac{3}{2}}}.$$

On introducing n in (39) and solving, it is found that

$$(40) \qquad n\,dt = \frac{r}{a}\frac{dr}{\sqrt{a^2e^2 - (a-r)^2}}.$$

In order to normalize the integral which appears in the right member of (40), let the auxiliary E be introduced by the equation

$$(41) \qquad \begin{cases} a - r = ae\cos E, & \text{whence} \\ \quad r = a(1 - e\cos E). \end{cases}$$

This angle E is called the *eccentric anomaly*. Then (40) becomes

$$n\,dt = (1 - e\cos E)dE,$$

the integral of which is

$$n(t - T) = E - e\sin E.$$

The quantity $n(t - T)$ is the angle which would have been described by the radius vector if it had moved uniformly with the average rate. It is usually denoted by M and is called the *mean anomaly*. Therefore

$$(42) \qquad n(t - T) = M = E - e\sin E.$$

The M can at once be found when $(t - T)$ is given, after which equation (42) must be solved for E. Then r and v can be found from (41) and the polar equation of the ellipse. Equation (42), known as *Kepler's equation*, is transcendental in E, and the solution for this quantity cannot be expressed in a finite number of terms. Since it is very desirable to have the solution as short as possible astronomers have devoted much attention to this equation, and several hundred methods of solving it have been discovered.

94. Geometrical Derivation of Kepler's Equation. Construct the ellipse in which the body moves, and also its auxiliary circle AQB. The angle AFP equals the true anomaly, v; the angle

ACQ will be defined as the eccentric anomaly, E, and it will be shown that the relation between M and E is given by Kepler's equation.

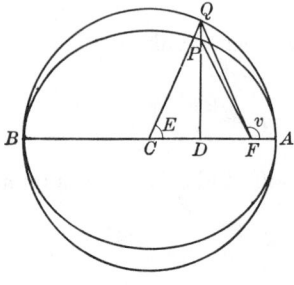

Fig. 28.

From the law of areas and the properties of the auxiliary circle, it follows that

$$\frac{M}{2\pi} = \frac{\text{area } AFP}{\text{area ellipse}} = \frac{\text{area } AFQ}{\text{area circle}}.$$

Area $AFQ = \text{area } ACQ - \text{area } FCQ = \dfrac{a^2 E}{2} - \dfrac{a}{2}\,ae \sin E.$

Therefore

$$\frac{M}{2\pi} = \frac{a^2}{2}\,\frac{(E - e \sin E)}{\pi a^2};$$

or,

$$\begin{cases} M = E - e \sin E, \\[2mm] FP = r = \dfrac{a(1 - e^2)}{1 + e \cos v} = \sqrt{\overline{PD}^2 + \overline{FD}^2} = a(1 - e \cos E), \end{cases}$$

which is the definition of the eccentric anomaly given in (41).

95. Solution of Kepler's Equation. It will be shown first that Kepler's equation always has one, and only one, real solution for every value of M and for every e such that $0 \lessgtr e < 1$. Write the equation in the form

$$\phi(E) \equiv E - e \sin E - M = 0.$$

Suppose M has some given value between $n\pi$ and $(n + 1)\pi$, where n is any integer; then there is but one real value of E satisfying this equation, and it lies between $n\pi$ and $(n + 1)\pi$. For, the function $\phi(E)$ when $E = n\pi$ is

$$\phi(n\pi) = n\pi - M < 0.$$

And $\phi(E)$ when $E = (n + 1)\pi$ is

$$\phi[(n + 1)\pi] = (n + 1)\pi - M > 0.$$

Consequently there is an odd number of real solutions for E which lie between $n\pi$ and $(n + 1)\pi$. But the derivative

$$\phi'(E) \equiv 1 - e \cos E$$

is always positive; therefore $\phi(E)$ increases continually with E and takes the value zero but once.

A convenient method of practically solving the equation is by means of an expansion due to Lagrange. Suppose z is defined as a function of w by the equation

$$(43) \qquad\qquad z = w + \alpha\phi(z),$$

where α is a parameter. Lagrange has shown that any function of z can be expressed in a power series in α, which converges for sufficiently small values of α, of the form*

$$(44) \quad \left\{ \begin{array}{l} F(z) = F(w) + \dfrac{\alpha}{1} \cdot \phi(w)F'(w) + \dfrac{\alpha^2}{1 \cdot 2} \dfrac{\partial}{\partial w}[\{\phi(w)\}^2 F'(w)] \\[2mm] \qquad + \cdots + \dfrac{\alpha^{n+1}}{(n + 1)!} \dfrac{\partial^n}{\partial w^n}[\{\phi(w)\}^{n+1} F'(w)] + \cdots. \end{array} \right.$$

This expansion can be applied to the solution of Kepler's equation by writing it

$$E = M + e \sin E,$$

which is of the same form as (43). The expansion of E as a series in e can be taken from (44) by putting $F(z) = E$, $\phi(z) = \sin E$, $w = M$, and $\alpha = e$. The result is

$$(45) \qquad E = M + \dfrac{e}{1} \sin M + \dfrac{e^2}{1 \cdot 2} \sin 2M + \cdots.$$

All the terms on the right except the first are expressed in radians and must be reduced to degrees by multiplying each of them by the number of degrees in a radian. The higher terms are considerably more complicated than those written, and the work of computing them increases very rapidly. In the planetary and satellite orbits the eccentricity is very small, and the series (45) converges with great rapidity, the first three terms giving quite an accurate value of E.

* Williamson's *Diff. Calc.*, p. 151.

96. Differential Corrections. A method will now be explained in one of its simplest applications, which is of great importance in many astronomical problems. Suppose an approximate value of E is determined by the first three terms of (45). Call it E_0; it is required to find the correct value of E.

Kepler's equation gives

$$M_0 = E_0 - e \sin E_0.$$

For a particular value of M, viz., M_0, the corresponding value of E, viz., E_0, is known. It is required to find the value of E corresponding to M, which differs only a little from M_0. The angle M is a function of E and may be written

$$M = M_0 + \Delta M_0 = f(E_0 + \Delta E_0).$$

On expanding the right member by Taylor's formula, this equation becomes

$$M = M_0 + \Delta M_0 = f(E_0) + f'(E_0)\Delta E_0 + \cdots.$$

By the definitions of the quantities, $M_0 = f(E_0)$; therefore this equation becomes

(46) $M - M_0 = f'(E_0)\Delta E_0 + \cdots = (1 - e \cos E_0)\Delta E_0 + \cdots.$

Since ΔE_0 is very small the squares and higher powers may be neglected,* and then equation (46) gives for the correction to be applied to E_0

(47) $$\Delta E_0 = \frac{M - M_0}{1 - e \cos E_0}.$$

With the more nearly correct value of E, $E_1 = E_0 + \Delta E_0$, and M_1 can be computed from Kepler's equation, and a second correction will be

$$\Delta E_1 = \frac{M - M_1}{1 - e \cos E_1}.$$

This process can be repeated until the value of E is found as near as may be desired.† In the planetary orbits two applications of

---

* If the higher terms in ΔE_0 were not neglected ΔE_0 could be expressed as a power series in $M - M_0$, of which the first term would be the right member of (47).

† For the proof of the convergence of a similar, but somewhat more laborious, process see Appell's *Mécanique* vol. i., p. 391.

the formulas will nearly always give results which are sufficiently accurate, and usually one correction will suffice.

97. Graphical Solution of Kepler's Equation. When the eccentricity is more than 0.2 the method of solving Kepler's equation given above is laborious because the first approximation will be very inexact. These high eccentricities occur in binary star and comet orbits, and are sometimes even so great as 0.9. In the case of binary star orbits it is usually sufficient to have a

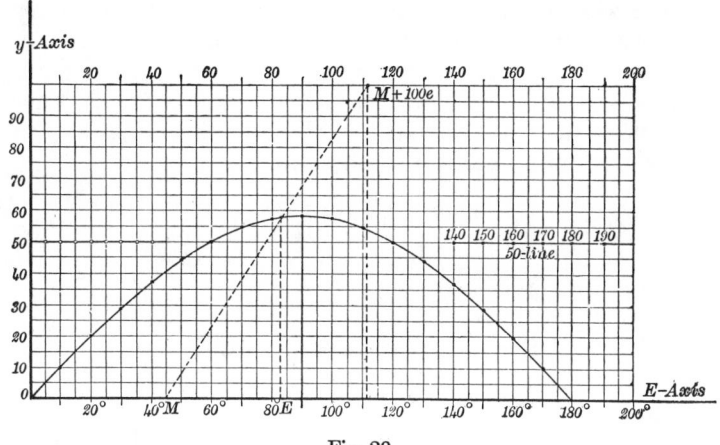

Fig. 29.

solution to within a tenth of one degree. In this work a rapid graphical method is of great practical value.

Consider Kepler's equation

$$E - e \sin E - M = 0,$$

where M is given and E is required. Take a rectangular system of axes and construct the sine curve and the straight line whose equations are

$$\begin{cases} y = \sin E, \\ y = \dfrac{1}{e}(E - M). \end{cases}$$

The abscissa of their point of intersection is the value of E satisfying the equation;* for, eliminating y, Kepler's equation results. The first curve is the familiar sine curve which can be constructed

* Due to J. J. Waterson, *Monthly Notices*, 1849–50, p. 169.

once for all; the second is a straight line making with the E-axis an angle whose tangent is $1/e$. Instead of drawing the straight line a straight-edge can be laid down making the proper slope with the axis. To facilitate the determination of its position construct a line with the degrees marked on it at an altitude of 100;* then place the bottom of the straight-edge at M and the top at $M + 100e$, and it follows that it will have the proper slope. If M is so near 180° that the straight-edge runs off from the diagram, the top can be placed at $M + 50e$ on the 50-line. As M becomes very near 180° the mean and eccentric anomalies become very nearly equal, exactly coinciding at $M = 180°$.

98. Recapitulation of Formulas. The equations for the computation of the polar coördinates, when the time is given, will now be given in the order in which they are used.

$$(48) \begin{cases} n = \dfrac{k\sqrt{1+m}}{a^{\frac{3}{2}}}, \\[2mm] M = n(t - T), \\[2mm] E_0 = M + e\sin M + \dfrac{e^2}{2}\sin 2M, \\[2mm] M_0 = E_0 - e\sin E_0, \\[2mm] \Delta E_0 = \dfrac{M - M_0}{1 - e\cos E_0}, \\[2mm] E_1 = E_0 + \Delta E_0, \\[2mm] r = a(1 - e\cos E) = \dfrac{a(1 - e^2)}{1 + e\cos v}; \end{cases}$$

whence

$$(49) \begin{cases} \cos v = \dfrac{\cos E - e}{1 - e\cos E}, \\[2mm] \sin v = \dfrac{\sqrt{1 - e^2}\sin E}{1 - e\cos E}, \\[2mm] 1 + \cos v = \dfrac{(1 - e)(1 + \cos E)}{1 - e\cos E}, \\[2mm] 1 - \cos v = \dfrac{(1 + e)(1 - \cos E)}{1 - e\cos E}. \end{cases}$$

* This device is due to C. A. Young.

The square root of the quotient of the last two equations gives a very convenient formula for the computation of v, viz.,

$$(50) \qquad \sqrt{\frac{1 - \cos v}{1 + \cos v}} = \tan \frac{v}{2} = \sqrt{\frac{1 + e}{1 - e}} \tan \frac{E}{2}.$$

The last equation of (48) and equation (50) give the polar co-ordinates when E is known.

99. The Development of E in Series. The equations which have been given are sufficient to enable one to compute the polar, and consequently the rectangular, coördinates at any epoch; yet, in some kinds of investigations, as in the theory of perturbations, it is necessary to have the developments of not only E, but also the polar coördinates, carried so far that the functions are represented by the series with the desired degree of accuracy.

The application of Lagrange's method of Art. 95 to Kepler's equation gives E as a power series in e whose coefficients are functions of M. This method has been used to get the first terms of the series and it can be continued as far as may be desired. It is very elegant in practice and is subject only to the difficulty of proving its legitimacy. But a direct treatment of Kepler's equation based on more elementary considerations is not difficult. The solution of

$$M = E - e \sin E$$

for E is $j\pi$ when $M = j\pi$, where $j = 0, 1, 2, \cdots$, whatever value e may have. Moreover, it has been shown that when e is less than unity the solution is unique for all values of M. When $e = 0$ the solution is $E = M$ for all values of M. If u is defined by the equation

$$E - M = u$$

Kepler's equation becomes

$$(51) \qquad u = e \sin (M + u),$$

which defines u in terms of M and e. For every value of M different from $j\pi$, for which the solution is already known, the right member of

$$e = \frac{u}{\sin (M + u)}$$

can be expanded as a converging power series in u. When this series is inverted u will be given as a power series in e whose coefficients are functions of M. Since u vanishes with e, it will have the form

(52) $$u = u_1 e + u_2 e^2 + u_3 e^3 + \cdots.$$

Instead of forming the series in u and then inverting, it is simpler to substitute (52) in (51) and to determine $u_1, u_2, \cdots$ by the condition that the result shall be an identity in e. The result of the substitution is

$$
\begin{cases}
u_1 e + u_2 e^2 + u_3 e^3 + \cdots = e \sin M \cos u + e \cos M \sin u \\[2mm]
\quad = e \sin M \left[1 - \dfrac{(u_1 e + u_2 e^2 \cdots)^2}{2!} + \dfrac{(u_1 e + \cdots)^4}{4!} - \cdots \right] \\[2mm]
\quad + e \cos M \left[(u_1 e + u_2 e^2 + \cdots) - \dfrac{(u_1 e \cdots)^3}{3!} + \cdots \right].
\end{cases}
$$

On equating coefficients of corresponding powers of e, it is found that

$$
(53)\quad
\begin{cases}
u_1 = \sin M, \\[2mm]
u_2 = u_1 \cos M = \dfrac{1}{2} \sin 2M, \\[2mm]
u_3 = -\dfrac{1}{2} u_1^2 \sin M + u_2 \cos M = \dfrac{3}{8} \sin 3M - \dfrac{1}{8} \sin M, \\[2mm]
\cdot \quad \cdot \quad \cdot \quad \cdot \quad \cdot \quad \cdot \quad \cdot \quad \cdot \quad \cdot \quad \cdot \quad \cdot \quad \cdot \quad \cdot
\end{cases}
$$

Some general properties of the solutions easily follow from these equations. It follows from (51) that if for any $M = M_0$ the solution for u, which is known to exist uniquely, is $u = u_0$, then the solution for $M = M_0 + 2j\pi$ (j any integer) is also $u = u_0$. Therefore u is a periodic function of M with the period 2π. Since this is true for all values of e, each u_j is separately periodic with the period 2π. If any M_0 and u_0 satisfy (51), then $-M_0$ and $-u_0$ also satisfy (51); therefore u is an odd function of M and the u_j are sines of multiples of M. If the sign of e is changed and π is added to M in (51), the equation is unchanged; therefore the u_j with odd subscripts involve only sines of odd multiples of M, and those with even subscripts only sines of even multiples of M.

It will be shown that the highest multiple of M appearing in u_j is jM. The general term of (53) is

$$u_j = \sin M \, P_j(u_1, u_2, \cdots, u_{j-1}) + \cos M \, Q_j(u_1, u_2, \cdots, u_{j-1}),$$

where P_j and Q_j are polynomials in $u_1, u_2, \cdots, u_{j-1}$. These quantities must enter in such powers that they are multiplied by e^{j-1}. Suppose the general terms of the polynomials P_j and Q_j

are, except for numerical coefficients which do not enter into the present argument, respectively

$$p_j = u_1^{j_1} \cdot u_2^{j_2} \cdots u_{j-1}^{j_{j-1}}, \qquad q_j = u_1^{k_1} \cdot u_2^{k_2} \cdots u_{j-1}^{k_{j-1}}.$$

The exponents of $u_1, \cdots, u_{j-1}$ are subject to the condition that p_j and q_j shall be multiplied by e^{j-1}. The term u_m carries with it the factor e^m, and therefore u_m^n carries the factor mn. Hence the exponents of $u_1, \cdots, u_{j-1}$ in p_j and q_j must satisfy

$$(54) \qquad \begin{cases} j_1 + 2j_2 + \cdots + (j-1)j_{j-1} = j - 1, \\ k_1 + 2k_2 + \cdots + (j-1)k_{j-1} = j - 1. \end{cases}$$

Now suppose that the highest multiples of M in u_m is mM for $m = 1, \cdots, j-1$. It follows from the properties of powers of the sines that the highest multiple in u_m^n is mnM. Since the highest multiple of the product of two or more sines is the sum of their highest multiples, the highest multiples in p_j and q_j are respectively

$$j_1 + 2j_2 + \cdots + (j-1)j_{j-1}, \qquad k_1 + 2k_2 + \cdots + (j-1)k_{j-1},$$

which are $j - 1$ by (54). But it follows from (53) that p_j is multiplied by $\sin M$ and q_j by $\cos M$; therefore the highest multiple appearing in u_j is jM. That is, u_j has the form

$$(55) \qquad \begin{cases} u_{2k} = + a_2^{(2k)} \sin 2M + \cdots + a_{2k}^{(2k)} \sin 2kM, \\ u_{2k+1} = a_1^{(2k+1)} \sin M + \cdots + a_{2k+1}^{(2k+1)} \sin (2k+1)M, \end{cases}$$

according as j is even or odd.

It is easy to develop a check on the accuracy of the computations. Since $E = M + u$, it follows that

$$\frac{\partial E}{\partial M} = 1 + \frac{\partial u}{\partial M} = 1 + \frac{\partial u_1}{\partial M}e + \frac{\partial u_2}{\partial M}e^2 + \cdots + \frac{\partial u_j}{\partial M}e^j + \cdots.$$

But it follows from Kepler's equation that

$$(56) \qquad \frac{\partial E}{\partial M} = \frac{1}{1 - e \cos E}.$$

Suppose $M = 0$; therefore $E = 0$ and for this value of M

$$\frac{\partial E}{\partial M} = \frac{1}{1-e} = 1 + e + e^2 + \cdots + e^j + \cdots.$$

Therefore, since the coefficient of e^j in this series is unity, for $M = 0$

$$(57) \quad \begin{cases} \dfrac{\partial u_{2k}}{\partial M} = 2a_2^{(2k)} + 4a_4^{(2k)} + \cdots + 2k a_{2k}^{(2k)} = 1, \\[2mm] \dfrac{\partial u_{2k+1}}{\partial M} = a_1^{(2k+1)} + \cdots + (2k+1)a_{2k+1}^{(2k+1)} = 1. \end{cases}$$

These equations constitute a valuable check on all the computations.

It is found from (56) that

$$\begin{cases} \dfrac{\partial^2 E}{\partial M^2} = \dfrac{-e \sin E}{(1 - e \cos E)^2}\dfrac{\partial E}{\partial M} = \dfrac{-e \sin E}{(1 - e \cos E)^3}, \\[3mm] \dfrac{\partial^3 E}{\partial M^3} = \dfrac{-e \cos E}{(1 - e \cos E)^4} + \dfrac{3e^2 \sin^2 E}{(1 - e \cos E)^5}. \end{cases}$$

For $M = 0$, the first of these equations is identically zero, but the second one becomes

$$\frac{\partial^3 E}{\partial M^3} = \frac{-e}{(1 - e)^4} = -\left[e + 4e^2 + \frac{4 \cdot 5}{1 \cdot 2}e^3 + \cdots \right.$$
$$\left. + \frac{4 \cdot 5 \cdots (n + 2)}{1 \cdot 2 \cdots (n - 1)} e^n + \cdots \right].$$

Then the conditions similar to (57) are

$$(58) \quad \begin{cases} -\dfrac{\partial^3 u_{2k}}{\partial M^3} = 2^3 a_2^{(2k)} + 4^3 a_4^{(2k)} + \cdots + (2k)^3 a_{2k}^{(2k)} \\[2mm] \qquad\qquad\qquad = \dfrac{4 \cdot 5 \cdots (2k + 2)}{1 \cdot 2 \cdots (2k - 1)}, \\[3mm] -\dfrac{\partial^3 u_{2k+1}}{\partial M^3} = 1^3 a_1^{(2k+1)} + 3^3 a_3^{(2k+1)} + \cdots \\[2mm] \qquad + (2k + 1)^3 a_{2k+1}^{(2k+1)} = \dfrac{4 \cdot 5 \cdots (2k + 3)}{1 \cdot 2 \cdots 2k}. \end{cases}$$

These equations constitute a check which is independent of that given in (57). In a similar way check formulas can be found from a consideration of all odd derivatives of E with respect to M.

Equations (57), (58), and similar ones for higher derivatives of E, are linear in the coefficients $a_j^{(k)}$, which it is desired to find; consequently, when the number of equations equals the number of unknowns, the latter are uniquely determined, at least if the determinant of the coefficients is not zero. It can be shown that the determinant is not zero.

For the purposes of illustration suppose $k = 0$. Then the second equation of (57) gives $a_1^{(1)} = 1$, whence $u_1 = \sin M$

agreeing with the result in (53). Suppose $k = 1$; then the first equation of (57) gives $2a_2^{(2)} = 1$, whence $u_2 = \frac{1}{2} \sin 2M$. As an illustration involving both (57) and (58), suppose $k = 1$ and consider the second equations of (57) and (58). They become in this case

$$\begin{cases} a_1^{(3)} + 3a_3^{(3)} = 1, \\ a_1^{(3)} + 27a_3^{(3)} = \dfrac{4\cdot 5}{1\cdot 2}; \end{cases}$$

whence $a_1^{(3)} = -\frac{1}{8}$, $a_3^{(3)} = +\frac{3}{8}$, agreeing with the results given in (53).

When the expansion is carried out by the method of Lagrange, or by that which has just been explained, the value of E to terms of the sixth order in e is found to be

$$(59) \quad \begin{cases} E = M + e \sin M + \dfrac{e^2}{2} \sin 2M \\[2mm] \quad + \dfrac{e^3}{3!\,2^2}\, (3^2 \sin 3M - 3 \sin M) \\[2mm] \quad + \dfrac{e^4}{4!\,2^3}\, (4^3 \sin 4M - 4 \cdot 2^3 \sin 2M) \\[2mm] \quad + \dfrac{e^5}{5!\,2^4}\, (5^4 \sin 5M - 5 \cdot 3^4 \sin 3M + 10 \sin M) \\[2mm] \quad + \dfrac{e^6}{6!\,2^5}\, (6^5 \sin 6M - 6 \cdot 4^5 \sin 4M + 15 \cdot 2^5 \sin 2M) \\[2mm] \quad + \;\cdot\;\;\cdot\;\;\cdot\;\;\cdot\;\;\cdot\;\;\cdot\;\;\cdot\;\;\cdot\;\;\cdot\;\;\cdot\;\;\cdot\;\;\cdot\;\;\cdot\;\;\cdot\;\cdot. \end{cases}$$

100. The Development of r and v in Series. The value of r in terms of e and M can be obtained by the method of Lagrange by letting $F(z) = \cos E$ and making use of the last equation of (48). This method has the disadvantage of being rather laborious.

It follows from Kepler's equation that *

$$\begin{cases} e\,\dfrac{\partial E}{\partial e} = \dfrac{e \sin E}{1 - e \cos E}, \\[2mm] dM = (1 - e \cos E)dE. \end{cases}$$

Therefore

$$e\,\frac{\partial E}{\partial e}\,dM = e \sin E\, dE.$$

---

* The method employed in this Art. is due to MacMillan.

The integral of this equation gives

(60) $$e \int_0^M \frac{\partial E}{\partial e}\, dM = - e \cos E + c,$$

which expresses $- e \cos E$ in terms of M very simply by substituting in the left member the explicit value of E given in (59). For example, the first terms are

$$- e \cos E = - c + e \int_0^M \left[\sin M + e \sin 2M \right.$$
$$\left. + \frac{3}{8} e^2 (3 \sin 3M - \sin M) + \cdots \right] dM$$
$$= - c - e \cos M - \frac{1}{2} e^2 \cos 2M - \frac{3}{8} e^3 (\cos 3M - \cos M) \cdots .$$

The last equation of (48) and (60) give for r the series

(61) $$\frac{r}{a} = 1 - e \cos E = 1 - c - e \cos M - \frac{1}{2} e^2 \cos 2M \cdots .$$

It remains to determine the constant c. Since r is measured from the focus of the ellipse, it follows that $r = a(1 - e)$ at $M = 0$; whence

$$1 - e = 1 - c - e - \frac{1}{2} e^2 + \cdots + b_j e^j + \cdots ,$$

where b_j is the coefficient of e^j in the series for $- e \cos E$ at $M = 0$. The two sides of this equation must be the same for all values of e for which (61) converges; therefore c must have the form

$$c = c_2 e^2 + c_3 e^3 + \cdots ,$$

where c_2, c_3, $\cdots$ are determined so that the right member will contain no terms in e^2, e^3, $\cdots$; that is, $- c_j + b_j = 0, j = 2, 3, \cdots$. Since $- e \cos E$, as defined by (60), is the integral of a sine series it contains no constant terms; therefore the b_j are the sums of the coefficients of the cosine terms. Now consider

$$\int_0^{2\pi} \frac{r}{a}\, dM = \int_0^{2\pi} \left[1 - c - e \cos M - \frac{e^2}{2} \cos 2M + \cdots \right] dM.$$

It was shown in Problem 4, p. 154, that the value of this integral is $2\pi (1 + \frac{1}{2} e^2)$. Therefore the coefficients of e^3, e^4, $\cdots$ contain no constant terms and the exact value of c is $- \frac{1}{2} e^2$.

The series for $\frac{r}{a}$ up to the sixth power of e is

$$(62) \begin{cases} \dfrac{r}{a} = 1 - e \cos M - \dfrac{e^2}{2} (\cos 2M - 1) \\[2mm] \qquad - \dfrac{e^3}{2\,!\,2^2} (3 \cos 3M - 3 \cos M) \\[2mm] \qquad - \dfrac{e^4}{3\,!\,2^3} (4^2 \cos 4M - 4 \cdot 2^2 \cos 2M) \\[2mm] \qquad - \dfrac{e^5}{4\,!\,2^4} (5^3 \cos 5M - 5 \cdot 3^3 \cos 3M + 10 \cos M) \\[2mm] \qquad - \dfrac{e^6}{5\,!\,2^5} (6^4 \cos 6M - 6 \cdot 4^4 \cos 4M + 15 \cdot 2^4 \cos 2M) \\[2mm] \qquad - \cdot \quad \cdot \quad \cdot \quad \cdot \quad \cdot \quad \cdot \quad \cdot \quad \cdot \quad \cdot \quad \cdot \quad \cdot \quad \cdot \quad \cdot \quad \cdot \quad \cdot \cdot. \end{cases}$$

The computation of the series for v will now be considered. It follows from the first two equations of (49) that

$$dv = \frac{\sqrt{1 - e^2}}{(1 - e \cos E)^2} dM,$$

which becomes as a consequence of Kepler's equation

$$(63) \qquad dv = \sqrt{1 - e^2} \left(\frac{dE}{dM} \right)^2 dM.$$

The quantity $\dfrac{dE}{dM}$ is found at once from (59), and the result squared and integrated gives, after $\sqrt{1 - e^2}$ has been expanded as a power series in e^2,

$$(64) \begin{cases} v = M + 2e \sin M + \tfrac{5}{4}e^2 \sin 2M \\[2mm] \qquad + \dfrac{e^3}{12} (13 \sin 3M - 3 \sin M) \\[2mm] \qquad + \dfrac{e^4}{96} (103 \sin 4M - 44 \sin 2M) \\[2mm] \qquad + \dfrac{e^5}{960} (1097 \sin 5M - 645 \sin 3M + 50 \sin M) \\[2mm] \qquad + \dfrac{e^6}{960} (1223 \sin 6M - 902 \sin 4M + 85 \sin 2M) \\[2mm] \qquad + \cdot \quad \cdot \quad \cdot \quad \cdot \quad \cdot \quad \cdot \quad \cdot \quad \cdot \quad \cdot \quad \cdot \quad \cdot \quad \cdot \quad \cdot \cdot. \end{cases}$$

When e is small, as in the planetary orbits, these series are very rapidly convergent; if e exceeds $0.6627 \cdots$ they diverge, as

Laplace first showed, for some values of M. This value of e is exceeded in the solar system only in the case of some of the comets' orbits, but developments of this sort are not employed in computing the perturbations of the comets.

101. Direct Computation of the Polar Coördinates.* It has been observed that there is considerable labor involved in finding the coördinates at any time in the case of elliptic motion. The question arises whether it may not be due partly to the fact that the final result is obtained by determining E as an intermediary function from Kepler's equation. The question also arises whether the coördinates cannot conveniently be found directly from the differential equations. It will be shown that the answer to the latter question is in the affirmative.

Equations (16) become in polar coördinates

$$\begin{cases} \dfrac{d^2r}{dt^2} - r\left(\dfrac{dv}{dt}\right)^2 + \dfrac{k^2(1+m)}{r^2} = 0, \\[2mm] \dfrac{d}{dt}\left(r^2\dfrac{dv}{dt}\right) = 0. \end{cases}$$

On integrating the second of these equations and eliminating $\dfrac{dv}{dt}$ from the first by means of the integral, the result is found to be

$$\begin{cases} \dfrac{d^2r}{dt^2} - \dfrac{h^2}{r^3} + \dfrac{k^2(1+m)}{r^2} = 0, \\[2mm] r^2\dfrac{dv}{dt} = h = k\sqrt{(1+m)a(1-e^2)}. \end{cases}$$

After eliminating $k^2(1+m)$ by the first equation of (48) and changing from the independent variable t to M by means of the second equation of (48), these equations become

$$(65) \quad \begin{cases} \dfrac{d^2r}{dM^2} - \dfrac{a^4(1-e^2)}{r^3} + \dfrac{a^3}{r^2} = 0, \\[2mm] \dfrac{dv}{dM} = \dfrac{a^2\sqrt{1-e^2}}{r^2}. \end{cases}$$

The first equation of (65) is independent of the second and can be integrated separately. It is satisfied by $r = a$ and $e = 0$, in which case the orbit is a circle. In order to get the elliptic orbit let

---

* This method was first published by the author in the *Astronomical Journal*, vol. 25 (1907).

(66) $$r = a(1 - \rho e),$$

where $a\rho e$ is the deviation from a circle. When the planet is at perihelion, $r = a(1 - e)$. Therefore $\rho = 1$ for $M = 0$. When the planet is at aphelion, $r = a(1 + e)$. Therefore $\rho = -1$ for $M = \pi$, and ρ varies between -1 and $+1$. Since $\frac{dr}{dM}$ is zero for M equal to 0 and π, it follows that $\frac{d\rho}{dM}$ is zero for M equal to 0 and π.

When (66) is substituted in (65), these equations become

$$\begin{cases} \dfrac{d^2\rho}{dM^2} + \dfrac{\rho - e}{(1 - \rho e)^3} = 0, \\[2ex] \dfrac{dv}{dM} - \dfrac{\sqrt{1 - e^2}}{(1 - \rho e)^2} = 0. \end{cases}$$

Since e is less than unity and ρ varies from -1 to $+1$, the second terms of these equations can be expanded as converging power series in e, giving

(67)
$$\begin{cases} \dfrac{d^2\rho}{dM^2} + \rho = \dfrac{1}{2} \displaystyle\sum_{i=1}^{\infty} (i + 1)[i - (i + 2)\rho^2]\rho^{i-1}e^i, \\[2ex] \dfrac{dv}{dM} = \sqrt{1 - e^2} \displaystyle\sum_{i=0}^{\infty} (i + 1)\rho^i e^i. \end{cases}$$

It has been shown that r, and hence ρ, is expansible as a power series in e. This fact also follows from the form of the first equation of (67) and the general principles of Differential Equations. Hence ρ can be written in the form

(68) $$\rho = \rho_0 + \rho_1 e + \rho_2 e^2 + \cdots,$$

where ρ_0, ρ_1, ρ_2, $\cdots$ are functions of M which remain to be determined. Since ρ is periodic with the period 2π for all e less than unity, each ρ_j separately is a sum of trigonometric terms. Since the motion is symmetrical with respect to the major axis of the orbit, and since $M = 0$ when the planet is at its perihelion, ρ is an even function of M. This is true for all values of e for which the series converges, and therefore each ρ_j is a sum of cosine terms.

A change in the sign of e is equivalent to changing the origin to the other focus of the ellipse. Hence if the sign of e is changed and π is added to M the value of r is unchanged; from (66) it follows that the sign of ρ is changed.   Since this is true for all values of e for which the series converges

$$\rho_j(M)e^j = -\rho_j(M + \pi)(-e)^j.$$

Therefore if j is even ρ_j is a sum of cosines of odd multiples of M, and if j is odd ρ_j is a sum of cosines of even multiples of M.   It is seen on referring to equations (68) and (66) that this is the same property as that which was established Art. 100.

It can easily be proved from the properties of the ρ_j and the second equation of (67) that v is expressible as a series of the form

(69) $$v = v_0 + v_1 e + v_2 e^2 + \cdots,$$

and that each v_j $(j > 1)$ is a sum of sines of integral multiples of M. A more detailed discussion shows that if j is even v_j is a sum of sines of even multiples of M, and if j is odd v_j is a sum of sines of odd multiples of M.

The solution can be directly constructed without any difficulty. The result of substituting (68) in the first of (67) is

$$\left[\frac{d^2\rho_0}{dM^2} + \frac{d^2\rho_1}{dM^2} e + \frac{d^2\rho_2}{dM^2} e^2 + \cdots \right] + [\rho_0 + \rho_1 e + \rho_2 e^2 + \cdots]$$
$$= [1 - 3\rho_0^2]e + [3\rho_0 - 6\rho_0\rho_1 - 6\rho_0^3]e^2 + \cdots.$$

On equating coefficients of corresponding powers of e in the left and right members of this equation, it is found that

(70)
$$\begin{cases}
(a) \quad \dfrac{d^2\rho_0}{dM^2} + \rho_0 = 0, \\[2mm]
(b) \quad \dfrac{d^2\rho_1}{dM^2} + \rho_1 = 1 - 3\rho_0^2, \\[2mm]
(c) \quad \dfrac{d^2\rho_2}{dM^2} + \rho_2 = 3\rho_0(1 - 2\rho_1 - 2\rho_0^2), \\[2mm]
\qquad \cdot \quad \cdot \quad \cdot \quad \cdot \quad \cdot \quad \cdot \quad \cdot \quad \cdot \quad \cdot \quad \cdot.
\end{cases}$$

Equations (70) can be integrated in the order in which they are written.   Two constants of integration arise at each step and they are to be determined so that $\rho = 1$ and $\dfrac{d\rho}{dM} = 0$ for $M = 0$ whatever may be the value of e.   It follows from (68) that these conditions are

$$\begin{cases}
\rho(0) = \rho_0(0) + \rho_1(0)e + \rho_2(0)e^2 + \cdots = 1, \\[2mm]
\dfrac{d\rho}{dM} = \dfrac{d\rho_0}{dM} + \dfrac{d\rho_1}{dM} e + \dfrac{d\rho_2}{dM} e^2 + \cdots = 0,
\end{cases}$$

where M is given the value 0 after the derivatives of the second equation have been formed. Since these equations hold for all values of e, it follows that

$$(71) \quad \begin{cases} \rho_0(0) = 1, & \rho_1(0) = 0, & \rho_2(0) = 0, & \cdots, \\ \dfrac{d\rho_0}{dM} = 0, & \dfrac{d\rho_1}{dM} = 0, & \dfrac{d\rho_2}{dM} = 0, & \cdots. \end{cases}$$

The general solution of equation (a) of (70) is (Art. 32)

$$\rho_0 = a_0 \cos M + b_0 \sin M,$$

where a_0 and b_0 are the constants of integration. It follows from (71) that $a_0 = 1$, $b_0 = C$, and therefore that

$$\rho_0 = \cos M.$$

The fact that b_0 is zero also follows from the general property that the ρ_j involve only cosines.

On substituting the value of ρ_0 in the right member of (b) of (70), this equation becomes

$$\frac{d^2\rho_1}{dM^2} + \rho_1 = -\tfrac{1}{2} - \tfrac{3}{2} \cos 2M.$$

This equation can be solved by the method of the variation of parameters (Art. 37). But since the part of the solution which comes from the right member will contain terms of the same form as the right member, it is simpler to substitute the expression

$$\rho_1 = a_1 \cos M + b_1 \sin M + c_1 + d_1 \cos 2M$$

in the differential equation and to determine c_1 and d_1 so that it will be satisfied. This leads to the solution

$$\rho_1 = a_1 \cos M + b_1 \sin M - \tfrac{1}{2} + \tfrac{1}{2} \cos 2M,$$

which is the general solution since it satisfies the differential equation and has the two arbitrary constants a_1 and b_1. On determining a_1 and b_1 by (71), the expression for ρ_1 becomes

$$\rho_1 = -\tfrac{1}{2} + \tfrac{1}{2} \cos 2M.$$

With the values of ρ_0 and ρ_1 which have been found equation (c) of (70) becomes

$$\frac{d^2\rho_2}{dM^2} + \rho_2 = -3 \cos 3M,$$

of which the general solution is

$$\rho_2 = a_2 \cos M + b_2 \sin M + \tfrac{3}{8} \cos 3M.$$

If a_2 and b_2 are determined by (71), the final expression for ρ_2 becomes

$$\rho_2 = \tfrac{3}{8}(- \cos M + \cos 3M).$$

This process of integration can be continued as far as may be desired. It follows from the results which have been found that

$$\frac{r}{a} = 1 - \rho e = 1 - (\rho_0 + \rho_1 e + \rho_2 e^2 + \cdots)e$$

$$= 1 - e \cos M - \tfrac{1}{2}e^2(\cos 2M - 1) - \tfrac{3}{8}e^3(\cos 3M - \cos M) \cdots,$$

which agrees with those given in (62).

When the values for ρ_0, ρ_1, $\cdots$ are substituted in the second equation of (67), the result is

$$\frac{dv}{dM} = 1 + 2e \cos M + \tfrac{5}{2}e^2 \cos 2M + \cdots,$$

and the integral of this equation is

$$v = c + M + 2e \sin M + \tfrac{5}{4}e^2 \sin 2M + \cdots.$$

Since $v = 0$ when $M = 0$, the arbitrary constant c is zero, and the result agrees with that given in (64).

The method which has just been developed is, for this special problem, perhaps not superior to that depending upon the solution of Kepler's equation. But if the conditions of the problem are modified a little, for example by adding the terms which would come from the oblateness of a planet when the body moves in the plane of its equator [equations (30), Chapter IV], Kepler's equation no longer holds and the method depending on it fails, while the one under consideration here can be applied without any modification except in the numerical values of the coefficients which depend upon the terms added to the differential equations. But additional terms in the differential equations change the period of the motion, if indeed it remains periodic, and in order to exhibit the periodicity explicitly some modifications of the methods of determining the constants of integration are in general necessary. This method of integrating in series is typical of those which are employed in the theories of perturbations and the more difficult parts of Celestial Mechanics, and for this reason it should be thoroughly mastered.

102. Position in Hyperbolic Orbits. There are close analogies between this problem and that of finding the position of a body in an elliptic orbit. But it follows from the polar equation of the hyperbola,

$$r = \frac{a(\epsilon^2 - 1)}{1 + \epsilon \cos v},$$

where a is its major semi-axis and ϵ its eccentricity, that in this case v can vary only from $-\pi + \cos^{-1}\left(\dfrac{1}{\epsilon}\right)$ to $+\pi - \cos^{-1}\left(\dfrac{1}{\epsilon}\right)$.

The integrals of areas and vis viva are respectively in the case of hyperbolic orbits

$$(72) \quad \begin{cases} r^2 \dfrac{dv}{dt} = k \sqrt{(1 + m)a(\epsilon^2 - 1)}, \\[2mm] \left(\dfrac{dr}{dt}\right)^2 + r^2 \left(\dfrac{dv}{dt}\right)^2 = k^2(1 + m)\left(\dfrac{2}{r} + \dfrac{1}{a}\right). \end{cases}$$

On eliminating v from the second of these equations by means of the first and solving, it is found that

$$a\nu \, dt = \frac{r dr}{\sqrt{(a + r)^2 - a^2 \epsilon^2}},$$

where

$$\nu = \frac{k \sqrt{1 + m}}{a^{\frac{3}{2}}}.$$

This equation can be integrated at once in terms of hyperbolic functions, but it is preferable to introduce first an auxiliary quantity F corresponding to the eccentric anomaly in elliptic orbits. Let

$$(73) \qquad a + r = \frac{a\epsilon}{2}(e^F + e^{-F}) = a\epsilon \cosh F;$$

then

$$\nu dt = \left\{-1 + \frac{\epsilon}{2}(e^F + e^{-F})\right\} dF = [-1 + \epsilon \cosh F]dF.$$

The integral of this equation is

$$(74) \quad M = \nu(t - T) = -F + \frac{\epsilon}{2}(e^F - e^{-F}) = -F + \epsilon \sinh F,$$

which gives t when F is known. The inverse problem of finding F when $\nu(t - T)$ is given is one of more difficulty. The most expeditious method would be, in general, to find an approximate value of F by some graphical process, and then a more exact value by differential corrections. The value of F satisfying (74) is the abscissa of the point of intersection of the line

$$y = \frac{1}{\epsilon}(F + M),$$

and the hyperbolic sine curve

$$y = \frac{e^F - e^{-F}}{2} = \sinh F.$$

The differential corrections could be computed in a manner analogous to that developed in the case of the elliptic orbits.

From (73) and the polar equation of the hyperbola, it follows that

$$r = \frac{a(\epsilon^2 - 1)}{1 + \epsilon \cos v} = a\left[-1 + \epsilon \cosh F\right]$$

and from this equation,

$$\tan \frac{v}{2} = \sqrt{\frac{\epsilon + 1}{\epsilon - 1}} \frac{\sqrt{-1 + \frac{1}{2}(e^F + e^{-F})}}{\sqrt{+1 + \frac{1}{2}(e^F + e^{-F})}} = \sqrt{\frac{\epsilon + 1}{\epsilon - 1}} \tanh \frac{F}{2},$$

which is a convenient formula for computing v when F has been found.

103. Position in Elliptic and Hyperbolic Orbits when e is Nearly Equal to Unity. The analytical solutions heretofore given have depended upon expansions in powers of e. If e is large, as in the case of some of the periodic comets' orbits, the convergence ceases or is so slow that the methods become impracticable. The graphical process, however, avoids this difficulty.

In order to obtain a workable analytical solution, the developments for elliptical orbits will be made in powers of $\frac{1 - e}{1 + e}$. The start is made from the equation of areas and the polar equation of the orbit which will be assumed to be an ellipse.

Let

$$\begin{cases} w = \tan \frac{v}{2}, \\ \lambda = \frac{1 - e}{1 + e}; \end{cases}$$

then the equation of areas becomes

$$\frac{n \sqrt{1 + e}}{2(1 - e)^{\frac{3}{2}}} dt = \frac{(1 + w^2)}{(1 + \lambda w^2)^2} dw.$$

When λ is very small the right member of this equation can be developed into a rapidly converging series in λ for all values of v not too near $180°$. Since the periodic comets are always invisible when near aphelion, there will seldom be occasion to consider the solution in this region. On expanding the right member and integrating, the result is found to be

$$
(75) \qquad \frac{n(1 + e)^{\frac{1}{2}}}{2(1 - e)^{\frac{3}{2}}} (t - T) = w + \frac{w^3}{3} - 2\lambda \left(\frac{w^3}{3} + \frac{w^5}{5} \right)
$$
$$
+ 3\lambda^2 \left(\frac{w^5}{5} + \frac{w^7}{7} \right) - 4\lambda^3 \left(\frac{w^7}{7} + \frac{w^9}{9} \right) + \cdots .
$$

When the orbit is a parabola $e = 1$ and $\lambda = 0$, and this equation reduces to (32), which is a cubic in w. Since the perihelion distance in an ellipse is $q = a(1 - e)$ and $n = \dfrac{k}{a^{\frac{3}{2}}}$, it follows that

$$
\frac{n \sqrt{1 + e}}{2(1 - e)^{\frac{3}{2}}} = \frac{k \sqrt{1 + e}}{2q^{\frac{3}{2}}} .
$$

It is desired to find the value of w for any value of t. If the eccentricity should become equal to unity, the left member keeping the same value, equation (75) would have the form

$$
(76) \qquad \frac{k(1 + e)^{\frac{1}{2}}}{2q^{\frac{3}{2}}} (t - T) = W + \tfrac{1}{3}W^3,
$$

where W would be the tangent of half the true anomaly in the resulting parabolic orbit. From this equation W can be determined by means of Barker's tables, or from equations (33). Suppose W has been found; then w can be expressed as a series in λ of which the coefficients are functions of W. For, assume the development

$$
(77) \qquad w = a_0 + a_1\lambda + a_2\lambda^2 + a_3\lambda^3 + \cdots ;
$$

substitute it in the right member of (75), which is equal to the right member of (76). The result of the substitution is

$$
W + \frac{W^3}{3} = a_0 + \frac{a_0{}^3}{3} + \left[a_1 + a_0{}^2a_1 - \tfrac{2}{3}a_0{}^3 - \tfrac{2}{5}a_0{}^5 \right]\lambda
$$
$$
+ \left[a_2 + a_0{}^2a_2 + a_0a_1{}^2 - 2a_0{}^2a_1 - 2a_0{}^4a_1 + \tfrac{3}{5}a_0{}^5 + \tfrac{3}{7}a_0{}^7 \right]\lambda^2
$$
$$
+ \left[a_3 + a_0{}^2a_3 + \frac{a_1{}^3}{3} - 2a_0{}^2a_2 - 2a_0{}^4a_2 - 2a_0a_1{}^2 \right.
$$
$$
\left. - 4a_0{}^3a_1{}^2 + 3a_0{}^4a_1 + 3a_0{}^6a_1 - \tfrac{4}{7}a_0{}^7 - \tfrac{4}{9}a_0{}^9 \right] \lambda^3
$$
$$
+ \quad \cdot \quad \cdot \quad \cdot \quad \cdot \quad \cdot \quad \cdot \quad \cdot \quad \cdot \quad \cdot \quad \cdot \quad \cdot \quad \cdot \quad \cdot \cdot .
$$

Since this equation is an identity in λ, the coefficients of corresponding powers of λ are equal. Hence

$$
\left\{
\begin{aligned}
&a_0 + \frac{a_0{}^3}{3} = W + \frac{W^3}{3}, \\[2mm]
&a_1(1 + a_0{}^2) = \tfrac{2}{3}a_0{}^3 + \tfrac{2}{5}a_0{}^5, \\[2mm]
&a_2(1 + a_0{}^2) = -a_0 a_1{}^2 + 2a_0{}^2 a_1 + 2a_0{}^4 a_1 - \tfrac{3}{5}a_0{}^5 - \tfrac{3}{7}a_0{}^7, \\[2mm]
&a_3(1 + a_0{}^2) = -\frac{a_1{}^3}{3} + 2a_0{}^2 a_2 + 2a_0{}^4 a_2 + 2a_0 a_1{}^2 + 4a_0{}^3 a_1{}^2 \\[2mm]
&\hspace{3cm} - 3a_0{}^4 a_1 - 3a_0{}^6 a_1 + \tfrac{4}{7}a_0{}^7 + \tfrac{4}{9}a_0{}^9, \\[2mm]
&\cdot \quad \cdot \quad \cdot \quad \cdot \quad \cdot \quad \cdot \quad \cdot \quad \cdot \quad \cdot \quad \cdot \quad \cdot \quad \cdot \quad \cdot
\end{aligned}
\right.
$$

There are three solutions for a_0, only one of which is real. On taking the real root of the first equation, it is found that

$$
\left\{
\begin{aligned}
&a_0 = W, \\[2mm]
&a_1 = \frac{\tfrac{2}{3}W^3 + \tfrac{2}{5}W^5}{1 + W^2}, \\[2mm]
&a_2 = \frac{\tfrac{11}{15}W^5 + \tfrac{439}{315}W^7 + \tfrac{33}{35}W^9 + \tfrac{37}{175}W^{11}}{(1 + W^2)^3}, \\[2mm]
&a_3 = \frac{\tfrac{292}{315}W^7 + \tfrac{7928}{2835}W^9 + \tfrac{10328}{2835}W^{11} + \tfrac{432}{175}W^{13} + \tfrac{6692}{7875}W^{15} + \tfrac{184}{1575}W^{17}}{(1 + W^2)^5}, \\[2mm]
&\cdot \quad \cdot \quad \cdot \quad \cdot \quad \cdot \quad \cdot \quad \cdot \quad \cdot \quad \cdot \quad \cdot \quad \cdot \quad \cdot \quad \cdot
\end{aligned}
\right.
$$

When the values of these coefficients are substituted in (77) the tangent of one-half the true anomaly is determined. The first term gives that which would come from a parabolic orbit, the remaining terms vanishing for $e = 1$. In the series (64) the first term in the right member would be the true anomaly if the orbit were a circle, the higher terms being the corrections to circular motion. In the series (77) the first term in the right member would give the tangent of one-half the true anomaly if the orbit were a parabola, the higher terms being the corrections to parabolic motion.

These equations apply equally to hyperbolic orbits in which the eccentricity is near unity if $1 - e$ and $1 + e$ are changed to $\epsilon - 1$ and $\epsilon + 1$ throughout, where ϵ is the eccentricity of the hyperbola.

XV. PROBLEMS.

1. Show how the cubic equation (32) can be solved approximately for $\tan \frac{v}{2}$ with great rapidity by the aid of a graphical construction.

2. Develop the equations for differential corrections to the approximate values found by the graphical method. Apply to a particular problem and verify the result.

3. If $e = 0.2$ and $M = 214°$, find E_0, M_0, E_1, M_1, E_2, and M_2.

Ans. $E_0 = 208° \, 39' \, 16''.6$, $M_0 = 214° \, 8' \, 58''.6$; $E_1 = 208° \, 31' \, 38''.4$,

$\qquad M_1 = 213° \, 59' \, 59''.8$; $E_2 = 208° \, 31' \, 38''.6$, $M_2 = 214° \, 00' \, 00''$.

4. Show from the curves employed in solving Kepler's equation that the solution is unique for all values of $e < 1$ and M.

5. In (50) the quadrant is not determined by the equation; show that corresponding values of $\frac{1}{2}v$ and $\frac{1}{2}E$ always lie in the same quadrant.

6. Express the rectangular coördinates $x = r \cos v$, $y = r \sin v$ in terms of the eccentric anomaly, and then, by means of the Lagrange expansion formula, in terms of M.

Ans.
$$
\begin{cases}
\dfrac{x}{a} = \cos M + \dfrac{e}{2} (\cos 2M - 3) + \dfrac{e^2}{2!\,2^2} (3 \cos 3M - 3 \cos M) \\[2mm]
\qquad\qquad + \dfrac{e^3}{3!\,2^3} (4^2 \cos 4M - 4 \cdot 2^2 \cos 2M) + \cdots. \\[4mm]
\dfrac{y}{a} = \sin M + \dfrac{e}{2} \sin 2M + \dfrac{e^2}{3!\,2^2} (3^2 \sin 3M - 15 \sin M) \\[2mm]
\qquad\qquad + \dfrac{e^3}{4!\,2^3} (4^3 \sin 4M - 10 \cdot 2^3 \sin 2M) + \cdots.
\end{cases}
$$

7. Show that the properties of E as a power series in e, which were established in Art. 99, follow from the Lagrange expansion.

8. Derive the first three terms of the series for r by the Lagrange formula.

9. Give a geometrical interpretation of F (Art. 102) corresponding to that of E in an elliptic orbit.

10. Express v as a power series in e by a method analogous to that used in Art. 103.

11. Show that the branch of the hyperbola which is convex to the sun is described by the body in purely imaginary time.

12. Add to the right members of equations (16) the terms $-\dfrac{3}{10}(1+m)b^2 e_1{}^2 \dfrac{x}{r^5}$ and $-\dfrac{3}{10}(1+m)b^2 e_1{}^2 \dfrac{y}{r^5}$, which come from the oblateness of the central body [equations (30), Chap. IV.], where e_1 is the eccentricity of a meridian section, and integrate by the method of Art. 101.

104. The Heliocentric Position in the Ecliptic System. Methods have been given for finding the positions in the orbits in the various cases which arise. The formulas will now be derived for determining the position referred to different systems of axes. The origin will first be kept fixed at the body with respect to which the motion of the second is given. Since most of the applications are in the solar system where the origin is at the center of the sun, the coördinates will be called *heliocentric.*

Positions of bodies in the solar system are usually referred to one of two systems of coördinates, the *ecliptic* system, or the *equatorial* system. The fundamental plane in the ecliptic system is the plane of the earth's orbit; in the equatorial system it is the plane of the earth's equator. The zero point of the fundamental circles in both systems is the *vernal equinox,* or the point at which the ecliptic cuts the equator from south to north, and is denoted by ♈. The polar coördinates in the ecliptic system are called *longitude* and *latitude;* and in the equatorial, *right ascension* and *declination.* When the origin is at the sun Roman letters are used to represent the coördinates, and when at the earth, Greek. Thus

Origin at sun. Origin at earth.

longitude	l	λ	measured eastward.
latitude	b	β	+ if north; − if south.
right ascension	a	α	measured eastward.
declination	d	δ	+ if north; − if south.
distance	r	ρ	

In practice a and d are very seldom used. Absolute positions of fundamental stars are given in the equatorial system, and the observed positions of comets are determined by comparison with them. In some theories relating to planets and comets, especially in considering the mutual perturbation of planets and their perturbations of comets, it is more convenient to use the ecliptic system; hence it is necessary to be able to transform the equations from one system to the other.

The *ascending node* is the projection on the ecliptic, from the sun, of the place at which the body crosses the plane of the ecliptic from south to north. It is measured from a fixed point in the ecliptic, the vernal equinox, and is denoted by ☊. The projection of the point where the body crosses the plane of the ecliptic from north to south is called the *descending node,* and is denoted by ☋.

The *inclination* is the angle between the plane of the orbit and the plane of the ecliptic, and is denoted by i. It has been the custom of some writers to take the inclination always less than 90°, and to define the direction of motion as *direct* or *retrograde*, according as it is the same as that of the earth or the opposite. Another method that has been used is to consider all motion direct and the inclination as varying from 0° to 180°. The latter method avoids the use of double signs in the formulas and is adopted here. [See Art. 86.] The node and inclination define the position of the plane of the orbit in space.

The distance from the ascending node to the perihelion point counted in the direction of the motion of the body in its orbit is ω, and defines the orientation of the orbit in its plane. The *longitude of the perihelion* is denoted by π, and is given by the equation

$$\pi = \Omega + \omega.$$

This element is not a longitude in the ordinary sense because it is counted in two different planes.

The problem of relative motion of two bodies was of the sixth order (Art. 85), and in the integration six arbitrary constants were introduced. There are six elements, therefore, which are independent functions of these constants. They are

a = major semi-axis, which defines the size of the orbit and the period of revolution.

e = the eccentricity, which defines the shape of the orbit.

Ω = longitude of ascending node, and

i = inclination to plane of the ecliptic, which together define the position of the plane of the orbit.

ω = longitude of the perihelion point measured from the node, or π = longitude of the perihelion, either defining the orientation of the orbit in its plane.

T = time of perihelion passage, defining, with the other elements, the position of the body in its orbit at any time.

The polar coördinates have been computed; hence the rectangular coördinates with the positive end of the x-axis directed to the perihelion point and the y-axis in the plane of the orbit are given by the equations

(78)
$$\begin{cases} x_0 = r \cos v, \\ y_0 = r \sin v, \\ z_0 = 0. \end{cases}$$

If the x-axis is rotated backward to the line of nodes, the coördinates in the new system are

$$(79) \quad \begin{cases} x = r \cos (v + \omega) = r \cos (v + \pi - \Omega), \\ y = r \sin (v + \omega) = r \sin (v + \pi - \Omega), \\ z = 0. \end{cases}$$

The longitude of the body in its orbit counted from the ascending node is called the *argument of the latitude* and is denoted by u. It is given by the equation

$$u = v + \omega;$$

hence u is known when v has been found.

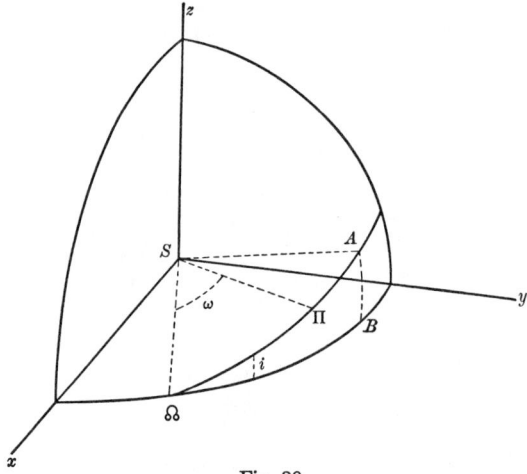

Fig. 30.

Let S represent the sun and Sxy the plane of the ecliptic; $S\Omega A$, the plane of the orbit; Ω, the ascending node; Π, the perihelion point; A, the projection of the position of the body; and angle $\Pi SA = v$. Then $\Omega A = \omega + v = u$.

Let the position of the body now be referred to a rectangular system of axes with the origin at the sun, the x-axis in the line of the nodes, and the y-axis in the plane of the ecliptic. Then equations (79) become

$$(80) \quad \begin{cases} x' = r \cos (v + \omega) = r \cos u, \\ y' = r \sin (v + \omega) \cos i = r \sin u \cos i, \\ z' = r \sin (v + \omega) \sin i = r \sin u \sin i. \end{cases}$$

But, in terms of the heliocentric latitude and longitude,

(81)
$$\begin{cases} x' = r \cos b \cos (l - \Omega), \\ y' = r \cos b \sin (l - \Omega), \\ z' = r \sin b. \end{cases}$$

Therefore, comparing (80) and (81), it is found that

(82)
$$\begin{cases} \cos b \cos (l - \Omega) = \cos u, \\ \cos b \sin (l - \Omega) = \sin u \cos i, \\ \sin b \qquad\quad = \sin u \sin i; \end{cases}$$

whence

(83)
$$\begin{cases} \tan (l - \Omega) = \tan u \cos i, \\ \tan b \qquad = \tan i \sin (l - \Omega). \end{cases}$$

Since $\cos b$ is always positive, equations (82) and (83) determine the heliocentric longitude and latitude, l and b, uniquely when Ω, i, and u are known.

105. Transfer of the Origin to the Earth. Let Ξ, H, Z be the geocentric coördinates of the center of the sun referred to a system of axes with the x-axis directed to the vernal equinox, and the y-axis in the plane of the ecliptic. Let P, Λ, and B* represent the geocentric distance, longitude, and latitude of the sun respectively. These quantities are given in the *Nautical Almanac* for every day in the year. The rectangular coördinates are expressed in terms of them by

(84)
$$\begin{cases} \Xi = P \cos B \cos \Lambda, \\ H = P \cos B \sin \Lambda, \\ Z = P \sin B. \end{cases}$$

The angle B is generally less than a second of arc, and unless great accuracy is required these equations may be replaced by

$$\begin{cases} \Xi = P \cos \Lambda, \\ H = P \sin \Lambda, \\ Z = 0. \end{cases}$$

Let ξ'', η'', and ζ'' be the geocentric, and x'', y'', and z'' the heliocentric, coördinates of the body with the x-axis directed toward the vernal equinox and the y-axis in the plane of the ecliptic. Therefore

* P, Λ, B = capital ρ, λ, β.

$$\begin{cases} \xi'' = x'' + \Xi, \\ \eta'' = y'' + H, \\ \zeta'' = z'' + Z. \end{cases}$$

In polar coördinates these equations are

$$\begin{cases} \rho \cos \beta \cos \lambda = r \cos b \cos l + P \cos B \cos \Lambda, \\ \rho \cos \beta \sin \lambda = r \cos b \sin l + P \cos B \sin \Lambda, \\ \rho \sin \beta \quad\;\; = r \sin b \quad\;\; + P \sin B. \end{cases}$$

From these equations λ and β can be found; but this system may be transformed into one which is more convenient by multiplying the first equation by $\cos \Lambda$, the second by $\sin \Lambda$, and adding the products; and then multiplying the first by $- \sin \Lambda$ and the second by $\cos \Lambda$, and adding the products. The results are

$$(85) \begin{cases} \rho \cos \beta \cos (\lambda - \Lambda) = r \cos b \cos (l - \Lambda) + P \cos B, \\ \rho \cos \beta \sin (\lambda - \Lambda) = r \cos b \sin (l - \Lambda), \\ \rho \sin \beta \quad\;\; = r \sin b \quad\;\;\;\; + P \sin B. \end{cases}$$

These equations give the geocentric distance, longitude, and latitude, ρ, λ, and β.

106. Transformation to Geocentric Equatorial Coördinates. Let ϵ represent the inclination of the plane of the ecliptic to the plane of the equator. Let ξ'', η'', and ζ'' be the geocentric coordinates of the body referred to the ecliptic system with the x-axis directed toward the vernal equinox. Then, the equatorial system can be obtained by rotating the ecliptic system around the x-axis in the negative direction through the angle ϵ, the relations between the coördinates in the two systems being

$$\begin{cases} \xi''' = \xi'', \\ \eta''' = \eta'' \cos \epsilon - \zeta'' \sin \epsilon, \\ \zeta''' = \eta'' \sin \epsilon + \zeta'' \cos \epsilon; \end{cases}$$

or, in polar coördinates,

$$(86) \begin{cases} \cos \delta \cos \alpha = \cos \beta \cos \lambda, \\ \cos \delta \sin \alpha = \cos \beta \sin \lambda \cos \epsilon - \sin \beta \sin \epsilon, \\ \sin \delta \quad\;\; = \cos \beta \sin \lambda \sin \epsilon + \sin \beta \cos \epsilon. \end{cases}$$

In order to solve these equations conveniently for δ and α the auxiliaries n and N will be introduced by the equations

$$(87) \quad \begin{cases} n \sin N = \sin \beta, \\ n \cos N = \cos \beta \sin \lambda, \end{cases}$$

in which n is a positive quantity. Then equations (86) become

$$\begin{cases} \cos \delta \cos \alpha = \cos \beta \cos \lambda, \\ \cos \delta \sin \alpha = n \cos (N + \epsilon), \\ \sin \delta \quad\;\; = n \sin (N + \epsilon); \end{cases}$$

whence

$$(88) \quad \begin{cases} n \sin N = \sin \beta, \\ n \cos N = \cos \beta \sin \lambda, \\ \tan \alpha = \dfrac{\cos (N + \epsilon) \tan \lambda}{\cos N}, \\ \tan \delta = \tan (N + \epsilon) \sin \alpha. \end{cases}$$

These equations, together with the first of (86), which is used in determining the quadrant in which α lies, give α and δ without ambiguity when λ and β are known.

If α and δ are given and λ and β are required, the equations from which they can be computed are found by interchanging α and δ with λ and β, and changing ϵ to $- \epsilon$ in (88). They are*

$$(89) \quad \begin{cases} m \sin M = \sin \delta, \\ m \cos M = \cos \delta \sin \alpha, \\ \tan \lambda = \dfrac{\cos (M - \epsilon) \tan \alpha}{\cos M}, \\ \tan \beta = \tan (M - \epsilon) \sin \lambda. \end{cases}$$

107. Direct Computation of the Geocentric Equatorial Coördinates. The geocentric equatorial coördinates, α and δ, can be found directly from the elements, i and Ω, and the argument of the latitude u, without first finding the ecliptic coördinates, λ and β.

In a system of axes with the x-axis directed to the node and the y-axis in the plane of the ecliptic, the equations for the heliocentric coördinates are

$$\begin{cases} x' = r \cos u, \\ y' = r \sin u \cos i, \\ z' = r \sin u \sin i. \end{cases}$$

* m and M are new auxiliaries, not being related to any of the quantities which these letters previously have represented.

If the system is rotated around the z-axis until the x-axis is directed toward the vernal equinox, the coördinates are

$$\begin{cases} x'' = x' \cos ☊ - y' \sin ☊, \\ y'' = x' \sin ☊ + y' \cos ☊, \\ z'' = z'; \end{cases}$$

or,

(90)
$$\begin{cases} x'' = r (\cos u \cos ☊ - \sin u \cos i \sin ☊), \\ y'' = r (\cos u \sin ☊ + \sin u \cos i \cos ☊), \\ z'' = r \sin u \sin i. \end{cases}$$

If the system is rotated now around the x-axis through the angle $-\epsilon$, the coördinates become

$$\begin{cases} x''' = x'', \\ y''' = y'' \cos \epsilon - z'' \sin \epsilon, \\ z''' = y'' \sin \epsilon + z'' \cos \epsilon; \end{cases}$$

or, in polar coördinates,

(91)
$$\begin{cases} x''' = r\{\cos u \cos ☊ - \sin u \cos i \sin ☊\}, \\ y''' = r\{(\cos u \sin ☊ + \sin u \cos i \cos ☊) \cos \epsilon \\ \qquad\qquad - \sin u \sin i \sin \epsilon\}, \\ z''' = r\{(\cos u \sin ☊ + \sin u \cos i \cos ☊) \sin \epsilon \\ \qquad\qquad + \sin u \sin i \cos \epsilon\}. \end{cases}$$

In order to facilitate the computation Gauss introduced the new auxiliaries A, a, B, b, C, and c by the equations

(92)
$$\begin{cases} \sin a \sin A = \cos ☊, \\ \sin a \cos A = - \sin ☊ \cos i, & \sin a > 0, \\ \sin b \sin B = \sin ☊ \cos \epsilon, & \sin b > 0, \\ \sin b \cos B = \cos ☊ \cos i \cos \epsilon - \sin i \sin \epsilon, \\ \sin c \sin C = \sin ☊ \sin \epsilon, & \sin c > 0, \\ \sin c \cos C = \cos ☊ \cos i \sin \epsilon + \sin i \cos \epsilon. \end{cases}$$

These constants depend upon the elements alone, so they need be computed but once for a given orbit. They are of particular advantage when the coördinates are to be computed for a large number of epochs, as in constructing an ephemeris. When these constants are substituted in (91), these equations for the heliocentric coördinates take the simple form

$$(93) \quad \begin{cases} x''' = r \sin a \sin (A + u), \\ y''' = r \sin b \sin (B + u), \\ z''' = r \sin c \sin (C + u), \end{cases}$$

from which x''', y''', and z''' can be found.

Then finally, the geocentric equatorial coördinates are defined by

$$(94) \quad \begin{cases} \rho \cos \delta \cos \alpha = x''' + X', \\ \rho \cos \delta \sin \alpha = y''' + Y', \\ \rho \sin \delta = z''' + Z', \end{cases}$$

where X', Y', and Z' are the rectangular geocentric coördinates of the sun referred to the equatorial system. They are given in the *Nautical Almanac* for every day in the year, and, therefore, these equations define ρ, α, and δ.

This completes the theory of the determination of the heliocentric and geocentric coördinates of a body, moving in any orbit, when either the ecliptic or the equatorial system is used.

XVI. PROBLEMS.

1. Interpret the angle N, equation (87), geometrically and show that n is simply a factor of proportionality.

2. Suppose the ascending node is taken always as that one which is less than 180°, and that the inclination varies from $- 90°$ to $+ 90°$; discuss the changes which will be made in the equations (78), $\cdots$, (93), and in particular, write the definitions of the Gaussian constants a, A, $\cdots\cdots$, C for this method of defining the elements.

3. Interpret the Gaussian constants, defined by (92), geometrically.

HISTORICAL SKETCH AND BIBLIOGRAPHY.

The Problem of Two Bodies for spheres of finite size was first solved by Newton about 1685, and is given in the *Principia*, Book I., Section 11. The demonstration is geometrical. The methods of the Calculus were cultivated with ardor in continental Europe at the beginning of the 18th century, but Newton's system of Mechanics did not find immediate acceptance; indeed, the French clung to the vortex theory of Descartes (1596–1650) until Voltaire, after his visit to London 1727, vigorously supported the Newtonian theory, 1728–1738. This, with the fact that the English continued to employ the geometrical methods of the *Principia*, delayed the analytical solution of the problem. It was probably accomplished by Daniel Bernouilli in the memoir for which he received the prize from the French Academy in 1734, and it was certainly solved in detail by Euler in 1744 in his *Theoria motuum planetarum et cometarum*. Since that time thè modifications have been chiefly in the choice of variables in which the problem has been expressed.

The solution of Kepler's equation naturally was first made by Kepler himself. The next was by Newton in the *Principia*. From a graphical construction involving the cycloid he was able to find very easily the approximate solution for the eccentric anomaly. A very large number of analytical and graphical solutions have been discovered, nearly every prominent mathematician from Newton until the middle of the last century having given the subject more or less attention. A bibliography containing references to 123 papers on Kepler's equation is given in the *Bulletin Astronomique*, Jan. 1900, and even this extended list is incomplete.

The transformations of coördinates involve merely the solutions of spherical triangles, the treatment of which in a perfectly general form the mathematical world owes to Gauss (1777–1855), and which was introduced into American Trigonometries by Chauvenet.

The Problem of Two Bodies is treated in every work on Analytical Mechanics. The reader will do well to consult further Tisserand's *Méc. Cél.*, vol. I., chapters VI. and VII.

CHAPTER VI.

THE DETERMINATION OF ORBITS.

108. General Consideration. In discussing the problem of two bodies [Arts. 86–88] it was shown how the constants of integration which arise when the differential equations are solved can be determined in terms of the original values of the coördinates and of the components of velocity; and then it was shown how the elements of the conic section orbit can be determined in terms of these constants. Consequently, it is natural to seek to determine the position and components of the observed body at some epoch. The difficulty arises from the fact that the observations, which are made from the moving earth, give only the direction of the object as seen by the observer, and furnish no direct information respecting its distance. An observation of apparent position simply determines the fact that the body is somewhere on one half of a defined line passing through the observer. The position of the body in space is therefore not given, and, of course, its components of velocity are not determined. It becomes necessary on this account to secure additional observations at other times. In the interval of time before the second observation is made the earth will have moved and the observed body will have gone to another place in its orbit. The second observation simply determines another line on which the body is located at another date. It is clear that the problem of finding the position of the body and the elements of its orbit from such data presents some difficulties.

The first question to settle is naturally the number of observations which are necessary in order that it shall be possible to determine the elements of the orbit. Since an orbit is defined by six elements, it follows that six independent quantities must be given by the observations in order that the elements may be determined. A single complete observation gives two quantities, the angular coördinates of the body. Therefore three complete observations are just sufficient, so far as these considerations are concerned, to define its orbit. It is at least certain that no smaller number will suffice. If the observed body is a comet whose orbit is a parabola, the eccentricity is unity and only five elements are to be found.   In this case two complete observations and one observation giving one of the two angular coördinates are enough.

109. Intermediate Elements.   The apparent positions of the observed body are usually obtained by measuring its angular distances and directions from neighboring fixed stars.   Since the stars are catalogued in right ascension and declination the results come out in these coördinates, but they can, of course, be changed to the ecliptic system, or any other, if it is desired.

Suppose the observations are made at the times t_1, t_2, and t_3, and let the corresponding coördinates be denoted by their usual symbols having the subscripts 1, 2, and 3 respectively.   The right ascensions and declinations are functions of the elements of the orbit and the dates of observation.   These relations may be represented by

$$(1) \quad \begin{cases} \alpha_1 = \varphi(\Omega, \ i, \ \omega, \ a, \ e, \ T; \ t_1), \\ \alpha_2 = \varphi(\Omega, \ i, \ \omega, \ a, \ e, \ T; \ t_2), \\ \alpha_3 = \varphi(\Omega, \ i, \ \omega, \ a, \ e, \ T; \ t_3), \\ \delta_1 = \psi(\Omega, \ i, \ \omega, \ a, \ e, \ T; \ t_1), \\ \delta_2 = \psi(\Omega, \ i, \ \omega, \ a, \ e, \ T; \ t_2), \\ \delta_3 = \psi(\Omega, \ i, \ \omega, \ a, \ e, \ T; \ t_3). \end{cases}$$

The problem consists in solving these six equations for the six unknown elements.   The functions φ and ψ are highly transcendental and involve the elements in a very complicated fashion.   In the case of an ellipse the position in the orbit is found by passing through Kepler's equation, in the hyperbola the process is similar, and in the parabola a cubic equation must be solved; and in all three cases the coördinates with respect to the earth are obtained by a number of trigonometrical transformations.   Hence it is clear that there is no direct solution of equations (1) by ordinary processes.

Although the ultimate object is to determine the elements of the orbit, the problem of finding other quantities which define the elements may be treated first.   These quantities may be considered as being *intermediate elements*.   It has been remarked that if the coördinates and the components of velocity are known at any epoch, the elements can be found.   Suppose it is desired to find the polar coördinates and their derivatives, which determine uniquely the rectangular coördinates and their derivatives, at the time of the second observation t_2. The equations corresponding to (1) become for this problem

$$(2) \quad \begin{cases} \alpha_1 = f(\alpha_2,\ \delta_2,\ \rho_2,\ \alpha_2',\ \delta_2',\ \rho_2';\ t_1,\ t_2), \\[4pt] \alpha_2 = \alpha_2, \\[4pt] \alpha_3 = f(\alpha_2,\ \delta_2,\ \rho_2,\ \alpha_2',\ \delta_2',\ \rho_2';\ t_2,\ t_3), \\[4pt] \delta_1 = g(\alpha_2,\ \delta_2,\ \rho_2,\ \alpha_2',\ \delta_2',\ \rho_2';\ t_1,\ t_2), \\[4pt] \delta_2 = \delta_2, \\[4pt] \delta_3 = g(\alpha_2,\ \delta_2,\ \rho_2,\ \alpha_2',\ \delta_2',\ \rho_2';\ t_2,\ t_3), \end{cases}$$

where

$$\alpha_2' = \frac{d\alpha}{dt}, \qquad \delta_2' = \frac{d\delta}{dt}, \qquad \rho_2' = \frac{d\rho}{dt} \quad \text{at} \quad t = t_2.$$

Since α_2 and δ_2 are observed quantities only the first, third, fourth, and sixth equations are to be solved for the four unknowns ρ_2, α_2', δ_2', and ρ_2'. The problem is therefore reduced to the solution of four simultaneous equations, and they are moreover much simpler than (1). These equations can be put in a manageable form, and this is, in fact, one of the methods of treating the problem. It was first developed and applied to the actual determination of orbits by Laplace in 1780, and it has been somewhat extended and modified as to details by many later writers.

As another set of intermediate elements the three coördinates at two epochs may be taken. Suppose the times t_1 and t_3 are chosen for this purpose. Then the fundamental equations corresponding to (1) can be written in the form

$$(3) \quad \begin{cases} \alpha_1 = \alpha_1, \\[4pt] \alpha_2 = F(\alpha_1,\ \delta_1,\ \rho_1,\ \alpha_3,\ \delta_3,\ \rho_3;\ t_1,\ t_2,\ t_3), \\[4pt] \alpha_3 = \alpha_3, \\[4pt] \delta_1 = \delta_1, \\[4pt] \delta_2 = G(\alpha_1,\ \delta_1,\ \rho_1,\ \alpha_3,\ \delta_3,\ \rho_3;\ t_1,\ t_2,\ t_3), \\[4pt] \delta_3 = \delta_3. \end{cases}$$

In this case the equations are reduced to two in the two unknowns ρ_1 and ρ_3, and they also can be solved. This is the line of attack on the problem laid out by Lagrange in 1778, taken up independently and carried out differently by Gauss in 1801, and followed more or less closely by many later writers. In spite of the hundreds of papers which have been written on the theory of the determination of orbits, very little that is really new or theoretically important has been added to the work of Laplace and Gauss unless more than three observations are used.

110. Preparation of the Observations. Whatever method it may be proposed to follow, the observations as obtained by the practical astronomer require certain slight corrections which should be made before the computation of the orbit is undertaken.

The attractions of the moon and the sun upon the equatorial bulge of the earth cause a small periodic oscillation and a slow secular change in the position of the plane of its equator. Since the equinoxes are the places where the equator and ecliptic intersect, the vernal equinox undergoes small periodic oscillations (the nutation) and slowly changes its position along the ecliptic (the precession). It is obviously necessary to have all the observations referred to the same coördinate system, and it is customary to use the mean equinox and position of the equator at the beginning of the year in which the observations are made.

The observed places are also affected by the aberration of light due to the revolution of the earth around the sun and to its rotation on its axis. Since the rotation is very slow compared to the revolution, the aberration due to the former is relatively small and generally may be neglected, especially if the observations are not very precise.

Suppose α_0 and δ_0 are the observed right ascension and declination of the body at any time. Then the right ascension and declination referred to the mean equinox of the beginning of the year, and corrected for the annual aberration, are

$$(4) \quad \begin{cases} \alpha = \alpha_0 - 15f - g \sin (G+\alpha_0) \tan \delta_0 - h \sin (H+\alpha_0) \sec \delta_0, \\ \delta = \delta_0 - i \cos \delta_0 - g \cos (G + \alpha_0) - h \cos (H + \alpha_0) \sin \delta_0, \end{cases}$$

where f, g, h, G, H, and i are auxiliary quantities, called the *Independent Star-Numbers*, which are given in the American Ephemeris and Nautical Almanac for every day of the year. In practice these numbers are to be taken from the Ephemeris. They depend upon the motions of the earth, but their derivation belongs to the domain of Spherical and Practical Astronomy, and cannot be taken up here.* The corrections to α_0 and δ_0 furnished by equations (4) are expressed in seconds of arc.

---

* Chauvenet, *Spherical and Practical Astronomy*, vol. i., chap. xi.

The corrections for the diurnal aberration are

(5)
$$\begin{cases} \Delta\alpha = -\,0''.322 \cos\varphi \cos(\theta - \alpha_0) \sec\delta_0, \\ \Delta\delta = -\,0''.322 \cos\varphi \sin(\theta - \alpha_0) \sin\delta_0, \end{cases}$$

where φ is the latitude of the observer, and $\theta - \alpha_0$ is the hour angle of the object at the time of the observation. The second of these corrections cannot exceed the small quantity $0''.322$, and the first is also small unless δ_0 is near $\pm\,90°$.

111. Outline of the Laplacian Method of Determining an Orbit. Before entering on the details which are necessary for the determination of the elements of an orbit by either of the two methods which are in common use, a brief exposition of the general lines of argument used in them will be given. From these outlines the plan of attack can be understood, and then the bearings of the detailed investigations will be fully appreciated.

In order to keep to the central thought suppose only three complete observations are available for the determination of the orbit. Let the dates of the observations be t_1, t_2, and t_3, and hence at these times the right ascensions and declinations of the observed body as seen from the earth are known. For the sake of definiteness in the terminology let C represent the observed body revolving around the sun, S, and observed from the earth E; ξ, η, ζ the rectangular coördinates of C with respect to E; x, y, z the rectangular coördinates of C with respect to S; X, Y, Z the rectangular coördinates of S with respect to E; ρ the distance from E to C; r the distance from S to C; R the distance from E to S. Then

(6)
$$\begin{cases} \xi = \rho \cos\delta \cos\alpha = \rho\lambda, \\ \eta = \rho \cos\delta \sin\alpha = \rho\mu, \\ \zeta = \rho \sin\delta \quad\;\; = \rho\nu. \end{cases}$$

The quantities λ, μ, and ν, which are the direction cosines of the line from E to C, are known at t_1, t_2, and t_3. The distance ρ is entirely unknown.

First Step. The first step is to determine the values of the first and second derivatives of λ, μ, ν, X, Y, and Z at some time near the dates of observation, say at t_2. It will be sufficient at present to show that it can be done with considerable approximation without discussing the best method of doing it. The value of the first derivative of λ during the interval t_1 to t_2 averages

$$\lambda'_{12} = \frac{\lambda_2 - \lambda_1}{t_2 - t_1},$$

and this is very nearly the value of λ' at the middle of the interval unless λ' happens to be changing very rapidly. The approximation is better the shorter the interval. In a similar manner λ'_{23} is formed. When the interval $t_2 - t_1$ equals the interval $t_3 - t_2$ the value of λ' at t_2 is very nearly

$$\lambda'_2 = \tfrac{1}{2}[\lambda'_{12} + \lambda'_{23}].$$

If the intervals are not equal, adjustment for the disparity can of course be made.

In a similar manner it follows from the definition of a derivative that the second derivative of λ at t_2, in case the two intervals are equal, is approximately

$$\lambda''_2 = \frac{\lambda'_{23} - \lambda'_{12}}{\tfrac{1}{2}(t_3 - t_1)}.$$

The first and second derivatives of μ and ν are given approximately by similar formulas, and it is to be understood that when the intervals are as short as they generally are in practice the approximations, especially as obtained by the more refined methods which will be considered in the detailed discussion, are very close. The American Ephemeris gives the values of X, Y, and Z for every day in the year, and from these data the values of their first and second derivatives can be found. As a matter of fact only the first derivatives of these coördinates will be required.

Second Step. The second step is to impose the condition that C moves around S in accordance with the law of gravitation. It will be assumed that C is not sensibly disturbed by the attractions of other bodies. Hence its coördinates satisfy the differential equations

(7)
$$\begin{cases} \dfrac{d^2x}{dt^2} = -\dfrac{k^2x}{r^3}, \\[2mm] \dfrac{d^2y}{dt^2} = -\dfrac{k^2y}{r^3}, \\[2mm] \dfrac{d^2z}{dt^2} = -\dfrac{k^2z}{r^3}. \end{cases}$$

But it also follows from the relations of C, E, and S that

(8)
$$\begin{cases} x = \rho\lambda - X, \\ y = \rho\mu - Y, \\ z = \rho\nu - Z. \end{cases}$$

On substituting these expressions for x, y, and z in equations (7), they become

(9)
$$\begin{cases} (\rho\lambda)'' - X'' = \dfrac{-k^2(\rho\lambda - X)}{r^3}, \\[2mm] (\rho\mu)'' - Y'' = \dfrac{-k^2(\rho\mu - Y)}{r^3}, \\[2mm] (\rho\nu)'' - Z'' = \dfrac{-k^2(\rho\nu - Z)}{r^3}. \end{cases}$$

But since E also revolves around S in accordance with the law of gravitation, it follows that

$$X'' = -\frac{k^2X}{R^3},$$

$$Y'' = -\frac{k^2Y}{R^3},$$

$$Z'' = -\frac{k^2Z}{R^3}.$$

Therefore equations (9) become

(10)
$$\begin{cases} \lambda\rho'' + 2\lambda'\rho' + \left[\lambda'' + \dfrac{k^2\lambda}{r^3}\right]\rho = -k^2X\left[\dfrac{1}{R^3} - \dfrac{1}{r^3}\right], \\[3mm] \mu\rho'' + 2\mu'\rho' + \left[\mu'' + \dfrac{k^2\mu}{r^3}\right]\rho = -k^2Y\left[\dfrac{1}{R^3} - \dfrac{1}{r^3}\right], \\[3mm] \nu\rho'' + 2\nu'\rho' + \left[\nu'' + \dfrac{k^2\nu}{r^3}\right]\rho = -k^2Z\left[\dfrac{1}{R^3} - \dfrac{1}{r^3}\right]. \end{cases}$$

The unknown quantities in these equations are ρ'', ρ', ρ, and r, the first three of which enter linearly.

Third Step. The third step is to determine the distance of C from E and S by means of equations (10) and a geometrical condition which the three bodies must satisfy. In order to solve equations (10) for ρ, let

(11)
$$D = \begin{vmatrix} \lambda, & \lambda', & \lambda'' + \dfrac{k^2\lambda}{r^3} \\[3mm] \mu, & \mu', & \lambda'' + \dfrac{k^2\mu}{r^3} \\[3mm] \nu, & \nu', & \lambda'' + \dfrac{k^2\nu}{r^3} \end{vmatrix} = \begin{vmatrix} \lambda, & \lambda', & \lambda'' \\[2mm] \mu, & \mu', & \mu'' \\[2mm] \nu, & \nu', & \nu'' \end{vmatrix}.$$

The second form of the determinant D is obtained by multiplying the first column by $\dfrac{k^2}{r^3}$ and subtracting the product from the third column. The determinant which is obtained by replacing the elements of the third column of D by the right member of (10) will also be needed. If the common factor $\left[\dfrac{1}{R^3} - \dfrac{1}{r^3}\right]$ is omitted, this determinant is

$$(12) \qquad D_1 = -k^2 \begin{vmatrix} \lambda, & \lambda', & X \\ \mu, & \mu', & Y \\ \nu, & \nu', & Z \end{vmatrix}.$$

The determinants D and D_1 involve only known quantities.

The solution of equations (10) for ρ is

$$(13) \qquad \rho = \dfrac{D_1}{D}\left[\dfrac{1}{R^3} - \dfrac{1}{r^3}\right].$$

To this equation in the two unknown quantities ρ and r must be added the equation

$$(14) \qquad r^2 = \rho^2 + R^2 - 2\rho R \cos\psi,$$

which expresses the fact that the three bodies C, S, and E form a triangle. The angle ψ is the angle at E between R and ρ, and this equation also has only the unknowns ρ and r. The problem of solving (13) and (14) for ρ and r is that which constitutes the third step. The solution of this problem gives the coördinates of C by means of equations (8) which involve only ρ as an unknown.

Fourth Step. The fourth step is the determination of the components of velocity of C. It follows from (8) that

$$(15) \qquad \begin{cases} x' = \rho'\lambda + \rho\lambda' - X', \\ y' = \rho'\mu + \rho\mu' - Y', \\ z' = \rho'\nu + \rho\nu' - Z'. \end{cases}$$

The only unknown in the right members of these equations is ρ' which can be determined from (10). The expression for it is

$$(16) \qquad \begin{cases} \rho' = +\dfrac{D_2}{2D}\left[\dfrac{1}{R^3} - \dfrac{1}{r^3}\right], \\[2ex] D_2 = -k^2 \begin{vmatrix} \lambda, & X, & \lambda'' \\ \mu, & Y, & \mu'' \\ \nu, & Z, & \nu'' \end{vmatrix}. \end{cases}$$

Therefore x', y', and z' become known.

Fifth Step. The fifth and last step is to determine the elements of the orbit from the position and components of velocity of the body. This is the problem which was solved in chap. v.

112. Outline of the Gaussian Method of Determining an Orbit. *First Step.* The first step in the Gaussian method is to impose the condition that C moves in a plane passing through S. Since S is the origin for the coördinates x, y, and z, this condition is

$$\left\{ \begin{array}{l} Ax_1 + By_1 + Cz_1 = 0, \\ Ax_2 + By_2 + Cz_2 = 0, \\ Ax_3 + By_3 + Cz_3 = 0, \end{array} \right.$$

where A, B, C are constants which depend upon the position of the plane of motion. The result of eliminating the unknown constants A, B, and C is the equation

$$(17) \qquad \begin{vmatrix} x_1, & y_1, & z_1 \\ x_2, & y_2, & z_2 \\ x_3, & y_3, & z_3 \end{vmatrix} = 0.$$

The determinant (17) can be expanded with respect to the elements of the three columns giving the three equations

$$(18) \quad \left\{ \begin{array}{l} (y_2z_3 - z_2y_3)x_1 - (y_1z_3 - z_1y_3)x_2 + (y_1z_2 - z_1y_2)x_3 = 0, \\ (x_2z_3 - z_2x_3)y_1 - (x_1z_3 - z_1x_3)y_2 + (x_1z_2 - z_1x_2)y_3 = 0, \\ (x_2y_3 - y_2x_3)z_1 - (x_1y_3 - y_1x_3)z_2 + (x_1y_2 - y_1x_2)z_3 = 0. \end{array} \right.$$

Evidently these three equations are but different forms of the same one; but when the nine parentheses are determined from additional principles and x_1, x_2, $\cdots$ are expressed in terms of the geocentric coördinates by (8), they become independent in the unknowns ρ_1, ρ_2, and ρ_3. The parentheses are the projections of twice the triangles formed by S and the positions of C taken in twos upon the three fundamental planes. Since in each equation the three areas are projected upon the same plane the triangles themselves can be used instead of their projections. If [1, 2], [1, 3], and [2, 3] represent the triangles formed by S and C at the times t_1t_2, t_1t_3, and t_2t_3 respectively, equations (18) become

$$(19) \quad \left\{ \begin{array}{l} [2, 3]\,x_1 - [1, 3]\,x_2 + [1, 2]\,x_3 = 0, \\ [2, 3]\,y_1 - [1, 3]\,y_2 + [1, 2]\,y_3 = 0, \\ [2, 3]\,z_1 - [1, 3]\,z_2 + [1, 2]\,z_3 = 0. \end{array} \right.$$

Second Step. The second step consists in developing the ratios of the triangles as power series in the time-intervals. This is done by integrating equations (7) as power series in the time-intervals, and then substituting the results for $t = t_1$, t_2, t_3 in the coefficients of (18) or (19). Inasmuch as these series are based upon equations (7) the condition that C shall move about S in accordance with the law of gravitation has been imposed. In order not to prolong the discussion at this point (for the details see Art. 127) the results will be given at once. For the purpose of simplifying the writing, let

(20)
$$\begin{cases} k(t_2 - t_1) = \theta_3, \\ k(t_3 - t_2) = \theta_1, \\ k(t_3 - t_1) = \theta_2, \\ \theta_2 = \theta_1 + \theta_3. \end{cases}$$

In this notation the ratios of the triangles [2, 3] and [1, 2] to [1, 3] are found to be

(21)
$$\begin{cases} \dfrac{[2, 3]}{[1, 3]} = \dfrac{\theta_1}{\theta_2} \left[1 + \dfrac{1}{6} \dfrac{\theta_2{}^2 - \theta_1{}^2}{r_2{}^3} + \cdots \right], \\ \dfrac{[1, 2]}{[1, 3]} = \dfrac{\theta_3}{\theta_2} \left[1 + \dfrac{1}{6} \dfrac{\theta_2{}^2 - \theta_3{}^2}{r_2{}^3} + \cdots \right]. \end{cases}$$

Third Step. The third step consists in developing equations for the determination of ρ_1, ρ_2, and ρ_3. The results of substituting equations (8) and (21) in (19) are

(22)
$$\begin{cases} \theta_1 \left[1 + \dfrac{1}{6} \dfrac{\theta_2{}^2 - \theta_1{}^2}{r_2{}^3} + \cdots \right] (\lambda_1 \rho_1 - X_1) - \theta_2 (\lambda_2 \rho_2 - X_2) \\ \qquad + \theta_3 \left[1 + \dfrac{1}{6} \dfrac{\theta_2{}^2 - \theta_3{}^2}{r_2{}^3} + \cdots \right] (\lambda_3 \rho_3 - X_3) = 0, \\[2ex] \theta_1 \left[1 + \dfrac{1}{6} \dfrac{\theta_2{}^2 - \theta_1{}^2}{r_2{}^3} + \cdots \right] (\mu_1 \rho_1 - Y_1) - \theta_2 (\mu_2 \rho_2 - Y_2) \\ \qquad + \theta_3 \left[1 + \dfrac{1}{6} \dfrac{\theta_2{}^2 - \theta_3{}^2}{r_2{}^3} + \cdots \right] (\mu_3 \rho_3 - Y_3) = 0, \\[2ex] \theta_1 \left[1 + \dfrac{1}{6} \dfrac{\theta_2{}^2 - \theta_1{}^2}{r_2{}^3} + \cdots \right] (\nu_1 \rho_1 - Z_1) - \theta_2 (\nu_2 \rho_2 - Z_2) \\ \qquad + \theta_3 \left[1 + \dfrac{1}{6} \dfrac{\theta_2{}^2 - \theta_3{}^2}{r_2{}^3} + \cdots \right] (\nu_3 \rho_3 - Z_3) = 0. \end{cases}$$

These equations involve the unknowns ρ_1, ρ_2, ρ_3, and r_2, the first three of which enter linearly. Since r_2 enters only as it is multiplied by the small quantities θ_1^2, θ_2^2, or θ_3^2, it might be supposed that in a first approximation these terms could be neglected, after which ρ_1, ρ_2, and ρ_3 would be determined by linear equations. A detailed discussion of the determinants which are involved shows, however, that it is necessary to retain the terms in r_2 even in the first approximation.

The solution of equations (22) for ρ_2 has the form

$$(23) \qquad \Delta\rho_2 = P + \frac{Q}{r_2^3},$$

where Δ is the determinant of the coefficients of ρ_1, ρ_2, and ρ_3, and P and Q are functions of the known quantities λ_1, λ_2, $\cdots$, X_1, Y_1, $\cdots$.

Since S, E, and C form a triangle at t_2 the quantities ρ_2 and r_2 satisfy the equation

$$(24) \qquad r_2^2 = \rho_2^2 + R_2^2 - 2\rho_2 R_2 \cos \psi_2.$$

The solution of any two equations of (22) for ρ_1 and ρ_3 in terms of ρ_2 and r_2 has the form

$$(25) \quad \begin{cases} M\rho_1 = P_1\rho_2 \left[1 - \frac{1}{6} \frac{\theta_2^2 - \theta_1^2}{r_2^3} + \cdots \right] + Q_1, \\[2ex] M\rho_3 = P_3\rho_2 \left[1 - \frac{1}{6} \frac{\theta_2^2 - \theta_3^2}{r_2^3} + \cdots \right] + Q_3, \end{cases}$$

where M, P_1, P_3 are functions of known quantities, and Q_1 and Q_3 involve only r_2 as an unknown.

Fourth Step. The fourth step consists in determining ρ_1 and ρ_3. The quantities ρ_2 and r_2 are found first by solving (23) and (24), which is exactly the same as the third step of the Laplacian method, and then ρ_1 and ρ_3 are given by (25).

Fifth Step. The fifth step consists in determining the elements from the known positions of C at the times t_1 and t_3. These two positions and that of C define the plane of the orbit without further work. Gauss solved the problem of determining the remaining elements by developing two equations involving only two unknowns. One equation was derived from the ratio of the triangle formed by S and C at t_1 and t_3 to the area of the sector contained between r_1, r_3, and the arc of the orbit described in the interval $t_1 t_3$. The other equation was derived from Kepler's equation at the epochs t_1 and t_3. The formulas are complex, but the method of solving the two equations is a rapid process of successive approximations. After the equations are solved the elements are uniquely determined without any trouble. Later methods have been devised which avoid many of the complexities of that due to Gauss.

I. THE LAPLACIAN METHOD OF DETERMINING ORBITS.

113. Determination of the First and Second Derivatives of the Angular Coördinates from Three Observations. It was found in the outline [Art. 111] of this method of determining orbits that the first and second derivatives of the angular coördinates, or of the direction cosines λ, μ, and ν will be required.

Let $k(t - t_0) = \tau$ and then equations (7) become

$$(26) \quad \begin{cases} \dfrac{d^2x}{d\tau^2} = -\dfrac{x}{r^3}, \\[2mm] \dfrac{d^2y}{d\tau^2} = -\dfrac{y}{r^3}, \\[2mm] \dfrac{d^2z}{d\tau^2} = -\dfrac{z}{r^3}. \end{cases}$$

Suppose $x = x_0$, $y = y_0$, $z = z_0$, $\dfrac{dx}{d\tau} = x_0'$, $\dfrac{dy}{d\tau} = y_0'$, $\dfrac{dz}{d\tau} = z_0'$ at $\tau = 0$. The solution of equations (26) can be expanded as power series in τ which will converge if the value of τ is not too great.* They will have the form

$$(27) \quad \begin{cases} x = x_0 + x_0'\tau + \dfrac{1}{2}\left(\dfrac{d^2x}{d\tau^2}\right)_0 \tau^2 + \cdots + \dfrac{1}{n!}\left(\dfrac{d^n x}{d\tau^n}\right)_0 \tau^n + \cdots, \\[3mm] y = y_0 + y_0'\tau + \dfrac{1}{2}\left(\dfrac{d^2y}{d\tau^2}\right)_0 \tau^2 + \cdots + \dfrac{1}{n!}\left(\dfrac{d^n y}{d\tau^n}\right)_0 \tau^n + \cdots, \\[3mm] z = z_0 + z_0'\tau + \dfrac{1}{2}\left(\dfrac{d^2z}{d\tau^2}\right)_0 \tau^2 + \cdots + \dfrac{1}{n!}\left(\dfrac{d^n z}{d\tau^n}\right)_0 \tau^n + \cdots, \end{cases}$$

where the subscript 0 on the parentheses indicates that the derivatives are taken for $\tau = 0$. The second derivatives can be replaced by the right members of (26) for $\tau = 0$; the third derivatives can be replaced by the first derivatives of the right members of (26), and so on. All the derivatives in this way will be expressed in terms of x_0, y_0, z_0, x_0', y_0', and z_0'.

* For the determination of the exact realm of convergence see a paper by F. R. Moulton in *The Astronomical Journal*, vol. 23 (1903).

It is important to know for how great intervals the series (27) are of practical value. The limits are smaller the smaller the perihelion distance and the greater the eccentricity, and moreover they depend upon the position of the body in its orbit at $\tau = 0$. For a small planet whose mean distance is 2.65, which is about the average for these bodies, and the eccentricity of whose orbit does not exceed 0.4, which is much greater than that of most of them, the series (27) always converge for an interval of less than 160 days. If the orbit is a parabola whose perihelion distance is unity the series (27) converge if the interval of time does not exceed 54 days. Of course, the series are not of practical value in their whole range of convergence. In practice in the case of small planets an interval of 90 days is nearly always small enough to secure rapid convergence of (27), and in the case of the orbits of comets 20 days is rarely too great an interval.

The coördinates of the earth also are expansible as series of the form of (27), and the rapid convergence holds for very long intervals because of the small eccentricity of the earth's orbit. Hence it follows from equations (8) that ρ, λ, μ, and ν can be expanded as power series of the type of (27). The range of usefulness of these expansions is the same as that of the series for x, y, and z.

It will be sufficient to consider the series for λ because those in μ and ν are symmetrically similar. The series for λ for a general value of τ and for τ_1, τ_2, and τ_3, which correspond to t_1, t_2, and t_3 respectively, are

$$(28) \quad \begin{cases} \lambda = c_0 + c_1\tau + c_2\tau^2 + \cdots, \\ \lambda_1 = c_0 + c_1\tau_1 + c_2\tau_1^2 + \cdots, \\ \lambda_2 = c_0 + c_1\tau_2 + c_2\tau_2^2 + \cdots, \\ \lambda_3 = c_0 + c_1\tau_3 + c_2\tau_3^2 + \cdots, \end{cases}$$

where c_0, c_1, c_2, $\cdots$ are constants. If these equations are terminated after the terms of the second degree the coefficients c_0, c_1, and c_2 are determined in terms of the observed quantities λ_1, λ_2, and λ_3, and the time-intervals τ_1, τ_2, and τ_3. If more observations are available more coefficients can be determined; the number which can be determined equals the number of observations.

The simplest way of expressing λ in terms of τ with known coefficients is to set equal to zero the eliminant of 1, c_0, c_1, and c_2 in (28), which is

(29)
$$\begin{vmatrix} \lambda, & 1, & \tau, & \tau^2 \\ \lambda_1, & 1, & \tau_1, & \tau_1^2 \\ \lambda_2, & 1, & \tau_2, & \tau_2^2 \\ \lambda_3, & 1, & \tau_3, & \tau_3^2 \end{vmatrix} = 0.$$

The expansion of this determinant with respect to the elements of the first column is

(30) $$A_0\lambda - A_1\lambda_1 + A_2\lambda_2 - A_3\lambda_3 = 0,$$

where

$$A_0 = \begin{vmatrix} 1, & \tau_1, & \tau_1^2 \\ 1, & \tau_2, & \tau_2^2 \\ 1, & \tau_3, & \tau_3^2 \end{vmatrix} = -(\tau_2 - \tau_1)(\tau_3 - \tau_2)(\tau_1 - \tau_3),$$

and where A_1, A_2, and A_3 are obtained from A_0 by permuting τ with τ_1, τ_2, and τ_3 respectively. The determinant A_0 is distinct from zero if τ_1, τ_2, and τ_3 are distinct. Hence equation (30) becomes

(31)
$$\lambda = \frac{(\tau - \tau_2)(\tau - \tau_3)}{(\tau_1 - \tau_2)(\tau_1 - \tau_3)}\lambda_1 + \frac{(\tau - \tau_3)(\tau - \tau_1)}{(\tau_2 - \tau_3)(\tau_2 - \tau_1)}\lambda_2$$
$$+ \frac{(\tau - \tau_1)(\tau - \tau_2)}{(\tau_3 - \tau_1)(\tau_3 - \tau_2)}\lambda_3.$$

It follows from the form of (31) that this equation gives λ exactly at τ_1, τ_2, and τ_3; for other small values of τ it gives λ approximately. The exact value of λ is given by an infinite series,

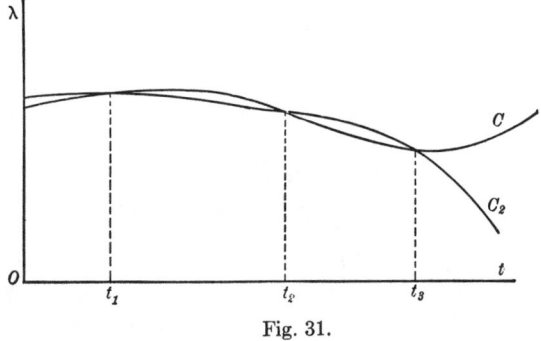

Fig. 31.

the first equation of (28), within the range of its convergence. Geometrically considered this series defines a curve, marked C in Fig. 31. The second degree polynomial (31) defines another curve C_2. These two curves intersect at τ_1, τ_2, and τ_3, but in general do not intersect elsewhere. For small values of τ the two curves nearly coincide, and the approximate value of λ can be found from the polynomial near the origin.

The first and second derivatives of λ are found from (31) to be given approximately by

$$(32) \begin{cases} \lambda' = \dfrac{2\tau - (\tau_2 + \tau_3)}{(\tau_1 - \tau_2)(\tau_1 - \tau_3)} \lambda_1 + \dfrac{2\tau - (\tau_3 + \tau_1)}{(\tau_2 - \tau_3)(\tau_2 - \tau_1)} \lambda_2 \\ \qquad + \dfrac{2\tau - (\tau_1 + \tau_2)}{(\tau_3 - \tau_1)(\tau_3 - \tau_2)} \lambda_3, \\ \lambda'' = \dfrac{2}{(\tau_1 - \tau_2)(\tau_1 - \tau_3)} \lambda_1 + \dfrac{2}{(\tau_2 - \tau_3)(\tau_2 - \tau_1)} \lambda_2 \\ \qquad + \dfrac{2}{(\tau_3 - \tau_1)(\tau_3 - \tau_2)} \lambda_3. \end{cases}$$

There are similar expressions in μ and ν.

114. Determination of the Derivatives from more than Three Observations. If the observations were perfectly exact and near together, the more there were available the more exactly could λ be determined for small values of τ, and the more of its derivatives could be determined. Suppose there are four observations. Then λ is defined by a third degree polynomial analogous to (31) which reduces to λ_1, λ_2, λ_3, and λ_4 for $\tau = \tau_1, \tau_2, \tau_3$, and τ_4 respectively. The explicit expression for λ is

$$(33) \begin{cases} \lambda = + \dfrac{(\tau - \tau_2)(\tau - \tau_3)(\tau - \tau_4)}{(\tau_1 - \tau_2)(\tau_1 - \tau_3)(\tau_1 - \tau_4)} \lambda_1 \\ \qquad + \dfrac{(\tau - \tau_3)(\tau - \tau_4)(\tau - \tau_1)}{(\tau_2 - \tau_3)(\tau_2 - \tau_4)(\tau_2 - \tau_1)} \lambda_2 \\ \qquad + \dfrac{(\tau - \tau_4)(\tau - \tau_1)(\tau - \tau_2)}{(\tau_3 - \tau_4)(\tau_3 - \tau_1)(\tau_3 - \tau_2)} \lambda_3 \\ \qquad + \dfrac{(\tau - \tau_1)(\tau - \tau_2)(\tau - \tau_3)}{(\tau_4 - \tau_1)(\tau_4 - \tau_2)(\tau_4 - \tau_3)} \lambda_4, \end{cases}$$

from which the first, second, and third, but not higher, derivatives can be found.

It is obvious from this how to proceed for any number of observations. The process is unique and does not become excessively laborious unless the number of observations is considerable. The number of derivatives which can be determined, at least approximately, is one less than the number of observations, but no derivative higher than the third will in any case be used. If the observations extend over a long period so that the convergence of (28) fails or becomes slow for the largest values of τ, it is necessary to omit some of them in the discussion. Usually, owing to the errors in the observations, four or five will give λ and its first two derivatives as accurately as any greater number.

115. The Approximations in the Determination of the Values of λ, μ, ν and their Derivatives. In the applications it is important to know the character of the approximations which are made, and whether all the quantities employed are determined with the same degree of accuracy. It is obvious no exact numerical answers can be given to these questions because the orbits under consideration are undetermined. But it has been insisted that the values of τ must not be too great in order that the series (28) shall converge rapidly. Consequently, the values of τ at the times of the observations can be considered as small quantities, and the degree of the approximation can be described in terms of the lowest powers of the τ_j which occur in the neglected terms. This gives a definite meaning to the order of approximation, and experience shows that it is a satisfactory measure of the accuracy of the results when the time-intervals are limited as described in Art. 113.

Suppose first that only three observations have been made. The approximations in the determination of λ and its derivatives arise from the fact that the higher terms of (28) are neglected. The coefficients c_0, c_1, and c_2 are determined by

$$(34) \quad \begin{cases} c_0 + c_1\tau_1 + c_2\tau_1{}^2 = \lambda_1 - c_3\tau_1{}^3 - c_4\tau_1{}^4 - \cdots, \\ c_0 + c_1\tau_2 + c_2\tau_2{}^2 = \lambda_2 - c_3\tau_2{}^3 - c_4\tau_2{}^4 - \cdots, \\ c_0 + c_1\tau_3 + c_2\tau_3{}^2 = \lambda_3 - c_3\tau_3{}^3 - c_4\tau_3{}^4 - \cdots. \end{cases}$$

The errors of lowest degree in the τ_j come from neglecting the terms in the right members which are multiplied by the unknown constant c_3. Let the errors be denoted by Δc_0, Δc_1, and Δc_2. Then

$$\begin{vmatrix} 1, & \tau_1, & \tau_1{}^2 \\ 1, & \tau_2, & \tau_2{}^2 \\ 1, & \tau_3, & \tau_3{}^2 \end{vmatrix} \Delta c_0 = - \begin{vmatrix} c_3\tau_1{}^3 + c_4\tau_1{}^4 + \cdots, & \tau_1, & \tau_1{}^2 \\ c_3\tau_2{}^3 + c_4\tau_2{}^4 + \cdots, & \tau_2, & \tau_2{}^2 \\ c_3\tau_3{}^3 + c_4\tau_3{}^4 + \cdots, & \tau_3, & \tau_3{}^2 \end{vmatrix}$$

$$= - c_3 \begin{vmatrix} \tau_1{}^3, & \tau_1, & \tau_1{}^2 \\ \tau_2{}^3, & \tau_2, & \tau_2{}^2 \\ \tau_3{}^3, & \tau_3, & \tau_3{}^2 \end{vmatrix} - c_4 \begin{vmatrix} \tau_1{}^4, & \tau_1, & \tau_1{}^2 \\ \tau_2{}^4, & \tau_2, & \tau_2{}^2 \\ \tau_3{}^4, & \tau_3, & \tau_3{}^2 \end{vmatrix} - \cdots,$$

and similar expressions for Δc_1 and Δc_2. These easily reduced by the elementary rules for s. minants, and it is found that

$$
(35) \begin{cases}
\Delta c_0 = -c_3\tau_1\tau_2\tau_3 - c_4\tau_1\tau_2\tau_3(\tau_1 + \tau_2 + \tau_3) \\
\Delta c_1 = +c_3(\tau_1\tau_2 + \tau_2\tau_3 + \tau_3\tau_1) \\
\qquad\qquad + c_4(\tau_1 + \tau_2)(\tau_2 + \tau_3)(\tau_3 + {}_{\,\prime} + \cdots, \\
\Delta c_2 = -c_3(\tau_1 + \tau_2 + \tau_3) \\
\qquad\qquad - c_4({\tau_1}^2 + {\tau_2}^2 + {\tau_3}^2 + \tau_1\tau_2 + \tau_2\tau_3 + \tau_3\tau_1) + \cdots.
\end{cases}
$$

It follows from these equations that c_0, c_1, and c_2 are determined up to the third, second, and first orders respectively.

Now consider the first equation of (28). Since c_1 is multiplied by τ and c_2 by τ^2, each of the first three terms in the series for λ is determined up to the third order in the τ_j. On taking the first and second derivatives, it is seen that λ' and λ'' are determined up to the second and first orders respectively. Consequently, λ in general is determined by the first terms of (28) more accurately than its first derivative, and its first derivative in general is determined more accurately than its second derivative. These facts must be remembered in the applications.

116. Choice of the Origin of Time. The origin of time has not been specified as yet except that it has been supposed that it is near the dates of the observations so that τ_1, τ_2, and τ_3 will be small. Any epoch t_0 which satisfies this condition can be used as an origin, and the problem at once arises of determining what one is most advantageous.

The choice of the origin of time which has been universally made is the date of the second observation. That is, $t_0 = t_2$ and therefore $\tau_2 = 0$. The value of λ is exactly known at $\tau = \tau_2 = 0$, and the derivative of λ at $t = t_2$ is

$$
\lambda_2' = c_1 + 2c_2\tau_2 + \cdots = c_1,
$$

which is subject to the error Δc_1, which, by (35), is in this case $c_3\tau_3\tau_1$. And similarly, the error in λ_2'' is $\Delta c_2 = -c_3[\tau_1 + \tau_3]$. The error in λ_2' is of the second order while that in λ_2'' is of the first order. In general, an error of the first order is more serious than one of the second order. But it should be noticed that when $t_0 = t_2$ the quantities τ_1 and τ_3 are opposite in sign; and if the intervals between the successive observations are equal, $\tau_1 + \tau_3 = 0$ and the error in λ_2'' is also of the second order. Conᶜequently, when $t_0 = t_2$ it is advantageous to have the successive observations separated by as nearly equal time-intervals as possible. But unfavorable weather and other circumstances generally cause the observations to be unequally spaced.

Suppose the epoch of the first observation is taken as the origin of time. The quantity λ_1 is exactly known. The error in λ_1' is $\Delta c_1 = c_3\tau_2\tau_3$, which is of the second order as before, but is approximately twice as great numerically as that in λ_2' because τ_3 now represents k times the whole interval between the first and third observations. The error in λ_1'' is $\Delta c_2 = -c_3(\tau_2 + \tau_3)$ which is much larger than before because τ_3 now depends on the whole interval covered by the observations, and because τ_2 and τ_3 in this case are both positive. It follows from this that it is not advantageous to use the time of the first observation as the origin of time; and for similar reasons the epoch of the third observation is to be rejected.

The question now arises what should be taken for the origin of time when the epoch of the second observation is not midway between those of the other two. Since in general the error in λ is only of the third order and that in λ' is only of the second, while λ'' is subject to an error of the first order, it is clear that the origin of time should be so chosen, if possible, as to make the first order error in λ'' vanish. It follows from the second equation of (35) that this result will be secured if

$$(36) \quad \begin{cases} \tau_1 + \tau_2 + \tau_3 = k(t_1 - t_0) + k(t_2 - t_0) + k(t_3 - t_0) = 0, \\ \text{whence } t_0 = \tfrac{1}{3}(t_1 + t_2 + t_3). \end{cases}$$

The best choice of the origin of time is therefore given by the second of (36), and this value of t_0 becomes the date of the second observation when the successive observations are equally distant from one another. With this choice of t_0 the errors in λ' and λ'' are of the second order, while λ is known up to the third order.

117. The Approximations when there are Four Observations. When there are four observations the equations which correspond to the last three of (28) are

$$(37) \quad \begin{cases} c_0 + c_1\tau_1 + c_2\tau_1{}^2 + c_3\tau_1{}^3 = \lambda_1 - (c_4\tau_1{}^4 + \cdots), \\ c_0 + c_1\tau_2 + c_2\tau_2{}^2 + c_3\tau_2{}^3 = \lambda_2 - (c_4\tau_2{}^4 + \cdots), \\ c_0 + c_1\tau_3 + c_2\tau_3{}^2 + c_3\tau_3{}^3 = \lambda_3 - (c_4\tau_3{}^4 + \cdots), \\ c_0 + c_1\tau_4 + c_2\tau_4{}^2 + c_3\tau_4{}^3 = \lambda_4 - (c_3\tau_4{}^4 + \cdots). \end{cases}$$

The determinant of the coefficients of c_0, c_1, c_2, and c_3 is

$$\delta = \begin{vmatrix} 1, & \tau_1, & \tau_1^2, & \tau_1^3 \\ 1, & \tau_2, & \tau_2^2, & \tau_2^3 \\ 1, & \tau_3, & \tau_3^2, & \tau_3^3 \\ 1, & \tau_4, & \tau_4^2, & \tau_4^3 \end{vmatrix} = (\tau_2 - \tau_1)(\tau_3 - \tau_1)(\tau_4 - \tau_1)(\tau_3 - \tau_2) \\ \times (\tau_4 - \tau_2)(\tau_4 - \tau_3),$$

which is not zero since the dates of the observations are distinct.

The errors of lowest order in c_0, c_1, c_2, and c_3 are determined from (37); when only the first terms in the right members are known they contain c_4 as a factor. Let these errors be represented by Δc_0, Δc_1, Δc_2, and Δc_3; their orders in the τ_j are required. The expression for Δc_0 is

$$\Delta c_0 = \frac{-c_4}{\delta} \begin{vmatrix} \tau_1^4, & \tau_1, & \tau_1^2, & \tau_1^3 \\ \tau_2^4, & \tau_2, & \tau_2^2, & \tau_2^3 \\ \tau_3^4, & \tau_3, & \tau_3^2, & \tau_3^3 \\ \tau_4^4, & \tau_4, & \tau_4^2, & \tau_4^3 \end{vmatrix}.$$

When the factors τ_1, τ_2, τ_3, and τ_4 are removed from this determinant it is identical with δ except the columns are permuted. Three permutations of columns bring it to the form of δ; hence

(38) $$\Delta c_0 = + c_4 \tau_1 \tau_2 \tau_3 \tau_4.$$

The expression for Δc_1 is

$$\Delta c_1 = \frac{-c_4}{\delta} \begin{vmatrix} 1, & \tau_1^4, & \tau_1^2, & \tau_1^3 \\ 1, & \tau_2^4, & \tau_2^2, & \tau_2^3 \\ 1, & \tau_3^4, & \tau_3^2, & \tau_3^3 \\ 1, & \tau_4^4, & \tau_4^2, & \tau_4^3 \end{vmatrix}.$$

If τ_2 is put equal to τ_1 in this determinant it vanishes because then two lines become the same. Therefore it is divisible by $\tau_2 - \tau_1$. Similarly, it is divisible by $\tau_3 - \tau_1$, $\tau_4 - \tau_1$, $\tau_3 - \tau_2$, $\tau_4 - \tau_2$, and $\tau_4 - \tau_3$; that is, it is divisible by δ. All the elements of each column are of the same degree; and since every term of the expansion of a determinant has a factor from each column, the terms of the expansion are all of the same degree. The degree of this determinant is nine, because this is the sum of the degrees of its columns. Hence Δc_1 is of the third degree because δ is of the sixth degree. Moreover, it is symmetrical in τ_1, $\cdots$, τ_4 because both δ and the numerator determinant are symmetrical in these quantities. Each term of the expansion contains τ_j only to the first degree because τ_j occurs in the numerator determinant to the fourth degree as the highest, and in δ to the third degree. The numerical coefficient of each term in the expansion is the same, because of the symmetry, and it can be determined by the consideration of a single term. It is found by considering the product of the main diagonal elements that it is $+ 1$. Analogous discussions can be made for Δc_2 and Δc_3, and it is found in this way that

$$(39) \quad \begin{cases} \Delta c_1 = -\ c_4[\tau_1\tau_2\tau_3 + \tau_2\tau_3\tau_4 + \tau_3\tau_4\tau_1 + \tau_4\tau_1\tau_2], \\ \Delta c_2 = +\ c_4[\tau_1\tau_2 + \tau_1\tau_3 + \tau_1\tau_4 + \tau_2\tau_3 + \tau_2\tau_4 + \tau_3\tau_4], \\ \Delta c_3 = -\ c_4[\tau_1 + \tau_2 + \tau_3 + \tau_4]. \end{cases}$$

It follows from (38) and (39) that when there are four observations λ, λ', λ'', and λ''' are determined up to small quantities of the fourth, third, second, and first order respectively. Ordinarily λ''' is not needed, though it becomes useful when the solution is double, as it may be, in determining which of them belongs to the physical problem. In this latter case it is advantageous to make Δc_3 vanish by determining t_0 so that

$$(40) \quad \begin{cases} \tau_1 + \tau_2 + \tau_3 + \tau_4 = 0, \quad \text{whence} \\ t_0 = \tfrac{1}{4}(t_1 + t_2 + t_3 + t_4). \end{cases}$$

If the solution of the problem is made to depend only on λ, λ', and λ'', it is most advantageous to choose t_0 so that Δc_2 shall vanish, for then all the quantities are determined up to the third order. This condition becomes

$$(41) \quad \begin{cases} \tau_1\tau_2 + \tau_1\tau_3 + \tau_1\tau_4 + \tau_2\tau_3 + \tau_2\tau_4 + \tau_3\tau_4 = 0, \quad \text{whence} \\ 6t_0{}^2 - 3(t_1 + t_2 + t_3 + t_4)t_0 + t_1t_2 \\ \qquad + t_1t_3 + t_1t_4 + t_2t_3 + t_2t_4 + t_3t_4 = 0. \end{cases}$$

The values of t_0' determined by this quadratic equation are of no practical value unless they are real. The discriminant of the quadratic is

$$9(t_1 + t_2 + t_3 + t_4)^2 - 24(t_1t_2 + t_1t_3 + t_1t_4 + t_2t_3 + t_2t_4 + t_3t_4)$$
$$= H = 3(t_1 - t_2)^2 + 3(t_1 - t_3)^2 + 3(t_1 - t_4)^2$$
$$\qquad + 3(t_2 - t_3)^2 + 3(t_2 - t_4)^2 + 3(t_3 - t_4)^2 > 0.$$

Therefore the solutions are always real, and are explicitly

$$(42) \quad t_0 = \frac{3(t_1 + t_2 + t_3 + t_4) \pm \sqrt{H}}{12}.$$

In order to get a concrete idea of the nature of the results suppose the intervals between the successive observations are equal to T. Then (42) gives

(43) $$t_0 = \tfrac{1}{4}(t_1 + t_2 + t_3 + t_4) \pm \tfrac{1}{6}\sqrt{15}\,T.$$

The first term on the right is the mean epoch of the observations, and the two values of t_0 are at the distance $\tfrac{1}{6}\sqrt{15}\,T$ either side of this time. Since the interval between the mean epoch and t_2 or t_3 is $\tfrac{1}{2}T$, it follows that t_0 is between t_1 and t_2 and distant $(\tfrac{1}{6}\sqrt{15} - \tfrac{1}{2})T = \tfrac{1}{6}T$ approximately from t_2, or symmetrically situated between t_3 and t_4. In practice it will be most convenient to choose $t_0 = t_2$ or $t_0 = t_3$, for then λ is given exactly, the coefficients of (33) are as simple as possible, and (41) is nearly satisfied.

The discussion when there are five or more observations can be carried out in a similar manner. For each additional observation one additional coefficient in the series (28) can be determined, and those which were determined previously become known to one order higher in the τ_j. In each case one additional order of accuracy in the determination of λ'' can be secured by properly selecting t_0, but it is simplest to let t_0 equal the date of the observation which is nearest the mean epoch of all of the observations.

118. The Fundamental Equations. The fundamental equations of the method of Laplace are (10), where λ, μ, ν, λ', μ', ν', λ'', μ'', ν'' are given by (31) and (32) and corresponding equations in μ and ν. The solution of equations (10) for ρ, ρ', and ρ'' is

(44)
$$\begin{cases} \rho = \dfrac{D_1}{D}\left[\dfrac{1}{R^3} - \dfrac{1}{r^3}\right], \\[2mm] \rho' = \dfrac{D_2}{2D}\left[\dfrac{1}{R^3} - \dfrac{1}{r^3}\right], \\[2mm] \rho'' = \dfrac{1}{D}\left[D_3 - \dfrac{D_1}{r^3}\right]\left[\dfrac{1}{R^3} - \dfrac{1}{r^3}\right], \end{cases}$$

where

(45)
$$\begin{cases} D = + \begin{vmatrix} \lambda, & \lambda', & \lambda'' \\ \mu, & \mu', & \mu'' \\ \nu, & \nu', & \nu'' \end{vmatrix}, & D_1 = - \begin{vmatrix} \lambda, & \lambda', & X \\ \mu, & \mu', & Y \\ \nu, & \nu', & Z \end{vmatrix}, \\[6mm] D_2 = - \begin{vmatrix} \lambda, & X, & \lambda'' \\ \mu, & Y, & \mu'' \\ \nu, & Z, & \nu'' \end{vmatrix}, & D_3 = - \begin{vmatrix} X, & \lambda', & \lambda'' \\ Y, & \mu', & \mu'' \\ Z, & \nu', & \nu'' \end{vmatrix}. \end{cases}$$

These determinants are subject to small errors because of the fact that the higher terms of equations (28) have been neglected. After ρ and ρ' have been approximately determined corrections can be made for these omissions. The determinants are also subject to small errors because they have been developed under the tacit assumption that the observations were made from the position of the center of the earth instead of from one or more points on its surface. After the approximate distances have been determined the observations can be corrected for the effects of the observer's position on the surface of the earth.

119. The Equations for the Determination of r and ρ. Consider the triangle formed by S, E, and C. Let ψ represent the angle at E and φ that at C. Then it follows that

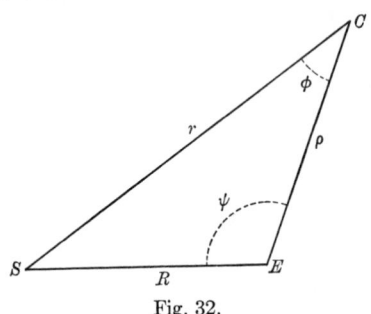

Fig. 32.

$$(46) \quad \begin{cases} R \cos \psi = X\lambda + Y\mu + Z\nu, \\[2mm] \rho = R \dfrac{\sin (\psi + \varphi)}{\sin \varphi}, \\[2mm] r = R \dfrac{\sin \psi}{\sin \varphi}. \end{cases}$$

When equations (46) are substituted in the first equation of (44) the result is

$$R \sin \psi \cos \varphi + \left[R \cos \psi - \frac{D_1}{DR^3} \right] \sin \varphi = \frac{-D_1}{DR^3 \sin^3 \psi} \sin^4 \varphi.$$

In order to simplify this expression let

$$(47) \quad \begin{cases} N \sin m = R \sin \psi, \\[2mm] N \cos m = R \cos \psi - \dfrac{D_1}{DR^3}, \\[2mm] M = \dfrac{- NDR^3 \sin^3 \psi}{D_1}, \end{cases}$$

where the sign of N will be so cnosen that M shall be positive. With this determination of the sign of N the first two equations of (47) uniquely determine N and m, and the equation in φ becomes simply

(48) $\sin^4 \varphi = M \sin (\varphi + m)$.*

The quantities M and m are known and M is positive.

Now consider the solution of (48) for φ. Since $\rho = 0$, $r = R$ is a solution of the problem, it follows from (48) that $\varphi = \pi - \psi$ is a solution of (48). This solution belongs to the position of the observer and is to be rejected. It follows from Fig. 32 that the φ belonging to the physical problem, which must exist if the computation is made from good observations, satisfies the inequality

(49) $\varphi < \pi - \psi$.

The solutions of (48) are the intersections of the curves defined by the equations

(50) $\begin{cases} y_1 = \sin^4 \varphi, \\ y_2 = M \sin (\varphi + m). \end{cases}$

For m negative and near zero and M somewhat less than unity these curves have the relation shown in Fig. 33.

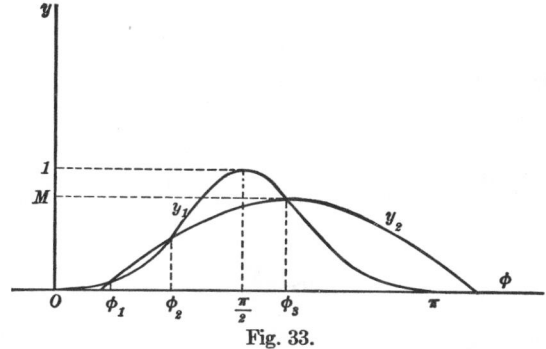

Fig. 33.

Consider first the case where $\dfrac{D_1}{D}$ is positive. Since both ρ and r must be positive, it follows from the first of (44) that in this case $r > R$. Since ψ is less than $180°$, it follows from (47) that N is negative, and that m is in the third or fourth quadrant.

In case m is in the fourth quadrant the ascending branch of the curve y_2 crosses the φ-axis in the first quadrant, and, if $M < 1$, the relations of the curves are as indicated in Fig. 33. If m is

---

* Hill, vol. I., pp. 2–4.

near 180° there are three solutions, φ_1, φ_2, and φ_3, one of which is $\pi - \psi$ and belongs to the position of the observer. If $\varphi_3 = \pi - \psi$, both φ_1 and φ_2 fulfill all the conditions of the problem and it can not be determined which belongs to the orbit of the observed body without additional information. However, it might happen that φ_1 would give so great values of r and ρ that it would be known from practical observational considerations that the body would be invisible; it would be known in this case that φ_2, which would give a smaller r, belongs to the physical problem. If $\varphi_2 = \pi - \psi$, it follows from (49) that φ_1 belongs to the problem. The case $\varphi_1 = \pi - \psi$ cannot occur for then the physical problem could have no solution. If, for a fixed M, the ascending branch of the curve y_2 moves to the right the roots φ_1 and φ_2 approach coincidence; and as it moves farther to the right φ_3 alone remains real. This case, which corresponds to m far from 180° in the fourth quadrant or in the third quadrant, cannot arise, for then the problem would have no solution. Therefore, *if $\dfrac{D_1}{D}$ is positive, then*

$r > R$, m is in the fourth quadrant, and there are one or two possible solutions of the physical problem according as φ_2 or φ_3 equals $\pi - \psi$.

Now suppose $\dfrac{D_1}{D}$ is negative. In this case $r < R$ and m is in the first or second quadrant. If m is in the first quadrant the descending branch of the curve y_2 crosses the φ-axis in the second

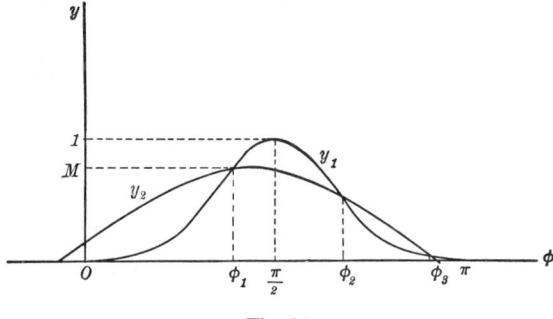

Fig. 34.

quadrant, and for a small m and $M < 1$ the relations are as shown in Fig 34. In this case the solution of the problem is unique or double according as φ_2 or φ_3 equals $\pi - \psi$. If m is in the second quadrant the descending branch of the curve y_2 crosses the φ-axis in the first quadrant, φ_2 and φ_3 are not real, and the problem has no solution. Therefore, *if $\dfrac{D_1}{D}$ is negative, then $r < R$, m is in the first quadrant, and there are one or two possible solutions of the physical problem according as φ_2 or φ_3 equals $\pi - \psi$.*

120. The Condition for a Unique Solution. The solution of the physical problem is unique whether $\dfrac{D_1}{D}$ is positive or negative if $\varphi_2 = \pi - \psi$, and otherwise it is double. Suppose $\varphi = \pi - \psi + \epsilon$, where ϵ is a small positive number. When $\dfrac{D_1}{D}$ is positive, it is seen from Fig. 33 that if $\varphi_2 = \pi - \psi$ the difference $y_1 - y_2$ is positive for $\varphi = \varphi_2 + \epsilon$; and, when $\dfrac{D_1}{D}$ is negative, it is seen from Fig. 34 that $y_1 - y_2$ is negative for $\varphi = \varphi_2 + \epsilon = \pi - \psi + \epsilon$.

It follows from (50) that y_1 and y_2 can be expanded as power series in ϵ when $\varphi = \pi - \psi + \epsilon$. The first two terms of the difference are

(51)
$$
\begin{aligned}
y_1 - y_2 = {} & [\sin^4(\pi - \psi) - M \sin(\pi - \psi + m)] \\
& + [4 \sin^3(\pi - \psi) \cos(\pi - \psi) \\
& \quad - M \cos(\pi - \psi + m)]\epsilon + \cdots.
\end{aligned}
$$

The term independent of ϵ is zero because $\varphi = \pi - \psi$ is a solution of (48). A reduction of the coefficient of ϵ by equations (47) and (48) gives

$$
y_1 - y_2 = \frac{MR}{N}\left[1 + \frac{3D_1}{DR^4}\cos\psi\right]\epsilon + \cdots.
$$

Therefore the condition that the solution of the physical problem shall be unique is

(52)
$$
\begin{cases}
\dfrac{1}{N}\left[1 + \dfrac{3D_1}{DR^4}\cos\psi\right] > 0 & \text{if } \dfrac{D_1}{D} > 0, \\[2ex]
\dfrac{1}{N}\left[1 + \dfrac{3D_1}{DR^4}\cos\psi\right] < 0 & \text{if } \dfrac{D_1}{D} < 0.
\end{cases}
$$

This function is completely determined by the observations, and consequently it is known without solving (48) whether the solution of the problem is unique or double.

The limit of the inequalities (52) is

(53)
$$
1 + \frac{3D_1}{DR^4}\cos\psi = 0.
$$

On eliminating $\cos \psi$ and $\dfrac{D_1}{D}$ by the first equations of (44) and (46), it is found that

$$(54) \qquad \rho^2 = r^2 + \frac{2}{3}\frac{R^5}{r^3} - \frac{5}{3}R^2.$$

The minimum value of the right member of this equation, considered as a function of r, is zero; therefore for each value of r there is a unique positive value of ρ. All points defined by pairs of values of r and ρ which satisfy (54) are on the boundary of the regions where the inequalities (52) are satisfied. These boundary surfaces are evidently surfaces of revolution around the line joining the earth and the sun. The section of these surfaces by a plane through the line SE is shown in Fig. 35.*

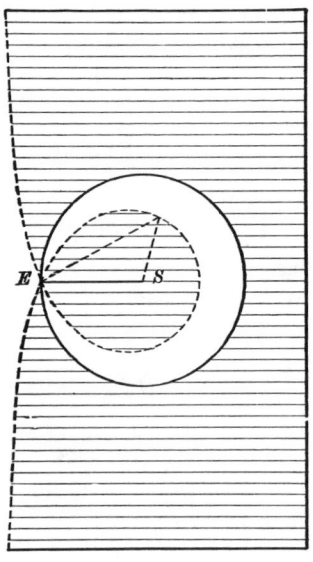

Fig. 35.

The surfaces defined by (54) divide space into four parts, two of which in the diagram are shaded, and two of which are plain. The function (52) has the same sign throughout each of these regions and changes sign when the boundary surface is crossed

* This figure was first given by Charlier, *Meddelande från Lunds Observatorium*, No. 45.

at any ordinary point. This is a special case of a general proposition which will be proved.

Suppose x_0, y_0, z_0 is an ordinary point on the surface defined by $F(x, y, z) = 0$. Consider the value of F at $x_0 + \Delta x$, $y_0 + \Delta y$, $z_0 + \Delta z$, where Δx, Δy, and Δz are small. The value of the function at this point is

$$F(x_0 + \Delta x, y_0 + \Delta y, z_0 + \Delta z)$$
$$= F(x_0, y_0, z_0) + \frac{\partial F}{\partial x}\Delta x + \frac{\partial F}{\partial y}\Delta y + \frac{\partial F}{\partial z}\Delta z + \cdots.$$

The first term in the right member of this equation is zero because x_0, y_0, z_0 is on the surface. Now suppose the point $x_0 + \Delta x$, $\cdots$ is on the perpendicular to the surface at x_0, y_0, z_0. Then

$$\Delta x = \frac{p\dfrac{\partial F}{\partial x}}{\sqrt{\left(\dfrac{\partial F}{\partial x}\right)^2 + \left(\dfrac{\partial F}{\partial y}\right)^2 + \left(\dfrac{\partial F}{\partial z}\right)^2}},$$

$$\Delta y = \frac{p\dfrac{\partial F}{\partial y}}{\sqrt{\left(\dfrac{\partial F}{\partial x}\right)^2 + \left(\dfrac{\partial F}{\partial y}\right)^2 + \left(\dfrac{\partial F}{\partial z}\right)^2}},$$

$$\Delta z = \frac{p\dfrac{\partial F}{\partial z}}{\sqrt{\left(\dfrac{\partial F}{\partial x}\right)^2 + \left(\dfrac{\partial F}{\partial y}\right)^2 + \left(\dfrac{\partial F}{\partial z}\right)^2}},$$

where p is the distance from x_0, y_0, z_0 to $x_0 + \Delta x$, $y_0 + \Delta y$, $z_0 + \Delta z$, because the factors by which p is multiplied are the direction cosines of the normal to the surface. On one side of the surface p is positive, and on the other side it is negative. The expression for the value of the function F at the point $x_0 + \Delta x$, $\cdots$ becomes

$$F(x_0 + \Delta x, y_0 + \Delta y, z_0 + \Delta z)$$
$$= p\left[\left(\frac{\partial F}{\partial x}\right)^2 + \left(\frac{\partial F}{\partial y}\right)^2 + \left(\frac{\partial F}{\partial z}\right)^2\right]^{\frac{1}{2}} + p^2[\ ] + \cdots.$$

For p very small the sign of the function is determined by the sign of the first term on the right whose coefficient is not zero. Since x_0, y_0, z_0 is by hypothesis an ordinary point of the surface, not all of the first partial derivatives of F are zero, and consequently the sign of the function changes with the change of sign of p. That is, the function changes sign when the surface for which it is zero is crossed; and it does not change sign at any other finite point because the function is continuous.

In order to find in which of the four regions of Fig. 35 the solution is unique, and in which it is double, consider a point on the line SE to the left of E. At such a point $r = \rho + R$, $\psi = \pi$, and it follows that

$$1 + \frac{3D_1}{DR^4} \cos \psi = 1 - \frac{3D_1}{DR^4} = 1 - \frac{3(\rho + R)^3}{R[\rho^2 + 3\rho R + 3R^2]},$$

which is clearly negative for ρ very large. Since in this case $r > R$ it follows that $\frac{D_1}{D} > 0$, $N < 0$ and the first inequality of (52) is the one under consideration. Since the inequality is satisfied the solution of the problem is unique if the observed body is in the unshaded area to the left of E. If the surface is crossed into the larger shaded area at a point for which $r > R$ the function changes sign while the sign of N is unchanged. Then the first inequality of (52) is not satisfied and the solution of the physical problem is double. In this region the function (53) is positive and N is negative. If the surface is crossed into the smaller unshaded area the function (53) becomes negative, N becomes positive, and the second inequality of (52), which is now in question, is satisfied. Therefore the solution is unique in this unshaded area. It is shown similarly that it is double in the smaller shaded area.

121. Use of a Fourth Observation in Case of a Double Solution. Suppose $\varphi_3 = \pi - \psi$ so that there are two solutions of (48) which correspond to the conditions of the physical problem. One method of determining which solution actually belongs to the physical problem, in case there are four observations, is obviously to develop (48), using the fourth observation instead of one of the original three. In general, this will make the result unique.

A better method of resolving the ambiguous case can be developed from equations (44). Eliminate r from the second and third equations of (44) by means of the first. The results are

$$\begin{cases} \rho' = \dfrac{D_2}{2D_1}\rho = P\rho, \qquad P = \dfrac{D_2}{2D_1}, \\[2ex] \rho'' = \dfrac{\rho}{D_1}\left[D_3 - \dfrac{D_1}{R^3} + D\rho \right]. \end{cases}$$

The derivative of the first of these equations is

$$\rho'' = P'\rho + P\rho' = (P' + P^2)\rho,$$

which equated to the right member of the second equation gives

$$(55) \qquad D_3 - \frac{D_1}{R^3} + D\rho = D_1(\dot{P}' + P^2).$$

Since this equation is linear ρ is uniquely determined unless D is zero. The determinant D will be examined in Art. 124. Equation (55) must be based upon not less than four observations, for P' involves λ''', μ''', and ν''' which cannot be determined, even approximately, from three observations.

122. The Limits on m and M. In an actual problem of the determination of an orbit the constants m and M are subject to the condition that equation (48) shall have three real roots between .0 and π. The limits imposed by this condition can be determined from the conditions that it shall have double roots; for, suppose M is fixed and that m varies. In the first case, represented in Fig. 33, there are three real solutions of (48) until, the curve y_2 moving to the right, φ_1 and φ_2 become equal; and in the second case, represented in Fig. 34, there are three real solutions of (48) until, the curve y_2 moving to the left, φ_2 and φ_3 become equal. The two cases are not essentially different for φ_1 in the first case corresponds exactly to φ_3 in the second. Similarly, if m remains fixed and M, starting from a small value, increases there are three real solutions of (56) until either φ_2 and φ_3 or φ_1 and φ_2, in the first and second cases respectively, become equal. When the limits are passed for which two values of φ which satisfy (48) are equal, there is only one real solution between 0 and π.

The conditions that (48) shall have a double root are

$$(56) \qquad \begin{cases} \sin^4 \varphi = M \sin (\varphi + m), \\ 4 \sin^3 \varphi \cos \varphi = M \cos (\varphi + m). \end{cases}$$

The solution of the quotient of these equations for $\tan \varphi$ is

$$(57) \qquad \tan \varphi = \frac{-3 \pm \sqrt{9 - 16 \tan^2 m}}{2 \tan m}.$$

It follows at once that m is subject to the condition

$$9 - 16 \tan^2 m \geqq 0$$

in order that the double root shall be real. Hence

$$(58) \qquad 323° \ 8' \gtreqless m \leqq 360°, \qquad 0 \gtreqless m \leqq 36° \ 52',$$

the first range for m belonging to the first case, represented in Fig. 33, and the second to the second case, represented in Fig. 34.

For each m there are two values of φ defined by (57) between 0 and π. In the first case, in which $\tan m$ is negative, $\tan \varphi$ is positive whether the upper or the lower sign is used before the radical, and it is smallest when the upper sign is used. Therefore the value of φ defined by (57) when the upper sign is used is that one for which $\varphi_1 = \varphi_2$ in Fig. 33, and the one determined when the lower sign is used is that one for which $\varphi_2 = \varphi_3$. When m has the limiting value for which the radical vanishes $\varphi_1 = \varphi_2 = \varphi_3$. The discussion is analogous in the second case in which $\tan m$ is positive.

The limiting values of φ, defined by (57), which correspond to the limiting values of m as given in (58), are respectively

$$(59) \qquad \varphi = 116° \ 34', \qquad \varphi = 63° \ 26',$$

and for both of these values of φ the value of M defined by (56) is $M = 1.431$. This is the maximum M for which (48) can have three real roots between 0 and π. In order that the three roots shall be real for this M the value of m must be $36° \ 52'$ or $323° \ 8'$, and the three roots are then equal.

Consider the first case and suppose m starts from $323° \ 8'$ and increases to $360°$. The two values of φ defined by (57) start from $63° \ 26'$. One goes to 0 and the other to $90°$. The two corresponding values of M start from 1.431, and one goes to 0 and the other to unity. For each value of m between the limits (58) there are two limits between which M must lie in order that (48) shall have three real solutions. In constructing a table of the solutions of (48) depending on the two independent parameters, M and m, these limits should be observed in order to reduce the work as much as possible.

123. Differential Corrections. Suppose the approximate solution of (48) has been found from the graphs of y_1 and y_2, or by numerical trials, or from the tables of the roots of this equation. Let φ_0 represent the approximate solution and $\varphi_0 + \Delta\varphi$ the exact solution. The problem is to find $\Delta\varphi$.

Let

$$(60) \qquad \sin^4 \varphi_0 - M \sin (\varphi_0 + m) = \eta,$$

where η will be a small quantity if φ_0 is an approximate solution of (48). If $\varphi_0 + \Delta\varphi$ is substituted in (48) in place of φ, the result expanded as a power series in $\Delta\varphi$ becomes

$$- \eta = [4 \sin^3 \varphi_0 \cos \varphi_0 - M \cos (\varphi_0 + m)]\Delta\varphi + [\ ] (\Delta\varphi)^2 + \cdots.$$

This power series can be inverted, giving $\Delta\varphi$ as a power series in η. The result is

$$(61) \quad \Delta\varphi = \frac{-\eta}{4 \sin^3 \varphi_0 \cos \varphi_0 - M \cos (\varphi_0 + m)} + [\ ] \eta^2 + \cdots.$$

The only exception is when the coefficient of $\Delta\varphi$ in the power series in $\Delta\varphi$ is zero. This is the second of equations (56), the conditions for a double root. In this case the expression for $\Delta\varphi$ proceeds in powers of $\pm \sqrt{\eta}$. In practice difficulty arises if the coefficient of $\Delta\varphi$ is small without being zero, for then φ_0 must be very close to the true value of φ before the method of differential corrections can be applied.

The higher terms of (61) can be computed without any difficulty, but they rapidly become more complex. It is simpler in practice to neglect them and to repeat the process with successive improved values of φ_0.

It is possible to develop a more convenient method for computing the differential corrections by making use of the fact that the work is done with logarithms. After m and M have been computed from the observational data the approximate solution of (48) can be determined from the diagram. The curve y_1 can be drawn accurately once for all. The better known sine curve, in this case flattened or stretched vertically by the factor M, can be drawn free hand with sufficient accuracy to enable one to get a very approximate estimate of the value of φ. Let it be φ_0. The logarithms of the right and left members of (48) will be computed and they will of course be found to be unequal. Let

$$4 \log \sin \varphi_0 - \log M - \log \sin (\varphi_0 + m) = \epsilon.$$

In the successive approximations only the first and third of these logarithms will be changed. The tables give the logarithms of the trigonometric functions. Let the tabular difference for the logarithm of $\sin \varphi_0$ and $\sin (\varphi_0 + \delta\varphi)$ be ϵ_1, where $\delta\varphi$ is some convenient increment to φ_0, and let ϵ_2 be the corresponding tabular difference for $\sin (\varphi_0 + m)$. These quantities are taken down from the margins of the tables when the logarithms of $\sin \varphi_0$ and $\sin (\varphi_0 + m)$ are taken out. Then the correction $\Delta\varphi$ is given by the equation

$$(62) \qquad\qquad \Delta\varphi = \frac{\delta\varphi \cdot \epsilon}{4\epsilon_1 - \epsilon_2},$$

where the result is expressed in the units used for $\delta\varphi$. This method is so convenient in practice that a very few minutes suffices in any case to find the solution of (48) with all the accuracy which may be desired. In the first approximation, where the error is in general large, one degree could be taken for $\delta\varphi$. In the later approximations $10''$ is a convenient increment because the tabular differences of the logarithms for differences of $10''$ are given on the margins of the tables.*

124. Discussion of the Determinant D. The determinant D, equation (45), enters into the determination of the constants M and m, and the solution becomes indeterminate in form if it is zero. Consequently it is important to find under what circumstances it vanishes.

Suppose the determination of the orbit is being based on only three observations. Then the values of λ, λ', and λ'', which occur in D, are given by (31) and (32). There are corresponding expressions for μ, μ', μ''; ν, ν', ν''. After they are substituted in (45) the determinant D can be factored into the product of two determinants. In order to simplify the notation let

$$(63) \quad P_1 = \frac{(\tau - \tau_2)(\tau - \tau_3)}{(\tau_1 - \tau_2)(\tau_1 - \tau_3)}, \qquad P_2 = \frac{(\tau - \tau_1)(\tau - \tau_3)}{(\tau_2 - \tau_1)(\tau_2 - \tau_3)},$$

$$P_3 = \frac{(\tau - \tau_1)(\tau - \tau_2)}{(\tau_3 - \tau_1)(\tau_3 - \tau_2)},$$

and denote the derivatives of these functions with respect to τ by accents. Then

$$(64) \quad \left\{ \begin{array}{l} D = \Delta_1 \Delta_2, \\[2ex] \Delta_1 = \begin{vmatrix} P_1, & P_1', & P_1'' \\ P_2, & P_2', & P_2'' \\ P_3, & P_3', & P_3'' \end{vmatrix}, \\[4ex] \Delta_2 = \begin{vmatrix} \lambda_1, & \lambda_2, & \lambda_3 \\ \mu_1, & \mu_2, & \mu_3 \\ \nu_1, & \nu_2, & \nu_3 \end{vmatrix}. \end{array} \right.$$

Consequently D can vanish only if Δ_1 or Δ_2 is zero.

*The solution of (48) depends on the *two* parameters M and m; if there were but one the relations between it and φ could easily be tabulated. In spite of the two parameters Leuschner has extended a table originally due to Oppolzer from which the solution can be read directly with considerable approximation. It is table xvi. in the third (Buchholz) edition of Klinkerfues' *Theoretische Astronomie*.

It will be shown first that Δ_1 is a constant which is distinct from zero. Since it is formally of the third degree in τ, necessary and sufficient conditions that it shall be independent of τ are that $\Delta_1' = 0$ for all values of τ. The derivative of a determinant is the sum of the determinants which are obtained by replacing successively the columns of the original determinant by their derivatives. Hence Δ_1' is a sum of three determinants. Since the derivative of the first column is identical with the second column, the first of these determinants is zero for all values of τ. Since the derivative of the second column is identical with the third, the second determinant is zero. The derivative of the third column is zero, and therefore the third determinant is zero. Hence Δ_1' is identically zero and Δ_1 is a constant. Its value, which is easily found for $\tau = 0$, is

(65)
$$\Delta_1 = \frac{2\begin{vmatrix} \tau_2\tau_3, & \tau_2 + \tau_3, & 1 \\ \tau_3\tau_1, & \tau_3 + \tau_1, & 1 \\ \tau_1\tau_2, & \tau_1 + \tau_2, & 1 \end{vmatrix}}{(\tau_1 - \tau_2)^2(\tau_2 - \tau_3)^2(\tau_3 - \tau_1)^2}$$

$$= \frac{2}{(\tau_2 - \tau_1)(\tau_3 - \tau_2)(\tau_3 - \tau_1)}.$$

This determinant is distinct from zero and independent of the choice of the epoch t_0.

In order to interpret Δ_2 multiply the first, second, and third columns by ρ_1, ρ_2, and ρ_3 respectively. Then, in the notation of equations (6), the determinant Δ_2 becomes

$$\rho_1\rho_2\rho_3\Delta_2 = \begin{vmatrix} \xi_1, & \xi_2, & \xi_3 \\ \eta_1, & \eta_2, & \eta_3 \\ \zeta_1, & \zeta_2, & \zeta_3 \end{vmatrix}$$

The right member of this equation is numerically the expression for six times the volume of the tetrahedron formed by the earth and the three positions of C with respect to E. The volume of this tetrahedron is zero only if the three positions of C lie in a plane passing through the fourth point E. This, of course, is referring the position of C to E as an origin. A simpler way of expressing the same result is, *the determinant Δ_2 (and therefore D) is zero only if the three apparent positions of C as observed from E lie on an arc of a great circle.*

It follows from (44) that if D is zero, D_1 and D_2 are also zero unless $R = r$. In general, the expressions for ρ and ρ' become indeterminate when D is zero, and they are poorly determined when D is small. One case in which Δ_2 and D are always zero is that in which C moves in the plane of the earth's orbit. But in this case there are only four elements to be determined, and since each observation gives a single coördinate (the longitude) four observations are required.

An expression for Δ_2 can be obtained by means of equations (6). After some simple reductions it is found that

$$(66) \quad \Delta_2 = \cos \delta_1 \cos \delta_2 \cos \delta_3 [\sin (\alpha_2 - \alpha_1) \tan \delta_3$$
$$+ \sin (\alpha_3 - \alpha_2) \tan \delta_1 + \sin (\alpha_1 - \alpha_3) \tan \delta_2].$$

125. Reduction of the Determinants D_1 and D_2. The expressions for D_1 and D_2, equations (45), become as a consequence of equations (31) and (32) and corresponding expressions for μ, μ', ν, and ν'

$$D_1 = - \begin{vmatrix} P_1\lambda_1 + P_2\lambda_2 + P_3\lambda_3, & P_1'\lambda_1 + P_2'\lambda_2 + P_3'\lambda_3, & X \\ P_1\mu_1 + P_2\mu_2 + P_3\mu_3, & P_1'\mu_1 + P_2'\mu_2 + P_3'\mu_3, & Y \\ P_1\nu_1 + P_2\nu_2 + P_3\nu_3, & P_1'\nu_1 + P_2'\nu_2 + P_3'\nu_3, & Z \end{vmatrix},$$

$$D_2 = + \begin{vmatrix} P_1\lambda_1 + P_2\lambda_2 + P_3\lambda_3, & P_1''\lambda_1 + P_2''\lambda_2 + P_3''\lambda_3, & X \\ P_1\mu_1 + P_2\mu_2 + P_3\mu_3, & P_1''\mu_1 + P_2''\mu_2 + P_3''\mu_3, & Y \\ P_1\nu_1 + P_2\nu_2 + P_3\nu_3, & P_1''\nu_1 + P_2''\nu_2 + P_3''\nu_3, & Z \end{vmatrix}.$$

If the first column of D_1 is multiplied by $\dfrac{P_3'}{P_3}$ and subtracted from the second column, the result is

$$- \frac{1}{P_3} \begin{vmatrix} P_\lambda, & (P_1'P_3 - P_1P_3')\lambda_1 + (P_2'P_3 - P_2P_3')\lambda_2, & X \\ P_\mu, & (P_1'P_3 - P_1P_3')\mu_1 + (P_2'P_3 - P_2P_3')\mu_2, & Y \\ P_\nu, & (P_1'P_3 - P_1P_3')\nu_1 + (P_2'P_3 - P_2P_3')\nu_2, & Z \end{vmatrix},$$

where

$$\begin{cases} P_\lambda = P_1\lambda_1 + P_2\lambda_2 + P_3\lambda_3, \\ P_\mu = P_1\mu_1 + P_2\mu_2 + P_3\mu_3, \\ P_\nu = P_1\nu_1 + P_2\nu_2 + P_3\nu_3. \end{cases}$$

This determinant is the sum of the two determinants

$$- \frac{1}{P_3} \begin{vmatrix} P_1\lambda_1 + P_2\lambda_2, & (P_1'P_3 - P_1P_3')\lambda_1 + (P_2'P_3 - P_2P_3')\lambda_2, & X \\ P_1\mu_1 + P_2\mu_2, & (P_1'P_3 - P_1P_3')\mu_1 + (P_2'P_3 - P_2P_3')\mu_2, & Y \\ P_1\nu_1 + P_2\nu_2, & (P_1'P_3 - P_1P_3')\nu_1 + (P_2'P_3 - P_2P_3')\nu_2, & Z \end{vmatrix}$$

and

$$-\begin{vmatrix} \lambda_3, & (P_1'P_3 - P_1P_3')\lambda_1 + (P_2'P_3 - P_2P_3')\lambda_2, & X \\ \mu_3, & (P_1'P_3 - P_1P_3')\mu_1 + (P_2'P_3 - P_2P_3')\mu_2, & Y \\ \nu_3, & (P_1'P_3 - P_1P_3')\nu_1 + (P_2'P_3 - P_2P_3')\nu_2, & Z \end{vmatrix}$$

The terms in λ_2, μ_2, and ν_2 can be eliminated in a similar manner from the second column of the first of these determinants. Then each of the determinants is the sum of two others, and the reduced expression for D_1 becomes

$$D_1 = - (P_1P_2' - P_1'P_2) \begin{vmatrix} \lambda_1, & \lambda_2, & X \\ \mu_1, & \mu_2, & Y \\ \nu_1, & \nu_2, & Z \end{vmatrix}$$

$$- (P_2P_3' - P_2'P_3) \begin{vmatrix} \lambda_2, & \lambda_3, & X \\ \mu_2, & \mu_3, & Y \\ \nu_2, & \nu_3, & Z \end{vmatrix}$$

$$- (P_3P_1' - P_3'P_1) \begin{vmatrix} \lambda_3, & \lambda_1, & X \\ \mu_3, & \mu_1, & Y \\ \nu_3, & \nu_1, & Z \end{vmatrix}.$$

The coefficients of these determinants are needed for $\tau = 0$. It is found from (63) that

$$P_1P_2' - P_1'P_2 = \frac{+ \tau_3^2}{(\tau_2 - \tau_1)(\tau_3 - \tau_2)(\tau_3 - \tau_1)},$$

$$P_2P_3' - P_2'P_3 = \frac{+ \tau_1^2}{(\tau_2 - \tau_1)(\tau_3 - \tau_2)(\tau_3 - \tau_1)},$$

$$P_3P_1' - P_3'P_1 = \frac{+ \tau_2^2}{(\tau_2 - \tau_1)(\tau_3 - \tau_2)(\tau_3 - \tau_1)}.$$

Then the expression for D_1 reduces to

(67)
$$\begin{cases} D_1 = -\frac{\tau_3^2}{P} \begin{vmatrix} \lambda_1, & \lambda_2, & X \\ \mu_1, & \mu_2, & Y \\ \nu_1, & \nu_2, & Z \end{vmatrix} - \frac{\tau_1^2}{P} \begin{vmatrix} \lambda_2, & \lambda_3, & X \\ \mu_2, & \mu_3, & Y \\ \nu_2, & \nu_3, & Z \end{vmatrix} \\ \qquad\qquad - \frac{\tau_2^2}{P} \begin{vmatrix} \lambda_3, & \lambda_1, & X \\ \mu_3, & \mu_1, & Y \\ \nu_3, & \nu_1, & Z \end{vmatrix}, \\ P = (\tau_2 - \tau_1)(\tau_3 - \tau_2)(\tau_3 - \tau_1). \end{cases}$$

In a similar manner the expression for D_2 reduces to

$$
(68) \quad
\begin{aligned}
D_2 = \frac{2\tau_3}{P}
\begin{vmatrix}
\lambda_1, & \lambda_2, & X \\
\mu_1, & \mu_2, & Y \\
\nu_1, & \nu_2, & Z
\end{vmatrix}
+ \frac{2\tau_1}{P}
\begin{vmatrix}
\lambda_2, & \lambda_3, & X \\
\mu_2, & \mu_3, & Y \\
\nu_2, & \nu_3, & Z
\end{vmatrix} \\
+ \frac{2\tau_2}{P}
\begin{vmatrix}
\lambda_3, & \lambda_1, & X \\
\mu_3, & \mu_1, & Y \\
\nu_3, & \nu_1, & Z
\end{vmatrix}
\end{aligned}
$$

Each of the determinants in the expressions for D_1 and D_3 can be developed in a form similar to (66).

126. Correction for the Time Aberration. Since the velocity of light is finite, the body C at any instant is apparently where it was at some preceding instant. This introduces a slight error in the data which must be corrected, if accurate results are desired, after the approximate distances have been determined. Since the velocity of light is very great and the apparent motions of the heavenly bodies are in general slow, it will not be necessary to know the distance of C with a high degree of accuracy in order to correct for the finite velocity of light.

Let E_1, E_2, and E_3 be the positions of the observer at t_1, t_2, and t_3 respectively. Let the observed directions of C at these

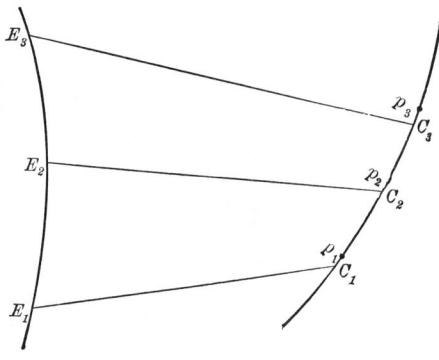

Fig. 36.

epochs be E_1C_1, E_2C_2, and E_3C_3. In the time required for the light to go from C to E the former will have moved forward in its orbit to the positions p_1, p_2, and p_3, which are its true places at the epochs t_1, t_2, and t_3. If the distances are known the observed coördinates can easily be corrected for these slight motions; but this changes all the observed data of the problem and makes it necessary to recompute all the determinants.

A second method, which is more convenient in practice, is to correct the times of the observations. The body C passed through the points C_1, C_2, and C_3, not at t_1, t_2, and t_3, but at these epochs diminished by the time required for light to move from C_1, C_2, and C_3 to E_1, E_2, and E_3 respectively. In order to make these corrections to the epochs it is necessary to know $E_1C_1 = \rho_1$, $E_2C_2 = \rho_2$, $E_3C_3 = \rho_3$. It will be supposed that (48), (46), and (44) have been solved and that ρ and ρ' are known. Then the values of ρ_1, ρ_2, and ρ_3 are given with sufficient approximations for present purposes by

$$
(69) \quad
\begin{cases}
\rho_1 = \rho + \rho'\tau_1, \\
\rho_2 = \rho + \rho'\tau_2, \\
\rho_3 = \rho + \rho'\tau_3.
\end{cases}
$$

Let V represent the velocity of light. Then the epochs at which C was at C_1, C_2, and C_3 are

$$
(70) \quad
\begin{cases}
\tau_1 - \Delta\tau_1 = \tau_1 - \dfrac{\rho_1}{V} = \tau_1 - \dfrac{(\rho + \rho'\tau_1)}{V}, \\[2ex]
\tau_2 - \Delta\tau_2 = \tau_2 - \dfrac{\rho_2}{V} = \tau_2 - \dfrac{(\rho + \rho'\tau_2)}{V}, \\[2ex]
\tau_3 - \Delta\tau_3 = \tau_3 - \dfrac{\rho_3}{V} = \tau_3 - \dfrac{(\rho + \rho'\tau_3)}{V}.
\end{cases}
$$

Now consider the correction to D, D_1, and D_2. In D only the factor Δ_1 is altered. But in the applications only the ratios of D to D_1 and D_2 are used, and the latter contain Δ_1 as a factor. Therefore the only change required is to replace τ_1, τ_2, and τ_3 by $\tau_1 - \Delta\tau_1$, $\tau_2 - \Delta\tau_2$, and $\tau_3 - \Delta\tau_3$ respectively in the numerators of the coefficients of the determinants in (67) and (68).

127. Development of x, y, and z in Series. In order to determine the corrections which should be added to λ' and λ'', so as to determine the elements of the orbit with greater accuracy, it is necessary to have x, y, and z developed as power series in τ. These quantities satisfy the differential equations

$$(71) \quad \begin{cases} \dfrac{d^2x}{d\tau^2} = -\dfrac{x}{r^3} = -ux, \qquad \left(u = \dfrac{1}{r^3}\right), \\[2mm] \dfrac{d^2y}{d\tau^2} = -\dfrac{y}{r^3} = -uy \\[2mm] \dfrac{d^2z}{d\tau^2} = -\dfrac{z}{r^3} = -uz. \end{cases}$$

It is shown in the theory of differential equations that the solutions of differential equations of this type are expansible as power series of the form

$$\begin{cases} x = x_0 + x_0'\tau + \tfrac{1}{2}x_0''\tau^2 + \tfrac{1}{6}x_0'''\tau^3 + \tfrac{1}{24}x_0^{iv}\tau^4 + \tfrac{1}{120}x_0^v\tau^5 + \cdots, \\[1mm] y = y_0 + y_0'\tau + \tfrac{1}{2}y_0''\tau^2 + \tfrac{1}{6}y_0'''\tau^3 + \tfrac{1}{24}y_0^{iv}\tau^4 + \tfrac{1}{120}y_0^v\tau^5 + \cdots, \\[1mm] z_0 = z_0 + z_c'\tau + \tfrac{1}{2}z_0''\tau^2 + \tfrac{1}{6}z_0'''\tau^3 + \tfrac{1}{24}z_0^{iv}\tau^4 + \tfrac{1}{120}z_0^v\tau^r + \cdots. \end{cases}$$

It is found from (71) and its successive derivatives that

$$(72) \quad \begin{cases} x_0'' = -u_0 x_0, \\[1mm] x_0''' = -u_0'x_0 - u_0 x_0', \\[1mm] x_0^{iv} = (-u_0'' + u_0^2)x_0 - 2u_0'x_0', \\[1mm] x_0^v = (-u_0''' + 4u_0 u_0')x_0 - (3u_0'' - u_0^2)x_0'. \end{cases}$$
$$\cdots \cdots \cdots \cdots \cdots \cdots \cdots \cdots$$

The coefficients of the series for y and z differ only in that y_0, y_0' and z_0, z_0' appear in place of x_0, x_0' respectively. Therefore

$$(73) \quad \begin{cases} x = f x_0 + g x_0', \\[1mm] y = f y_0 + g y_0', \\[1mm] z = f z_0 + g z_0', \\[1mm] f = 1 - \tfrac{1}{2}u_c\tau^2 - \tfrac{1}{6}u_0'\tau^3 - \tfrac{1}{24}(u_0'' - u_0^2)\tau^4 \\[1mm] \qquad\qquad\qquad\quad - \tfrac{1}{120}(u_0''' - 4u_0u_0')\tau^5 + \cdots, \\[1mm] g = \tau - \tfrac{1}{6}u_0\tau^3 - \tfrac{1}{12}u_0'\tau^4 - \tfrac{1}{120}(3u_0'' - u_0^2)\tau^5 + \cdots. \end{cases}$$

In order to have f and g in a form for practical use the derivatives of u must be expressed in terms of x_0, y_0, z_0, x_0', y_0', and z_0'. Lagrange has done this very elegantly by introducing p and q by the equations

$$(74) \quad \begin{cases} r^2 p = \dfrac{1}{2}\dfrac{dr^2}{d\tau} = xx' + yy' + zz', \\[2mm] r^2 q = \dfrac{1}{2}\dfrac{d^2r^2}{d\tau^2} = x'^2 + y'^2 + z'^2 - r^2u. \end{cases}$$

Then it is found that

$$\begin{cases} u' = -\dfrac{3}{r^4}\dfrac{dr}{d\tau} = -\dfrac{3}{r^4}\dfrac{1}{2r}\dfrac{dr^2}{d\tau} = -3up, \\[2ex] p' = +\dfrac{1}{2r^2}\dfrac{d^2r^2}{d\tau^2} - \dfrac{1}{r^3}\dfrac{dr^2}{d\tau}\dfrac{dr}{d\tau} = q - 2p^2, \\[2ex] q' = -\dfrac{1}{r^3}\dfrac{d^2r^2}{d\tau^2}\dfrac{dr}{d\tau} + \dfrac{1}{2r^2}\dfrac{d^3r^2}{d\tau^3} = -up - 2pq. \end{cases}$$

By means of these equations and their successive derivatives the coefficients in the series for f and g can be expressed as polynomials in u, p, and q. The expressions for f and g become

(75)
$$\begin{cases} f = 1 - \tfrac12 u_0\tau^2 + \tfrac12 u_0 p_0\tau^3 + \tfrac{1}{24}(3u_0q_0 - 15u_0p_0^2 + u_0^2)\tau^4 \\[1ex] \qquad + \tfrac18(7u_0p_0^3 - 3u_0p_0q_0 - u_0^2p_0)\tau^5 + \cdots, \\[1ex] g = \tau - \tfrac16 u_0\tau^3 + \tfrac14 u_0 p_0\tau^4 + \tfrac{1}{120}(9u_0q_0 - 45u_0p_0^2 \\[1ex] \qquad\qquad\qquad\qquad\qquad\qquad + u_0^2)\tau^5 + \cdots. \end{cases}$$

The derivatives of x, y, and z can be determined from equations (73) and (75). For example

(76)
$$\begin{cases} x''' = f'''x_0 + g'''x_0', \\[1ex] x^{iv} = f^{iv}x_0 + g^{iv}x_0', \\[1ex] \quad .\quad .\quad .\quad .\quad .\quad .\quad . \end{cases}$$

128. Computation of the Higher Derivatives of λ, μ, ν. The values of λ, λ', and λ'' determined by equations (31) and (32) are only approximate because c_3, c_4, $\cdots$ were unknown. But after the higher derivatives become known these coefficients are obtainable, and the approximate values can be corrected.

The third derivatives of equations (8) are

(77)
$$\begin{cases} \rho'''\lambda + 3\rho''\lambda' + 3\rho'\lambda'' + \rho\lambda''' = x''' + X''', \\[1ex] \rho'''\mu + 3\rho''\mu' + 3\rho'\mu'' + \rho\mu''' = y''' + Y''', \\[1ex] \rho'''\nu + 3\rho''\nu' + 3\rho'\nu'' + \rho\nu''' = z''' + Z'''. \end{cases}$$

The left members of these equations involve the four unknowns ρ''', λ''', μ''', and ν''', the first and second derivatives having been determined approximately by equations (31), (32), and (44); but the unknowns are not independent because λ, μ, ν, and their derivatives satisfy the relations

$$\begin{cases} \lambda^2 + \mu^2 + \nu^2 = 1, \\[1ex] \lambda\lambda' + \mu\mu' + \nu\nu' = 0, \\[1ex] \lambda\lambda'' + \mu\mu'' + \nu\nu'' + \lambda'^2 + \mu'^2 + \nu'^2 = 0, \\[1ex] \lambda\lambda''' + \mu\mu''' + \nu\nu''' + 3(\lambda'\lambda'' + \mu'\mu'' + \nu'\nu'') = 0. \end{cases}$$

Consequently if equations (77) are multiplied by λ, μ, and ν respectively and added, the result is

$$
(78) \quad
\begin{aligned}
\rho''' = {} & 3\rho'(\lambda'^2 + \mu'^2 + \nu'^2) + 3\rho(\lambda'\lambda'' + \mu'\mu'' + \nu'\nu'') \\
& + (x''' + X''')\lambda + (y''' + Y''')\mu + (z''' + Z''')\nu,
\end{aligned}
$$

which uniquely defines ρ'''. Then λ'', μ''', and ν''' are determined by (77) because x''', y''', z''' are given by (76) and X''', Y''', and Z''' can be found from the Ephemeris.

The quantities λ^{iv}, μ^{iv}, and ν^{iv} can be computed in a similar way by taking the derivatives of (77) and reducing by means of the relations among λ, μ, and ν.

129. Improvement of the Values of x, y, z, x', y', z'. After D, D_1, and D_2 have been found from (65), (66), (67), and (68) equation (48) can be solved, and then x, y, z and their first derivatives can be determined from (8) and their first derivatives. These results are only approximate because of the errors to which λ, μ, ν, λ', μ', and ν' are subject, and the problem is to correct them after λ''', μ''', $\cdots$ have been determined.

It follows from the first equation of (28) that

$$
c_3 = \tfrac{1}{6}\lambda''', \qquad c_4 = \tfrac{1}{24}\lambda^{iv}, \qquad \cdots\cdots.
$$

Then equations (35) give

$$
\left\{
\begin{aligned}
\Delta c_0 = {} & -\tfrac{1}{6}\lambda'''\tau_1\tau_2\tau_3 - \tfrac{1}{24}\lambda^{iv}\tau_1\tau_2\tau_3(\tau_1 + \tau_2 + \tau_3) + \cdots, \\
\Delta c_1 = {} & +\tfrac{1}{6}\lambda'''(\tau_1\tau_2 + \tau_2\tau_3 + \tau_3\tau_1) \\
& + \tfrac{1}{24}\lambda^{iv}(\tau_1 + \tau_2)(\tau_2 + \tau_3)(\tau_3 + \tau_1) + \cdots, \\
\Delta c_2 = {} & -\tfrac{1}{6}\lambda'''(\tau_1 + \tau_2 + \tau_3) \\
& -\tfrac{1}{24}\lambda^{iv}(\tau_1{}^2 + \tau_2{}^2 + \tau_3{}^2 + \tau_1\tau_2 + \tau_2\tau_3 + \tau_3\tau_1) + \cdots,
\end{aligned}
\right.
$$

and the expression for λ becomes

$$
(79) \quad
\begin{aligned}
\lambda = {} & c_0 + \Delta c_0 + (c_1 + \Delta c_1)\tau + (c_2 + \Delta c_2)\tau^2 \\
& + \tfrac{1}{6}\lambda'''\tau^3 + \tfrac{1}{24}\lambda^{iv}\tau^4 + \cdots,
\end{aligned}
$$

where c_0, c_1, and c_2 are the approximate values of the coefficients of the series which are obtained from (31) and (32) by putting τ equal to zero. There are corresponding equations for μ and ν. With these more nearly correct values of λ, λ', λ'', $\cdots$, the determinants D, D_1, and D_2 are computed from (45), φ is determined from (48), ρ and ρ' from (44), and x, y, z, x', y', z' from (8) and their first derivatives. Then still higher derivatives of λ, μ, ν can be computed and still more nearly exact values of λ, λ', and λ'' determined, or the elements can be determined from x, y, z, x', y', z' by the methods of chap. v.

There are two principal objections to the method of Laplace. One is that it is necessary to recompute all determinants and auxiliaries at each stage of the approximation, each of which costs a very considerable amount of labor. The other is that the method depends upon the motion of the observer through the equations by means of which X'', Y'', and Z'' were eliminated from (9). Obviously all that is really fundamental in the problem is that C shall have been observed from definite known places and that it shall move about the sun in accordance with the law of gravitation.

130. The Modifications of Harzer and Leuschner. The method of Laplace for determining orbits has not been found very satisfactory in practice. The reason seems to be that the conditions that the first and third observations shall be exactly satisfied are not directly imposed as they are, for example, in the method of Gauss. To remedy this defect Harzer proposed* the plan of so determining x, y, z, x', y', z' by differential corrections, after their approximate values have been found, that the three observations shall be exactly fulfilled. If more than three observations are under consideration, they cannot in general be exactly satisfied, and the adjustments are then made by the method of least squares.

It will be sufficient here to sketch the method of making the differential corrections. The right ascensions and declinations are expressed in terms of the coördinates and components of velocity at t_0 by

$$\begin{cases} \rho\lambda = f x_0 + g x_0' + X, \\ \rho\mu = f y_0 + g y_0' + Y, \\ \rho\nu = f z_0 + g z_0' + Z, \end{cases}$$

which are obtained by substituting equations (73) in equations (8). The right ascension and declination enter through λ, μ, and ν of equations (6). The result can be indicated

$$\begin{cases} \alpha = F(x_0, y_0, z_0, x_0', y_0', z_0'), \\ \delta = G(x_0, y_0, z_0, x_0', y_0', z_0'). \end{cases}$$

* *Astronomische Nachrichten*, Nos. 3371–2 (1896).

From these equations the variations in α and δ, which are the known differences between the observations and the approximate theory, are expressed in terms of the variations in $x_0, \cdots, z_0'$, which are required. The relations are

$$
\begin{cases}
\Delta\alpha = \dfrac{\partial F}{\partial x_0}\Delta x_0 + \dfrac{\partial F}{\partial y_0}\Delta y_0 + \dfrac{\partial F}{\partial z_0}\Delta z_0 + \dfrac{\partial F}{\partial x_0'}\Delta x_0' + \dfrac{\partial F}{\partial y_0'}\Delta y_0' + \dfrac{\partial F}{\partial z_0'}\Delta z_0', \\[2mm]
\Delta\delta = \dfrac{\partial G}{\partial x_0}\Delta x_0 + \dfrac{\partial G}{\partial y_0}\Delta y_0 + \dfrac{\partial G}{\partial z_0}\Delta z_0 + \dfrac{\partial G}{\partial x_0'}\Delta x_0' + \dfrac{\partial G}{\partial y_0'}\Delta y_0' + \dfrac{\partial G}{\partial z_0'}\Delta z_0'.
\end{cases}
$$

In forming the partial derivatives it must be remembered that $x_0, \cdots, z_0'$ enter through f and g as well as explicitly. When these equations are written for three dates they become equal to the number of arbitraries, viz., $\Delta x_0, \cdots, \Delta z_0'$, and consequently determine them uniquely provided the determinant of their coefficients is distinct from zero. The circumstances under which it vanishes have not been investigated. If there are more than three observations the number of equations exceeds the number of arbitraries and the method of least squares is employed.

When the date of the second observation is taken as the origin of time and the number of observations is only three, the number of equations of condition reduces to four which in general can be satisfied by suitably determining $\Delta\rho_0$, $\Delta x_0'$, $\Delta y_0'$, and $\Delta z_0'$. This is the procedure adopted by Leuschner* to abbreviate the method of Harzer. In its simplified form the method has been found very convenient in practice and has led to highly satisfactory results.

II. The Gaussian Method of Determining Orbits.

131. The Equation for ρ_2. Equations (19) are fundamental in the method of Gauss. If the geocentric coördinates are introduced by equations (8), equations (19) become

$$
(80)\quad
\begin{cases}
[2, 3]\rho_1\lambda_1 - [1, 3]\rho_2\lambda_2 + [1, 2]\rho_3\lambda_3 \\
\qquad\qquad = [2, 3]X_1 - [1, 3]X_2 + [1, 2]X_3, \\[1mm]
[2, 3]\rho_1\mu_1 - [1, 3]\rho_2\mu_2 + [1, 2]\rho_3\mu_3 \\
\qquad\qquad = [2, 3]Y_1 - [1, 3]Y_2 + [1, 2]Y_3, \\[1mm]
[2, 3]\rho_1\nu_1 - [1, 3]\rho_2\nu_2 + [1, 2]\rho_3\nu_3 \\
\qquad\qquad = [2, 3]Z_1 - [1, 3]Z_2 + [1, 2]Z_3.
\end{cases}
$$

The left members of these equations are linear in the three unknowns ρ_1, ρ_2, and ρ_3. Their solution for ρ_2 is

* *Publications of the Lick Observatory*, vol. vii., Part 1 (1902).

$$(81) \quad \begin{cases} \rho_2 = \dfrac{D}{\Delta}, \\[2mm] \Delta = -[2,3][1,3][1,2] \begin{vmatrix} \lambda_1, & \lambda_2, & \lambda_3 \\ \mu_1, & \mu_2, & \mu_3 \\ \nu_1, & \nu_2, & \nu_3 \end{vmatrix} = -[2,3][1,3][1,2]\Delta_2, \\[6mm] D = [2,3][1,2] \begin{vmatrix} \lambda_1, & [2,3]X_1 - [1,3]X_2 + [1,2]X_3, & \lambda_3 \\ \mu_1, & [2,3]Y_1 - [1,3]Y_2 + [1,2]Y_3, & \mu_3 \\ \nu_1, & [2,3]Z_1 - [1,3]Z_2 + [1,2]Z_3, & \nu_3 \end{vmatrix}. \end{cases}$$

The determinant D is the sum of three determinants

$$(82) \quad \begin{cases} D = [2,3]^2[1,2]D^{(1)} - [2,3][1,3][1,2]D^{(2)} + [2,3][1,2]^2 D^{(3)}, \\[3mm] D^{(1)} = \begin{vmatrix} \lambda_1, & X_1, & \lambda_3 \\ \mu_1, & Y_1, & \mu_3 \\ \nu_1, & Z_1, & \nu_3 \end{vmatrix}, \qquad D^{(2)} = \begin{vmatrix} \lambda_., & X_2, & \lambda_3 \\ \mu_1, & Y_2, & \mu_3 \\ \nu_1, & Z_2, & \nu_3 \end{vmatrix}, \\[9mm] D^{(3)} = \begin{vmatrix} \lambda_1, & X_3, & \lambda_3 \\ \mu_1, & Y_3, & \mu_3 \\ \nu_1, & Z_3, & \nu_3 \end{vmatrix}. \end{cases}$$

Consequently the first equation of (81) becomes

$$(83) \qquad \Delta_2 \rho_2 = -\frac{[2,3]}{[1,3]}D^{(1)} + D^{(2)} - \frac{[1,2]}{[1,3]}D^{(3)}.$$

Suppose t_2 is taken as the origin of time. Then it follows from equations (73) that

$$\begin{cases} x_1 = f_1 x_2 + g_1 x_2', & y_1 = f_1 y_2 + g_1 y_2', & z_1 = f_1 z_2 + g_1 z_2', \\ x_3 = f_3 x_2 + g_3 x_2', & y_3 = f_3 y_2 + g_3 y_2', & z_3 = f_3 z_2 + g_3 z_2'. \end{cases}$$

The expressions for the ratios of triangles then become

$$(84) \quad \begin{cases} \dfrac{[2,3]}{[1,3]} = \dfrac{x_2 y_3 - x_3 y_2}{x_1 y_3 - x_3 y_1} = \dfrac{+ g_3}{f_1 g_3 - f_3 g_1}, \\[4mm] \dfrac{[1,2]}{[1,3]} = \dfrac{x_1 y_2 - x_2 y_1}{x_1 y_3 - x_3 y_1} = \dfrac{- g_1}{f_1 g_3 - f_3 g_1}. \end{cases}$$

The numerators and denominators of the expressions for the right members of these equations are found from (75) to be expansible as power series in τ_1 and τ_3. But in order to simplify (83) it is convenient to let

$$(85) \quad \begin{cases} k(t_3 - t_1) = \tau_3 - \tau_1 = 2\tau, \\ \tau_1 = -\tau + \epsilon, \qquad \tau_3 = +\tau + \epsilon, \end{cases}$$

where ϵ is in general small compared to τ, and will be supposed to be of the order of τ^2. Then the expressions for the ratios of the triangles become

$$(86) \quad \begin{cases} \dfrac{[2, 3]}{[1, 3]} = \dfrac{+ g_3}{f_1 g_3 - f_3 g_1} = \dfrac{1}{2} + \dfrac{\epsilon}{2\tau} + \dfrac{1}{4} u\tau^2 P + \dfrac{\tau\epsilon}{12} uQ, \\[2ex] \dfrac{[1, 2]}{[1, 3]} = \dfrac{- g_1}{f_1 g_3 - f_3 g_1} = \dfrac{1}{2} - \dfrac{\epsilon}{2\tau} + \dfrac{1}{4} u\tau^2 P - \dfrac{\tau\epsilon}{12} uQ, \\[2ex] P = 1 - \dfrac{\epsilon^2}{\tau^2} - 2p\epsilon + \dfrac{1}{12}(7u - 15p^2 + 3q)\tau^2 + \cdots, \\[2ex] Q = 1 - \dfrac{\epsilon^2}{\tau^2} + \dfrac{3}{2} p \dfrac{\tau^2}{\epsilon} - 3p\epsilon + \dfrac{1}{60}(37u - 765p^2 + 153q)\tau^2 \\[2ex] \qquad\qquad + \dfrac{1}{4} p(3u + 14p^2 - 6q)\dfrac{\tau^4}{\epsilon} + \cdots, \end{cases}$$

where all terms up to the sixth order have been written. The quantity u is defined by $u = \dfrac{1}{r_2{}^3}$ and p and q are defined in (74).

On making use of equations (86), equation (83) becomes

$$(87) \qquad \Delta_2 \rho_2 = K + \frac{\tau^2}{4r_2^3} PK_1 + \frac{\tau\epsilon}{12r_2^3} QK_2 + \frac{\epsilon}{2\tau} K_2$$

where

$$K = -\frac{1}{2} \begin{vmatrix} \lambda_1, & X_1, & \lambda_3 \\ \mu_1, & Y_1, & \mu_3 \\ \nu_1, & Z_1, & \nu_3 \end{vmatrix} + \begin{vmatrix} \lambda_1, & X_2, & \lambda_3 \\ \mu_1, & Y_2, & \mu_3 \\ \nu_1, & Z_2, & \nu_3 \end{vmatrix} - \frac{1}{2} \begin{vmatrix} \lambda_1, & X_3, & \lambda_3 \\ \mu_1, & Y_3, & \mu_3 \\ \nu_1, & Z_3, & \nu_3 \end{vmatrix},$$

$$K_1 = - \begin{vmatrix} \lambda_1, & X_1, & \lambda_3 \\ \mu_1, & Y_1, & \mu_3 \\ \nu_1, & Z_1, & \nu_3 \end{vmatrix} - \begin{vmatrix} \lambda_1, & X_3, & \lambda_3 \\ \mu_1, & Y_3, & \mu_3 \\ \nu_1, & Z_3, & \nu_3 \end{vmatrix},$$

$$K_2 = - \begin{vmatrix} \lambda_1, & X_1, & \lambda_3 \\ \mu_1, & Y_1, & \mu_3 \\ \nu_1, & Z_1, & \nu_3 \end{vmatrix} + \begin{vmatrix} \lambda_1, & X_3, & \lambda_3 \\ \mu_1, & Y_3, & \mu_3 \\ \nu_1, & Z_3, & \nu_3 \end{vmatrix}.$$

The right members of the expressions for K, K_1, and K_2 add, giving the simpler expressions

$$(88)\quad\begin{cases} K = -\dfrac{1}{2}\begin{vmatrix} \lambda_1, & X_1 + X_3 - 2X_2, & \lambda_3 - \lambda_1 \\ \mu_1, & Y_1 + Y_3 - 2Y_2, & \mu_3 - \mu_1 \\ \nu_1, & Z_1 + Z_3 - 2Z_2, & \nu_3 - \nu_1 \end{vmatrix}, \\[3mm] K_1 = -\begin{vmatrix} \lambda_1, & X_1 + X_3, & \lambda_3 - \lambda_1 \\ \mu_1, & Y_1 + Y_3, & \mu_3 - \mu_1 \\ \nu_1, & Z_1 + Z_3, & \nu_3 - \nu_1 \end{vmatrix}, \\[3mm] K_2 = +\begin{vmatrix} \lambda_1, & X_3 - X_1, & \lambda_3 - \lambda_1 \\ \mu_1, & Y_3 - Y_1, & \mu_3 - \mu_1 \\ \nu_1, & Z_3 - Z_1, & \nu_3 - \nu_1 \end{vmatrix} \end{cases}$$

Consider equation (87). The determinant Δ_2 by which the left member is multiplied is given in terms of the α_i and δ_i by (66), which appeared in the method of Laplace. It can also be written, by properly combining columns, in the form

$$\Delta_2 = \begin{vmatrix} \lambda_1, & \lambda_2, & \lambda_3 \\ \mu_1, & \mu_2, & \mu_3 \\ \nu_1, & \nu_2, & \nu_3 \end{vmatrix} = \begin{vmatrix} \lambda_1, & \lambda_1 + \lambda_3 - 2\lambda_2, & \lambda_3 - \lambda_1 \\ \mu_1, & \mu_1 + \mu_3 - 2\mu_2, & \mu_3 - \mu_1 \\ \nu_1, & \nu_1 + \nu_3 - 2\nu_2, & \nu_3 - \nu_1 \end{vmatrix}$$

If λ_i, μ_i, ν_i are replaced by the series (28), taking $\tau_2 = 0$, the second column is of the second order and the third column is of the first order in the time-intervals. Therefore Δ_2 is of the third order.

Since the left member of (87) is of the third order the right member also must be of the third order. The second column of the expression tor K, the first equation of (88), is of the second order, and the third column is of the first order. Therefore K is of the third order. The determinant K_1 is of the first order and K_2 is of the second order. The former is multiplied by τ^2, which is of the second order, and the latter by $\tau\epsilon$, which is of the third order. In a preliminary determination of an orbit the terms of higher order may be omitted, after which (87) becomes

$$(87a)\qquad \Delta_2\rho_2 = K + \frac{\tau^2 K_1}{4r_2{}^3} + \frac{\epsilon^2}{2\tau} K_2.$$

This equation is of the same form as the first of (44), and involves the two unknowns ρ_2 and r_2. They are expressible in terms of a single unknown φ by means of equations (46) affected with the subscript 2. The resulting equation has exactly the same form as (48), and its solution gives approximate values of ρ_2 and r_2.

132. The Equations for ρ_1 and ρ_3. Equations (80) are linear in ρ_1 and ρ_3, and these quantities can be determined from any two of the three equations. The two to be used in practice are those for which the determinant of the coefficients of ρ_1 and ρ_3 is the greatest, for they will best determine these quantities.

The solution of the first two equations of (80) for ρ_1 and ρ_2 if they are written first in determinant form, and if they are then expanded as a sum of determinants, is

$$(89) \begin{cases} \begin{vmatrix} \lambda_1, & \lambda_3 \\ \mu_1, & \mu_3 \end{vmatrix} \rho_1 = \begin{vmatrix} X_1, & \lambda_3 \\ Y_1, & \mu_3 \end{vmatrix} - \frac{[1,3]}{[2,3]} \begin{vmatrix} X_2, & \lambda_3 \\ Y_2, & \mu_3 \end{vmatrix} \\ \qquad + \frac{[1,2]}{[2,3]} \begin{vmatrix} X_3, & \lambda_3 \\ Y_3, & \mu_3 \end{vmatrix} + \rho_2 \frac{[1,3]}{[2,3]} \begin{vmatrix} \lambda_2, & \lambda_3 \\ \mu_2, & \mu_3 \end{vmatrix}, \\ \begin{vmatrix} \lambda_1, & \lambda_3 \\ \mu_1, & \mu_3 \end{vmatrix} \rho_3 = \frac{[2,3]}{[1,2]} \begin{vmatrix} \lambda_1, & X_1 \\ \mu_1, & Y_1 \end{vmatrix} - \frac{[1,3]}{[1,2]} \begin{vmatrix} \lambda_1, & X_2 \\ \mu_1, & Y_2 \end{vmatrix} \\ \qquad + \begin{vmatrix} \lambda_1, & X_3 \\ \mu_1, & Y_3 \end{vmatrix} + \rho_2 \frac{[1,3]}{[1,2]} \begin{vmatrix} \lambda_1, & \lambda_2 \\ \mu_1, & \mu_2 \end{vmatrix}. \end{cases}$$

The solution of the first and third equations of (80) differs from this only in that the μ_i are replaced by the ν_i, and the Y_i by the Z_i; and the solution of the second and third equations of (80) can be obtained from (89) by changing the λ_i, μ_i, X_i, and Y_i to μ_i, ν_i, Y_i, and Z_i respectively.

After ρ_1, ρ_2, and ρ_3 have been computed the correction of the time for the time-aberration can be computed. The method was explained in Art. 126.

133. Improvement of the Solution. The results so far obtained are only approximate because only the first term of P was retained while the term in Q was entirely neglected. Having found an approximate solution it is easy to correct it. The values of ρ_1, ρ_2, and ρ_3 are known, and the corresponding values of r can be found at each of the three epochs from

$$r^2 = \rho^2 + R^2 - 2\rho R \cos \psi,$$

which expresses the fact that S, E, and C form triangles at the dates of the three observations. After r_1, r_2, and r_3 have been found the first and second derivatives of r at $t = t_2$ can be found by the method of Art. 113.   Then equations (74) define p and q after which more approximate values of P and Q can be determined.

134. The Method of Gauss for Computing the Ratios of the Triangles.   Equation (83), which is fundamental in determining ρ_2 and r_2, involves two ratios of triangles.   It follows from (86) that they can be written in the form

$$(90) \quad \begin{cases} \dfrac{[2, 3]}{[1, 3]} = \dfrac{1}{2} + \dfrac{\epsilon}{2\tau} + \dfrac{P_1}{r_2{}^3}, \\[2mm] \dfrac{[1, 2]}{[1, 3]} = \dfrac{1}{2} - \dfrac{\epsilon}{2\tau} + \dfrac{P_2}{r_2{}^3}. \end{cases}$$

Consequently, if the ratios of the triangles can be determined P_1 and P_2 can be found from these equations.   One of the important features of the method of Gauss is a convenient means of determining the ratios of the triangles.   In order to apply this method it is necessary to find the inclination and node of the orbit and the argument of the latitude at the dates of the observations.

Since the geocentric coördinates are all known after ρ_1, ρ_2, ρ_3 have been determined, the heliocentric coördinates can be computed.   Suppose ecliptic coördinates are used and that the

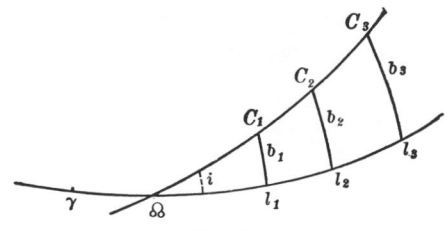

Fig. 37.

longitudes and latitudes, as well as the distances, are known at t_1, t_2, and t_3.   The inclination is less or greater than $90°$ according as l_3 is greater or less than l_1.   Then it follows from the spherical triangles $C_1 \Omega l_1$ and $C_3 \Omega l_3$ that

$$\begin{cases} \tan i \sin (l_1 - \Omega) = \tan b_1, \\ \tan i \sin (l_3 - \Omega) = \tan b_3. \end{cases}$$

But $l_3 - \Omega = (l_3 - l_1) + (l_1 - \Omega)$;   therefore these equations become

$$(91) \quad \begin{cases} \tan i \sin (l_1 - \Omega) = \tan b_1, \\[2mm] \tan i \cos (l_1 - \Omega) = \dfrac{\tan b_3 - \tan b_1 \cos (l_3 - l_1)}{\sin (l_3 - l_1)}, \end{cases}$$

which determine i and Ω uniquely since the quadrant of i is already known from the sign of $l_3 - l_1$.

The longitude of C from the node is called the argument of the latitude. It follows from Fig. 37 that

$$(92) \quad \begin{cases} \cos (l_j - \Omega) \cos b_j = \cos u_j, \qquad (j = 1, 2, 3), \\[2mm] \sin (l_j - \Omega) \cos b_j = \sin u_j \cos i, \\[2mm] \qquad\qquad \sin b_j = \sin u_j \sin i, \end{cases}$$

which uniquely define u_1, u_2, and u_3.

Let A equal the area of the sector contained between the radii r_1 and r_2 and the orbit. Then the ratio of the area of the sector to the area of the triangle contained between r_1 and r_2 is

$$(93) \qquad \eta = \frac{A}{[r_1,\ r_2]} = \frac{k \sqrt{p}\ (t_2 - t_1)}{r_1 r_2 \sin (u_2 - u_1)},$$

where p now represents the parameter of the conic. Suppose the corresponding ratios for $t_3 - t_1$ and $t_3 - t_2$ have been found; then the ratios of the triangles are known. The method of Gauss depends upon the determination of these ratios. Each of these quantities is defined by two simultaneous equations in two unknown quantities.

135. The First Equation of Gauss. The polar equation of the conic gives

$$\begin{cases} \dfrac{p}{r_1} = 1 + e \cos v_1, \\[3mm] \dfrac{p}{r_2} = 1 + e \cos v_2; \end{cases}$$

whence

$$(94) \quad \begin{aligned} p \frac{r_1 + r_2}{r_1 r_2} &= 2 + e(\cos v_1 + \cos v_2) \\[2mm] &= 2 + 2e \cos \left(\frac{v_2 + v_1}{2} \right) \cos \left(\frac{v_2 - v_1}{2} \right). \end{aligned}$$

Since $v_2 - v_1 = u_2 - u_1$ is known, the only unknown in the right member of this equation is $e \cos \left(\dfrac{v_2 + v_1}{2} \right)$, which will now be eliminated. From the equations of Art. 98 it follows that

$$(95) \begin{cases} \sqrt{r_1} \cos \dfrac{v_1}{2} = \sqrt{a(1-e)} \, \cos \dfrac{E_1}{2}, \\[2ex] \sqrt{r_1} \sin \dfrac{v_1}{2} = \sqrt{a(1+e)} \, \sin \dfrac{E_1}{2}, \\[2ex] \sqrt{r_2} \cos \dfrac{v_2}{2} = \sqrt{a(1-e)} \, \cos \dfrac{E_2}{2}, \\[2ex] \sqrt{r_2} \sin \dfrac{v_2}{2} = \sqrt{a(1+e)} \, \sin \dfrac{E_2}{2}. \end{cases}$$

From these equations it is found that

$$\begin{cases} \sqrt{r_1 r_2} \cos \left(\dfrac{v_2 - v_1}{2} \right) = a \cos \left(\dfrac{E_2 - E_1}{2} \right) - ae \cos \left(\dfrac{E_2 + E_1}{2} \right), \\[2ex] \sqrt{r_1 r_2} \cos \left(\dfrac{v_2 + v_1}{2} \right) = a \cos \left(\dfrac{E_2 + E_1}{2} \right) - ae \cos \left(\dfrac{E_2 - E_1}{2} \right). \end{cases}$$

On eliminating $e \cos \left(\dfrac{E_2 + E_1}{2} \right)$ and solving for $e \cos \left(\dfrac{v_2 + v_1}{2} \right)$, it is found that

$$e \cos \left(\frac{v_2 + v_1}{2} \right) = \frac{p}{\sqrt{r_1 r_2}} \cos \left(\frac{E_2 - E_1}{2} \right) - \cos \left(\frac{v_2 - v_1}{2} \right).$$

As a consequence of this equation (94) reduces to

$$p = \frac{2 r_1 r_2 \sin^2 \left(\dfrac{v_2 - v_1}{2} \right)}{r_1 + r_2 - 2 \sqrt{r_1 r_2} \cos \left(\dfrac{v_2 - v_1}{2} \right) \cos \left(\dfrac{E_2 - E_1}{2} \right)}.$$

On eliminating p from this equation and (93) the equation

$$(96) \quad \eta^2 = \frac{k^2 (t_2 - t_1)^2 \sec^2 \left(\dfrac{v_2 - v_1}{2} \right)}{2 r_1 r_2 \left\{ r_1 + r_2 - 2 \sqrt{r_1 r_2} \cos \left(\dfrac{v_2 - v_1}{2} \right) \cos \left(\dfrac{E_2 - E_1}{2} \right) \right\}}$$

is obtained.   In order to simplify it let

$$(97) \begin{cases} v_2 - v_1 = u_2 - u_1 = 2f, \\[1.5ex] E_2 - E_1 = 2g, \\[1.5ex] m = \dfrac{k(t_2 - t_1)}{(2 \sqrt{r_1 r_2} \cos f)^{\frac{3}{2}}}, \\[2.5ex] l = \dfrac{r_1 + r_2}{4 \sqrt{r_1 r_2} \cos f} - \dfrac{1}{2}. \end{cases}$$

Then the expression for η^2 reduces to

$$(98) \qquad \eta^2 = \frac{m^2}{l + \sin^2\dfrac{g}{2}},$$

in which η and g are the unknowns.   This is the first equation in the method of Gauss.

136. The Second Equation of Gauss.   An independent equation involving η and g will now be derived.   It will be made to depend upon Kepler's equation, thus insuring its independence of (98) which was derived without reference to Kepler's equation. The first equations are

$$\begin{cases} M_1 = \dfrac{k(t_1 - T)}{a^{\frac{3}{2}}} = E_1 - e \sin E_1, \\[2ex] M_2 = \dfrac{k(t_2 - T)}{a^{\frac{3}{2}}} = E_2 - e \sin E_2; \end{cases}$$

whence

$$\frac{k(t_2 - t_1)}{a^{\frac{3}{2}}} = 2g - 2e \sin g \cos\left(\frac{E_2 + E_1}{2}\right).$$

The quantities a and $e \cos\left(\dfrac{E_2 + E_1}{2}\right)$ must be eliminated in order to reduce this equation to the required type.   On making use of the first equation following (95), it is found that

$$(99) \qquad \frac{k(t_2 - t_1)}{a^{\frac{3}{2}}} = 2g - \sin 2g + 2\,\frac{\sqrt{r_1 r_2}}{a} \sin g \cos f.$$

It remains to eliminate a.   By Art. 98

$$\begin{cases} \dfrac{r_1}{a} = 1 - e \cos E_1, \\[2ex] \dfrac{r_2}{a} = 1 - e \cos E_2; \end{cases}$$

whence

$$\frac{r_1 + r_2}{a} = 2 - 2e \cos g \cos\left(\frac{E_2 + E_1}{2}\right)$$

On eliminating $e \cos\left(\dfrac{E_2 + E_1}{2}\right)$ by the first equation following (95) this equation becomes

$$\frac{1}{a} = \frac{2 \sin^2 g}{r_1 + r_2 - 2\sqrt{r_1 r_2}\cos g \cos f},$$

which becomes as a consequence of the expression for η^2

(100) $$\frac{1}{a} = \left(\frac{2\eta \sin g \cos f}{k(t_2 - t_1)} \right)^2 r_1 r_2.$$

On eliminating a between (99) and (100), it is found that

(101) $$\frac{\eta^3}{m^2} - \frac{\eta^2}{m^2} = \frac{2g - \sin 2g}{\sin^3 g},$$

which is the second equation in η and g. There are similar equations for the time-intervals $t_3 - t_1$ and $t_3 - t_2$.

137. Solution of (98) and (101). It follows from the definition of η that it is positive if the heliocentric motion in the orbit is less than $180°$ in the interval $t_2 - t_1$. It will be supposed in what follows that the observations are so close together that this condition is fulfilled.

Let

(102) $$\begin{cases} \sin^2 \dfrac{g}{2} = x, \\ \dfrac{2g - \sin 2g}{\sin^3 g} = X. \end{cases}$$

On eliminating η from (98) and (101) and making use of (102), it is found that

(103) $$m = (l + x)^{\frac{1}{2}} + X(l + x)^{\frac{3}{2}}.$$

The quantity X must now be expressed in terms of x, after which (103) will involve this quantity alone as an unknown. This will be done by first expressing X in terms of g, and then g in terms of x. The following are well-known expansions of the trigonometrical functions:

$$\begin{cases} \sin 2g = 2g - \tfrac{4}{3}g^3 + \tfrac{4}{15}g^5 - \tfrac{8}{315}g^7 + \tfrac{8}{5670}g^9 - \cdots, \\ \sin^3 g = g^3 - \tfrac{1}{2}g^5 + \tfrac{13}{120}g^7 - \tfrac{41}{3024}g^9 + \cdots; \end{cases}$$

whence

(104) $$\begin{aligned} X &= \frac{\tfrac{4}{3} - \tfrac{4}{15}g^2 + \tfrac{8}{315}g^4 - \tfrac{8}{5670}g^6 + \cdots}{1 - \tfrac{1}{2}g^2 + \tfrac{13}{120}g^4 - \tfrac{41}{3024}g^6 + \cdots} \\ &= \tfrac{4}{3}(1 + \tfrac{3}{10}g^2 + \tfrac{17}{280}g^4 + \tfrac{29}{2800}g^6 + \cdots). \end{aligned}$$

From the first of (102) it follows that

$$\begin{cases} g = 2 \sin^{-1}(x^{\frac{1}{2}}) = 2x^{\frac{1}{2}} + \tfrac{1}{3}x^{\frac{3}{2}} + \tfrac{3}{20}x^{\frac{5}{2}} + \tfrac{5}{56}x^{\frac{7}{2}} + \cdots, \\ g^2 = 4x + \tfrac{4}{3}x^2 + \tfrac{32}{45}x^3 + \tfrac{16}{35}x^4 + \cdots, \\ g^4 = 16x^2 + \tfrac{32}{3}x^3 + \tfrac{112}{15}x^4 + \cdots, \\ g^6 = 64x^3 + 64x^4 + \cdots. \end{cases}$$

Then (104) becomes

$$X = \frac{4}{3} \left[1 + \frac{6}{5} x + \frac{6 \cdot 8}{5 \cdot 7} x^2 + \frac{6 \cdot 8 \cdot 10}{5 \cdot 7 \cdot 9} x^3 + \cdots \right],$$

or

$$X = \frac{1}{\frac{3}{4} \left[1 + \frac{6}{5} x + \frac{6 \cdot 8}{5 \cdot 7} x^2 + \frac{6 \cdot 8 \cdot 10}{5 \cdot 7 \cdot 9} x^3 + \cdots \right]^{-1}}$$

$$= \frac{1}{\frac{3}{4} - \frac{9}{10} \left[x - \frac{2}{35} x^2 - \frac{52}{1575} x^3 + \cdots \right]}.$$

Let

(105)
$$\xi = \frac{2}{35} x^2 + \frac{52}{1575} x^3 + \cdots.$$

If $\frac{1}{2} g$ is a small quantity of the first order, x is of the second order and ξ is of the fourth order.

From (98) it is found that

(106)
$$x = \frac{m^2}{\eta^2} - l.$$

Let

(107)
$$h = \frac{m^2}{\frac{5}{6} + l + \xi};$$

then (101) may be written

$$\eta - 1 = \frac{m^2 X}{\eta^2} = \frac{\frac{10}{9} h}{\eta^2 - h},$$

from which it is found that

(108)
$$\eta^3 - \eta^2 - h\eta - \frac{h}{9} = 0.$$

If ξ were known h would be determined by (107) and η by (108), which has but one real positive root. In the first approximation compute h assuming that the small quantity ξ is zero; then find the real positive root of (108). Or, instead of computing the root, use may be made of the tables which have been constructed by Gauss, giving the real positive values of η for values of h ranging from 0 to 0.6.* The value of x is then computed by (106) and the value of ξ by (105).† With this value of ξ, h, and η are recomputed, and the process is repeated until the desired degree of precision is attained. Experience has shown that this method of computing

* This table is XIII. in Watson's *Theoretical Astronomy*, and VIII. in Oppolzer's *Bahnbestimmung*.

† The value of ξ with argument x is given in Watson's *Theoretical Astronomy*, Table XIV., and in Oppolzer's *Bahnbestimmung*, Table IX.

the ratio of the sector to the triangle converges very rapidly, even when the time-interval is considerable.

The species of conic section is decided at this point, the orbit being an ellipse, parabola, or hyperbola according as x is positive, zero, or negative; for, $x = \sin^2 \dfrac{g}{2} = \sin^2 \dfrac{1}{4} (E_2 - E_1)$, and E_2 and E_1 are real in ellipses, zero in parabolas, and imaginary in hyperbolas.

Gauss has introduced a transformation which facilitates the computation of l which was defined in the last equation of (97).‡ Let

$$\sqrt[4]{\frac{r_2}{r_1}} = \tan (45° + \omega'), \qquad 0° \leqq \omega' \leqq 45°;$$

whence

$$\frac{r_1 + r_2}{\sqrt{r_1 r_2}} = \sqrt{\frac{r_2}{r_1}} + \sqrt{\frac{r_1}{r_2}} = \tan^2 (45° + \omega') + \cot^2 (45° + \omega'),$$

or

$$\frac{r_1 + r_2}{\sqrt{r_1 r_2}} = 2 + 4 \tan^2 2\omega'.$$

Then the last equation of (97) becomes

$$l = \frac{\sin^2 \dfrac{f}{2} + \tan^2 2\omega'}{\cos f}.$$

138. Determination of the Elements a, e, and ω.

After g has been found by the method of Art. 137 it is easy to obtain the elements a, e, and ω. The major semi-axis a is defined by the last equation on page 240, or by the preceding equation for the longer time-interval $t_3 - t_1$,

$$(109) \qquad a = \frac{r_1 + r_3 - 2\sqrt{r_1 r_3} \cos g \cos f}{2 \sin^2 g}.$$

The parameter of the orbit p is determined by equation (93). Since

$$(110) \qquad p = a(1 - e^2) \quad \text{or} \quad p = a(e^2 - 1)$$

according as the orbit is an ellipse or hyperbola, e is determined when a and p are known.

If the angle v is computed from the perihelion point it is related to the heliocentric distances and e and p by the polar equation of the conic,

‡ *Theoria Motus*, Art. 86.

$$(111) \qquad r_i = \frac{p}{1 + e \cos v_i} \qquad (i = 1, 2, 3).$$

Either of these equations determines a value of v since r is known at t_1, t_2, and t_3, and then ω is determined by

$$(112) \qquad \omega = u_i - v_i.$$

139. Second Method of Determining a, e, and ω. The method of Gauss depends upon the complicated formulas of Arts. 135 and 136. If the higher terms of P and Q, equations (86), give sufficiently accurate values of the ratios of the triangles, there is another method * which is simpler and especially advantageous when the intervals between the observations are not very great. The data which will be used in the solution are r_1, u_1; r_2, u_2; r_3, u_3, the heliocentric coördinates at t_1, t_2, and t_3.

The elements i and Ω can be computed by equations (91), which are valid for any orbit. The difficulties all arise in finding a, e, ω. Let the parameter p be adopted as an element in place of the major semi-axis a. It is more convenient in that it does not become infinite when e equals unity, and it is involved alone in the equation of areas,

$$k \sqrt{p}\, dt = r^2 dv = r^2 du.$$

The integral of this equation is

$$(113) \qquad k \sqrt{p}(t_3 - t_1) = \int_{u_1}^{u_3} r^2 du.$$

If r^2 were expressed in terms of u the integral in the right member could be found, when the value of p would be given. It will be shown from the knowledge of the value of r^2 when $u = u_1$, u_2, u_3, viz., $r^2 = r_1^2$, r_2^2, r_3^2, that r^2 can be expressed in terms of u with sufficient accuracy to give a very close approximation to the value of p.

For values of u not too remote from u_2 the function r^2 can be expanded in a converging series of the form

$$(114) \quad r^2 = r_2^2 + c_1(u - u_2) + c_2(u - u_2)^2 + c_3(u - u_2)^3 + \cdots.$$

In an unknown orbit the coefficients of the series (114) are unknown, but it will now be shown how a sufficient number to define p with the desired degree of accuracy can be easily found. By hypothesis, the radii and arguments of latitude are known at the epochs t_1, t_2, and t_3. Hence (114) becomes at t_1 and t_3

* F. R. Moulton; *The Astronomical Journal*, vol. XXII., No. 510 (1901).

$$(115) \quad \begin{cases} r_1^2 = r_2^2 + c_1(u_1 - u_2) + c_2(u_1 - u_2)^2 \\ \qquad\qquad + c_3(u_1 - u_2)^3 + c_4(u_1 - u_2)^4 + \cdots, \\ r_3^2 = r_2^2 + c_1(u_3 - u_2) + c_2(u_3 - u_2)^2 \\ \qquad\qquad + c_3(u_3 - u_2)^3 + c_4(u_3 - u_2)^4 + \cdots. \end{cases}$$

For abbreviation let

$$(116) \quad \begin{cases} \sigma_1 = u_3 - u_2, \\ \sigma_2 = u_3 - u_1, \\ \sigma_3 = u_2 - u_1, \\ \epsilon_1 = c_3(u_1 - u_2)^3 + c_4(u_1 - u_2)^4 + \cdots, \\ \epsilon_3 = c_3(u_3 - u_2)^3 + c_4(u_3 - u_2)^4 + \cdots. \end{cases}$$

Then equations (115) can be written

$$\begin{cases} -c_1\sigma_3 + c_2\sigma_3^2 = r_1^2 - r_2^2 - \epsilon_1, \\ +c_1\sigma_1 + c_2\sigma_1^2 = r_3^2 - r_2^2 - \epsilon_3. \end{cases}$$

On solving for c_1 and c_3, it is found that

$$\begin{cases} c_1 = \dfrac{-(r_1^2 - \epsilon_1)\sigma_1^2 + (r_3^2 - \epsilon_3)\sigma_3^2 - r_2^2(\sigma_3^2 - \sigma_1^2)}{\sigma_1\sigma_2\sigma_3}, \\[2mm] c_2 = \dfrac{(r_1^2 - \epsilon_1)\sigma_1 + (r_3^2 - \epsilon_3)\sigma_3 - r_2^2\sigma_2}{\sigma_1\sigma_2\sigma_3}; \end{cases}$$

and, on substituting the values of ϵ_1 and ϵ_3,

$$(117) \quad \begin{cases} c_1 = \dfrac{-r_1^2\sigma_1^2 + r_3^2\sigma_3^2 - r_2^2(\sigma_3^2 - \sigma_1^2)}{\sigma_1\sigma_2\sigma_3} \\ \qquad\qquad\qquad - c_3\sigma_1\sigma_3 - c_4\sigma_1\sigma_3(\sigma_1 - \sigma_3) - \cdots, \\[2mm] c_2 = \dfrac{r_1^2\sigma_1 + r_3^2\sigma_3 - r_2^2\sigma_2}{\sigma_1\sigma_2\sigma_3} \\ \qquad\qquad\qquad - c_3(\sigma_1 - \sigma_3) + c_4(3\sigma_1\sigma_3 - \sigma_2^2) - \cdots. \end{cases}$$

Having obtained these expressions for the coefficients of the second and third terms of (114), let this series be substituted for r^2 in (113) and the result integrated. On making use of (116), it is easily found that

$$\begin{cases} k\sqrt{p}(t_3 - t_1) = r_2^2\sigma_2 + \dfrac{c_1}{2}(\sigma_1^2 - \sigma_3^2) + \dfrac{c_2}{3}(\sigma_1^3 + \sigma_3^3) \\ \qquad\qquad\qquad + \dfrac{c_3}{4}(\sigma_1^4 - \sigma_3^4) + \dfrac{c_4}{5}(\sigma_1^5 + \sigma_3^5) + \cdots. \end{cases}$$

On substituting the values of c_1 and c_2 given in (117), this equation becomes

$$(118) \quad \begin{cases} \sqrt{p}(t_3 - t_1) = \dfrac{r_2^2 \sigma_2^3}{6 \sigma_1 \sigma_3} + \dfrac{r_1^2 \sigma_2}{6 \sigma_3}(2\sigma_3 - \sigma_1) \\[2mm] \qquad + \dfrac{r_3^2 \sigma_2}{6 \sigma_1}(2\sigma_1 - \sigma_3) + \dfrac{c_3 \sigma_2^3}{12}(\sigma_3 - \sigma_1) \\[2mm] \qquad - \dfrac{c_4 \sigma_2^3}{30}\{4(\sigma_3 - \sigma_1)^2 + \sigma_1 \sigma_3\} - \cdots. \end{cases}$$

If the second observation divides the whole interval into two nearly equal parts, as generally will be the case in practice, σ_1 and σ_3 will be nearly equal. Let

$$\sigma_1 - \sigma_3 = \epsilon, \quad \text{and} \quad \sigma_1 + \sigma_3 = \sigma_2;$$

whence

$$\begin{cases} \sigma_1 = \dfrac{\sigma_2 + \epsilon}{2}, \\[3mm] \sigma_3 = \dfrac{\sigma_2 - \epsilon}{2}, \end{cases}$$

where ϵ is in general a very small quantity. On substituting these expressions in the last terms of (118) this equation becomes

$$(119) \quad \begin{cases} k\sqrt{p}(t_3 - t_1) = \dfrac{r_2^2 \sigma_2^3}{6 \sigma_1 \sigma_3} + \dfrac{r_1^2 \sigma_2}{6 \sigma_3}(2\sigma_3 - \sigma_1) \\[2mm] \qquad + \dfrac{r_3^2 \sigma_2}{6 \sigma_1}(2\sigma_1 - \sigma_3) - \dfrac{c_3 \sigma_2^3 \epsilon}{12} \\[2mm] \qquad - \dfrac{c_4}{120} \sigma_2^3 (\sigma_2^2 + 15\epsilon^2) - \cdots. \end{cases}$$

It is found in a similar way on integrating between the limits corresponding to t_2 and t_1 that

$$(120) \quad \begin{cases} k\sqrt{p}(t_2 - t_1) = \dfrac{r_2^2 \sigma_3(3\sigma_1 + \sigma_3)}{6 \sigma_1} + \dfrac{r_1^2 \sigma_3(3\sigma_1 + 2\sigma_3)}{6 \sigma_2} \\[2mm] \qquad - \dfrac{r_3^2 \sigma_3^3}{6 \sigma_1 \sigma_2} + \dfrac{c_3 \sigma_3^3}{12}(2\sigma_1 + \sigma_3) \\[2mm] \qquad + \dfrac{c_4}{30} \sigma_3^3 (5\sigma_1^2 - 5\sigma_1 \sigma_3 - 4\sigma_3^2) + \cdots. \end{cases}$$

For the intervals of time which are used in determining an orbit these series converge very rapidly, and an approximate value of p, which is generally as accurate as is desired, can be obtained by taking only the first three terms* in the right member of (119). By considering equations (119) and (120) simultaneously and neglecting terms in c_4 and of higher order, it is possible to determine both p and c_3. But not much increase in accuracy is obtained because the term in c_3 in (119) is multiplied by the small quantity ϵ, while that in c_4 does not carry this factor. Suppose the value of p has been computed; it will be shown how ω and e can be found.

The polar equation of the conic gives

$$(121) \quad \begin{cases} e \cos(u_1 - \omega) = \dfrac{p - r_1}{r_1}, \\[2mm] e \cos(u_3 - \omega) = \dfrac{p - r_3}{r_3}. \end{cases}$$

Now $u_3 - \omega = (u_3 - u_1) + (u_1 - \omega)$. On substituting this expression for $u_3 - \omega$ in the second equation of (121), expanding, and reducing by the first, it is found that

$$(122) \quad \begin{cases} e \sin(u_1 - \omega) = \dfrac{r_3(p - r_1)\cos(u_3 - u_1) - r_1(p - r_3)}{r_1 r_3 \sin(u_3 - u_1)}, \\[2mm] e \cos(u_1 - \omega) = \dfrac{p - r_1}{r_1}. \end{cases}$$

Since e is positive these equations define e and ω uniquely. When p and e are known, a is defined by $p = a(1 - e^2)$ or $p = a(e^2 - 1)$ according as the orbit is an ellipse or an hyperbola.

If the elements a, e, and ω have not been found with sufficient approximation it is now possible to correct them. It follows from (114) that

$$c_3 = \frac{1}{6}\frac{\partial^3(r^2)}{\partial u_2^3}, \qquad c_4 = \frac{1}{24}\frac{\partial^4(r^2)}{\partial u_2^4},$$

and since

$$r^2 = \frac{p^2}{[1 + e\cos(u - \omega)]^2},$$

it is found that

---

* For conditions and rapidity of convergence see the original paper in the *Astronomical Journal*, No. 510. It is shown there that the elements of asteroid orbits will be given by the first three terms of (119) correct to the sixth decimal place if the whole interval covered by the observations is not more than 40 days, and in the case of comets' orbits, if the interval is not more than 10 days. When the two corrective terms defined by (123) are added the corresponding intervals are 100 days and 20 days.

$$(123) \begin{cases} \dfrac{c_3}{p^2} = \dfrac{- e \sin{(u - \omega)}}{3[1 + e \cos{(u - \omega)}]^3} + \dfrac{3e^2 \sin{(u - \omega)} \cos{(u - \omega)}}{[1 + e \cos{(u - \omega)}]^4} \\ \qquad\qquad\qquad\qquad + \dfrac{4e^3 \sin^3{(u - \omega)}}{[1 + e \cos{(u - \omega)}]^5}, \\ \dfrac{c_4}{p^2} = \dfrac{- e \cos{(u - \omega)}}{12[1 + e \cos{(u - \omega)}]^3} - \dfrac{e^2 \sin^2{(u - \omega)}}{[1 + e \cos{(u - \omega)}]^4} \\ \qquad\quad + \dfrac{3e^2 \cos^2{(u - \omega)}}{4[1 + e \cos{(u - \omega)}]^4} + \dfrac{6e^3 \sin^2{(u - \omega)} \cos{(u - \omega)}}{[1 + e \cos{(u - \omega)}]^5} \\ \qquad\qquad\qquad\qquad + \dfrac{5e^4 \sin^4{(u - \omega)}}{[1 + e \cos{(u - \omega)}]^6}. \end{cases}$$

With the values of c_3 and c_4 computed from these equations the higher terms of (119) can be added, thus determining a more accurate value of p, after which e and ω can be recomputed by (122). Besides being very brief this method has the advantage of being the same for all conics.

140. Computation of the Time of Perihelion Passage. The methods of computing the time of perihelion passage depend upon whether the body moves in a parabola, ellipse, or hyperbola, and are based on the formulas of chap. v.

Parabolic Case. Equation (32), of chap. v., is

$$(124) \qquad k(t - T) = \sqrt{2}\, q^{\frac{3}{2}} \left[\tan{\frac{v}{2}} + \frac{1}{3} \tan^3{\frac{v}{2}} \right],$$

where $2q = p$. Since $u = v + \omega$, and u_1, u_2, and u_3 are known, this equation determines T.

Elliptic Case. The first two equations of (49), chap. v., give

$$(125) \begin{cases} \sin E = \dfrac{\sqrt{1 - e^2}\, \sin v}{1 + e \cos v}, \\ \cos E = \dfrac{e + \cos v}{1 + e \cos v}, \end{cases}$$

which uniquely define E. Then Kepler's equation

$$(126) \qquad M = n(t - T) = E - e \sin E$$

determines T by using v and the corresponding E at t_1, t_2, or t_3.

Hyperbolic Case. The quantity F is defined by

$$(127) \qquad \tanh{\frac{F}{2}} = \sqrt{\frac{e - 1}{e + 1}} \tan{\frac{v}{2}},$$

after which T is given by

(128) $$\frac{k\sqrt{1+m}}{a^{\frac{3}{2}}}(t-T) = -F + e \sinh F.$$

141. Direct Derivation of Equations Defining Orbits.

The motion of an observed body must satisfy both geometrical and dynamical conditions. Altogether the simplest mode of procedure is to write out at once these conditions. They will involve directly or indirectly many of the equations of the methods of Laplace and Gauss, for these methods both rest in the end on the essentials of the problem.

Let the notation of Art. 111 be adopted. Think of the sun as an origin. Then obviously the x-coördinate of C equals the x-coördinate of the observer plus the x-coördinate of C with respect to the observer. Similar equations are of course true in the two other coördinates. These relations are explicitly

(129) $$\begin{cases} -\lambda_i\rho_i + x_i = -X_i, & (i = 1,\ 2,\ 3), \\ -\mu_i\rho_i + y_i = -Y_i, \\ -\nu_i\rho_i + z_i = -Z_i. \end{cases}$$

These equations are subject to no errors of parallax because the coördinates of the observer have been used. Moreover, they contain all the geometrical relations which exist among the bodies S, E, and C at t_1, t_2, and t_3.

The next condition to be applied is that C shall move about S according to the law of gravitation. This is equivalent to stating that its coördinates can be developed in series of the form of (73). On making use of this notation, equations (129) become

(130) $$\begin{cases} -\lambda_1\rho_1 + f_1x_0 + g_1x_0' = -X_1, \\ -\lambda_2\rho_2 + f_2x_0 + g_2x_0' = -X_2, \\ -\lambda_3\rho_3 + f_3x_0 + g_3x_0' = -X_3; \\ -\mu_1\rho_1 + f_1y_0 + g_1y_0' = -Y_1, \\ -\mu_2\rho_2 + f_2y_0 + g_2y_0' = -Y_2, \\ -\mu_3\rho_3 + f_3y_0 + g_3y_0' = -Y_3; \\ -\nu_1\rho_1 + f_1z_0 + g_1z_0' = -Z_1, \\ -\nu_2\rho_2 + f_2z_0 + g_2z_0' = -Z_2, \\ -\nu_3\rho_3 + f_3z_0 + g_3z_0' = -Z_3. \end{cases}$$

If the date of the second observation is taken as the origin of time, as is convenient in practice, $f_2 = 1$ and $g_2 = 0$.

Equations (130) contain fully the geometrical and dynamical conditions of the problem and are valid for all classes of conics. Since they are only the necessary conditions no artificial difficulties or exceptional cases have been introduced; and if in a special case they should fail no other mode of approach could succeed.

The right members of equations (130) are entirely known; the unknowns in the left members are ρ_1, ρ_2, ρ_3, x_0, x_0', y_0, y_0', z_0, and z_0'. That is, the number of unknowns exactly equals the number of equations. The quantities ρ_1, ρ_2, and ρ_3 enter linearly, but x_0, $\cdots$, z_0' occur not only explicitly but also in the higher terms of the f_i and the g_i. The solution of (130) for ρ_1, ρ_2, and ρ_3 is

$$(131) \quad \begin{cases} \Delta_2 \rho_1 = + A_1 - \dfrac{(f_1 g_3 - f_3 g_1)}{g_3} A_2 - \dfrac{g_1}{g_3} A_3, \\[2ex] \Delta_2 \rho_2 = \dfrac{- g_3}{f_1 g_3 - f_3 g_1} B_1 + B_2 + \dfrac{g_1}{f_1 g_3 - f_3 g_1} B_3, \\[2ex] \Delta_2 \rho_3 = \dfrac{- g_3}{g_1} C_1 + \dfrac{(f_1 g_3 - f_3 g_1)}{g_1} C_2 + C_3, \end{cases}$$

where

$$(132) \quad \begin{cases} \Delta_2 = \begin{vmatrix} \lambda_1, & \lambda_2, & \lambda_3 \\ \mu_1, & \mu_2, & \mu_3 \\ \nu_1, & \nu_2, & \nu_3 \end{vmatrix}, \qquad A_i = \begin{vmatrix} X_i, & \lambda_2, & \lambda_3 \\ Y_i, & \mu_2, & \mu_3 \\ Z_i, & \nu_2, & \nu_3 \end{vmatrix}, \\[4ex] B_i = \begin{vmatrix} \lambda_1, & X_i, & \lambda_3 \\ \mu_1, & Y_i, & \mu_3 \\ \nu_1, & Z_i, & \nu_3 \end{vmatrix}, \qquad C_i = \begin{vmatrix} \lambda_1, & \lambda_2, & X_i \\ \mu_1, & \mu_2, & Y_i \\ \nu_1, & \nu_2, & Z_i \end{vmatrix}. \end{cases}$$

In order to complete the discussion the coefficients of the determinants in the right members of these equations must be developed, as they were in (86); and since Δ_2 is of the third order, terms of the right member of the third order must be retained even in the first approximation. When applied to the second of (131), this leads to an equation of the form of the first of (44). The details of this and the completion of the solution of equations (130) will be called out in the questions which follow Art. 142.

142. Formulas for Computing an Approximate Orbit. For convenience in use the formulas for the computation of an approximate orbit are collected here in the order in which they are used. The numbers attached are those occurring in the text.

Preparation of the data. The observed right ascensions and declinations, α_0 and δ_0, are corrected for precession, aberration, etc., by

$$(4) \quad \begin{cases} \alpha = \alpha_0 - 15f - g \sin (G + \alpha_0) \tan \delta_0 - h \sin (H + \alpha_0) \sec \delta_0, \\ \delta = \delta_0 - i \cos \delta_0 - g \cos (G + \alpha_0) - h \cos (H + \alpha_0) \sin \delta_0. \end{cases}$$

The direction cosines are given by

$$(6) \quad \begin{cases} \lambda_j = \cos \delta_j \cos \alpha_j, \quad (j = 1, 2, 3), \\ \mu_j = \cos \delta_j \sin \alpha_j, \\ \nu_j = \sin \delta_j. \end{cases}$$

The Method of Laplace. Take $t_0 = t_2$ unless the intervals between the successive observations are very unequal, when $t_0 = \frac{1}{3}(t_1 + t_2 + t_3)$. It will be supposed that $t_0 = t_2$. Suppose X, Y, and Z are tabulated in the Ephemeris for t_a, t_b, t_c where t_b is near t_0. Then compute X, Y, and Z at t_0 from formulas of the type*

$$(31) \quad \begin{aligned} X = {} & \frac{(t_0 - t_b)(t_0 - t_c)}{(t_a - t_b)(t_a - t_c)} X_a + \frac{(t_0 - t_a)(t_0 - t_c)}{(t_b - t_a)(t_b - t_c)} X_b \\ & + \frac{(t_0 - t_a)(t_0 - t_b)}{(t_c - t_a)(t_c - t_b)} X_c. \end{aligned}$$

$$(26) \quad k(t_j - t_2) = \tau_j, \quad (j = 1, 2, 3; \ \tau_2 = 0).$$

$$(67) \quad P = -\tau_1 \tau_3 (\tau_3 - \tau_1).$$

$$(64, 65) \quad D = \frac{2}{P} \begin{vmatrix} \lambda_1, & \lambda_2, & \lambda_3 \\ \mu_1, & \mu_2, & \mu_3 \\ \nu_1, & \nu_2, & \nu_3 \end{vmatrix}.$$

$$(67) \quad D_1 = -\frac{\tau_3^2}{P} \begin{vmatrix} \lambda_1, & \lambda_2, & X \\ \mu_1, & \mu_2, & Y \\ \nu_1, & \nu_2, & Z \end{vmatrix} - \frac{\tau_1^2}{P} \begin{vmatrix} \lambda_2, & \lambda_3, & X \\ \mu_2, & \mu_3, & Y \\ \nu_2, & \nu_3, & Z \end{vmatrix}.$$

$$(68) \quad D_2 = +\frac{2\tau_3}{P} \begin{vmatrix} \lambda_1, & \lambda_2, & X \\ \mu_1, & \mu_2, & Y \\ \nu_1, & \nu_2, & Z \end{vmatrix} + \frac{2\tau_1}{P} \begin{vmatrix} \lambda_2, & \lambda_3, & X \\ \mu_2, & \mu_3, & Y \\ \nu_2, & \nu_3, & Z \end{vmatrix}.$$

---

*These equations are very simple because t_a, t_b, and t_c differ by intervals of one day, but there are other methods of interpolation which are even simpler.

$$(46) \qquad R \cos \psi = X\lambda + Y\mu + Z\nu, \quad (0 < \psi \leqq \pi).$$

$$(47) \qquad \begin{cases} N \sin m = R \sin \psi, \\[2mm] N \cos m = R \cos \psi - \dfrac{D_1}{DR^3}, \\[2mm] \qquad M = -\dfrac{NDR^3}{D_1} \sin^3 \psi > 0. \end{cases}$$

$$(48) \qquad \sin^4 \varphi = M \sin (\varphi + m).$$

$$(46) \qquad r = R\frac{\sin \psi}{\sin \varphi}, \qquad \rho = R\frac{\sin (\psi + \varphi)}{\sin \varphi}.$$

$$(44) \qquad \frac{d\rho}{d\tau} = \rho' = \frac{D_2}{2D}\left[\frac{1}{R^3} - \frac{1}{r^3}\right].$$

$$(8) \qquad x = \rho\lambda - X, \qquad y = \rho\mu - Y, \qquad z = \rho\nu - Z$$

Compute λ', μ', ν' from equations of the type

$$(32) \qquad \begin{aligned} \lambda' = &\frac{-(\tau_2 + \tau_3)\lambda_1}{(\tau_1 - \tau_2)(\tau_1 - \tau_3)} - \frac{(\tau_3 + \tau_1)\lambda_2}{(\tau_2 - \tau_3)(\tau_2 - \tau_1)} \\[2mm] &- \frac{(\tau_1 + \tau_2)\lambda_3}{(\tau_3 - \tau_1)(\tau_3 - \tau_2)}. \end{aligned}$$

Compute X', Y', and Z' from equations of the type

$$(32) \qquad \begin{aligned} kX' = &\frac{2t_2 - (t_b + t_c)}{(t_a - t_b)(t_a - t_c)} X_a + \frac{2t_2 - (t_c + t_a)}{(t_b - t_a)(t_b - t_c)} X_b \\[2mm] &+ \frac{2t_2 - (t_a + t_b)}{(t_c - t_a)(t_c - t_b)} X_c. \end{aligned}$$

$$(8) \qquad \begin{cases} x' = \rho'\lambda + \rho\lambda' - X', \\[2mm] y' = \rho'\mu + \rho\mu' - Y', \\[2mm] z' = \rho'\nu + \rho\nu' - Z'. \end{cases}$$

where differentiation is with respect to τ.

At this point the correction for the time aberration may be made by equations (70), and the approximate values of x, y, z, x', y', and z' may be improved by the methods of Arts. 128 and 129; or, the elements may be computed at once from the formulas given in chap. v. The formulas for the determination of the elements will be given and the numbers of the equations refer to chap. v

The integrals of areas in the equator system are

$$(10) \quad \begin{cases} xy' - yx' = b_1, \\ yz' - zy' = b_2, \\ zx' - xz' = b_3. \end{cases}$$

If ϵ represents the obliquity of the ecliptic, the corresponding constants in the ecliptic system are

$$\begin{cases} a_1 = b_1 \cos \epsilon - b_3 \sin \epsilon, \\ a_2 = b_2, \\ a_3 = b_1 \sin \epsilon + b_3 \cos \epsilon, \end{cases}$$

and i and Ω are defined by (chap. v.)

$$(15) \quad \begin{cases} a_1 = \sqrt{a_1{}^2 + a_2{}^2 + a_3{}^2} \cos i, \\ a_2 = \pm \sqrt{a_1{}^2 + a_2{}^2 + a_3{}^2} \sin i \sin \Omega, \\ a_3 = \mp \sqrt{a_1{}^2 + a_2{}^2 + a_3{}^2} \sin i \cos \Omega. \end{cases}$$

The major axis and parameter are defined by

$$(24) \quad x'^2 + y'^2 + z'^2 = k^2 \left(\frac{2}{r} - \frac{1}{a} \right),$$

where differentiation is with respect to τ.

$$(22) \quad k^2 p = k^2 a(1 - e^2) = a_1{}^2 + a_2{}^2 + a_3{}^2$$

It follows from Fig. 37, p. 237, that

$$\begin{cases} \sin i \sin u = \sin b = \dfrac{\bar{z}}{r}, \\ \cos i \sin u = \cos b \sin (l - \Omega) = \dfrac{\bar{y}}{r} \cos \Omega - \dfrac{\bar{x}}{r} \sin \Omega, \\ \cos u = \cos b \cos (l - \Omega) = \dfrac{\bar{y}}{r} \sin \Omega + \dfrac{\bar{x}}{r} \cos \Omega, \end{cases}$$

which define u, where $(\bar{x}, \bar{y}, \bar{z})$ are from equations, p. 256. The angle v is given by

$$r = \frac{p}{1 + e \cos v}.$$

and
$$\omega = u - v.$$

use $e \sin v = \sqrt{\rho}\, r^1$ to determine quadrant.

If the orbit is a parabola, T is defined by

$$(32) \quad k(t - T) = \tfrac{1}{2} p^{\frac{3}{2}} \left[\tan \frac{v}{2} + \frac{1}{3} \tan^3 \frac{v}{2} \right].$$

If the orbit is an ellipse, E, n, and T are defined by

(50) $$\tan\frac{E}{2} = \sqrt{\frac{1 - e}{1 + e}}\tan\frac{v}{2},$$

(30) $$n = \frac{k}{a^{\frac{3}{2}}},$$

(42) $$n(t - T) = E - e\sin E.$$

The corresponding equations for hyperbolic orbits are

(73) $$a + r = ae\cosh F,$$

(74) $$n(t - T) = -F + e\sinh F.$$

The Method of Gauss. The observed data are corrected by (4) and the direction cosines are given by (6). The coördinates of the sun at t_1, t_2, and t_3 can be computed from equations of the type

(31) $$X_i = \frac{(t_i - t_b)(t_i - t_c)}{(t_a - t_b)(t_a - t_c)}X_a + \frac{(t_i - t_a)(t_i - t_c)}{(t_b - t_a)(t_b - t_c)}X_b$$
$$+ \frac{(t_i - t_a)(t_i - t_b)}{(t_c - t_a)(t_c - t_b)}X_c$$

where X_a, X_b, X_c are taken from the Ephemeris and t_b is the time nearest to t_i for which X is given. Then

(64) $$\Delta_2 = \begin{vmatrix} \lambda_1, & \lambda_2, & \lambda_3 \\ \mu_1, & \mu_2, & \mu_3 \\ \nu_1, & \nu_2, & \nu_3 \end{vmatrix},$$

(88) $$\begin{cases} K = -\tfrac{1}{2}\begin{vmatrix} \lambda_1, & X_1 + X_3 - 2X_2, & \lambda_3 \\ \mu_1, & Y_1 + Y_3 - 2Y_2, & \mu_3 \\ \nu_1, & Z_1 + Z_3 - 2Z_2, & \nu_3 \end{vmatrix}, \\[4ex] K_1 = -\begin{vmatrix} \lambda_1, & X_1 + X_3, & \lambda_3 \\ \mu_1, & Y_1 + Y_3, & \mu_3 \\ \nu_1, & Z_1 + Z_3, & \nu_3 \end{vmatrix}. \end{cases}$$

On neglecting the last term of (87), which is very small, and comparing the result with the first of (44), it is seen that the explicit formulas for determining r_2 and ρ_2 are

(46) $$R_2\cos\psi_2 = X_2\lambda_2 + Y_2\mu_2 + Z_2\nu_2, \qquad (0 < \psi_2 \leqq \pi),$$

(47)
$$\begin{cases} N \sin m = R_2 \sin \psi_2, \\[2mm] N \cos m = R_2 \cos \psi_2 - \dfrac{K}{\Delta_2}, \\[2mm] M = \dfrac{N}{\tau^2} \dfrac{4\Delta_2 R^3 \sin^3 \psi}{K_1} > 0. \end{cases}$$

(48)
$$\sin^4 \varphi = M \sin (\varphi + m).$$

(46)
$$\begin{cases} r_2 = \dfrac{R_2 \sin \psi_2}{\sin \varphi}, \\[2mm] \rho_2 = R_2 \dfrac{\sin (\psi_2 + \varphi)}{\sin \varphi}. \end{cases}$$

Then ρ_1 and ρ_3 are given by

(89)
$$\begin{cases} \begin{vmatrix} \lambda_1, & \lambda_3 \\ \mu_1, & \mu_3 \end{vmatrix} \rho_1 = \begin{vmatrix} X_1, & \lambda_3 \\ Y_1, & \mu_3 \end{vmatrix} - \dfrac{[1,3]}{[2,3]} \begin{vmatrix} X_2, & \lambda_3 \\ Y_2, & \mu_3 \end{vmatrix} \\[4mm] \qquad\quad + \dfrac{[1,2]}{[2,3]} \begin{vmatrix} X_3, & \lambda_3 \\ Y_3, & \mu_3 \end{vmatrix} + \rho_2 \dfrac{[1,3]}{[1,2]} \begin{vmatrix} \lambda_1, & \lambda_2 \\ \mu_1, & \mu_2 \end{vmatrix}, \\[4mm] \begin{vmatrix} \lambda_1, & \lambda_3 \\ \mu_1, & \mu_3 \end{vmatrix} \rho_3 = \dfrac{[2,3]}{[1,2]} \begin{vmatrix} \lambda_1, & X_1 \\ \mu_1, & Y_1 \end{vmatrix} - \dfrac{[1,3]}{[1,2]} \begin{vmatrix} \lambda_1, & X_2 \\ \mu_1, & Y_2 \end{vmatrix} \\[4mm] \qquad\quad + \begin{vmatrix} \lambda_1, & X_3 \\ \mu_1, & Y_3 \end{vmatrix} - \rho_2 \dfrac{[1,3]}{[2,3]} \begin{vmatrix} \lambda_1, & \lambda_2 \\ \mu_1, & \mu_2 \end{vmatrix}, \end{cases}$$

(or by formulas obtained from these by cyclical permutation of the letters λ, μ, ν and X, Y, Z), where

(85)
$$2\tau = \tau_3 - \tau_1, \qquad 2\epsilon = \tau_3 + \tau_1,$$

and

(86)
$$\begin{cases} \dfrac{[1,3]}{[2,3]} = \dfrac{1}{\dfrac{1}{2} + \dfrac{\epsilon}{2\tau} + \dfrac{\tau^2}{4r_2^3}}, \\[6mm] \dfrac{[1,2]}{[2,3]} = \dfrac{1 - \dfrac{\epsilon}{\tau} + \dfrac{\tau^2}{2r_2^3}}{1 + \dfrac{\epsilon}{\tau} + \dfrac{\tau^2}{2r_2^3}}, \\[6mm] \dfrac{[2,3]}{[1,2]} = \dfrac{1 + \dfrac{\epsilon}{\tau} + \dfrac{\tau^2}{2r_2^3}}{1 - \dfrac{\epsilon}{\tau} + \dfrac{\tau^2}{2r_2^3}}, \\[6mm] \dfrac{[1,3]}{[1,2]} = \dfrac{1}{\dfrac{1}{2} - \dfrac{\epsilon}{2\tau} + \dfrac{\tau^2}{4r_2^3}}. \end{cases}$$

$$(8) \quad \begin{cases} x_j = \rho_j \lambda_j - X_j, & (j = 1, 2, 3), \\ y_j = \rho_j \mu_j - Y_j, \\ z_j = \rho_j \nu_j - Z_j. \end{cases}$$

At this point the correction for the time aberration may be made; the first two derivatives of r_2^2 may be computed from the values r_1^2, r_2^2, and r_3^2 by applying the formulas (32) to this case; p and q may be computed from (74) and more approximate values of P and Q may be determined from (86); and then the computation may be repeated beginning with equations (46); or, the method of Gauss of Art. 134 may be used to improve the accuracy of the expressions for the ratios of the triangles; or, the elements may be computed without further approximation of the intermediate quantities. The formulas for the computation of the elements will be given. Let the rectangular coördinates in the ecliptic system be $\bar{x}_i$, $\bar{y}_i$, $\bar{z}_i$, and the obliquity of the ecliptic ϵ, which will not be confused with the ϵ defined in (85). Then

$$\begin{cases} \bar{x}_j = x_j, & (j = 1, 2, 3), \\ \bar{y}_j = + y_j \cos \epsilon + z_j \sin \epsilon, \\ \bar{z}_j = - y_j \sin \epsilon + z_j \cos \epsilon. \end{cases}$$

$$(17) \quad \begin{cases} A\bar{x}_1 + B\bar{y}_1 + C\bar{z}_1 = 0, \\ A\bar{x}_2 + B\bar{y}_2 + C\bar{z}_2 = 0, \\ A\bar{x}_3 + B\bar{y}_3 + C\bar{z}_3 = 0, \end{cases}$$

from which

$$A : B : C = \begin{vmatrix} \bar{y}_1, & \bar{z}_1 \\ \bar{y}_2, & \bar{z}_2 \end{vmatrix} : \begin{vmatrix} \bar{z}_1, & \bar{x}_1 \\ \bar{z}_2, & \bar{x}_2 \end{vmatrix} : \begin{vmatrix} \bar{x}_1, & \bar{y}_1 \\ \bar{x}_2, & \bar{y}_2 \end{vmatrix}.$$

Then, from equations corresponding to (11), (14), and (15) of chap. v.,

$$(15) \quad \begin{cases} \cos i = \dfrac{C}{\sqrt{A^2 + B^2 + C^2}}, \\ \sin \Omega \sin i = \dfrac{\pm A}{\sqrt{A^2 + B^2 + C^2}}, \\ \cos \Omega \sin i = \dfrac{\mp B}{\sqrt{A^2 + B^2 + C^2}}, \end{cases}$$

which define Ω and i.

It follows from Fig. 37 that the arguments of the latitude are defined by

$$\begin{cases} \sin i \sin u_j = \dfrac{\bar{z}_j}{r_j}, \qquad (j = 1, 2, 3), \\[2mm] \cos i \sin u_j = \dfrac{\bar{y}_j}{r_j} \cos \Omega - \dfrac{\bar{x}_j}{r_j} \sin \Omega, \\[2mm] \qquad \cos u_j = \dfrac{\bar{y}_j}{r_j} \sin \Omega + \dfrac{\bar{x}_j}{r_j} \cos \Omega. \end{cases}$$

(116) $\sigma_1 = u_3 - u_2, \qquad \sigma_2 = u_3 - u_1, \qquad \sigma_3 = u_2 - u_1.$

(119) $k \sqrt{p}(t_3 - t_1) = \dfrac{r_2^2 \sigma_2^3}{6\sigma_1 \sigma_3} + \dfrac{r_1^2 \sigma_2 (2\sigma_3 - \sigma_1)}{6\sigma_3} + \dfrac{r_3^2 \sigma_2 (2\sigma_1 - \sigma_3)}{6\sigma_1},$

defines p.

(122) $\begin{cases} e \sin (u_1 - \omega) = \dfrac{r_3(p - r_1) \cos \sigma_2 - r_1(p - r_3)}{r_1 r_3 \sin \sigma_2}, \\[2mm] e \cos (u_1 - \omega) = \dfrac{p - r_1}{r_1} \end{cases}$

define e and ω. Hence a can be determined from p and e.

Since $v_j = u_j - \omega$ $(j = 1, 2, 3)$, the time of perihelion passage is determined precisely as in the method of Laplace by equations (of chap. v.) (32), [(50), (30), (42)], [(73), (74)] in the parabolic, elliptic, and hyperbolic cases respectively.

XVII. PROBLEMS.

1. Take three observations of an asteroid not separated from one another by more than 15 days, or three of a comet not separated from one another by more than 6 days, and compute the elements of the orbit by both the method of Laplace and also that of Gauss.

2. Prove that the apparent motion of C cannot be permanently along a great circle unless it moves in the plane of the ecliptic.

3. Apply formulas (31) and (32) on a definite closed function, as for example $x = \sin t$.

4. By means of the equation

$$r^2 = R^2 + \rho^2 - 2Rr \cos \psi$$

eliminate ρ from the first equation of (44) and discuss the result by the methods of the Theory of Algebraic Equations, and show that the solutions agree qualitatively with those obtained in Art. 119.

5. Discuss the determinants D, D_1, and D_2 when there are four observations.

6. Express Δ_2 when there are three observations in terms of the α_i and the δ_i in such a manner that the fact it is of the third order will be explicitly exhibited.

7. Develop the explicit formulas, using the λ_i, μ_i, and ν_i and the determinant notation, for the differential corrections of the method of Harzer and Leuschner.

8. Give a geometrical interpretation of the vanishing of the coefficients of ρ_1 and ρ_3 in equations (89).

9. Suppose three positions of C are known as in Art. 139. Show (a) that the three equations

$$r_i = \frac{p}{1 + e \cos (u_i - \omega)}, \qquad (i = 1, 2, 3),$$

define p, e, and ω without using the intervals of time in which the arcs are described; (b) write out the explicit formulas for computing p, e, and ω; (c) compare their length with that of (119) and (122); and (d) show that p is not well determined as it depends upon ratios of small quantities of the third order.

10. Suppose $f_2 = 1$, $g_2 = 0$ and regard (130) as linear equations in ρ_1, ρ_2, ρ_3, x_0, x_0', y_0, y_0', z_0, z_0'. Show that the determinant of the coefficients is

$$\Delta = - g_1 g_3 (f_1 g_3 - f_3 g_1) \begin{vmatrix} \lambda_1, & \lambda_2, & \lambda_3 \\ \mu_1, & \mu_2, & \mu_3 \\ \nu_1, & \nu_2, & \nu_3 \end{vmatrix}$$

11. Show that on using the expansions of equations (86) the second equation of (131) becomes (87).

12. Having found ρ_2 from the equation corresponding to (87), and ρ_1 and ρ_3 from (131), show that x_0, x_0', y_0, y_0', z_0, z_0' can be determined from equations (130). (Then the elements can be determined as in the Laplacian Method.)

HISTORICAL SKETCH AND BIBLIOGRAPHY.

The first method of finding the orbit of a body (comet moving in a parabola) from three observations was devised by Newton, and is given in the *Principia*, Book III., Prop. XLI. The solution depends upon a graphical construction, which, by successive approximations, leads to the elements. One of the earliest applications of the method was by Halley to the comet which has since borne his name. Newton seems to have had trouble with the problem of determining orbits, for he said, " This being a problem of very great difficulty, I tried many methods of resolving it." Newton's success in basing his discussion on the fundamental elements of the problem was fully explained by Laplace in his memoir on the subject

The first complete solution which did not depend upon a graphical construction was given by Euler in 1744 in his *Theoria Motuum Planetarum et Cometarum*. Some important advances were made by Lambert in 1761. Up to this time the methods were for the most part based upon one or the other of two assumptions, which are only approximately true, viz., that in the interval $t_3 - t_1$ the observed body describes a straight line with uniform speed, or that the radius at the time of the second observation divides the chord joining the end positions into segments which are proportional to the intervals between the observations. In attempting to improve on the second of these assumptions Lambert made the discovery of the relation among the radii, chord, time-interval, and major axis mentioned in Art. 92. He later made the determination depend upon the curvature of the apparent orbit, which is closely related to the determinant Δ_2, and in this direction approached the best modern methods. He had an unusual grasp of the physics and geometry of the problem, and really anticipated many of the ideas which were carried out by his successors in better and more convenient ways.

Lagrange wrote three memoirs on the theory of orbits, two in 1778 and one in 1783. They are printed together in his collected works, vol. iv., pp. 439–532. As one would expect, with Lagrange came generality, precision, and mathematical elegance. He determined the geocentric distance of C at the time of the second observation by an equation of the eighth degree, which is nothing else than (87) with r_2 eliminated by means of the equation which expresses the fact that S, E, and C form a triangle at t_2. He developed the expressions for the heliocentric coördinates as power series in the time-intervals [eqs. (73)], and laid the foundation for the development of expressions for intermediate elements in power series. These developments have been completed and put in form for numerical applications by Charlier, *Meddelande från Lunds Astronomiska Observatorium*, No. 46. The original work of Lagrange was not put in a form adapted to the needs of the computer, and has not been used in practice.

In 1780 Laplace published an entirely new method in *Mémoires de l'Académie Royale des Sciences de Paris* (Collected Works, vol. x., pp. 93–146). This method, the fundamental ideas of which have been given in this chapter, has been the basis for a great deal of later work. Among the developments in this line may be mentioned a memoir by Villarceau (*Annales de l'Observatoire de Paris*, vol. iii.), the work of Harzer (*Astronomische Nachrichten*, vol. 141), and its simplification by Leuschner (*Publications of the Lick Observatory*, vol. vii., Part i.). The approximations beyond the first are not conveniently carried out in the original method of Laplace, but the method of differential corrections devised by Harzer and simplified by Leuschner has proved very satisfactory in practice.

Olbers published his classical *Abhandlung über die leichteste und bequemste Methode, die Bahn eines Kometen zu berechnen*, in 1797. This method has not been surpassed for computing parabolic orbits and is in very general use even at the present time. It is given in nearly every treatise on the theory of determining orbits.

The discovery of Ceres in 1801 and its loss after having been observed only a short time drew the attention of a brilliant young German mathematician, Gauss, to the problem of determining the elements of the orbit of a heavenly body from observations made from the earth. The problem was quickly solved, and an application of the method led to the recovery of Ceres. Gauss elaborated and perfected his work, and in 1809 brought it out in his *Theoria Motus Corporum Coelestium*. This work, written by a man at once a master of mathematics and highly skilled as a computer, is so filled with valuable ideas and is so exhaustive that it remains a classic treatise on the subject to this day. The later treatises all are under the greatest obligations to the work of Gauss.

In the *Memoirs of the National Academy of Science*, vol. IV. (1888), Gibbs published a method of considerable originality in which the first approximation to the ratios of the triangles was obtained more exactly by including all three geocentric distances as unknown from the beginning. The method is also distinguished by the fact that it was developed by the calculus of vector analysis.

The works to be consulted are:

The *Theoria Motus* of Gauss.

Watson's *Theoretical Astronomy* (now out of print).

Oppolzer's *Bahnbestimmung*, an exhaustive treatise.

Tisserand's *Leçons sur la Détermination des Orbites*, written in the characteristically clear French style.

Bauschinger's *Bahnbestimmung*, a recent book of great excellence by one of the best authorities on the subject of the theory of orbits.

Klinkerfues' *Theoretische Astronomie* (third edition by Buchholz), an excellent work and the most exhaustive one yet issued.

CHAPTER VII.

THE GENERAL INTEGRALS OF THE PROBLEM OF n BODIES.

143. The Differential Equations of Motion. Suppose the bodies are homogeneous in spherical layers; then they will attract each other as though their masses were at their centers. Let m_1, m_2, $\cdots$, m_n represent their masses. Let the coördinates of m_i referred to a fixed system of axes be x_i, y_i, z_i $(i = 1, \cdots, n)$. Let $r_{i,j}$ represent the distance between the centers of m_i and m_j. Let k^2 represent a constant depending upon the units employed. Then the components of force on m_1 parallel to the x-axis are

$$-\frac{k^2 m_1 m_2}{r^2_{1,2}} \cdot \frac{(x_1 - x_2)}{r_{1,2}}, \quad \cdots, \quad -\frac{k^2 m_1 m_n}{r^2_{1,n}} \cdot \frac{(x_1 - x_n)}{r_{1,n}},$$

and the total force is their sum. Therefore

$$m_1 \frac{d^2 x_1}{dt^2} = -k^2 m_1 \sum_{j=2}^{n} m_j \frac{(x_1 - x_j)}{r^3_{1,j}},$$

and there are corresponding equations in y and z.

There are similar equations for each body, and the whole system of equations is

$$(1) \quad \begin{cases} m_i \dfrac{d^2 x_i}{dt^2} = -k^2 m_i \sum\limits_{j=1}^{n} m_j \dfrac{(x_i - x_j)}{r^3_{i,j}}, \\[2ex] m_i \dfrac{d^2 y_i}{dt^2} = -k^2 m_i \sum\limits_{j=1}^{n} m_j \dfrac{(y_i - y_j)}{r^3_{i,j}}, \\[2ex] m_i \dfrac{d^2 z_i}{dt^2} = -k^2 m_i \sum\limits_{j=1}^{n} m_j \dfrac{(z_i - z_j)}{r^3_{i,j}}; \quad (i = 1, \cdots, n; j \neq i). \end{cases}$$

Each of these equations involves all of the $3n$ variables x_i, y_i, and z_i, and the system must, therefore, be solved simultaneously. There are $3n$ equations each of the second order, so that the problem is of order $6n$.

Equations (1) can be put in a simple and elegant form by the introduction of the potential function, which in this problem will be denoted by U instead of V. The constant k^2 will be included in the potential. In chap. iv the potential, V, was defined by

261

the integral $V = \int \dfrac{dm}{\rho}$.   In this case the system is composed of discrete masses, and the potential is

(2) $$U = \tfrac{1}{2}k^2 \sum_{i=1}^{n} \sum_{j=1}^{n} \frac{m_i m_j}{r_{i,j}}, \qquad (i \neq j).$$

The partial derivative of U with respect to x_i is

$$\frac{\partial U}{\partial x_i} = k^2 m_i \frac{\partial}{\partial x_i} \sum_{j=1}^{n} \frac{m_j}{r_{i,j}} = -k^2 m_i \sum_{j=1}^{n} m_j \frac{(x_i - x_j)}{r^3{}_{i,j}}, \qquad (i \neq j),$$

and there are similar equations in y_i and z_i. Therefore equations (1) can be written in the form

(3)
$$\begin{cases} m_i \dfrac{d^2 x_i}{dt^2} = \dfrac{\partial U}{\partial x_i}, \\[2mm] m_i \dfrac{d^2 y_i}{dt^2} = \dfrac{\partial U}{\partial y_i}, \\[2mm] m_i \dfrac{d^2 z_i}{dt^2} = \dfrac{\partial U}{\partial z_i}, \qquad (i = 1, \cdots, n). \end{cases}$$

144. The Six Integrals of the Motion of the Center of Mass. The function U is independent of the choice of the coördinate axes since it depends upon the mutual distances of the bodies alone. Therefore, if the origin is displaced parallel to the x-axis in the negative direction through a distance α, the x-coördinate of every body will be increased by the quantity α, but the potential function will not be changed. Let the fact that U is a function of all the x-coördinates be indicated by writing

$$U = U(x_1, x_2, \cdots, x_n).$$

After the origin is displaced the x-coördinates become

$$x_i' = x_i + \alpha, \quad (i = 1, \cdots, n).$$

The partial derivative of U with respect to α is

$$\frac{\partial U}{\partial \alpha} = \frac{\partial U}{\partial x_1'} \frac{\partial x_1'}{\partial \alpha} + \frac{\partial U}{\partial x_2'} \frac{\partial x_2'}{\partial \alpha} + \cdots + \frac{\partial U}{\partial x_n'} \frac{\partial x_n'}{\partial \alpha}$$

But $\dfrac{\partial x_i'}{\partial \alpha} = 1,\ (i = 1, \cdots, n)$, and $\dfrac{\partial U}{\partial \alpha} = 0$, because U does not involve α explicitly. Therefore, on dropping the accents and writing the corresponding equations in y_i and z_i for displacements β and γ, it is found that

$$\begin{cases} \dfrac{\partial U}{\partial \alpha} = \displaystyle\sum_{i=1}^{n} \dfrac{\partial U}{\partial x_i} = 0, \\[2ex] \dfrac{\partial U}{\partial \beta} = \displaystyle\sum_{i=1}^{n} \dfrac{\partial U}{\partial y_i} = 0, \\[2ex] \dfrac{\partial U}{\partial \gamma} = \displaystyle\sum_{i=1}^{n} \dfrac{\partial U}{\partial z_i} = 0. \end{cases}$$

Therefore equations (3) give

$$\begin{cases} \displaystyle\sum_{i=1}^{n} m_i \dfrac{d^2 x_i}{dt^2} = 0, \\[2ex] \displaystyle\sum_{i=1}^{n} m_i \dfrac{d^2 y_i}{dt^2} = 0, \\[2ex] \displaystyle\sum_{i=1}^{n} m_i \dfrac{d^2 z_i}{dt^2} = 0. \end{cases}$$

These equations are at once integrable, and the result of integration is

$$(4) \quad \begin{cases} \displaystyle\sum_{i=1}^{n} m_i \dfrac{dx_i}{dt} = \alpha_1, \\[2ex] \displaystyle\sum_{i=1}^{n} m_i \dfrac{dy_i}{dt} = \beta_1, \\[2ex] \displaystyle\sum_{i=1}^{n} m_i \dfrac{dz_i}{dt} = \gamma_1, \end{cases}$$

where α_1, β_1, γ_1 are the constants of integration. On integrating again, it follows that

$$(5) \quad \begin{cases} \displaystyle\sum_{i=1}^{n} m_i x_i = \alpha_1 t + \alpha_2, \\[2ex] \displaystyle\sum_{i=1}^{n} m_i y_i = \beta_1 t + \beta_2, \\[2ex] \displaystyle\sum_{i=1}^{n} m_i z_i = \gamma_1 t + \gamma_2. \end{cases}$$

Let $\displaystyle\sum_{i=1}^{n} m_i = M$, and $\bar{x}$, $\bar{y}$, and $\bar{z}$ represent the coördinates of the center of mass of the n bodies; then, by Art. 19,

$$(6) \quad \begin{cases} \sum_{i=1}^{n} m_i x_i = M\bar{x}, \\[2mm] \sum_{i=1}^{n} m_i y_i = M\bar{y}, \\[2mm] \sum_{i=1}^{n} m_i z_i = M\bar{z}. \end{cases}$$

Therefore, equations (5) become

$$(7) \quad \begin{cases} M\bar{x} = \alpha_1 t + \alpha_2, \\ M\bar{y} = \beta_1 t + \beta_2, \\ M\bar{z} = \gamma_1 t + \gamma_2; \end{cases}$$

that is, the coördinates of the center of mass vary directly as the time. From this it can be inferred that the center of mass moves with uniform speed in a straight line. Or otherwise, the velocity of the center of mass is

$$(8) \quad \bar{V} = \sqrt{\left(\frac{d\bar{x}}{dt}\right)^2 + \left(\frac{d\bar{y}}{dt}\right)^2 + \left(\frac{d\bar{z}}{dt}\right)^2} = \frac{1}{M}\sqrt{\alpha_1^2 + \beta_1^2 + \gamma_1^2},$$

which is a constant; and on eliminating t from equations (7), it is found that

$$(9) \quad \frac{M\bar{x} - \alpha_2}{\alpha_1} = \frac{M\bar{y} - \beta_2}{\beta_1} = \frac{M\bar{z} - \gamma_2}{\gamma_1},$$

which are the symmetrical equations of a straight line in space of three dimensions. Equations (8) and (9) give the theorem:

If n bodies are subject to no forces except their mutual attractions, their center of mass moves in a straight line with uniform speed. The special case $\bar{V} = 0$ will arise if $\alpha_1 = \beta_1 = \gamma_1 = 0$. Since it is impossible to know any fixed point in space it is impossible to determine the six constants.

The origin might now be transferred to the center of mass of the system, as it was in the Problem of Two Bodies, or, to the center of one of the bodies, as it will be in Art. 148, and the order of the problem reduced six units.

145. The Three Integrals of Areas. The potential function is not changed by a rotation of the axes. Suppose the system of coördinates is rotated around the z-axis through the angle $-\phi$, and call the new coördinates x_i', y_i', and z_i'. They are related to the old by the equations

$$
(10) \quad
\begin{cases}
x_i' = x_i \cos \phi - y_i \sin \phi, \\
y_i' = x_i \sin \phi + y_i \cos \phi, \\
z_i' = z_i, \qquad (i = 1, \cdots, n).
\end{cases}
$$

Since the function U is not changed by the rotation it does not contain ϕ explicitly; therefore

$$
(11) \quad \frac{\partial U}{\partial \phi} = \sum_{i=1}^{n} \frac{\partial U}{\partial x_i'} \frac{\partial x_i'}{\partial \phi} + \sum_{i=1}^{n} \frac{\partial U}{\partial y_i'} \frac{\partial y_i'}{\partial \phi} + \sum_{i=1}^{n} \frac{\partial U}{\partial z_i'} \frac{\partial z_i'}{\partial \phi} = 0.
$$

But from (10) it follows that

$$
\frac{\partial x_i'}{\partial \phi} = - y_i', \qquad \frac{\partial y_i'}{\partial \phi} = x_i', \qquad \frac{\partial z_i'}{\partial \phi} = 0, \qquad (i = 1, \cdots, n);
$$

therefore (11) becomes

$$
\sum_{i=1}^{n} \left[x_i' \frac{\partial U}{\partial y_i'} - y_i' \frac{\partial U}{\partial x_i'} \right] = 0.
$$

On dropping the accents, which are of no further use, it is found as a consequence of (3) that

$$
\begin{cases}
\sum_{i=1}^{n} m_i \left[x_i \frac{d^2 y_i}{dt^2} - y_i \frac{d^2 x_i}{dt^2} \right] = 0, \text{ and similarly,} \\
\sum_{i=1}^{n} m_i \left[y_i \frac{d^2 z_i}{dt^2} - z_i \frac{d^2 y_i}{dt^2} \right] = 0, \\
\sum_{i=1}^{n} m_i \left[z_i \frac{d^2 x_i}{dt^2} - x_i \frac{d^2 z_i}{dt^2} \right] = 0.
\end{cases}
$$

Each term of these sums can be integrated separately, giving

$$
(12) \quad
\begin{cases}
\sum_{i=1}^{n} m_i \left[x_i \frac{dy_i}{dt} - y_i \frac{dx_i}{dt} \right] = c_1, \\
\sum_{i=1}^{n} m_i \left[y_i \frac{dz_i}{dt} - z_i \frac{dy_i}{dt} \right] = c_2, \\
\sum_{i=1}^{n} m_i \left[z_i \frac{dx_i}{dt} - x_i \frac{dz_i}{dt} \right] = c_3.
\end{cases}
$$

The parentheses are the projections of the areal velocities of the various bodies upon the three fundamental planes (Art. 16). As it is impossible to determine any fixed point in space, so also it is impossible to determine any fixed direction in space; consequently it is impossible to determine practically the constants c_1, c_2, c_3. Yet, in this case it is customary to assume that the fixed stars, on the average, do not revolve in space, so that, by observing them, these constants can be determined. It is evident, however, that there is no more reason for assuming that the stars do not revolve than there is for assuming that they are not drifting through space, each being a pure assumption without any possibility of proof or disproof. But it is to be noted that, if these assumptions are granted, the constants c_1, c_2, and c_3 can be determined easily with a high degree of precision, while in the present state of observational Astronomy the constants of equations (4) cannot be found with any considerable accuracy.

Let A_i, B_i, and C_i represent the projections of the areas described by the line from the origin to the body m_i upon the xy, yz, and zx-planes respectively; then (12) can be written

$$
\begin{cases}
\displaystyle\sum_{i=1}^{n} m_i \frac{dA_i}{dt} = c_1, \\[2mm]
\displaystyle\sum_{i=1}^{n} m_i \frac{dB_i}{dt} = c_2, \\[2mm]
\displaystyle\sum_{i=1}^{n} m_i \frac{dC_i}{dt} = c_3,
\end{cases}
$$

the integrals of which are

$$
(13) \quad
\begin{cases}
\displaystyle\sum_{i=1}^{n} m_i A_i = c_1 t + c_1{}', \\[2mm]
\displaystyle\sum_{i=1}^{n} m_i B_i = c_2 t + c_2{}', \\[2mm]
\displaystyle\sum_{i=1}^{n} m_i C_i = c_3 t + c_3{}'.
\end{cases}
$$

Hence the theorem:

The sums of the products of the masses and the projections of the areas described by the corresponding radii are proportional to the time; or, from (12), the sums of the products of the masses and the rates of the projections of the areas are constants.

It is possible, as was first shown by Laplace, to direct the axes so that two of the constants in equations (12) shall be zero, while the third becomes $\sqrt{c_1{}^2 + c_2{}^2 + c_3{}^2}$. This is the plane of maximum sum of the products of the masses and the rates of the projections of areas. Its relations to the original fixed axes are defined by the constants c_1, c_2, c_3, and its position is, therefore, always the same. On this account it was called the *invariable* *plane* by Laplace. At present the invariable plane of the solar system is inclined to the ecliptic by about 2°, and the longitude of its ascending node is about 286°. These figures are subject to some uncertainty because of our imperfect knowledge regarding the masses of some of the planets. If the position of the plane were known with exactness it would possess some practical advantages over the ecliptic, which undergoes considerable variations, as a fundamental plane of reference. It has been of great value in certain theoretical investigations.*

146. The Energy Integral.† On multiplying equations (3) by $\frac{dx_i}{dt}, \frac{dy_i}{dt}, \frac{dz_i}{dt}$ respectively, adding, and summing with respect to i, it is found that

$$
(14) \quad
\begin{aligned}
\sum_{i=1}^{n} m_i & \left\{ \frac{d^2x_i}{dt^2}\frac{dx_i}{dt} + \frac{d^2y_i}{dt^2}\frac{dy_i}{dt} + \frac{d^2z_i}{dt^2}\frac{dz_i}{dt} \right\} \\
& = \sum_{i=1}^{n} \left\{ \frac{\partial U}{\partial x_i}\frac{dx_i}{dt} + \frac{\partial U}{\partial y_i}\frac{dy_i}{dt} + \frac{\partial U}{\partial z_i}\frac{dz_i}{dt} \right\}.
\end{aligned}
$$

The potential U is a function of the $3n$ variables x_i, y_i, z_i, alone; therefore the right member of (14) is the total derivative of U with respect to t. Upon integrating both members of this equation, it is found that

$$
(15) \quad \frac{1}{2}\sum_{i=1}^{n} m_i \left\{ \left(\frac{dx_i}{dt}\right)^2 + \left(\frac{dy_i}{dt}\right)^2 + \left(\frac{dz_i}{dt}\right)^2 \right\} = U + h.
$$

The left member of this equation is the kinetic energy of the whole system, and the right member is the potential function plus a constant.

Let the potential energy of one configuration of a system with respect to another configuration be defined as the amount of work required to change it from the one to the other. If two bodies attract each other according to the law of the inverse squares, the force existing between them is $\frac{k^2 m_i m_j}{r^2_{i,\,j}}$. The amount of work done in changing their distance apart from $r^{(0)}_{i,\,j}$ to $r_{i,\,j}$ is

$$
(16) \quad W_{i,\,j} = k^2 m_i\, m_j \int_{r^{(0)}_{i,\,j}}^{r_{i,\,j}} \frac{dr_{i,\,j}}{r^2_{i,\,j}} = k^2 m_i\, m_j \left[\frac{1}{r^{(0)}_{i,\,j}} - \frac{1}{r_{i,\,j}} \right].
$$

* See memoirs by Jacobi, *Journal de Math.*, vol. ix.; Tisserand, *Méc. Cél.* vol. i., chap. xxv.; Poincaré, *Les Méthodes Nouvelles de la Méc. Cél.*, vol. i., p. 39.

† This is very frequently called the *Vis Viva* integral.

If the bodies are at an infinite distance from one another at the start, then $r_{i,j}^{(0)} = \infty$, and (16) becomes

$$- W_{i,j} = \frac{k^2 m_i m_j}{r_{i,j}} ;$$

hence

$$U = -\frac{1}{2} \sum_{i=1}^{n} \sum_{j=1}^{n} W_{i,j}.$$

Therefore, U is the negative of the potential energy of the whole system with respect to the infinite separation of the bodies as the original configuration.   Hence (15) gives the theorem:

In a system of n bodies subject to no forces except their mutual attractions the sum of the kinetic and potential energies is a constant.

147. The Question of New Integrals. Ten of the whole $6n$ integrals which are required in order to solve the problem completely have been found.   These ten integrals are the only ones known, and the question arises whether any more of certain types exist.

In a profound memoir in the *Acta Mathematica*, vol. XI., Bruns has demonstrated that, when the rectangular coördinates are chosen as dependent variables, there are no new algebraic integrals. This does not, of course, exclude the possibility of algebraic integrals when other variables are used.   Poincaré has demonstrated in his prize memoir in the *Acta Mathematica*, vol. XIII., and again with some additions in *Les Méthodes Nouvelles de la Mécanique Céleste*, chap. v., that the Problem of Three Bodies admits no new uniform transcendental integrals, even when the masses of two of the bodies are very small compared to that of the third.   In this theorem the dependent variables are the elements of the orbits of the bodies, which continually change under their mutual attractions.   It does not follow that integrals of the class considered by Poincaré do not exist when other dependent variables are employed.   In fact, Levi-Civita has shown the existence of this class of integrals in a special problem, which comes under Poincaré's theorem, when suitable variables are used (*Acta Mathematica*, vol. XXX.).   The practical importance of the theorems of Bruns and Poincaré have often been overrated by those who have forgotten the conditions under which they have been proved to hold true.

XVIII. PROBLEMS.

1. Write equations (1) when the force varies inversely as the nth power of the distance. For what values of n do the equations all become independent? The Problem of n Bodies can be completely solved for this law of force; show that the orbits with respect to the center of mass of the system are all ellipses with this point as center. Show that the orbit of any body with respect to any other is also a central ellipse, and that the same is true for the motion of any body with respect to the center of mass of any subgroup of the whole system. Show that the periods are all equal.

2. What will be the definition of the potential function when the force varies inversely as the nth power of the distance?

3. Derive the equations immediately preceding (4) directly from equations (1).

4. Prove that the theorem regarding the motion of the center of mass holds when the force varies as any power of the distance.

5. Derive the equations immediately preceding (12) directly from equations (1), and show that they hold when the force varies as any power of the distance.

6. Any plane through the origin can be changed into any other plane through the origin by a rotation around each of two of the coördinate axes. Transform equations (12) by successive rotations around two of the axes, and show that the angles of rotation can be so chosen that two of the constants, to which the functions of the new coördinates similar to (12) are equal, are zero, and that the third is $\sqrt{c_1{}^2 + c_2{}^2 + c_3{}^2}$. (This is the method used by Laplace to prove the existence of the invariable plane.)

7. Why are equations (13) not to be regarded as integrals of the differential equations (1), thus making the whole number of integrals thirteen?

148. Transfer of the Origin to the Sun.

Nothing is known of the absolute motions of the planets because the observations furnish information regarding only their relative positions, or their positions with respect to the sun. It is true that it is known that the solar system is moving toward the constellation Hercules, but it must be remembered that this motion is only with respect to certain of the stars. The problem for the student of Celestial Mechanics is to determine the relative positions of the members of the solar system; or, in particular, to determine the positions of the planets with respect to the sun. To do this it is advantageous to transfer the origin to the sun, and to employ the resulting differential equations.

Suppose m_n is the sun and take its center as the origin, and let the coördinates of the body m_i referred to the new system be x_i', y_i', z_i'. Then the old coördinates are expressed in terms of the new by the equations

$$x_i = x_i' + x_n, \quad y_i = y_i' + y_n, \quad z_i = z_i' + z_n, \quad (i=1, \cdots, n-1).$$

Since the differences of the old variables are equal to the corresponding differences of the new, it follows that

$$\frac{\partial U}{\partial x_i} = \frac{\partial U}{\partial x_i'}, \quad \frac{\partial U}{\partial y_i} = \frac{\partial U}{\partial y_i'}, \quad \frac{\partial U}{\partial z_i} = \frac{\partial U}{\partial z_i'}, \quad (i = 1, \cdots, n-1).$$

As a consequence of these transformations equations (3) become

$$(17) \quad \begin{cases} \dfrac{d^2 x_i'}{dt^2} + \dfrac{d^2 x_n}{dt^2} = \dfrac{1}{m_i}\dfrac{\partial U}{\partial x_i'}, \\[2ex] \dfrac{d^2 y_i'}{dt^2} + \dfrac{d^2 y_n}{dt^2} = \dfrac{1}{m_i}\dfrac{\partial U}{\partial y_i'}, \\[2ex] \dfrac{d^2 z_i'}{dt^2} + \dfrac{d^2 z_n}{dt^2} = \dfrac{1}{m_i}\dfrac{\partial U}{dz_i'}, \quad (i = 1, \cdots, n-1). \end{cases}$$

Since the origin is at $x_n' = y_n' = z_n' = 0$, the first equation of (1) gives, on putting $i = n$,

$$(18) \quad \begin{aligned} \frac{d^2 x_n}{dt^2} &= \frac{k^2 m_1 x_1'}{r^3_{1,\,n}} + \frac{k^2 m_2 x_2'}{r^3_{2,\,n}} + \cdots + \frac{k^2 m_{n-1} x'_{n-1}}{r^3_{n-1,\,n}} \\[1ex] &= k^2 \sum_{j=1}^{n-1} \frac{m_j x_j'}{r^3_{j,\,n}}. \end{aligned}$$

This equation, with the corresponding ones in y and z, substituted in (17) completes the transformation to the new variables; but it will be advantageous to combine the terms in another manner so that those which come from the attraction of the sun shall be separate from the others. The differential equations will be written for the body m_1, from which the others can be formed by permuting the subscripts.

The potential function can be broken up into the sum

$$U = k^2 m_n \sum_{i=1}^{n-1} \frac{m_i}{r_{i,\,n}} + \tfrac{1}{2}k^2 \sum_{i=1}^{n-1} \sum_{j=1}^{n-1} \frac{m_i m_j}{r_{i,\,j}}, \quad (i \neq j);$$

or,

$$(19) \quad U = k^2 m_n \sum_{j=1}^{n-1} \frac{m_j}{r_{j,\,n}} + U'.$$

On substituting equations (18) and (19) in equations (17), the latter become

$$(20) \begin{cases} \dfrac{d^2x_1'}{dt^2} + k^2(m_1 + m_n)\dfrac{x_1'}{r^3_{1,\,n}} = \dfrac{1}{m_1}\dfrac{\partial U'}{\partial x_1'} - k^2\sum_{j=2}^{n-1}\dfrac{m_j x_j'}{r^3_{j,\,n}}, \\[2ex] \dfrac{d^2y_1'}{dt^2} + k^2(m_1 + m_n)\dfrac{y_1'}{r^3_{1,\,n}} = \dfrac{1}{m_1}\dfrac{\partial U'}{\partial y_1'} - k^2\sum_{j=2}^{n-1}\dfrac{m_j y_j'}{r^3_{j,\,n}}, \\[2ex] \dfrac{d^2z_1'}{dt^2} + k^2(m_1 + m_n)\dfrac{z_1'}{r^3_{1,\,n}} = \dfrac{1}{m_1}\dfrac{\partial U'}{\partial z_1'} - k^2\sum_{j=2}^{n-1}\dfrac{m_j z_j'}{r^3_{j,\,n}}. \end{cases}$$

Let

$$R_{1,\,j} = k^2\left\{\frac{1}{r_{1,\,j}} - \frac{x_1'x_j' + y_1'y_j' + z_1'z_j'}{r^3_{j,\,n}}\right\};$$

then, equations (20) can be written in the form

$$(21) \begin{cases} \dfrac{d^2x_1'}{dt^2} + k^2(m_1 + m_n)\dfrac{x_1'}{r^3_{1,\,n}} = \sum_{j=2}^{n-1} m_j \dfrac{\partial R_{1,\,j}}{\partial x_1'}, \\[2ex] \dfrac{d^2y_1'}{dt^2} + k^2(m_1 + m_n)\dfrac{y_1'}{r^3_{1,\,n}} = \sum_{j=2}^{n-1} m_j \dfrac{\partial R_{1,\,j}}{\partial y_1'}, \\[2ex] \dfrac{d^2z_1'}{dt^2} + k^2(m_1 + m_n)\dfrac{z_1'}{r^3_{1,\,n}} = \sum_{j=2}^{n-1} m_j \dfrac{\partial R_{1,\,j}}{\partial z_1'}. \end{cases}$$

Let the accents, which have become useless, be dropped, and, in order to derive the general equations corresponding to (21), let

$$(22) \qquad R_{i,\,j} = k^2\left\{\frac{1}{r_{i,\,j}} - \frac{x_ix_j + y_iy_j + z_iz_j}{r^3_{j,\,n}}\right\}, \qquad (i \neq j).$$

Then, the general equations for relative motion are

$$(23) \begin{cases} \dfrac{d^2x_i}{dt^2} + k^2(m_i + m_n)\dfrac{x_i}{r^3_{i,\,n}} = \sum_{j=1}^{n-1} m_j \dfrac{\partial R_{i,\,j}}{\partial x_i}, \qquad (i \neq j), \\[2ex] \dfrac{d^2y_i}{dt^2} + k^2(m_i + m_n)\dfrac{y_i}{r^3_{i,\,n}} = \sum_{j=1}^{n-1} m_j \dfrac{\partial R_{i,\,j}}{\partial y_i}, \\[2ex] \dfrac{d^2z_i}{dt^2} + k^2(m_i + m_n)\dfrac{z_i}{r^3_{i,\,n}} = \sum_{j=1}^{n-1} m_j \dfrac{\partial R_{i,\,j}}{\partial z_i}, \end{cases}$$

in which $i = 1, \cdots, n - 1$.

149. Dynamical Meaning of the Equations. In order to understand easily the meaning of the equations, suppose that there are but three bodies, m_1, m_2, and m_n Suppose m_n is the sun, let its mass equal unity, and let the distances from it to m_1 and m_2 be r_1 and r_2 respectively. Then equations (23) are, in full,

$$(24) \begin{cases} \dfrac{d^2x_1}{dt^2} + k^2(1+m_1)\dfrac{x_1}{r_1^3} = k^2m_2\dfrac{\partial}{\partial x_1}\left\{\dfrac{1}{r_{1,\,2}} - \dfrac{x_1x_2+y_1y_2+z_1z_2}{r_2^3}\right\}, \\[2ex] \dfrac{d^2y_1}{dt^2} + k^2(1+m_1)\dfrac{y_1}{r_1^3} = k^2m_2\dfrac{\partial}{\partial y_1}\left\{\dfrac{1}{r_{1,\,2}} - \dfrac{x_1x_2+y_1y_2+z_1z_2}{r_2^3}\right\}, \\[2ex] \dfrac{d^2z_1}{dt^2} + k^2(1+m_1)\dfrac{z_1}{r_1^3} = k^2m_2\dfrac{\partial}{\partial z_1}\left\{\dfrac{1}{r_{1,2}} - \dfrac{x_1x_2+y_1y_2+z_1z_2}{r_2^3}\right\}; \\[2ex] \dfrac{d^2x_2}{dt^2} + k^2(1+m_2)\dfrac{x_2}{r_2^3} = k^2m_1\dfrac{\partial}{\partial x_2}\left\{\dfrac{1}{r_{2,\,1}} - \dfrac{x_2x_1+y_2y_1+z_2z_1}{r_1^3}\right\}, \\[2ex] \dfrac{d^2y_2}{dt^2} + k^2(1+m_2)\dfrac{y_2}{r_2^3} = k^2m_1\dfrac{\partial}{\partial y_2}\left\{\dfrac{1}{r_{2,\,1}} - \dfrac{x_2x_1+y_2y_1+z_2z_1}{r_1^3}\right\}, \\[2ex] \dfrac{d^2z_2}{dt^2} + k^2(1+m_2)\dfrac{z_2}{r_2^3} = k^2m_1\dfrac{\partial}{\partial z_2}\left\{\dfrac{1}{r_{2,\,1}} - \dfrac{x_2x_1+y_2y_1+z_2z_1}{r_1^3}\right\}. \end{cases}$$

If m_2 were zero the first three equations would be independent of the second three, and they would then be the equations for the relative motion of the body m_1 with respect to $m_n = 1$, and could be integrated. All the variations from the purely elliptical motion arise from the presence of the right members, which are, in the first three equations, the partial derivatives of $R_{1,\,2}$ with respect to the variables x_1, y_1, and z_1 respectively. On this account $m_2R_{1,\,2}$ is called the *perturbative function*.

The partial derivatives of the first terms of the right members of the first three equations are respectively

$$- k^2m_2\frac{(x_1 - x_2)}{r^3_{1,\,2}}, \quad - k^2m_2\frac{(y_1 - y_2)}{r^3_{1,\,2}}, \quad - k^2m_2\frac{(z_1 - z_2)}{r^3_{1,\,2}},$$

which are the components of acceleration of m_1 due to the attraction of m_2. The partial derivatives of the second terms are

$$- k^2m_2\frac{x_2}{r_2^3}, \quad - k^2m_2\frac{y_2}{r_2^3}, \quad - k^2m_2\frac{z_2}{r_2^3},$$

which are the negatives of the components of the acceleration of the sun due to the attraction of m_2. Therefore the right members of the first three equations of (24) are the differences of the components of acceleration of m_1 and of the sun due to the attraction of m_2. Similarly, the right members of the last three equations are the differences of the components of the acceleration of m_2 and of the sun due to the attraction of m_1. If two bodies are subject to equal parallel accelerations their relative positions will not be changed. The differences of their accelerations are due to the disturbing forces, and measure these disturbances. The right members of (24) are, therefore, exactly those parts of the accelerations due to the disturbing forces.

If there are $n - 2$ disturbing bodies the right members are the sums of terms depending upon the bodies $m_2, \cdots, m_{n-1}$ similar to the right members of (24), which depend upon m_2 alone; or, in other words, the whole resultants of the disturbing accelerations are equal to the sums of the parts arising from the action of the separate disturbing bodies.

150. The Order of the System of Equations. The order of the system of equations (23) is $6n - 6$, instead of $6n$ as (1) was in the case of absolute motion. In the absolute motion ten integrals were found which reduced the problem to order $6n - 10$. Six of these related to the motion of the center of mass, three to the areal velocities, and one to the energy of the system. In the present case but four integrals, the three integrals of areas and the energy integral, can be found, which leaves the problem of order $6n - 10$ also.

The problem can be reduced to the order $6n - 6$ by using the integrals for the center of mass directly. In particular, consider the differential equations for the bodies $m_1, m_2, \cdots, m_{n-1}$. In the original equations they involve the coördinates of m_n, but these quantities can be eliminated by means of equations (5).

If the origin is taken at the center of mass

$$\sum_{i=1}^{n} m_i x_i = 0, \qquad \sum_{i=1}^{n} m_i y_i = 0, \qquad \sum_{i=1}^{n} m_i z_i = 0,$$

and the elimination becomes particularly simple. Or, because of these linear homogeneous relations, the n variables of each set can be expressed linearly and homogeneously in terms of $n - 1$ new variables. Thus

$$x_1 = a_{11}\xi_1 + a_{12}\xi_2 + \cdots + a_{1,\,n-1}\xi_{n-1},$$
$$x_2 = a_{21}\xi_1 + a_{22}\xi_2 + \cdots + a_{2,\,n-1}\xi_{n-1},$$
$$\cdot \quad \cdot \quad \cdot \quad \cdot \quad \cdot \quad \cdot \quad \cdot \quad \cdot \quad \cdot \quad \cdot \quad \cdot \quad \cdot$$
$$x_n = a_{n1}\xi_1 + a_{n2}\xi_2 + \cdots + a_{n,\,n-1}\xi_{n-1},$$

and similar sets of equations for y and z. The coefficients a_{ij} are arbitrary constants except that they must be so chosen that every determinant of the matrix of the substitutions shall be distinct from zero; for, otherwise, a linear relation would exist among the ξ_i. These constants can be so chosen that the transformed equations preserve a symmetrical form. This method was employed by Jacobi in an important memoir entitled, *Sur l'élimination des nœuds dans le problème des trois corps* (*Journal de Math.* vol. IX., 1844), and by Radau in a memoir entitled, *Sur une transformation des équations différentielles de la Dynamique* (*Annales de l'École Normale*, 1st series, vol. V.).

XIX. PROBLEMS.

1. Make the transformation $x_i = x_i' + x_n$ in the integrals (12) and (15), and eliminate x_n, y_n, z_n, $\dfrac{dx_n}{dt}$, $\dfrac{dy_n}{dt}$, and $\dfrac{dz_n}{dt}$ by means of equations (4) and (5). Prove that the resulting expressions are four integrals of equations (23).

2. Derive equations (23) directly by taking the origin at m_n, without first making use of the fixed axes.

3. The equations (23) are not symmetrical, since each body requires a different perturbative function $R_{i, j}$ in the right members. Construct the corresponding system of differential equations where the motion of m_{n-1} is referred to a rectangular system of axes with the origin at m_n; the motion of m_{n-2} to a parallel system of axes with origin at the center of mass of m_n and m_{n-1}; the motion of m_{n-3} to a parallel system of axes with the origin at the center of mass of m_n, m_{n-1}, and m_{n-2}, and continue in this way. Show that the results are the symmetrical equations.

$$\frac{\mu_n}{\mu_{n-1}} m_{n-1} \frac{d^2 x_{n-1}}{dt^2} = \frac{\partial U}{\partial x_{n-1}}, \qquad \mu_n = m_n, \qquad \mu_{n-1} = m_{n-1} + m_n,$$

$$\frac{\mu_{n-1}}{\mu_{n-2}} m_{n-2} \frac{d^2 x_{n-2}}{dt^2} = \frac{\partial U}{\partial x_{n-2}}, \qquad \mu_{n-2} = m_{n-2} + m_{n-1} + m_n,$$

$$\vdots$$

$$\frac{\mu_2}{\mu_1} m_1 \frac{d^2 x_1}{dt^2} = \frac{\partial U}{\partial x_1}, \qquad \mu_1 = m_1 + m_2 + \cdots + m_n,$$

and similar equations in y and z, where

$$U = k^2 m_n \left(\frac{m_{n-1}}{r_{n,\, n-1}} + \frac{m_{n-2}}{r_{n,\, n-2}} + \cdots + \frac{m_1}{r_{n,\, 1}} \right)$$

$$+ k^2 m_{n-1} \left(\frac{m_{n-2}}{r_{n-1,\, n-2}} + \frac{m_{n-3}}{r_{n-1,\, n-3}} + \cdots + \frac{m_1}{r_{n-1,\, 1}} \right)$$

$$\vdots$$

$$+ k^2 m_2 \frac{m_1}{r_{1,\, 2}}.$$

(These equations are the same as found by Radau from a different standpoint in the memoir cited in Art. 150. They have been employed by Tisserand in a very elegant demonstration of Poisson's theorem of the invariability of the major axes of the planets' orbits up to perturbations of the second order inclusive with respect to the masses. Poincaré has generally used this system in his researches in the Problem of Three Bodies.)

4. Derive the differential equations corresponding to (23) in polar co-ordinates.

$$\left\{\begin{array}{l}\dfrac{d^2 r_{j,\,n}}{dt^2} - r_{j,\,n}\cos^2\phi_i\left(\dfrac{d\theta_j}{dt}\right)^2 - r_{j,\,n}\left(\dfrac{d\phi_j}{dt}\right)^2 = -\dfrac{k^2(m_j + m_n)}{r^2_{j,\,n}} + \sum\limits_{i=1}^{n-1} m_i\,\dfrac{\partial R_{j,\,i}}{\partial r_{i,\,n}},\\[4mm]
\dfrac{d}{dt}\left(r^2_{j,\,n}\cos^2\phi_i\dfrac{d\theta_j}{dt}\right) = \sum\limits_{i=1}^{n-1} m_i\,\dfrac{\partial R_{j,\,i}}{\partial\theta_j},\\[4mm]
\dfrac{d}{dt}\left(r^2_{j,\,n}\dfrac{d\phi_j}{dt}\right) + r^2_{j,\,n}\sin\phi_i\cos\phi_i\left(\dfrac{d\theta_j}{dt}\right)^2 = \sum\limits_{i=1}^{n-1} m_i\cdot\dfrac{\partial R_{j,\,i}}{\partial\phi_j},\\[4mm]
\hspace{4cm}(j = 1,\,\cdots,\,n-1),\quad(i \neq j).\end{array}\right.$$

HISTORICAL SKETCH AND BIBLIOGRAPHY.

The investigations in the Problem of n Bodies are of two classes; first, those which lead to general theorems holding in every system; and second, those which give good approximations for a certain length of time in particular systems, such as the solar system. Investigations of the second class are known as theories of perturbations, the discussion of which will be given in another chapter.

The first general theorems are regarding the motion of the center of mass, and were given by Newton in the *Principia*. The ten integrals and the theorems to which they lead were known by Euler. The next general result was the proof of the existence and the discussion of the properties of the invariable plane by Laplace in 1784. In the winter semester of 1842–43 Jacobi gave a course of lectures in the University of Königsberg on Dynamics. In this course he gave the results of some very important investigations on the integration of the differential equations which arise in Mechanics. In all cases where the forces depend upon the coördinates alone, and where a potential function exists, conditions which are fulfilled in the Problem of n Bodies, he proved that if all the integrals except two have been found the last two can always be found. He also showed, in extending some investigations of Sir William Rowan Hamilton, that the problem is reducible to that of solving a partial differential equation whose order is one-half as great as that of the original system. Jacobi's lectures are published in the supplementary volume to his collected works. They are of great importance in themselves, as well as being an absolutely necessary prerequisite to the reading of the epoch-making memoirs of Poincaré, and they should be accessible to every student of Celestial Mechanics.

It is a question of the highest interest whether the motions of the members of such a system as the sun and planets are purely periodic. Newcomb has shown in an important memoir published in the *Smithsonian Contributions to Knowledge*, December 1874, that the differential equations can be formally satisfied by purely periodic series. He did not, however, prove the convergence of these series; and, indeed, Poincaré has shown in *Les Méthodes Nouvelles*, chaps. IX. and XII., that they are in general divergent.

As was stated in Art. 147, Bruns has proved in the *Acta Mathematica,* vol. XI., that, using rectangular coördinates, there are no new algebraic integrals; and Poincaré, in the *Acta Mathematica,* vol. XIII., that, using the elements as variables, there are no new uniform transcendental integrals, even when the masses of all the bodies except one are very small.

For further reading regarding the general differential equations in different sets of variables the student will do well to consult Tisserand's *Mécanique Céleste,* vol. I. chapters III., IV., and V.

CHAPTER VIII.

THE PROBLEM OF THREE BODIES.

151. Problem Considered. There are a number of important results in the Problem of Three Bodies which have been established with mathematical rigor if the initial coördinates and the components of velocity fulfill certain special conditions. While these special cases have not been found in nature, there are nevertheless some applications of the results obtained, and the processes employed are mathematically elegant and lead to most interesting conclusions. This chapter will contain such of these results as fall within the scope of this work, reserving the theories of perturbations, by means of which the positions of the heavenly bodies are predicted, to subsequent chapters.

The first part of the chapter will be devoted to a discussion of some of the properties of motion of an infinitesimal body when it is attracted by two finite bodies which revolve in circles around their center of mass, and will include the proof of the existence of certain particular solutions in which the distances of the infinitesimal body from the finite bodies are constants. The second part of the chapter will be devoted to an exposition of a method of finding particular solutions of the motion of three finite bodies such that the ratios of their mutual distances are constants. These solutions include the former, but the discoverable properties of motion are so much fewer, and are obtained with so much more difficulty, that it is advisable to divide the discussion into two parts.

The particular solutions of the Problem of Three Bodies which will be discussed here were given for the first time by Lagrange in a prize memoir in 1772. The method adopted here is radically different from that employed by him, and lends itself much more readily to a generalization to the case where a larger number of bodies is involved. But, on the other hand, the reduction of the order of the problem by one unit, which was a very interesting feature of Lagrange's memoir, is not accomplished by this method. However, as it has not been possible to make any use of this reduction, it has not been of any practical importance.

Mathematically speaking, an *infinitesimal body* is one that is attracted by finite masses but does not attract them. Physically speaking, it is a body of such a small mass that it will disturb the motion of finite bodies less than an arbitrarily assigned amount, however small, during any arbitrarily assigned time, however long. To actually determine a small mass fulfilling these conditions it is only necessary to make it so small that its whole attraction, which is always greater than its disturbing force, on one of the large bodies, if placed at the minimum distance possible, would move the large body less than the assigned small distance in the assigned time.

Motion of the Infinitesimal Body.

152. The Differential Equations of Motion. Suppose the system consists of two finite bodies revolving in circles around their common center of mass, and of an infinitesimal body subject to their attraction. Let the unit of mass be so chosen that the sum of the masses of the finite bodies shall be unity; then they can be represented by $1 - \mu$ and μ, where the notation is so chosen that $\mu \leqq \frac{1}{2}$. Let the unit of distance be so chosen that the constant distance between the finite bodies shall be unity. Let the unit of time be so chosen that k^2 shall equal unity. Let the origin of coördinates be taken at the center of mass of the finite bodies, and let the direction of the axes be so chosen that the $\xi\eta$-plane is the plane of their motion. Let the coördinates of $1-\mu$, μ, and the infinitesimal body be ξ_1, η_1, 0; ξ_2, η_2, 0; and ξ, η, ζ respectively, and

$$r_1 = \sqrt{(\xi - \xi_1)^2 + (\eta - \eta_1)^2 + \zeta^2},$$
$$r_2 = \sqrt{(\xi - \xi_2)^2 + (\eta - \eta_2)^2 + \zeta^2}.$$

Then the differential equations of motion for the infinitesimal body are

$$(1) \quad \begin{cases} \dfrac{d^2\xi}{dt^2} = -(1 - \mu)\dfrac{(\xi - \xi_1)}{r_1{}^3} - \mu\dfrac{(\xi - \xi_2)}{r_2{}^3}, \\[2mm] \dfrac{d^2\eta}{dt^2} = -(1 - \mu)\dfrac{(\eta - \eta_1)}{r_1{}^3} - \mu\dfrac{(\eta - \eta_2)}{r_2{}^3}, \\[2mm] \dfrac{d^2\zeta}{dt^2} = -(1 - \mu)\dfrac{\zeta}{r_1{}^3} - \mu\dfrac{\zeta}{r_2{}^3}. \end{cases}$$

As a consequence of the way the units have been chosen the mean angular motion of the finite bodies is

$$n = k\frac{\sqrt{(1 - \mu) + \mu}}{a^{\frac{3}{2}}} = 1.$$

Let the motion of the bodies be referred to a new system of axes having the same origin as the old, and rotating in the $\xi\eta$-plane in the direction in which the finite bodies move with the uniform angular velocity unity. The coördinates in the new system are defined by the equations

(2)
$$\begin{cases} \xi = x \cos t - y \sin t, \\ \eta = x \sin t + y \cos t, \\ \zeta = z, \end{cases}$$

and similar equations for the letters with subscripts 1 and 2. On computing the second derivatives of (2) and substituting in (1), it is found that

(3)
$$\begin{cases}
\left\{ \dfrac{d^2x}{dt^2} - 2\dfrac{dy}{dt} - x \right\} \cos t - \left\{ \dfrac{d^2y}{dt^2} + 2\dfrac{dx}{dt} - y \right\} \sin t \\[2ex]
\qquad = - \left\{ (1-\mu)\dfrac{(x-x_1)}{r_1^3} + \mu\dfrac{(x-x_2)}{r_2^3} \right\} \cos t \\[2ex]
\qquad\qquad + \left\{ (1-\mu)\dfrac{(y-y_1)}{r_1^3} + \mu\dfrac{(y-y_2)}{r_2^3} \right\} \sin t, \\[3ex]
\left\{ \dfrac{d^2x}{dt^2} - 2\dfrac{dy}{dt} - x \right\} \sin t + \left\{ \dfrac{d^2y}{dt^2} + 2\dfrac{dx}{dt} - y \right\} \cos t \\[2ex]
\qquad = - \left\{ (1-\mu)\dfrac{(x-x_1)}{r_1^3} + \mu\dfrac{(x-x_2)}{r_2^3} \right\} \sin t \\[2ex]
\qquad\qquad - \left\{ (1-\mu)\dfrac{(y-y_1)}{r_1^3} + \mu\dfrac{(y-y_2)}{r_2^3} \right\} \cos t, \\[3ex]
\dfrac{d^2z}{dt^2} = -(1-\mu)\dfrac{z}{r_1^3} - \mu\dfrac{z}{r_2^3}.
\end{cases}$$

Multiply the first two equations by $\cos t$ and $\sin t$ respectively, then by $-\sin t$ and $\cos t$, and add; the results are

$$\begin{cases}
\dfrac{d^2x}{dt^2} - 2\dfrac{dy}{dt} = x - (1-\mu)\dfrac{(x-x_1)}{r_1^3} - \mu\dfrac{(x-x_2)}{r_2^3}, \\[2ex]
\dfrac{d^2y}{dt^2} + 2\dfrac{dx}{dt} = y - (1-\mu)\dfrac{(y-y_1)}{r_1^3} - \mu\dfrac{(y-y_2)}{r_2^3}, \\[2ex]
\dfrac{d^2z}{dt^2} \quad = \quad -(1-\mu)\dfrac{z}{r_1^3} - \mu\dfrac{z}{r_2^3}.
\end{cases}$$

The position of the axes can be so taken at the origin of time that the x-axis will continually pass through the centers of the finite bodies; then $y_1 = 0$, $y_2 = 0$, and the equations become

$$(4) \quad \begin{cases} \dfrac{d^2x}{dt^2} - 2\dfrac{dy}{dt} = x - (1 - \mu)\dfrac{(x - x_1)}{r_1{}^3} - \mu\dfrac{(x - x_2)}{r_2{}^3}, \\[2ex] \dfrac{d^2y}{dt^2} + 2\dfrac{dx}{dt} = y - (1 - \mu)\dfrac{y}{r_1{}^3} - \mu\dfrac{y}{r_2{}^3}, \\[2ex] \dfrac{d^2z}{dt^2} \quad\ = \quad - (1 - \mu)\dfrac{z}{r_1{}^3} - \mu\dfrac{z}{r_2{}^3}. \end{cases}$$

These are the differential equations of motion of the infinitesimal body referred to axes rotating so that the finite bodies always lie on the x-axis. They have the important property that they do not involve explicitly the independent variable t because the coördinates of the finite bodies have become constants as a consequence of the particular manner in which the axes are rotated. On the other hand, in equations (1) the quantities ξ_1, ξ_2, η_1, and η_2 are functions of t.

The general problem of determining the motion of the infinitesimal body is of the sixth order; if it moves in the plane of motion of the finite bodies, the problem is of the fourth order.

153. Jacobi's Integral. Equations (4) admit an integral which was first given by Jacobi in *Comptes Rendus de l'Académie des Sciences de Paris*, vol. III., p. 59, and which has been discussed by Hill in the first of his celebrated papers on the Lunar Theory, *The American Journal of Mathematics*, vol. I., p. 18, and again by Darwin in his memoir on Periodic Orbits in *Acta Mathematica*, vol. XXI., p. 102. Let

$$(5) \qquad U = \tfrac{1}{2}(x^2 + y^2) + \frac{(1 - \mu)}{r_1} + \frac{\mu}{r_2};$$

then equations (4) can be written in the form

$$(6) \quad \begin{cases} \dfrac{d^2x}{dt^2} - 2\dfrac{dy}{dt} = \dfrac{\partial U}{\partial x}, \\[2ex] \dfrac{d^2y}{dt^2} + 2\dfrac{dx}{dt} = \dfrac{\partial U}{\partial y}, \\[2ex] \dfrac{d^2z}{dt^2} \quad\ = \dfrac{\partial U}{\partial z}. \end{cases}$$

If these equations are multiplied by $2\dfrac{dx}{dt}$, $2\dfrac{dy}{dt}$, and $2\dfrac{dz}{dt}$ respectively, and added, the resulting equation can be integrated, since U is a function of x, y, and z alone, and give

$$
\left(\frac{dx}{dt}\right)^2 + \left(\frac{dy}{dt}\right)^2 + \left(\frac{dz}{dt}\right)^2 = V^2 = 2U - C
$$
(7)
$$
\equiv x^2 + y^2 + \frac{2(1-\mu)}{r_1} + \frac{2\mu}{r_2} - C.
$$

Five integrals more are required in order completely to solve the problem. If the infinitesimal body moved in the xy-plane only three would remain to be found, the last two of which could be obtained by Jacobi's *last multiplier*,* if the first one were found. Thus it appears that only one new integral is needed for the complete solution of this special problem in the plane.† But Bruns has proved in *Acta Mathematica*, vol. XI., that, when rectangular coördinates are used, no new algebraic integrals exist; and Poincaré has proved in *Les Méthodes Nouvelles de la Mécanique Céleste*, vol. I., chap. V., that when the elements of the orbits are used as variables, there are no new uniform transcendental integrals, even when the mass of one of the finite bodies is very small compared to that of the other (see Art. 147). These demonstrations are entirely outside the scope of this work and cannot be reproduced here.

154. The Surfaces of Zero Relative Velocity.‡ Equation (7) is a relation between the square of the velocity and the coördinates of the infinitesimal body referred to the rotating axes. Therefore, when the constant of integration C has been determined numerically by the initial conditions, equation (7) determines the velocity with which the infinitesimal body will move, if at all, at all points of the rotating space; and conversely, for a given velocity, equation (7) gives the locus of those points of relative space where alone the infinitesimal body can be. In particular, if V is put equal to zero in this equation it will define the surfaces at which the velocity will be zero. On one side of these surfaces the velocity will be real and on the other side imaginary; or, in other words, it is

* Developed in *Vorlesungen über Dynamik*, supplementary volume to Jacobi's collected works.

† Hill put his special equations in such a form that they would be reduced to quadratures if a single variable were expressed in terms of the time, *American Journal of Mathematics*, vol. I., p. 16.

‡ First discussed by Hill in his *Lunar Theory*, *The American Journal of Mathematics*, vol. I.; and again, for motion in the xy-plane, by Darwin in his *Periodic Orbits*, in *Acta Mathematica*, vol. XXI.

possible for the body to move on one side, and impossible for it to move on the other. The general proposition that a function changes sign as the surface at which it is zero is crossed (at least at a regular point of the surface) was proved in Art. 120. While it will not be possible to say in any except very particular cases what the orbit will be, yet this partition of relative space will show in what portions the infinitesimal body can move and in what portions it can not.

The equation of the surfaces of zero relative velocity is

$$(8) \quad \begin{cases} x^2 + y^2 + \dfrac{2(1 - \mu)}{r_1} + \dfrac{2\mu}{r_2} = C, \\[2mm] r_1 = \sqrt{(x - x_1)^2 + y^2 + z^2}, \\[2mm] r_2 = \sqrt{(x - x_2)^2 + y^2 + z^2}. \end{cases}$$

Since only the squares of y and z occur the surfaces defined by (8) are symmetrical with respect to the xy and xz-planes, and, when $\mu = \frac{1}{2}$, with respect to the yz-plane also. The surfaces for $\mu \neq \frac{1}{2}$ can be regarded as being deformations of those for $\mu = \frac{1}{2}$. It follows from the way in which z enters that a line parallel to the z-axis pierces the surfaces in two (or no) real points. Moreover, the surfaces are contained within a cylinder whose axis is the z-axis and whose radius is $\sqrt{C}$, to which certain of the folds are asymptotic at $z^2 = \infty$; for, as z^2 increases the equation approaches as a limit

$$x^2 + y^2 = C.$$

155. Approximate Forms of the Surfaces. From the properties of the surfaces given in the preceding article and from the shapes of the curves in which the surfaces intersect the reference planes, a general idea of their form can be obtained. The equation of the curves of intersection of the surfaces with the xy-plane is obtained by putting z equal to zero in the first of (8), and is

$$(9) \quad x^2 + y^2 + \frac{2(1 - \mu)}{\sqrt{(x - x_1)^2 + y^2}} + \frac{2\mu}{\sqrt{(x - x_2)^2 + y^2}} = C.$$

For large values of x and y which satisfy this equation the third and fourth terms are relatively unimportant, and the equation may be written

$$x^2 + y^2 = C - \frac{2(1 - \mu)}{\sqrt{(x - x_1)^2 + y^2}} - \frac{2\mu}{\sqrt{(x - x_2)^2 + y^2}} = C - \epsilon,$$

where ϵ is a small quantity. This is the equation of a circle whose radius is $\sqrt{C - \epsilon}$; therefore, one branch of the curve in the xy-plane is an approximately circular oval within the asymptotic cylinder. It is also to be noted that the larger C is, the larger are the values of x and y which satisfy the equation, the smaller is ϵ, the more nearly circular is the curve, and the more nearly does it approach its asymptotic cylinder.

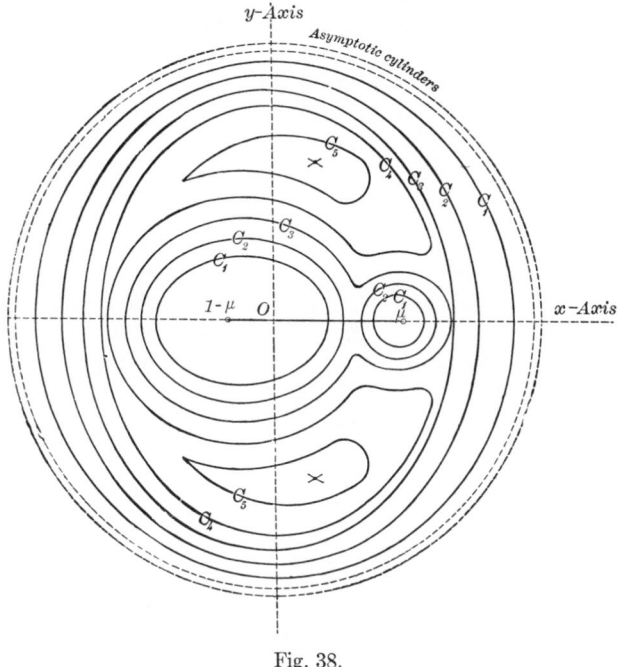

Fig. 38.

For small values of x and y satisfying (9) the first and second terms are relatively unimportant, and the equation may be written

$$\frac{1 - \mu}{r_1} + \frac{\mu}{r_2} = \frac{C}{2} - \frac{x^2 + y^2}{2} = \frac{C}{2} - \epsilon.$$

This is the equation of the *equipotential curves** for the two centers of force, $1 - \mu$ and μ. For large values of C they consist of closed ovals around each of the bodies $1 - \mu$ and μ; for smaller values of C these ovals unite between the bodies forming a dumb-

*Thomson and Tait's *Natural Philosophy*, Part II., Art. 508.

bell shaped figure in which the ends are of different size except when $\mu = \frac{1}{2}$; and for still smaller values of C the handle of the dumb-bell enlarges until the figure becomes an oval enclosing both of the bodies

From the foregoing considerations it follows that the approximate forms of the curves in which the surfaces intersect the xy-plane are as given in Fig. 38. The curves C_1, C_2, C_3, C_4, C_5 are in the order of decreasing values of the constant C. They were not drawn from numerical calculations and are intended to show only qualitatively the character of the curves.

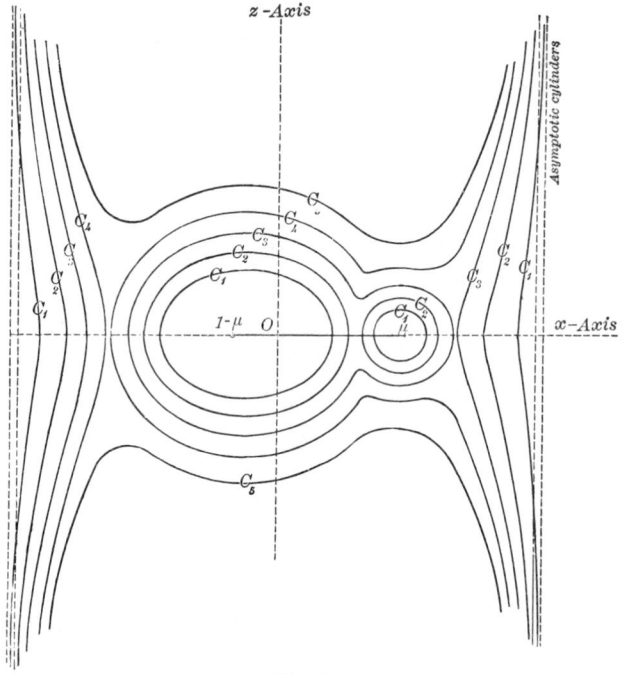

Fig. 39.

The equation of the curves of intersection of the surfaces and the xz-plane is obtained by putting y equal to zero in equation (8), and is

$$(10) \quad x^2 + \frac{2(1 - \mu)}{\sqrt{(x - x_1)^2 + z^2}} + \frac{2\mu}{\sqrt{(x - x_2)^2 + z^2}} = C.$$

For large values of x and z satisfying this equation the second and third terms are relatively unimportant, and it may be written

$$x^2 = C - \epsilon,$$

which is the equation of a symmetrical pair of straight lines parallel to the z-axis. The larger C is, the larger is the value of x which, for a given value of z, satisfies the equation, and, therefore, the smaller is ϵ. Hence, the larger C the closer the lines are to the asymptotic cylinder.

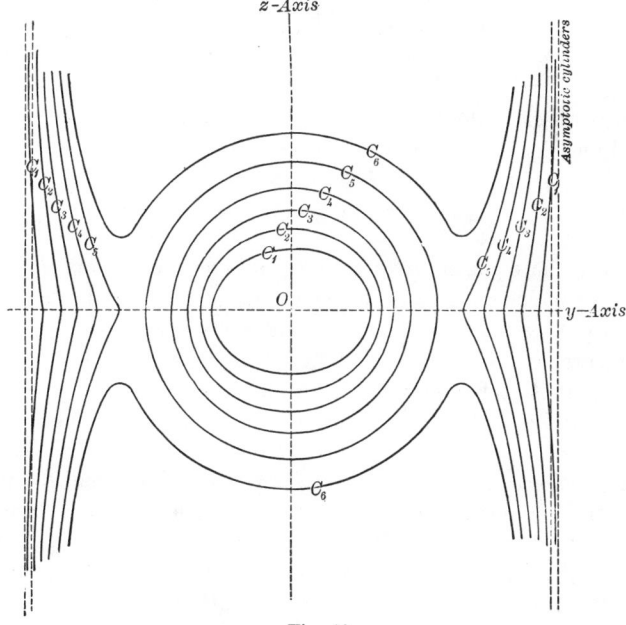

Fig. 40.

For small values of x and z satisfying equation (10) the first term is relatively unimportant, and the equation may be written

$$\frac{1 - \mu}{r_1} + \frac{\mu}{r_2} = \frac{C}{2} - \epsilon.$$

This is again the equation of the equipotential curves and has the same properties as before. Hence, the forms of the curves in the xz-plane are qualitatively like those given in Fig. 39. Again, the curves $C_1, \cdots, C_5$ are in the order of decreasing values of the constant C, and were not drawn from numerical calculations.

The equation of the curves of intersection of the surfaces and the yz-plane is obtained by putting x equal to zero in equation (8), and is

$$(11) \qquad y^2 + \frac{2(1 - \mu)}{\sqrt{x_1^2 + y^2 + z^2}} + \frac{2\mu}{\sqrt{x_2^2 + y^2 + z^2}} = C.$$

For large values of y and z satisfying this equation the second and third terms are relatively unimportant, and it may be written

$$y^2 = C - \epsilon,$$

which is the equation of a pair of lines near the asymptotic cylinder, approaching it as C increases.

If $1 - \mu$ is much greater than μ, the numerical value of x_2 is much greater than that of x_1; hence, for small values of y and z satisfying (11), this equation may be written

$$\frac{2(1 - \mu)}{r_1} = C - \epsilon,$$

which is the equation of a circle which becomes larger as C decreases. Hence, the forms of the curves in the yz-plane are qualitatively as given in Fig. 40. Again, the curves $C_1, \cdots, C_5$ are in the order of decreasing values of the constant C.

From these three sections of the surfaces it is easy to infer their forms for the different values of C. They may be roughly described as consisting of, for large values of C, a closed fold approximately spherical in form around each of the finite bodies, and of curtains hanging from the asymptotic cylinder symmetrically with respect to the xy-plane; for smaller values of C, the folds expand and coalesce (Fig. 38, curve C_3); for still smaller values of C the united folds coalesce with the curtains, the first points of contact being in every case in the xy-plane; and for sufficiently small values of C the surfaces consist of two parts symmetrical with respect to the xy-plane but not intersecting it (Figs. 39, curve C_5, and 40, curve C_6).

156. The Regions of Real and Imaginary Velocity. Having determined the forms of the surfaces, it remains to find in what regions of relative space the motion is real and in what it is imaginary. The equation for the square of the velocity is

$$V^2 = x^2 + y^2 + \frac{2(1 - \mu)}{r_1} + \frac{2\mu}{r_2} - C.$$

Suppose C is so large that the ovals and curtains are all separate.

The motion will be real in those portions of relative space for which the right member of this equation is positive. If it is positive in one point in a closed fold it will be positive in every other point within it, for the function changes sign only at a surface of zero relative velocity.

It is evident from the equation that x and y can be taken so large that the right member will be positive, however great C may be; therefore, *the motion is real outside of the curtains.* It is also clear that a.point can be chosen so near to either $1 - \mu$ or μ, that is, either r_1 or r_2 may be taken so small, that the right member will be positive, however great C may be; therefore, *the motion is real within the folds around the finite bodies.*

If the value of C were so large that the folds around the finite bodies were closed, and if the infinitesimal body should be within one of these folds at the origin of time, it would always remain there since it could not cross a surface of zero velocity. If the earth's orbit is supposed to be circular and the mass of the moon infinitesimal, it is found that the constant C, determined by the motion of the moon, is so large that the fold around the earth is closed with the moon within it. Therefore the moon cannot recede indefinitely from the earth. It was in this manner, and with these approximations, that Hill proved that the moon's distance from the earth has a superior limit.*

157. Method of Computing the Surfaces. Actual points on the surfaces can be found most readily by first determining the curves in the xy-plane, and then finding by methods of approximation the values of z which satisfy (7). Besides, the curves in the xy-plane are of most interest because the first points of contact as the various folds coalesce occur in this plane, and, indeed, on the x-axis, as can be·seen from the symmetries of the surfaces.

The equation of the curves in the xy-plane is

$$x^2 + y^2 + \frac{2(1 - \mu)}{\sqrt{(x - x_1)^2 + y^2}} + \frac{2\mu}{\sqrt{(x - x_2)^2 + y^2}} = C.$$

If this equation is rationalized and cleared of fractions the result is a polynomial of the sixteenth degree in x and y. When the value of one of the variables is taken arbitrarily the corresponding values of the other can be found by solving this rationalized equation. This problem presents great practical difficulties

---

* *Lunar Theory, Am. Jour. Math.,* vol. i., p. 23.

because of the high degree of the equation, and these troubles are supplemented by the presence of foreign solutions which are introduced by the processes of rationalization.

The difficulty from foreign solutions can be avoided entirely, and the degree of the equation can be very much reduced by transforming to bi-polar coördinates. That is, points on the curves can be defined by giving their distances from two fixed points on the x-axis. This method could not be applied if the curves were not symmetrical with respect to the axis on which the poles lie. Let the centers of the bodies $1 - \mu$ and μ be taken as the poles; the distances from these points are r_1 and r_2 respectively. To complete the transformation it is only necessary to express $x^2 + y^2$ in terms of these quantities.

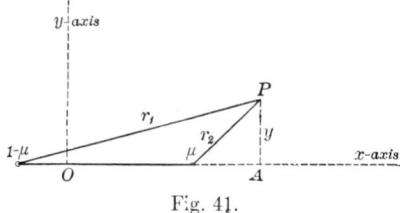

Fig. 41.

Let P be a point on one of the curves; then $OA = x$, $AP = y$, and, since O is the center of mass of $1 - \mu$ and μ, $\overline{O\mu} = 1 - \mu$, and $\overline{O(1 - \mu)} = -\mu$. It follows that

$$\begin{cases} y^2 = r_1^2 - (x + \mu)^2 = r_1^2 - x^2 - 2\mu x - \mu^2, \\ y^2 = r_2^2 - [x - (1 - \mu)]^2 = r_2^2 - x^2 + 2(1 - \mu)x - (1 - \mu)^2. \end{cases}$$

On eliminating the first power of x from these equations and solving for $x^2 + y^2$, it is found that

$$x^2 + y^2 = (1 - \mu)r_1^2 + \mu r_2^2 - \mu(1 - \mu).$$

As a consequence of this equation, (9) becomes

$$(12) \quad (1 - \mu)\left[r_1^2 + \frac{2}{r_1} \right] + \mu\left[r_2^2 + \frac{2}{r_2} \right] = C + \mu(1-\mu) = C'.$$

If an arbitrary value of r_2 is assumed r_1 can be computed from this equation; the points of intersection of the circles around $1 - \mu$ and μ as centers, with the computed and assumed values respectively of r_1 and r_2 as radii, will be points on the curves. To follow out this plan, let equation (12) be written in the form

$$(13) \quad \begin{cases} r_1{}^3 + ar_1 + b = 0, \\ a = -\dfrac{C'}{1-\mu} + \dfrac{\mu}{1-\mu}\left[r_2{}^2 + \dfrac{2}{r_2} \right], \\ b = 2. \end{cases}$$

Since $b = 2$ is positive there is at least one real negative root of the first of (13) whatever value a may have. But the only value of r_1 which has a meaning in this problem is real and positive; hence the condition for real positive roots must be considered.

It follows from (12) that C' is always greater than $\mu\left[r_2{}^2 + \dfrac{2}{r_2} \right]$ for all real positive values of r_1 and r_2; therefore a is always negative. It is shown in the Theory of Equations that a cubic equation of this form has three distinct real roots if $27b^2 + 4a^3 < 0$; or, since $b = 2$, if

$$(14) \qquad\qquad a + 3 < 0.$$

Suppose this inequality is satisfied. Then a convenient method of solving the cubic is

$$(15) \quad \begin{cases} \sin\theta = \dfrac{b}{2}\sqrt{\dfrac{27}{-a^3}}, \qquad \theta \leqq \dfrac{\pi}{2}, \\ r_{11} = 2\sqrt{\dfrac{-a}{3}}\sin\dfrac{\theta}{3}, \\ r_{12} = 2\sqrt{\dfrac{-a}{3}}\sin\left(60° - \dfrac{\theta}{3} \right), \\ r_{13} = -2\sqrt{\dfrac{-a}{3}}\sin\left(60° + \dfrac{\theta}{3} \right), \end{cases}$$

where r_{11}, r_{12}, r_{13} are the three roots of the cubic.

The limit of the inequality (14) is $a + 3 = 0$; or, in terms of the original quantities,

$$(16) \quad \begin{cases} r_2{}^3 + a'r_2 + b' = 0, \\ a' = -\dfrac{C'}{\mu} + \dfrac{3(1-\mu)}{\mu}, \\ b' = 2. \end{cases}$$

The solution of this equation gives the extreme values of r_2 for which (13) has real roots. Therefore, in the actual computation equation (16) should be solved first for r_{21} and r_{22}. The values of

r_2 to be substituted in (13) should be chosen at convenient intervals between these roots.

Equation (16) will not have real positive roots for all values of a', the condition for real positive roots being

$$a' + 3 \leqq 0;$$

the limiting value of which is, in the original quantities,

$$-\frac{C'}{\mu} + \frac{3(1-\mu)}{\mu} = -3;$$

whence

$$C' = 3.$$

Therefore C' must be equal to, or greater than, 3 in order that the curves shall have real points in the xy-plane. For $C' = 3$ the curves are just vanishing from the plane, and it follows at once that equation (12) is then satisfied by $r_1 = 1$, $r_2 = 1$; that is, the surfaces vanish from the xy-plane at the points which form equilateral triangles with $1 - \mu$ and μ.

158. Double Points of the Surfaces and Particular Solutions of the Problem of Three Bodies. It follows from the general forms of the surfaces that the double points which appear as C diminishes are all in the xy-plane. Therefore it is sufficient in this discussion to consider the equation of the curves in the xy-plane. There are three double points on the x-axis which appear when the ovals around the finite bodies touch each other and when they touch the exterior curve enclosing them both. There are two more which appear, as the surfaces vanish from the xy-plane, at the two points making equilateral triangles with the finite bodies.

These double points are of interest as critical points of the curves, and it will now be shown that they are connected with important dynamical properties of the system. Let the equation of the curves be written

$$(17) \quad F(x, y) \equiv x^2 + y^2 + \frac{2(1-\mu)}{r_1} + \frac{2\mu}{r_2} - C = 0.$$

The conditions for double points are

$$(18) \quad \begin{cases} \dfrac{1}{2}\dfrac{\partial F}{\partial x} \equiv x - (1-\mu)\dfrac{(x-x_1)}{r_1^3} - \mu\dfrac{(x-x_2)}{r_2^3} = 0, \\[2ex] \dfrac{1}{2}\dfrac{\partial F}{\partial y} \equiv y - (1-\mu)\dfrac{y}{r_1^3} - \mu\dfrac{y}{r_2^3} = 0. \end{cases}$$

The left members of these equations are the same as the right members of the equations (4) for $z = 0$. The expressions $\frac{1}{2}\frac{\partial F}{\partial x}$ and $\frac{1}{2}\frac{\partial F}{\partial y}$ are proportional to the direction cosines of the normal at all ordinary points of the curves; and since $\frac{dx}{dt}$ and $\frac{dy}{dt}$ are zero at the surfaces of zero velocity it follows from (4) that *the directions of acceleration, or the lines of effective force, are orthogonal to the surfaces of zero relative velocity.* Therefore, if the infinitesimal body is placed on a surface of zero relative velocity it will start in its motion in the direction of the normal. But at the double points the sense of the normal becomes ambiguous; hence, it might be surmised that if the infinitesimal body were placed at one of these points it would remain relatively at rest.

The conditions imposed by (17) and (18) are also the conditions that $\frac{d^2x}{dt^2}$ and $\frac{d^2y}{dt^2}$, or the components of acceleration, in equations (4) shall vanish. Hence, *if the infinitesimal body is placed at a double point with zero relative velocity, its coördinates will identically fulfill the differential equations of motion and it will remain forever relatively at rest, unless disturbed by forces exterior to the system under consideration.* These are particular solutions of the Problem of Three Bodies, and are special cases of the Lagrangian solutions.

Consider equations (18), the second of which is satisfied by $y = 0$. The double points on the x-axis, and the straight line solutions of the problem are given by the conditions

$$(19) \quad \begin{cases} x - (1-\mu)\frac{(x-x_1)}{[(x-x_1)^2]^{\frac{3}{2}}} - \mu\frac{(x-x_2)}{[(x-x_2)^2]^{\frac{3}{2}}} = 0, \\ y = 0, \\ z = 0. \end{cases}$$

The left member of the first equation considered as a function of x is positive for $x = +\infty$; it is negative for $x = x_2 + \epsilon$, where ϵ is a very small positive quantity; it is positive for $x = x_2 - \epsilon$; it is negative for $x = x_1 + \epsilon$; it is positive for $x = x_1 - \epsilon$; and it is negative for $x = -\infty$. Since the function is finite and continuous except when $x = +\infty$, x_2, x_1, or $-\infty$, it follows that the function changes sign three times by passing through zero, (a) once between $+\infty$ and x_2, (b) once between x_2 and x_1, and (c) once between x_1 and $-\infty$. Therefore, there are three positions on the line through $1 - \mu$ and μ at which the infinitesimal body will remain when given proper initial projection.

(a) Let the distance from μ to the double point on the x-axis between $+ \infty$ and x_2 be represented by ρ. Then $x - x_2 = \rho$, $x - x_1 = r_1 = 1 + \rho$, $x = 1 - \mu + \rho$; therefore the first equation of (19) becomes after clearing of fractions

(20) $\quad \rho^5 + (3 - \mu)\rho^4 + (3 - 2\mu)\rho^3 - \mu\rho^2 - 2\mu\rho - \mu = 0.$

This quintic equation has one variation in the sign of its coefficients, and hence only one real positive root. The value of this root depends upon μ. Consider the left member of the equation as a function of ρ and μ. For $\mu = 0$ the equation becomes

$$\rho^3(\rho^2 + 3\rho + 3) = 0,$$

which has three roots $\rho = 0$, and two others, coming from the second factor, which are complex. It follows from the theory of the solution of algebraic equations that, for μ different from zero but sufficiently small, three roots of the equation are expressible as power series in $\mu^{\frac{1}{3}}$, vanishing with this parameter.* The one of these three roots obtained by taking the real value of $\mu^{\frac{1}{3}}$ is real; the other two are complex. Therefore, the real root has the form

$$\rho = a_1\mu^{\frac{1}{3}} + a_2\mu^{\frac{2}{3}} + a_3\mu^{\frac{3}{3}} + \cdots.$$

On substituting this expression for ρ in (20) and equating to zero the coefficients of corresponding powers of $\mu^{\frac{1}{3}}$, it is found that

$$a_1 = \frac{3^{\frac{1}{3}}}{3}, \qquad a_2 = \frac{3^{\frac{1}{3}}}{9}, \qquad a_3 = -\frac{1}{27}, \qquad \cdots.$$

Hence

(21) $\quad \begin{cases} r_2 = \rho = \left(\dfrac{\mu}{3}\right)^{\frac{1}{3}} + \dfrac{1}{3}\left(\dfrac{\mu}{3}\right)^{\frac{2}{3}} - \dfrac{1}{9}\left(\dfrac{\mu}{3}\right)^{\frac{3}{3}} + \cdots, \\ r_1 = 1 + \rho. \end{cases}$

The corresponding value of C' is found by substituting these values of r_1 and r_2 in equation (12).

(b) Let the distance from μ to the double point on the x-axis between x_2 and x_1 be represented by ρ. Then in this case $x - x_2 = - \rho$, $x - x_1 = r_1 = 1 - \rho$, $x = (1 - \mu) - \rho$; therefore the first equation of (19) becomes

$$\rho^5 - (3 - \mu)\rho^4 + (3 - 2\mu)\rho^3 - \mu\rho^2 + 2\mu\rho - \mu = 0.$$

* See Harkness and Morley's *Theory of Functions*, chapter IV.

On solving as in (a), the values of r_2 and r_1 are found to be

$$(22) \quad \begin{cases} r_2 = \rho = \left(\dfrac{\mu}{3}\right)^{\frac{1}{3}} - \dfrac{1}{3}\left(\dfrac{\mu}{3}\right)^{\frac{2}{3}} - \dfrac{1}{9}\left(\dfrac{\mu}{3}\right)^{\frac{3}{3}} + \cdots, \\ r_1 = 1 - \rho. \end{cases}$$

The corresponding value of C' is found by substituting these values of r_1 and r_2 in equation (12).

(c) Let the distance from $1 - \mu$ to the double point on the x-axis between x_1 and $-\infty$ be represented by $1-\rho$. In this case $x - x_2 = -2 + \rho$, $x - x_1 = -1 + \rho$, $x = -\mu - 1 + \rho$, and the first equation of (19) becomes

$$(23) \quad \begin{aligned} \rho^5 - (7 + \mu)\rho^4 + (19 + 6\mu)\rho^3 - (24 + 13\mu)\rho^2 \\ + (12 + 14\mu)\rho - 7\mu = 0. \end{aligned}$$

When $\mu = 0$ this equation becomes

$$\rho^5 - 7\rho^4 + 19\rho^3 - 24\rho^2 + 12\rho = 0,$$

which has but one root $\rho = 0$. Therefore ρ can be expressed as a power series in μ which converges for sufficiently small values of this parameter, and vanishes with it. This root will have the form

$$\rho = c_1\mu + c_2\mu^2 + c_3\mu^3 + c_4\mu^4 + \cdots.$$

On substituting this expression for ρ in (23), and equating to zero the coefficients of the various powers of μ, it is found that

$$c_1 = \frac{7}{12}, \qquad c_2 = 0, \qquad c_3 = \frac{23 \times 7^2}{12^4}, \qquad \cdots.$$

Hence

$$(24) \quad \begin{cases} \rho = \dfrac{7}{12}\mu + \dfrac{23 \times 7^2}{12^4}\mu^3 + \cdots, \\ r_1 = 1 - \rho, \\ r_2 = 1 + r_1 = 2 - \rho \end{cases}$$

The corresponding value of C' is found by substituting these values of r_1 and r_2 in equation (12).

If the values of r_1 and r_2 given by the first three terms of the series (21), (22), and (24) are not sufficiently accurate, more nearly correct values should be found by differential corrections.

In order to find the double points *not* on the x-axis consider equations (18) again. They, or any two independent functions of them, define the double points. Since y is distinct from zero in this case the second equation may be divided by it, giving

$$1 - \frac{1 - \mu}{r_1^3} - \frac{\mu}{r_2^3} = 0.$$

Multiply this equation by $x - x_2$, and $x - x_1$, and subtract the products separately from the first of (18). The results are

$$\begin{cases} x_2 - (1 - \mu) \dfrac{(x_2 - x_1)}{r_1^3} = 0, \\[2mm] x_1 - \mu \dfrac{(x_1 - x_2)}{r_2^3} = 0, \\[2mm] z = 0. \end{cases}$$

But $x_2 = 1 - \mu$, $x_1 = -\mu$, and $x_2 - x_1 = 1$; therefore these equations reduce to

$$\begin{cases} 1 - \dfrac{1}{r_1^3} = 0, \\[2mm] -1 + \dfrac{1}{r_2^3} = 0, \\[2mm] z = 0. \end{cases}$$

The only real solutions are $r_1 = 1$, $r_2 = 1$, and the points form equilateral triangles with the finite bodies whatever their relative masses may be. As was shown in the last of Art. 157, they occur at the places where the surfaces vanish from the xy-plane.

XX. PROBLEMS.

1. The units defined in Art. 152 are called *canonical units;* what would the canonical unit of time be in days for the earth and sun?

2. Show on *à priori* grounds that, when the motion of the system is referred to axes rotating as in Art. 152, the differential equations should not involve the time explicitly.

3. Why cannot an integral corresponding to (7) be derived from equations (1) at once without any transformations? Prove that there is an integral of (1).

4. What are the surfaces of zero velocity for a body projected vertically upward against gravity? For a body moving subject to a central force varying inversely as the square of the distance?

5. Show by direct reductions from (13) and (14) that

$$(r_1 - r_{11})(r_1 - r_{12})(r_1 - r_{13}) \equiv r_1^3 + ar_1 + b = 0.$$

6. Prove that the solution of (16) gives the extreme values of r_2 for which (14) has real roots. *Hint.* Consider the graph of $y = r_2^3 + a'r_2 + b'$.

7. Impose the conditions on (12) that C' shall be a minimum and show that it is satisfied only for $r_1 = 1$, $r_2 = 1$, and that the minimum value of C' is 3.

8. Why are not the lines of effective force orthogonal to all of the surfaces of constant velocity?

9. Prove that the double point between μ and $1 - \mu$ is nearer μ than is the one between μ and $+ \infty$.

10. Prove that, as C' diminishes, the first double point to appear is the one between μ and $1 - \mu$; the second, the one between μ and $+ \infty$; the third, the one between $1 - \mu$ and $- \infty$; and the last, those which make equilateral triangles with the finite bodies.

11. If $\mu = \frac{1}{11}$, $1 - \mu = \frac{10}{11}$, find the values of r_1, r_2, and C' from (21), (22), (24), and (12).

$$
Ans. \begin{cases} (21) & r_2 = 0.340, & r_1 = 1.340, & C' = 3.535; \\ (22) & r_2 = 0.276, & r_1 = 0.724, & C' = 3.653; \\ (24) & r_2 = 1.947, & r_1 = 0.947, & C' = 3.173. \end{cases}
$$

12. From the approximate values of the last example find by the method of differential corrections more accurate values.

$$
Ans. \begin{cases} (21) & r_2 = 0.347, & r_1 = 1.347, & C' = 3.534; \\ (22) & r_2 = 0.282, & r_1 = 0.718, & C' = 3.653; \\ (23) & r_2 = 1.947, & r_1 = 0.947, & C' = 3.173. \end{cases}
$$

13. Considering the earth's orbit to be a circle, find the distance in miles from the earth to the double point which is opposite to the sun. Would an infinitesimal body at this point be eclipsed?

Ans. 930,240 miles.

159. Tisserand's Criterion for the Identity of Comets.* Comets sometimes pass near the planets in their revolutions around the sun, and then the elements of their orbits are greatly changed. The planet Jupiter is especially potent in producing these perturbations because of its great mass and because at its distance the attraction of the sun is much less than it is at the distances of the earth-like planets. Since a comet has no characteristic features by which it may be recognized with certainty, its identity might be in question if it were not followed visually during the time of the perturbations.

One way of testing the identity of two comets appearing at different epochs is to take the orbit of the earlier and to compute the perturbations which it undergoes, and then to compare the derived elements with those determined from the later obser-

* *Bulletin Astronomique*, vol. VI., p. 289, and *Méc. Cél.*, vol. IV., p. 203.

vations; or, the start may be made with the elements of the later comet, and by inverse processes the earlier elements may be computed and the comparison made. One or the other of these plans has been followed until recent years.

But the question arises if there is not some relation among the elements which remains unaltered by the perturbations. This is the question which Tisserand has answered in the affirmative in one of his characteristically elegant and important papers on Celestial Mechanics.

Let the eccentricity of Jupiter's orbit be supposed equal to zero, and the mass of the comet infinitesimal. While both of these assumptions are false they are very nearly fulfilled, and the error introduced will be inappreciable, especially as the comet will be near enough to Jupiter to suffer sensible disturbances only a very short time. Under these suppositions, and when the units are properly chosen, the integral

$$(7) \quad \left(\frac{dx}{dt}\right)^2 + \left(\frac{dy}{dt}\right)^2 + \left(\frac{dz}{dt}\right)^2 = x^2 + y^2 + \frac{2(1-\mu)}{r_1} + \frac{2\mu}{r_2} - C$$

holds true. This is an answer to the question; for, when the elements are known the velocity and coördinates can be computed at any time, and the motion referred to rotating axes by equations (2). Hence, to test the identity of two comets, compute the function (7) for each orbit and see if the constant C is the same for both. If the two values of C are the same, the probability is very strong that only one comet has been observed; if they are different, the two comets are certainly distinct bodies.

The process, just explained has the inconvenience of involving considerable computation. This can be largely avoided by expressing (7) in terms of the ordinary elements of the orbit. The first step is to express (7) in terms of coördinates measured from fixed axes. The equations of transformation are the inverse of equations (2), viz.,

$$\begin{cases} x = + \xi \cos t + \eta \sin t, \\ y = - \xi \sin t + \eta \cos t, \\ z = \zeta. \end{cases}$$

From these equations it is found that

$$x^2 + y^2 = \xi^2 + \eta^2,$$

$$\left(\frac{dx}{dt}\right)^2 + \left(\frac{dy}{dt}\right)^2 + \left(\frac{dz}{dt}\right)^2 = \left(\frac{d\xi}{dt}\right)^2 + \left(\frac{d\eta}{dt}\right)^2 + \left(\frac{d\zeta}{dt}\right)^2$$

$$+ \xi^2 + \eta^2 - 2\left(\xi\frac{d\eta}{dt} - \eta\frac{d\xi}{dt}\right).$$

Hence equation (7) becomes

$$(25) \quad \left(\frac{d\xi}{dt}\right)^2 + \left(\frac{d\eta}{dt}\right)^2 + \left(\frac{d\zeta}{dt}\right)^2 - 2\left(\xi\frac{d\eta}{dt} - \eta\frac{d\xi}{dt}\right)$$

$$= \frac{2(1-\mu)}{r_1} + \frac{2\mu}{r_2} - C.$$

Let r represent the distance of the comet from the origin, and i the angle between the plane of its instantaneous orbit and the $\xi\eta$-plane. Then equations (24), Art. 89, give

$$\begin{cases} \left(\frac{d\xi}{dt}\right)^2 + \left(\frac{d\eta}{dt}\right)^2 + \left(\frac{d\zeta}{dt}\right)^2 = \frac{2}{r} - \frac{1}{a}, \\ \left(\xi\frac{d\eta}{dt} - \eta\frac{d\xi}{dt}\right) = \sqrt{a(1-e^2)}\cos i. \end{cases}$$

Hence equation (25) becomes

$$(26) \quad \frac{2}{r} - \frac{1}{a} - 2\sqrt{a(1-e^2)}\cos i = \frac{2(1-\mu)}{r_1} + \frac{2\mu}{r_2} - C.$$

In the case of Jupiter and the sun μ is less than one-thousandth. Therefore the origin is very near the center of the sun, and r_1 is sensibly equal to r. In both instances the elements will be determined when the comet is far from both Jupiter and the sun so that $-\frac{2\mu}{r_1} + \frac{2\mu}{r_2}$ will be so small that it may be neglected without important error; then (26) reduces to the simple expression

$$\frac{1}{a} + 2\sqrt{a(1-e^2)}\cos i = C.$$

It will be noticed that the elements of this formula are the instantaneous elements for motion around a unit mass situated at the center of mass of the finite bodies. The actual elements used in Astronomy are the elements referred to the center of the sun, with the sun as the attracting mass. Nevertheless, on account of the small relative mass of Jupiter the two sets of elements are very nearly the same, and if the two orbits are of the same body, the equation

$$(27) \quad \frac{1}{a_1} + 2\sqrt{a_1(1 - e_1{}^2)}\cos i_1 = \frac{1}{a_2} + 2\sqrt{a_2(1 - e_2{}^2)}\cos i_2$$

must be fulfilled, where the elements are those in actual use by astronomers. Such is the criterion developed by Tisserand, and employed later by Schulhof and others.

160. Stability of Particular Solutions. Five particular solutions of the motion of the infinitesimal body have been found. If the infinitesimal body is displaced a very little from the exact points of the solutions and given a small velocity it will either oscillate around these respective points, at least for a considerable time, or it will rapidly depart from them. In the first case the particular solution from which the displacement is made is said to be *stable;* in the second case, it is said to be *unstable.*

The question of stability must be formulated mathematically. Consider the equations

$$(28) \quad \begin{cases} \dfrac{d^2x}{dt^2} - 2\dfrac{dy}{dt} = f(x,\ y), \\[2mm] \dfrac{d^2y}{dt^2} + 2\dfrac{dx}{dt} = g(x,\ y). \end{cases}$$

Suppose $x = x_0$, $y = y_0$, where x_0 and y_0 are constants, is a particular solution of (28). That is,

$$f(x_0,\ y_0) = 0, \qquad g(x_0,\ y_0) = 0.$$

Give the body a small displacement and a small velocity so that its coördinates and components of velocity are

$$(29) \quad \begin{cases} x = x_0 + x', \\[1mm] y = y_0 + y', \\[1mm] \dfrac{dx}{dt} = \dfrac{dx'}{dt}, \\[2mm] \dfrac{dy}{dt} = \dfrac{dy'}{dt}, \end{cases}$$

where x', y', $\dfrac{dx'}{dt}$, and $\dfrac{dy'}{dt}$ are initially very small. On making these substitutions in (28), the differential equations become

$$(30) \quad \begin{cases} \dfrac{d^2x'}{dt^2} - 2\dfrac{dy'}{dt} = f(x_0 + x',\ y_0 + y'), \\[2mm] \dfrac{d^2y'}{dt^2} + 2\dfrac{dx'}{dt} = g(x_0 + x',\ y_0 + y'). \end{cases}$$

When the right members are developed by Taylor's formula, they take the form

$$
\begin{cases}
f(x_0 + x',\ y_0 + y') = f(x_0,\ y_0) + \dfrac{\partial f}{\partial x} x' + \dfrac{\partial f}{\partial y} y' + \cdots, \\[2mm]
g(x_0 + x',\ y_0 + y') = g(x_0,\ y_0) + \dfrac{\partial g}{\partial x} x' + \dfrac{\partial g}{\partial y} y' + \cdots.
\end{cases}
$$

In the partial derivatives $x = x_0$ and $y = y_0$. The first terms in the right members are respectively zero; hence equations (30) become

$$
(31) \quad
\begin{cases}
\dfrac{d^2x'}{dt^2} - 2\dfrac{dy'}{dt} = \dfrac{\partial f}{\partial x'} x' + \dfrac{\partial f}{\partial y'} y' + \cdots, \\[3mm]
\dfrac{d^2y'}{dt^2} + 2\dfrac{dx'}{dt} = \dfrac{\partial g}{\partial x'} x' + \dfrac{\partial g}{\partial y'} y' + \cdots.
\end{cases}
$$

If x' and y' are taken very small on the start the influence of the higher powers in the right members will be inappreciable, at least for a considerable time. If the parts which involve second and higher degree terms in x' and y' are neglected, the differential equations reduce to the linear system

$$
(32) \quad
\begin{cases}
\dfrac{d^2x'}{dt^2} - 2\dfrac{dy'}{dt} = \dfrac{\partial f}{\partial x'} x' + \dfrac{\partial f}{\partial y'} y', \\[3mm]
\dfrac{d^2y'}{dt^2} + 2\dfrac{dx'}{dt} = \dfrac{\partial g}{\partial x'} x' + \dfrac{\partial g}{\partial y'} y'.
\end{cases}
$$

The solutions of a system of linear differential equations with constant coefficients can in general be expressed in terms of exponentials in the form

$$
\begin{cases}
x' = \alpha_1 e^{\lambda_1 t} + \alpha_2 e^{\lambda_2 t} + \alpha_3 e^{\lambda_3 t} + \alpha_4 e^{\lambda_4 t}, \\[2mm]
y' = \beta_1 e^{\lambda_1 t} + \beta_2 e^{\lambda_2 t} + \beta_3 e^{\lambda_3 t} + \beta_4 e^{\lambda_4 t},
\end{cases}
$$

where $\alpha_1, \cdots, \alpha_4$ are the constants of integration, and $\beta_1, \cdots, \beta_4$ are constants depending upon them and the constants involved in the differential equations. If $\lambda_1, \cdots, \lambda_4$ are pure imaginary numbers, then x' and y' are expressible in periodic functions, and the solution from which the start was made is said to be *stable;* if any of $\lambda_1, \cdots, \lambda_4$ are real or complex numbers, then x' and y' change indefinitely with t, and the solution is said to be *unstable.*

There are exceptional cases where the solution contains constant terms instead of exponentials; they are of course stable if all the exponentials are purely imaginary. There are other exceptional cases in which the solution contains exponentials multiplied by some power of t; these solutions are usually regarded as unstable.

161. Application of the Criterion for Stability to the Straight Line Solutions. The definitions and general methods of the last article will now be applied to the special cases which have arisen in the discussion of the motion of the infinitesimal body. The original differential equations were (Art. 152)

$$\begin{cases} \dfrac{d^2x}{dt^2} - 2\dfrac{dy}{dt} = x - (1-\mu)\dfrac{(x-x_1)}{r_1^3} - \mu\dfrac{(x-x_2)}{r_2^3} \equiv f(x, y, z), \\[2mm] \dfrac{d^2y}{dt^2} + 2\dfrac{dx}{dt} = y - (1-\mu)\dfrac{y}{r_1^3} - \mu\dfrac{y}{r_2^3} \equiv g(x, y, z), \\[2mm] \dfrac{d^2z}{dt^2} \quad\;\; = \quad - (1-\mu)\dfrac{z}{r_1^3} - \mu\dfrac{z}{r_2^3} \equiv h(x, y, z). \end{cases}$$

The straight line solutions occur for

$$x = x_{0i}, \qquad y = 0, \qquad z = 0,$$

where $i = 1, 2, 3$ according as the point lies between $+\infty$ and μ, μ and $1-\mu$, or $1-\mu$ and $-\infty$, and where these values of x, y, and z satisfy equation (19). Make the substitution

$$\begin{cases} x = x_{0i} + x', \qquad y = y', \qquad z = z', \\[2mm] \dfrac{dx}{dt} = \dfrac{dx'}{dt}, \qquad \dfrac{dy}{dt} = \dfrac{dy'}{dt}, \qquad \dfrac{dz}{dt} = \dfrac{dz'}{dt}. \end{cases}$$

Then it is found that

$$(33)\;\begin{cases} \dfrac{\partial f}{\partial x'}x' + \dfrac{\partial f}{\partial y'}y' + \dfrac{\partial f}{\partial z'}z' \equiv x' + \dfrac{2(1-\mu)x'}{[(x_{0i}-x_1)^2]^{\frac{3}{2}}} + \dfrac{2\mu x'}{[(x_{0i}-x_2)^2]^{\frac{3}{2}}}, \\[3mm] \dfrac{\partial g}{\partial x'}x' + \dfrac{\partial g}{\partial y'}y' + \dfrac{\partial g}{\partial z'}z' \equiv y' - \dfrac{(1-\mu)y'}{[(x_{0i}-x_1)^2]^{\frac{3}{2}}} - \dfrac{\mu y'}{[(x_{0i}-x_2)^2]^{\frac{3}{2}}}, \\[3mm] \dfrac{\partial h}{\partial x'}x' + \dfrac{\partial h}{\partial y'}y' + \dfrac{\partial h}{\partial z'}z' \equiv \quad - \dfrac{(1-\mu)z'}{[(x_{0i}-x_1)^2]^{\frac{3}{2}}} - \dfrac{\mu z'}{[(x_{0i}-x_2)^2]^{\frac{3}{2}}}. \end{cases}$$

Let

$$(34) \qquad A_i = \dfrac{1-\mu}{[(x_{0i}-x_1)^2]^{\frac{3}{2}}} + \dfrac{\mu}{[(x_{0i}-x_2)^2]^{\frac{3}{2}}}.$$

Then the equations corresponding to (32) become in this case

$$(35) \quad \begin{cases} \dfrac{d^2x'}{dt^2} - 2\dfrac{dy'}{dt} = (1 + 2A_i)x', \\[2mm] \dfrac{d^2y'}{dt^2} + 2\dfrac{dx'}{dt} = (1 - A_i)y', \\[2mm] \dfrac{d^2z'}{dt^2} \quad\quad = -A_i z'. \end{cases}$$

The last equation is independent of the first two and can be treated separately. The solution is (Art. 32)

$$(36) \quad z' = c_1 e^{\sqrt{-1}\,\sqrt{A_i}\,t} + c_2 e^{-\sqrt{-1}\,\sqrt{A_i}\,t}.$$

Therefore the motion parallel to the z-axis, for small displacements, is periodic with the period $\dfrac{2\pi}{\sqrt{A_i}}$.

Consider now the simultaneous equations

$$(37) \quad \begin{cases} \dfrac{d^2x'}{dt^2} - 2\dfrac{dy'}{dt} = (1 + 2A_i)x', \\[2mm] \dfrac{d^2y'}{dt^2} + 2\dfrac{dx'}{dt} = (1 - A_i)y'. \end{cases}$$

To find the solutions let

$$(38) \quad \begin{cases} x' = Ke^{\lambda t}, \\ y' = Le^{\lambda t}, \end{cases}$$

where K and L are constants. On substituting these expressions in equations (37) and dividing out $e^{\lambda t}$, it is found that

$$(39) \quad \begin{cases} [\lambda^2 - (1 + 2A_i)]K - 2\lambda L = 0, \\ 2\lambda K + [\lambda^2 - (1 - A_i)]L = 0. \end{cases}$$

In order that equations (38) shall be particular solutions of (37) equations (39) must be fulfilled. They are verified by $K = 0$, $L = 0$; but in this case $x' = 0$, $y' = 0$, and the solutions reduce to the straight line solutions. Equations (39) can be satisfied by values of K and L different from zero only if the determinant of the coefficients vanishes. This condition is

$$(40) \quad \begin{vmatrix} \lambda^2 - (1 + 2A_i), & -2\lambda \\ +2\lambda, & \lambda^2 - (1 - A_i) \end{vmatrix} = 0.$$

This equation is the condition upon λ that equations (38) may be a solution of (37). There are four roots of this biquadratic, each giving a particular solution, and the general solution is the sum of the four particular solutions multiplied by arbitrary constants; that is, if the four roots of (40) are λ_1, λ_2, λ_3, λ_4, the general solution is

(41)
$$\begin{cases} x' = K_1 e^{\lambda_1 t} + K_2 e^{\lambda_2 t} + K_3 e^{\lambda_3 t} + K_4 e^{\lambda_4 t}, \\ y' = L_1 e^{\lambda_1 t} + L_2 e^{\lambda_2 t} + L_3 e^{\lambda_3 t} + L_4 e^{\lambda_4 t}, \end{cases}$$

where the K_j are the arbitrary constants of integration, and the L_j are defined in terms of them respectively by either of the equations (39). The λ_j depend of course upon the subscript i on A, but the notation need not be burdened with this fact since the equations all have the same form whether i is 1, 2, or 3.

It remains to determine the character of the roots of the biquadratic (40). It follows from (34) and (21), (22), and (24) respectively that

(42)
$$\begin{cases} A_1 \equiv \dfrac{1-\mu}{(1+r_2)^3} + \dfrac{\mu}{r_2{}^3} = 4 - 2 \cdot 3 \left(\dfrac{\mu}{3}\right)^{\frac{1}{3}} + \cdots, \\[2mm] A_2 \equiv \dfrac{1-\mu}{(1-r_2)^3} + \dfrac{\mu}{r_2{}^3} = 4 + 2 \cdot 3 \left(\dfrac{\mu}{3}\right)^{\frac{1}{3}} + \cdots, \\[2mm] A_3 \equiv \dfrac{1-\mu}{(1-\rho)^3} + \dfrac{\mu}{(2-\rho)^3} = 1 + \tfrac{7}{8}\mu + \cdots. \end{cases}$$

It follows from (42) that, for small values of μ, the term of (40) which is independent of λ satisfies the inequality

$$1 + A_i - 2A_i{}^2 < 0, \qquad (i = 1,\, 2,\, 3);$$

and, indeed, this relation is true for values of μ up to the limit $\frac{1}{2}$, as can be verified easily.* Therefore the biquadratic has two real roots which are equal in numerical value and opposite in sign, and two conjugate pure imaginaries. It follows from the definitions given that the motion is unstable. If the infinitesimal body were displaced a very little from the points of solution it would in general depart to a comparatively great distance.

162. Particular Values of the Constants of Integration. The constants of integration will now be expressed in terms of the initial conditions, and it will be shown that the latter can be selected so that the motion will be periodic.

Suppose λ_1 and λ_2 are the real roots of equation (40); then $\lambda_1 = -\lambda_2$. The imaginary roots are

* H. C. Plummer gave a general proof in *Monthly Not. of Roy. Astr. Soc.,* vol. LXII. (1901).

$$\begin{cases} \lambda_3 = + \sqrt{-1}\,\sigma, \\ \lambda_4 = - \sqrt{-1}\,\sigma, \end{cases}$$

where σ is a real number. The L_j are expressed in terms of the K_j by equations (39), and are

$$(43) \qquad L_j = \frac{[\lambda_j^2 - (1 + 2A_i)]}{2\lambda_j} K_j = c_j K_j, \qquad \left(\begin{matrix} i = 1, 2, 3; \\ j = 1, 2, 3, 4 \end{matrix} \right).$$

Since the λ_j are equal in numerical value but opposite in sign in pairs, and the last two are imaginary, it follows that

$$(44) \qquad \begin{cases} c_1 = - c_2, \\ c_3 = + \sqrt{-1}\,c, \\ c_4 = - \sqrt{-1}\,c, \end{cases}$$

where c is a real constant depending on i.

Let x_0', y_0', $\dfrac{dx_0'}{dt}$, and $\dfrac{dy_0'}{dt}$ be the initial coördinates and components of velocity; then equations (41) give at $t = 0$

$$\begin{cases} x_0' = K_1 + K_2 + K_3 + K_4, \\ y_0' = c_1(K_1 - K_2) + \sqrt{-1}\,c(K_3 - K_4), \\ \dfrac{dx_0'}{dt} = \lambda_1(K_1 - K_2) + \sqrt{-1}\,\sigma(K_3 - K_4), \\ \dfrac{dy_0'}{dt} = c_1\lambda_1(K_1 + K_2) - c\sigma(K_3 + K_4). \end{cases}$$

The values of the constants of integration are found in terms of the initial coördinates and components of velocity by solving these equations.

The values of x' and y' increase in general without limit with the time, but if the initial conditions are such that $K_1 = K_2 = 0$ they become purely periodic. This case will now be considered. The initial coördinates, x_0', y_0', will determine K_3 and K_4, by means of which $\dfrac{dx_0'}{dt}$ and $\dfrac{dy_0'}{dt}$ are defined. Thus

$$\begin{cases} x_0' = K_3 + K_4, \\ y_0' = \sqrt{-1}\,c(K_3 - K_4); \end{cases}$$

whence

$$\begin{cases} K_3 = \dfrac{x_0'}{2} - \dfrac{\sqrt{-1}}{2c}y_0', \\[2mm] K_4 = \dfrac{x_0'}{2} + \dfrac{\sqrt{-1}}{2c}y_0'. \end{cases}$$

The equations (41) become

$$(45) \begin{cases} x' = \dfrac{x_0'}{2}(e^{\sqrt{-1}\sigma t} + e^{-\sqrt{-1}\sigma t}) - \dfrac{\sqrt{-1}}{2c}y_0'(e^{\sqrt{-1}\sigma t} - e^{-\sqrt{-1}\sigma t}) \\[2mm] \quad = x_0' \cos \sigma t + \dfrac{y_0'}{c}\sin \sigma t, \\[3mm] y' = \sqrt{-1}\dfrac{c}{2}x_0'(e^{\sqrt{-1}\sigma t} - e^{-\sqrt{-1}\sigma t}) + \dfrac{y_0'}{2}(e^{\sqrt{-1}\sigma t} + e^{-\sqrt{-1}\sigma t}) \\[2mm] \quad = - cx_0' \sin \sigma t + y_0' \cos \sigma t. \end{cases}$$

The equation of the orbit is found by eliminating t from these equations. Solve for $\cos \sigma t$ and $\sin \sigma t$; then square and add, and the result, after dividing out common factors, is

$$(46) \qquad \frac{x'^2}{\dfrac{c^2x_0'^2 + y_0'^2}{c^2}} + \frac{y'^2}{c^2x_0'^2 + y_0'^2} = 1.$$

This is the equation of an ellipse with the major and minor axes lying along the coördinate axes, and with the center at the origin. Since λ_3 is imaginary it follows from (43) and (44) that $c^2 > 1$; therefore the major axis of the ellipse is parallel to the y-axis. The eccentricity is given by

$$e^2 = \frac{c^2 - 1}{c^2},$$

which, for large values of c, is very near unity. The orbits have the remarkable property that their eccentricity is independent of the initial small displacements, depending only upon the distribution of the mass between the finite bodies, and upon the one of the three straight line solutions from which they spring.

It is obvious that this discussion is not completely rigorous because the terms of higher degree in the right members of the differential equations have been neglected. The linear terms alone do not give sufficient conditions for the existence of periodic orbits, and consequently when the discussion is thus restricted it answers only the question as to the stability of the solution. But in the present case periodic orbits actually exist about all three points for all $0 < \mu \leqq \frac{1}{2}$. Some special examples for $\mu = \frac{1}{11}$ were found by Darwin in his memoir in *Acta Mathematica*, vol. 21. The complete analysis for these orbits, including the much more difficult case in which the finite bodies describe elliptical orbits, was given by the author in the *Mathematische Annalen*, vol. LXXIII. (1912), pp. 441–479, and in the *Publications of the Carnegie Institution of Washington*, No. 161, Periodic Orbits, chapters V., VI., and VII.

163. Application to the Gegenschein. If the constants K_1 and K_2 are zero the infinitesimal body will revolve in an ellipse around the point of equilibrium. If these constants are not zero but small in numerical value compared to K_3 and K_4, the motion will be nearly in an ellipse for a considerable time, but will eventually depart very far from it. It would be possible to have any number of infinitesimal bodies revolving around the same point without disturbing one another.

Consider the motion of the earth around the sun. It is in a curve which is nearly a circle. One of the straight line solution points is exactly opposite to the sun, and if a meteor should pass near it with initial conditions approximately such as have been defined in the last article it would make one or more circuits around this point before pursuing its path into other regions. If a very great number were swarming around this point at one time they would appear from the earth as a hazy patch of light with its center at the anti-sun, and elongated along the ecliptic. This is the appearance of the gegenschein which was discovered independently by Brorsen, Backhouse, and Barnard in 1855, 1868, and 1875 respectively.

The crucial question seems to be whether or not there are enough meteors with the approximate initial conditions to explain the observed phenomena, but no certain answer can be given. However, it is certain that the meteors are exceedingly numerous, as many as 8,000,000 striking into the earth's atmosphere daily according to H. A. Newton; and it is only reasonable to suppose that they cause the zodiacal light which is very bright compared to the gegenschein. The suggestion that this may be the cause of the gegenschein was first made by Gyldèn in the closing paragraph of a memoir in the *Bulletin Astronomique*, vol. I., entitled, *Sur un Cas Particulier du Problème des Trois Corps.**

---

* See also a paper by F. R. Moulton in *The Astronomical Journal*, No. 483.

164. Application of the Criterion for Stability to the Equilateral Triangle Solutions. The particular solutions of the original differential equations in this case are $r_1 = 1$, $r_2 = 1$. The equations corresponding to (33) are

$$\begin{cases} \dfrac{\partial f}{\partial x'}\,x' + \dfrac{\partial f}{\partial y'}\,y' + \dfrac{\partial f}{\partial z'}\,z' = \tfrac{3}{4}x' + \dfrac{3\sqrt{3}}{4}\,(1 - 2\mu)y', \\[2mm] \dfrac{\partial g}{\partial x'}\,x' + \dfrac{\partial g}{\partial y'}\,y' + \dfrac{\partial g}{\partial z'}\,z' = \dfrac{3\sqrt{3}}{4}\,(1 - 2\mu)x' + \tfrac{9}{4}y', \\[2mm] \dfrac{\partial h}{\partial x'}\,x' + \dfrac{\partial h}{\partial y'}\,y' + \dfrac{\partial h}{\partial z'}\,z' = -\,z'; \end{cases}$$

and the differential equations up to terms of the second degree are

$$(47) \quad \begin{cases} \dfrac{d^2x'}{dt^2} - 2\dfrac{dy'}{dt} = \tfrac{3}{4}x' + \dfrac{3\sqrt{3}}{4}\,(1 - 2\mu)y', \\[2mm] \dfrac{d^2y'}{dt^2} + 2\dfrac{dx'}{dt} = \dfrac{3\sqrt{3}}{4}\,(1 - 2\mu)x' + \tfrac{9}{4}y', \\[2mm] \dfrac{d^2z'}{dt^2} \qquad\quad = -\,z'. \end{cases}$$

The last equation is independent of the first two, and its solution is

$$z' = c_1 \sin t + c_2 \cos t.$$

Therefore the motion parallel to the z-axis, for small displacements, is periodic with period 2π, the same as that of the revolution of the finite bodies.

To find the solutions of the first two equations let

$$(48) \quad \begin{cases} x' = Ke^{\lambda t}, \\ y' = Le^{\lambda t}. \end{cases}$$

On substituting these expressions in the first two equations of (47) and dividing out common factors, it is found that

$$(49) \quad \begin{cases} [\lambda^2 - \tfrac{3}{4}]K - \left[2\lambda + \dfrac{3\sqrt{3}}{4}\,(1 - 2\mu) \right]L = 0, \\[3mm] \left[2\lambda - \dfrac{3\sqrt{3}}{4}\,(1 - 2\mu) \right]K + [\lambda^2 - \tfrac{9}{4}]L = 0. \end{cases}$$

In order that solutions may be obtained other than $x' = 0$, $y' = 0$ the determinant of these equations must vanish. That is,

$$(50) \quad \begin{vmatrix} \lambda^2 - \frac{3}{4}, & -2\lambda - \frac{3\sqrt{3}}{4}(1 - 2\mu) \\ 2\lambda - \frac{3\sqrt{3}}{4}(1 - 2\mu), & \lambda^2 - \frac{9}{4} \end{vmatrix} \equiv \lambda^4 + \lambda^2 + \frac{27}{4}\mu(1 - \mu) = 0.$$

Let λ_1, λ_2, λ_3, λ_4 be the roots of this biquadratic. Then the general solutions of (47) are

$$\begin{cases} x' = K_1 e^{\lambda_1 t} + K_2 e^{\lambda_2 t} + K_3 e^{\lambda_3 t} + K_4 e^{\lambda_4 t}, \\ y' = L_1 e^{\lambda_1 t} + L_2 e^{\lambda_2 t} + L_3 e^{\lambda_3 t} + L_4 e^{\lambda_4 t}, \end{cases}$$

where K_1, K_2, K_3, K_4 are the constants of integration, and L_1, L_2, L_3, L_4 are constants related to them by either of equations (49).

It is found from (50) that

$$\begin{cases} \lambda_1 = -\lambda_2 = \sqrt{\dfrac{-1 + \sqrt{1 - 27\mu(1 - \mu)}}{2}}, \\ \lambda_3 = -\lambda_4 = \sqrt{\dfrac{-1 - \sqrt{1 - 27\mu(1 - \mu)}}{2}}. \end{cases}$$

The number μ never exceeds $\frac{1}{2}$, and if $1 - 27\mu(1 - \mu) \geqq 0$ the roots are pure imaginaries in conjugate pairs; if this inequality is not fulfilled they are complex quantities. The inequality may be written

$$1 - 27\mu(1 - \mu) = \epsilon,$$

where ϵ is a positive quantity whose limit is zero. The solution of this equation is

$$(51) \quad \mu = \tfrac{1}{2} \pm \sqrt{\frac{23 + 4\epsilon}{108}}.$$

Since μ represents the mass which is less than one-half the negative sign must be taken. At the limit $\epsilon = 0$, $\mu = .0385 \cdots$. Therefore if $\mu < .0385 \cdots$ the roots of (50) are pure imaginaries and the equilateral triangle solutions are stable; if $\mu > .0385 \cdots$ the roots of (50) are complex and the equilateral triangle solutions are unstable.

XXI. PROBLEMS.

1. If a comet approaching the sun in a parabola should be disturbed by Jupiter so that its orbit remained a parabola while its perihelion distance was doubled, what would be the relation between the new inclination and the old?

Ans. $$\cos i_2 = \frac{\sqrt{2}}{2} \cos i_1.$$

2. Prove that if a comet's orbit, whose inclination to Jupiter's orbit is zero, is changed by the perturbations of Jupiter from a parabola to an ellipse the parameter of the orbit is necessarily decreased. Investigate the changes in the parameters for changes in the major axes of the other species of conics.

3. Suppose a comet is moving in an ellipse in the plane of Jupiter's orbit, and that the perturbing action of Jupiter is inappreciable except for a short time when they are near each other. Prove that if the perturbation of Jupiter has increased the eccentricity, the period has been increased or decreased according as the product of the major semi-axis and the square root of the parameter in the original ellipse is greater or less than unity when expressed in the canonical units.

4. A particle placed midway between two equal fixed masses is in equilibrium. Investigate the character of the equilibrium by the method of Art. 161.

5. Suppose $1 - \mu$ and μ are the sun and earth respectively; find the period of oscillation parallel to the z-axis for an infinitesimal body slightly displaced from the xy-plane near the straight line solution point opposite to the sun with respect to the earth as an origin.
Ans. 183.304 mean solar days.

6. In the same case, find the period of oscillation in the xy-plane.
Ans. 139.6 mean solar days.

7. Prove that in general for small values of μ the periods of oscillation both parallel to the z-axis and in the xy-plane, are longest for the point opposite to μ with respect to $1 - \mu$ as origin; next longest for the point opposite to $1 - \mu$ with respect to μ as origin; and shortest for the point between $1 - \mu$ and μ.

8. Find the eccentricity of the orbit in the xy-plane opposite to the sun in the case of the sun and earth.

9. The differential equations (35) admit the integral

$$\left(\frac{dx'}{dt}\right)^2 + \left(\frac{dy'}{dt}\right)^2 + \left(\frac{dz'}{dt}\right)^2 = (1 + 2A_i)x'^2 + (1 - A_i)y'^2 - A_i z'^2 + C;$$

discuss the meaning of this integral after the manner of articles 154–159.

10. What can be said regarding the independence of equations (39) after the condition has been imposed that the determinant shall vanish?

11. If the explanation of the gegenschein given in Art. 163 is true what should be its maximum parallax in celestial latitude for an observer in latitude 45°?
Ans. Roughly 15'. (Too small to be observed with certainty in such an indefinite object.)

12. Suppose $\mu = \frac{1}{2}$ and reduce the problem of finding the motion of the infinitesimal body through the origin along the z-axis to elliptic integrals.

CASE OF THREE FINITE BODIES.

165. Conditions for Circular Orbits. The theorem of Lagrange that it is possible to start three finite bodies in such a manner that their orbits will be similar ellipses, all described in the same time, will be proved in this section. It will be established first for the special case in which the orbits are circles. It will be assumed that the three bodies are projected in the same plane. Take the origin at their center of mass and the $\xi\eta$-plane as the plane of motion. Then the differential equations of motion are (Art. 143)

(52)
$$\begin{cases} \dfrac{d^2\xi_i}{dt^2} = \dfrac{1}{m_i}\dfrac{\partial U}{\partial \xi_i}, \qquad (i = 1,\ 2,\ 3), \\[2mm] \dfrac{d^2\eta_i}{dt^2} = \dfrac{1}{m_i}\dfrac{\partial U}{\partial \eta_i}, \\[2mm] U = \dfrac{k^2 m_1 m_2}{r_{1,2}} + \dfrac{k^2 m_2 m_3}{r_{2,3}} + \dfrac{k^2 m_3 m_1}{r_{3,1}}. \end{cases}$$

The motion of the system is referred to axes rotating with the uniform angular velocity n by the substitution

(53)
$$\begin{cases} \xi_i = x_i \cos nt - y_i \sin nt, \qquad (i = 1,\ 2,\ 3), \\[1mm] \eta_i = x_i \sin nt + y_i \cos nt. \end{cases}$$

On making the substitution, and reducing as in Art. 152, it is found that

(54)
$$\begin{cases} \dfrac{d^2 x_i}{dt^2} - 2n\dfrac{dy_i}{dt} - n^2 x_i - \dfrac{1}{m_i}\dfrac{\partial U}{\partial x_i} = 0, \\[2mm] \dfrac{d^2 y_i}{dt^2} + 2n\dfrac{dx_i}{dt} - n^2 y_i - \dfrac{1}{m_i}\dfrac{\partial U}{\partial y_i} = 0. \end{cases}$$

If the bodies are moving in circles around the origin with the angular velocity n, their coördinates with respect to the rotating axes are constants. Since the first and second derivatives are then zero, equations (54) become

(55)
$$\begin{cases} -n^2 x_1 + k^2 m_2 \dfrac{(x_1 - x_2)}{r^3_{1,2}} + k^2 m_3 \dfrac{(x_1 - x_3)}{r^3_{1,3}} = 0, \\[3mm] -n^2 x_2 + k^2 m_1 \dfrac{(x_2 - x_1)}{r^3_{1,2}} + k^2 m_3 \dfrac{(x_2 - x_3)}{r^3_{2,3}} = 0, \\[3mm] -n^2 x_3 + k^2 m_1 \dfrac{(x_3 - x_1)}{r^3_{1,3}} + k^2 m_2 \dfrac{(x_3 - x_2)}{r^3_{2,3}} = 0, \end{cases}$$

$$(55) \quad \begin{cases} - n^2 y_1 + k^2 m_2 \dfrac{(y_1 - y_2)}{r^3_{1,\,2}} + k^2 m_3 \dfrac{(y_1 - y_3)}{r^3_{1,\,3}} = 0, \\[2mm] - n^2 y_2 + k^2 m_1 \dfrac{(y_2 - y_1)}{r^3_{1,\,2}} + k^2 m_3 \dfrac{(y_2 - y_3)}{r^3_{2,\,3}} = 0, \\[2mm] - n^2 y_3 + k^2 m_1 \dfrac{(y_3 - y_1)}{r^3_{1,\,3}} + k^2 m_2 \dfrac{(y_3 - y_2)}{r^3_{2,\,3}} = 0. \end{cases}$$

And conversely, if the masses and initial projections are such that these six equations are fulfilled the bodies move in circles around the origin with the uniform angular velocity n.

Since the origin is at the center of mass the coördinates satisfy

$$(56) \quad \begin{cases} m_1 x_1 + m_2 x_2 + m_3 x_3 = 0, \\ m_1 y_1 + m_2 y_2 + m_3 y_3 = 0. \end{cases}$$

If the first equation of (55) is multiplied by m_1, the second by m_2, and the products added, the sum becomes, as a consequence of the first equation of (56), the third of (55). In a similar manner the last equation of (55) can be derived from the others in y and the last of (56). Therefore the third and sixth equations of (55) can be suppressed, and equations (56) used in place of them, giving a somewhat simpler system of equations.

The units of time, space, and mass are so far arbitrary. It is possible, without loss of generality, to select them so that $r_{1,\,2} = 1$ and $k^2 = 1$. Then necessary and sufficient conditions for the existence of solutions in which the orbits are circles are

$$(57) \quad \begin{cases} m_1 x_1 + m_2 x_2 + m_3 x_3 = 0, \\[2mm] - n^2 x_1 + m_2 (x_1 - x_2) + m_3 \dfrac{(x_1 - x_3)}{r^3_{1,\,3}} = 0, \\[2mm] - n^2 x_2 + m_1 (x_2 - x_1) + m_3 \dfrac{(x_2 - x_3)}{r^3_{2,\,3}} = 0, \\[2mm] m_1 y_1 + m_2 y_2 + m_3 y_3 = 0, \\[2mm] - n^2 y_1 + m_2 (y_1 - y_2) + m_3 \dfrac{(y_1 - y_3)}{r^3_{1,\,3}} = 0, \\[2mm] - n^2 y_2 + m_1 (y_2 - y_1) + m_3 \dfrac{(y_2 - y_3)}{r^3_{2,\,3}} = 0. \end{cases}$$

166. Equilateral Triangle Solutions. There is a solution of the problem for every set of real values of the variables satisfying equations (57). It is easy to show that the equations are fulfilled if the bodies lie at the vertices of an equilateral triangle. Then $r_{1,2} = r_{2,3} = r_{1,3} = 1$, and equations (57) become

$$\left\{ \begin{array}{l} m_1 x_1 + m_2 x_2 + m_3 x_3 = 0, \\ (m_2 + m_3 - n^2) x_1 - m_2 x_2 - m_3 x_3 = 0, \\ (m_1 + m_3 - n^2) x_2 - m_1 x_1 - m_3 x_3 = 0, \\ m_1 y_1 + m_2 y_2 + m_3 y_3 = 0, \\ (m_2 + m_3 - n^2) y_1 - m_2 y_2 - m_3 y_3 = 0, \\ (m_1 + m_3 - n^2) y_2 - m_1 y_1 - m_3 y_3 = 0. \end{array} \right.$$

These equations are linear and homogeneous in x_1, x_2, $\cdots$, y_3. In order that they may have a solution different from $x_1 = x_2 = \cdots = y_3 = 0$, which is incompatible with $r_{1,2} = r_{2,3} = r_{1,3} = 1$, the determinant of their coefficients must vanish. On letting $M = m_1 + m_2 + m_3$, it is easily found that this condition is

$$m_3{}^2 (M - n^2)^4 = 0,$$

from which $n^2 = M$. Then two of the x_i and two of the y_i are arbitrary, and hence the equations have a solution compatible with $r_{i,j} = 1$. Therefore, *the equilateral triangular configuration with proper initial components of velocity is a particular solution of the Problem of Three Bodies; and, if the units are such that the mutual distances and k^2 are unity, the square of the angular velocity of revolution is equal to the sum of the masses of the three bodies.*

167. Straight Line Solutions. The last three equations of (57) are fulfilled by $y_1 = y_2 = y_3 = 0$, that is, if the bodies are all on the x-axis. Suppose they lie in the order m_1, m_2, m_3 from the negative end of the axis toward the positive. Then $x_3 > x_2 > x_1$ and $r_{1,2} = x_2 - x_1 = 1$, and the first three equations of (57) become

$$(58) \quad \left\{ \begin{array}{l} m_1 x_1 + m_2 (1 + x_1) + m_3 x_3 = 0, \\ m_2 + \dfrac{m_3}{(x_3 - x_1)^2} + n^2 x_1 = 0, \\ - m_1 + \dfrac{m_3}{(x_3 - x_1 - 1)^2} + n^2 (1 + x_1) = 0. \end{array} \right.$$

On eliminating x_3 and n^2, it is found that

$$(59) \quad m_2 + (m_1 + m_2) x_1 + \frac{m_3{}^3 (1 + x_1)}{(M x_1 + m_2)^2} - \frac{m_3{}^3 x_1}{(M x_1 + m_2 + m_3)^2} = 0.$$

If this equation is cleared of fractions a quintic equation in x_1 is obtained whose coefficients are all positive. Therefore there is no real positive root but there is at least one real negative root, and consequently at least one solution of the problem.

Instead of adopting x_1 as the unknown, $x_3 - x_2$, which will be denoted by A, may be used. The distance x_1 must be expressed in terms of this new variable. The relations among x_1, x_2, x_3, and A are

$$\left\{ \begin{array}{r} m_1x_1 + m_2x_2 + m_3x_3 = 0, \\ x_2 - x_1 = 1, \\ x_3 - x_2 = A; \end{array} \right.$$

whence

$$x_1 = -\frac{m_2 + m_3 + m_3A}{M}.$$

On substituting this expression for x_1 in (59), clearing of fractions, and dividing out common factors, the condition for the collinear solutions becomes

$$(60) \quad \begin{array}{l} (m_1 + m_2)A^5 + (3m_1 + 2m_2)A^4 + (3m_1 + m_2)A^3 \\ \quad - (m_2 + 3m_3)A^2 - (2m_2 + 3m_3)A - (m_2 + m_3) = 0. \end{array}$$

This is precisely Lagrange's quintic equation in A,* and has but one real positive root since the coefficients change sign but once. The only A valid in the problem for the chosen order of the masses is positive; hence the solution of (60) is unique and defines the distribution of the bodies in the straight line solution of the Problem of Three Bodies. It is evident that two more distinct straight line solutions will be obtained by cyclically permuting the order of the three bodies.

168. Dynamical Properties of the Solutions. Since the bodies revolve in circles with uniform angular velocity around the center of mass, the law of areas holds for each body separately; therefore *the resultant of all the forces acting upon each body is constantly directed toward the center of mass* (Art. 48).

Let the distances of m_1, m_2, and m_3 from their center of mass be a_1, a_2, and a_3 respectively. Then the centrifugal acceleration to which m_i is subject is $\alpha_i = \dfrac{V_i^2}{a_i}$, where V_i is the linear velocity of m_i. But this may be written $\alpha_i = n^2a_i$. The centripetal force

* See Lagrange's *Collected Works*, vol. VI., p. 277, and Tisserand's *Méc. Cél.*, vol. I., p. 155.

exactly balances the centrifugal; therefore the acceleration toward the center of mass is

$$\alpha_i = n^2 a_i;$$

that is, *the accelerations of the various bodies toward their common center of mass are directly proportional to their respective distances from this point.*

169. General Conic Section Solutions. The solutions of the problem of three bodies which have been discussed are characterized by the fact that their orbits are circles. It will be shown that corresponding to each of them there is a solution in which the orbits are conic sections of arbitrary eccentricity. These solutions are characterized by the fact that in them the ratios of the mutual distances of the bodies are constant, though the distances themselves are variable.

The differential equations of motion when the system is referred to fixed axes with the origin at the center of gravity of the system are

$$(61) \begin{cases} \dfrac{d^2\xi_1}{dt^2} = -\dfrac{m_2(\xi_1 - \xi_2)}{r^3_{1,\,2}} - \dfrac{m_3(\xi_1 - \xi_3)}{r^3_{1,\,3}}, \\[2ex] \dfrac{d^2\eta_1}{dt^2} = -\dfrac{m_2(\eta_1 - \eta_2)}{r^3_{1,\,2}} - \dfrac{m_3(\eta_1 - \eta_3)}{r^3_{1,\,3}}, \\[2ex] \dfrac{d^2\xi_2}{dt^2} = -\dfrac{m_1(\xi_2 - \xi_1)}{r^3_{1,\,2}} - \dfrac{m_3(\xi_2 - \xi_3)}{r^3_{2,\,3}}, \\[2ex] \dfrac{d^2\eta_2}{dt^2} = -\dfrac{m_1(\eta_2 - \eta_1)}{r^3_{1,\,2}} - \dfrac{m_3(\eta_2 - \eta_3)}{r^3_{2,\,3}}, \\[2ex] \dfrac{d^2\xi_3}{dt^2} = -\dfrac{m_1(\xi_3 - \xi_1)}{r^3_{1,\,3}} - \dfrac{m_2(\xi_3 - \xi_2)}{r^3_{2,\,3}}, \\[2ex] \dfrac{d^2\eta_3}{dt^2} = -\dfrac{m_1(\eta_3 - \eta_1)}{r^3_{1,\,3}} - \dfrac{m_2(\eta_3 - \eta_2)}{r^3_{2,\,3}}. \end{cases}$$

Suppose the coördinates of m_1, m_2, and m_3 at $t = t_0$ are respectively (x_1, y_1), (x_2, y_2), and (x_3, y_3), and let the respective distances from the origin be $r_1^{(0)}$, $r_2^{(0)}$, and $r_3^{(0)}$. Suppose the angles that $r_1^{(0)}$, $r_2^{(0)}$, and $r_3^{(0)}$ make with the ξ-axis are φ_1, φ_2, and φ_3. Then

$$(62) \begin{cases} x_1 = r_1^{(0)} \cos \varphi_1, & x_2 = r_2^{(0)} \cos \varphi_2, & x_3 = r_3^{(0)} \cos \varphi_3, \\[1ex] y_1 = r_1^{(0)} \sin \varphi_1, & y_2 = r_2^{(0)} \sin \varphi_2, & y_3 = r_3^{(0)} \sin \varphi_3. \end{cases}$$

Now let the coördinates of the bodies at any time t be (ξ_1, η_1), (ξ_2, η_2), and (ξ_3, η_3). Suppose the ratios of the mutual distances are constants; then the mutual distances at t are

$$\rho\, r_{1,\,2}, \qquad \rho\, r_{2,\,3}, \qquad \rho\, r_{1,\,3},$$

where ρ is the factor of proportionality. Since the shape of the figure formed by the three bodies is unaltered, it follows that

$$(63) \qquad r_1 = r_1^{(0)}\rho, \qquad r_2 = r_2^{(0)}\rho, \qquad r_1 = r_1^{(0)}\rho.$$

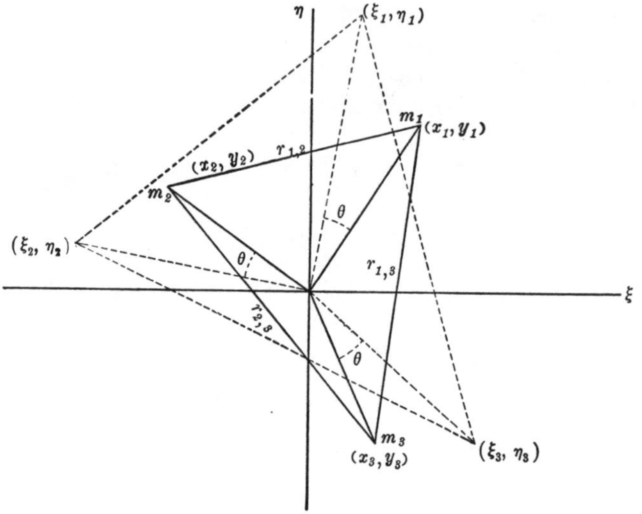

Fig. 42.

Moreover, the radii r_1, r_2, and r_3 will have turned through the same angle θ. Hence

$$(64)\;\begin{cases} \xi_1 = r_1^{(0)}\rho\,\cos\,(\theta + \varphi_1) = (x_1\cos\theta - y_1\sin\theta)\rho, \\[4pt] \eta_1 = r_1^{(0)}\rho\,\sin\,(\theta + \varphi_1) = (x_1\sin\theta + y_1\cos\theta)\rho, \\[4pt] \xi_2 = r_2^{(0)}\rho\,\cos\,(\theta + \varphi_2) = (x_2\cos\theta - y_2\sin\theta)\rho, \\[4pt] \eta_2 = r_2^{(0)}\rho\,\sin\,(\theta + \varphi_2) = (x_2\sin\theta + y_2\cos\theta)\rho, \\[4pt] \xi_3 = r_3^{(0)}\rho\,\cos\,(\theta + \varphi_3) = (x_3\cos\theta - y_3\sin\theta)\rho, \\[4pt] \eta_3 = r_3^{(0)}\rho\,\sin\,(\theta + \varphi_3) = (x_3\sin\theta + y_3\cos\theta)\rho. \end{cases}$$

If equations (61) are transformed by means of (64) they will involve only the two dependent variables ρ and θ, and they will be necessary conditions for the existence of solutions in which the ratios of the mutual distances are constants. It follows from the first two equations of (61) and (64) after multiplying the results of the transformation by $\cos \theta$ and $\sin \theta$ and adding, and then by $-\sin \theta$ and $\cos \theta$ and adding, that

(65)
$$
\begin{cases}
x_1 \dfrac{d^2\rho}{dt^2} - 2y_1 \dfrac{d\rho}{dt}\dfrac{d\theta}{dt} - x_1\rho \left(\dfrac{d\theta}{dt}\right)^2 - y_1\rho \dfrac{d^2\theta}{dt^2} \\
\qquad = -\left\{ \dfrac{m_2(x_1 - x_2)}{r^3_{1,\,2}} + \dfrac{m_3(x_1 - x_3)}{r^3_{1,\,3}} \right\} \dfrac{1}{\rho^2}, \\[2ex]
y_1 \dfrac{d^2\rho}{dt^2} + 2x_1 \dfrac{d\rho}{dt}\dfrac{d\theta}{dt} - y_1\rho \left(\dfrac{d\theta}{dt}\right)^2 + x_1\rho \dfrac{d^2\theta}{dt^2} \\
\qquad = -\left\{ \dfrac{m_2(y_1 - y_2)}{r^3_{1,\,2}} + \dfrac{m_3(y_1 - y_3)}{r^3_{1,\,3}} \right\} \dfrac{1}{\rho^2}.
\end{cases}
$$

Let

(66)
$$ \rho^2 \frac{d\theta}{dt} = \psi. $$

Then

$$ 2\frac{d\rho}{dt}\frac{d\theta}{dt} + \rho \frac{d^2\theta}{dt^2} = \frac{1}{\rho}\frac{d\psi}{dt}, \qquad \rho\left(\frac{d\theta}{dt}\right)^2 = \frac{\psi^2}{\rho^3}, $$

and equations (65) become

(67)
$$
\begin{cases}
\dfrac{d^2\rho}{dt^2} - \dfrac{y_1}{x_1\rho}\dfrac{d\psi}{dt} - \dfrac{\psi^2}{\rho^3} = -\dfrac{1}{x_1}\left\{ \dfrac{m_2(x_1-x_2)}{r^3_{1,\,2}} + \dfrac{m_3(x_1-x_3)}{r^3_{1,\,3}} \right\} \dfrac{1}{\rho^2}, \\[2ex]
\dfrac{d^2\rho}{dt^2} + \dfrac{x_1}{y_1\rho}\dfrac{d\psi}{dt} - \dfrac{\psi^2}{\rho^3} = -\dfrac{1}{y_1}\left\{ \dfrac{m_2(y_1-y_2)}{r^3_{1,\,2}} + \dfrac{m_3(y_1-y_3)}{r^3_{1,\,3}} \right\} \dfrac{1}{\rho^2}.
\end{cases}
$$

And the equations which are similarly derived from the last four equations of (61) and of (65) are

(68)
$$
\begin{cases}
\dfrac{d^2\rho}{dt^2} - \dfrac{y_2}{x_2\rho}\dfrac{d\psi}{dt} - \dfrac{\psi^2}{\rho^3} = -\dfrac{1}{x_2}\left\{ \dfrac{m_1(x_2 - x_1)}{r^3_{1,\,2}} + \dfrac{m_3(x_2 - x_3)}{r^3_{2,\,3}} \right\} \dfrac{1}{\rho^2}, \\[2ex]
\dfrac{d^2\rho}{dt^2} + \dfrac{x_2}{y_2\rho}\dfrac{d\psi}{dt} - \dfrac{\psi^2}{\rho^3} = -\dfrac{1}{y_2}\left\{ \dfrac{m_1(y_2 - y_1)}{r^3_{1,\,2}} + \dfrac{m_3(y_2 - y_3)}{r^3_{2,\,3}} \right\} \dfrac{1}{\rho^2}; \\[2ex]
\dfrac{d^2\rho}{dt^2} - \dfrac{y_3}{x_3\rho}\dfrac{d\psi}{dt} - \dfrac{\psi^2}{\rho^3} = -\dfrac{1}{x_3}\left\{ \dfrac{m_1(x_3 - x_1)}{r^3_{1,\,3}} + \dfrac{m_2(x_3 - x_2)}{r^3_{2,\,3}} \right\} \dfrac{1}{\rho^2}, \\[2ex]
\dfrac{d^2\rho}{dt^2} + \dfrac{x_3}{y_3\rho}\dfrac{d\psi}{dt} - \dfrac{\psi^2}{\rho^3} = -\dfrac{1}{y_3}\left\{ \dfrac{m_1(y_3 - y_1)}{r^3_{1,\,3}} + \dfrac{m_2(y_3 - y_2)}{r^3_{2,\,3}} \right\} \dfrac{1}{\rho^2}.
\end{cases}
$$

Equations (67) and (68) are necessary conditions for the existence of solutions in which the ratios of the distances of the bodies are constants. There are but two variables, ρ and ψ, to be determined. The first gives the dimensions of the system by means of (63), and the second its orientation by means of (66). In order that the solutions in question may exist these equations must be consistent. In pairs of two they define ρ and ψ when the initial conditions are specified. In order that for given initial conditions the ρ and ψ shall be identical as defined by each of the three pairs of differential equations, the coefficients of corresponding terms in ρ and ψ must be the same. This can be proved by considering the expansion of the solutions as power series in $t - t_0$ by the method of Art. 127. In order that the solutions shall be the same the coefficients of corresponding powers of $t - t_0$ must be identical; and in order that these conditions shall be satisfied the coefficients of corresponding terms in the differential equations must be identical. Therefore the conditions for the consistency of equations (67) and (68) are either

$$(69) \qquad \frac{y_1}{x_1} = \frac{y_2}{x_2} = \frac{y_3}{x_3},$$

or

$$(70) \qquad \frac{d\psi}{dt} = 0,$$

and the system of six equations

$$(71) \quad \begin{cases} \dfrac{m_2(x_1 - x_2)}{r^3_{1,2}} + \dfrac{m_3(x_1 - x_3)}{r^3_{1,3}} = n^2 x_1, \\[2mm] \dfrac{m_1(x_2 - x_1)}{r^3_{1,2}} + \dfrac{m_3(x_2 - x_3)}{r^3_{2,3}} = n^2 x_2, \\[2mm] \dfrac{m_1(x_3 - x_1)}{r^3_{1,3}} + \dfrac{m_2(x_3 - x_2)}{r^3_{2,3}} = n^2 x_3, \\[2mm] \dfrac{m_2(y_1 - y_2)}{r^3_{1,2}} + \dfrac{m_3(y_1 - y_3)}{r^3_{1,3}} = n^2 y_1, \\[2mm] \dfrac{m_1(y_2 - y_1)}{r^3_{1,2}} + \dfrac{m_3(y_2 - y_3)}{r^3_{2,3}} = n^2 y_2, \\[2mm] \dfrac{m_1(y_3 - y_1)}{r^3_{1,3}} + \dfrac{m_2(y_3 - y_2)}{r^3_{2,3}} = n^2 y_3, \end{cases}$$

where n^2 is the common constant value of the brackets in the right members of (67) and (68). And it follows from equations (71), as well as from the original definitions of the x_i and the y_i, that the center of mass equations

$$\begin{cases} m_1 x_1 + m_2 x_2 + m_3 x_3 = 0, \\ m_1 y_1 + m_2 y_2 + m_3 y_3 = 0, \end{cases}$$

are fulfilled.

Equations (69) are satisfied only if the three bodies are in a straight line at $t = t_0$. Since, by hypothesis, the shape of the configuration is constant, they always remain in a straight line in this case. The position of the axes can be so chosen at $t = t_0$ that $y_1 = y_2 = y_3 = 0$ and the conditions for the existence of the solutions reduce to the first three equations of (71). These equations are the same as (55) of Art. 165, and it was shown in Art. 167 that they have but three real solutions.

Suppose equations (69) are satisfied and that the bodies remain collinear; therefore the resultant of all the forces to which each one is subject is directed constantly toward the center of gravity of the system, and consequently the law of areas with respect to this point holds. Hence

$$r_1{}^2\frac{d\theta}{dt} = c_1, \qquad r_2{}^2\frac{d\theta}{dt} = c_2, \qquad r_3{}^2\frac{d\theta}{dt} = c_3,$$

where c_1, c_2, and c_3 are constants. It follows from (63) that $\rho^2\frac{d\theta}{dt} = \frac{c_1}{(r_1{}^{(0)})^2}$, and then from (66) that $\frac{d\psi}{dt} = 0$. Hence equations (66), (67), and (68) become in this case

$$(72) \quad \begin{cases} \dfrac{d^2\rho}{dt^2} - \dfrac{\psi^2}{\rho^3} = -n^2\dfrac{1}{\rho^2}, \\[2mm] \psi = c_0 = \text{constant}, \\[2mm] \dfrac{d\theta}{dt} = \dfrac{\psi}{\rho^2} = \dfrac{c_0}{\rho^2}. \end{cases}$$

These are the differential equations in polar coördinates for the Problem of Two Bodies. Except for differences of notation, they are the same as equations (65) of chap. v. Therefore ρ and θ satisfy the conditions of conic section motion under the law of gravitation, and it follows from (63) and the definition of θ that the three bodies describe similar conic sections having an arbitrary eccentricity. These solutions include the straight line solutions in which the orbits are circles as a special case.

Suppose equations (69) are not satisfied; then the bodies are not collinear. But if the bodies are not collinear equation (70) must hold in order that equations (67) and (68) may be compatible. It follows from equations (66) and (63) that the law of areas with respect to the origin holds for each body separately. It was shown in Art. 166 that equations (71) are satisfied if the bodies are at the vertices of an equilateral triangle. It is easy to show that, unless they are collinear, there is no other solution. In the case of the equilateral triangle solution equations (67) and (68) also reduce to (72), and the orbits are similar conic sections of arbitrary eccentricity.

XXII. PROBLEMS.

1. Take as an hypothesis that a solution exists in which the three bodies are always collinear. Prove that the law of areas holds for each body separately with respect to the center of mass of the system, with respect to either of the other bodies, and with respect to the center of mass of any two of the bodies.

2. Write the conditions that the accelerations to which the bodies are subject shall be directed toward their common center of mass and proportional to their respective distances.

Ans. Equations (55).

3. The resultant of the forces acting on each body always passes through a fixed point. Prove that the equilateral triangle configuration is the only solution of equations (55) unless the bodies lie in a straight line.

4. Suppose $m_1 = m_2 = m_3 = 1$, and that the bodies move according to the equilateral triangular solution. Find the radius of the circle in which a particle would revolve around one of them in the period in which they revolve around their center of mass.

Ans. $R = 3^{\frac{1}{3}}$.

5. Prove that the equilateral triangular circular solutions hold when the mutual attractions of the bodies vary as any power of the distance.

6. Find the number of collinear solutions when the force varies as any power of the distance.

7. Prove that when the force varies inversely as the fifth power one solution is that each of the bodies moves in a circle through their center of mass in such a way that the three bodies are always at the vertices of an equilateral triangle.

8. Prove that if the three bodies are placed at rest in any one of the configurations admitting circular solutions, they will fall to their center of mass in the same time in straight lines.

9. Find the distribution of mass among the three bodies for which the time of falling to their center of mass will be the least; the greatest.

10. Prove that if any four masses are placed at the vertices of a regular tetrahedron, the resultant of all the forces acting on each body passes through the center of mass of the four, and that the magnitudes of the accelerations are proportional to the respective distances of the bodies from their center of mass.

11. Prove that there are no circular solutions in the Problem of Four Bodies in which the bodies do not all move in the same plane.

12. Investigate the stability of the triangle and straight line solutions of the Problem of Three Bodies when all of the masses are finite.

HISTORICAL SKETCH AND BIBLIOGRAPHY.

The first particular solutions of the Problem of Three Bodies were found by Lagrange in his prize memoir, *Essai sur le Problème des Trois Corps*, which was submitted to the Paris Academy in 1772 (*Coll. Works*, vol. VI., p. 229, Tisserand's *Méc. Cél.* vol. I., chap. VIII.). The solutions which he found are precisely those given in the last part of this chapter. His method was to divide the problem into two parts; (a) the determination of the mutual distances of the bodies, (b) having solved (a), the determination of the plane of the triangle in space and the orientation of the triangle in the plane. He proved that if the part (a) were solved the part (b) could also be solved. To solve (a) it was necessary to derive three differential equations involving the three mutual distances alone as dependent variables. He found three equations, one of which was of the third order, and the remaining two of the second order each, making the whole problem of the *seventh* order. The reduction of the general problem of three bodies by the ten integrals leaves it of the *eighth* order; hence Lagrange's analysis reduced the problem by one unit. He found that he could integrate the differential equations completely by assuming that the ratios of the mutual distances were constants. The demonstration was repeated by Laplace in the *Mécanique Céleste*, vol. V., p. 310. In *l'Exposition du Système du Monde* he remarked that if the moon had been given to the earth by Providence to illuminate the night, as some have maintained, the end sought has been only imperfectly attained; for, if the moon were properly started in opposition to the sun it would always remain there relatively, and the whole earth would have either the full moon or the sun always in view. The demonstration upon which he based his remark was made under the assumption that there was no disturbing force. If there were disturbing forces the configuration would not be preserved unless the solution were stable, which it is not, as was proved by Liouville, *Journal de Mathématiques*, vol. VII., 1845.

A number of memoirs have appeared following more or less closely along the lines marked out by Lagrange. Among them may be mentioned one by Radau in the *Bulletin Astronomique*, vol. III., p. 113; by Lindstedt in the *Annales de l'École Normale*, 3rd series, vol. I., p. 85; by Allegret in the *Journal de Mathématiques*, 1875, p. 277; by Bour in the *Journal de l'École Polytechnique*, vol. XXXVI.; and by Mathieu in the *Journal de Mathématiques*, 1876, p. 345.

Jacobi, without a knowledge of the work of Lagrange, reduced the general Problem of Three Bodies to the seventh order in *Crelle's Journal*, 1843, p. 115 (*Coll. Works*, vol. IV., p. 478). It has never been reduced further.

Concerning the solutions of the problem of more than three bodies in which the ratios of the mutual distances are constants a number of papers have appeared, among which are one by Lehmann-Filhes in the *Astronomische Nachrichten*, vol. CXXVII., p. 137, one by F. R. Moulton in *The Transactions of the American Mathematical Society*, vol. I., p. 17, and one by W. R. Longley in *Bulletin of the American Mathematical Society*, vol. XIII., p. 324.

No new periodic solutions of the problem of three bodies were discovered after those of Lagrange until Hill developed his Lunar Theory, *The American Journal of Mathematics*, vol. I. (1878). These solutions of Hill are of immensely greater practical value than those of the Lagrangian type. It should be stated, however, that they are not strictly periodic solutions of any actual case, because a small part of the perturbing action of the sun was neglected.

The next important advance was made by Poincaré in a memoir in the *Bulletin Astronomique,* vol. I., in which he proved that when the masses of two of the bodies are small compared to that of the third, there is an infinite number of sets of initial conditions for which the motion is periodic. These ideas were elaborated and the results extended in a memoir crowned with the prize offered by the late King Oscar of Sweden. This memoir appeared in *Acta Mathematica,* vol. XIII. The methods employed by Poincaré are incomparably more profound and powerful than any previously used in Celestial Mechanics, and mark an epoch in the development of the science. The work of Poincaré was recast and extended in many directions, and published in three volumes entitled, *Les Méthodes Nouvelles de la Mécanique Céleste.* It is written with admirable directness and clearness, and is given in sufficient detail to make so profound a work as easily read as possible.

An important memoir on Periodic Orbits by Sir George Darwin appeared in *Acta Mathematica,* vol. XXI. (1899). In this investigation it was assumed that one of the three masses is infinitesimal and that the finite masses, having the ratio of ten to one, revolve in circles. A large number of periodic orbits, belonging to a number of families, were discovered by numerical experiments. The question of their stability was answered by using essentially the method employed by Hill in his discussion of the motion of the lunar perigee.

A considerable number of investigations in the domain of periodic orbits, employing analytical processes based on the methods of Poincaré, have been published by F. R. Moulton and his former students Daniel Buchanan, Thomas Buck, F. L. Griffin, Wm. R. Longley, and W. D. MacMillan. These papers have appeared in the *Transactions of the American Mathematical Society,* the *Proceedings of the London Mathematical Society,* the *Mathematische Annalen,* and the *Proceedings of the Fifth International Congress of Mathematicians.* Besides containing the analysis for a great variety of periodic orbits, they show the existence of infinite sets of closed orbits of ejection which form the boundaries between different classes of periodic orbits. These investigations are published under the title " Periodic Orbits " as *Publication* 161 of the Carnegie Institution of Washington.

CHAPTER IX.

170. Meaning of Perturbations. It was shown in chapter v. that if two spherical bodies move under the influence of their mutual attractions each describes a conic section with respect to their center of mass as a focus, and that the path of each body with respect to the other is a conic. The converse theorem is also true; that is, if the law of areas holds and if the orbit of one body is a conic with respect to the other as a focus, then if the force depends only on the distance it varies inversely as the square of the distance (see also Art. 58). If there is a resisting medium, or if either of the bodies is oblate, or if there is a third body attracting the two under consideration, or if there is any force acting upon the bodies other than that of the mutual attractions of the two spheres, their orbits will cease to be exact conic sections. Suppose the coördinates and components of velocity are given at a definite instant t_0; then, if the conditions of the two-body problem were precisely fulfilled, the orbits would be definite conics in which the bodies would move so as to fulfill the law of areas. The differences between the coördinates and the components of velocity in the actual orbits and those which the bodies would have had if the motion had been undisturbed are the *perturbations*. It is necessary to include the changes in the components of velocity as perturbations, for the paths described depend not only upon the relative positions of the bodies and the forces to which they are subject, but also upon the relative velocities with which they are moving.

Several methods of computing perturbations have been devised depending upon the somewhat different points of view which may be taken. Of these the two following are the ones most frequently used.

171. Variation of Coördinates. The simplest conception of perturbations is that the coördinates are directly perturbed. For example, if a planet is subject to the attraction of another planet the coördinates and components of velocity of the former at any time t differ by definite amounts from what they would have been if the sun had been the only source of attraction, and these differences are computed by appropriate devices. No attempt is made to get the equations of the curve described, and usually no general information as to what will happen in the course of a long time is secured. This method is applied only to comets and small planets.

172. Variation of the Elements. This method is variously called the *Variation of the Elements*, the *Variation of Parameters*, and the *Variation of the Constants of Integration*. According to this conception, a body subject to the law of gravitation is always moving in a conic section, but in one which changes at each instant. The variable conic is tangent to the actual orbit at every point

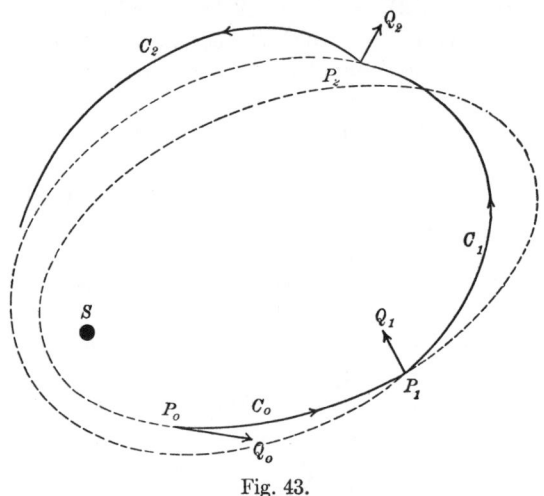

Fig. 43.

of it; and further, if the body were moving undisturbed in any one of the tangent conics it would have the same velocity at the point of tangency which it has in the actual orbit at that point. This conic is said to *osculate* with the actual orbit at the point of contact. The perturbations are the differences between the elements of the orbit on the start, and those of the osculating conic at any time. An obvious advantage of this method is that the elements change very slowly, since in most of the cases which actually arise in the solar system the perturbing forces are small. But if the perturbations were very large, as they are in some of the multiple star systems, this method would lose its relative advantages.

The conception of perturbations as being variations of the elements arises quite naturally in considering the factors which determine the elements of an orbit. It was shown in chap. v. that the initial positions of the two bodies and the directions of projection determine the plane of the orbit; that the initial positions and the velocities of projection determine the length of the major axis; and that the initial conditions, including the direction of projection and the velocities, determine the eccentricity and the line of the apsides.

Suppose a body m is projected from P_0, Fig. 43, in the direction Q_0 with the velocity V_0. Suppose there are no forces acting upon it except the attraction of S; then, in accordance with the results of the two-body problem, it follows that it will move in a conic section C_0 whose elements are uniquely determined. Suppose that when it arrives at P_1 it becomes subject to an instantaneous impulse of intensity f_1 in the direction P_1Q_1; this position and the new velocity and direction of motion determine a new conic C_1 in which the body will move until it is again disturbed by some external force. Suppose it becomes subject to the impulse f_2 in the direction P_2Q_2 when it arrives at P_2; it will move in the new conic C_2. This may be supposed to continue indefinitely. The body will be moving in conic sections which change from time to time when it is subject to the disturbing impulses. Suppose the instantaneous impulses become very small, and that the intervals of time between them become shorter and shorter. The general characteristics of the motion will remain the same. At the limit the impulses become a continually disturbing force, and the orbit a conic section which continually changes.

173. Derivation of the Elements from a Graphical Construction. It was shown in Art. 89 that the major semi-axis is given by the very simple equation

$$(1) \qquad V^2 = k^2(S + m)\left(\frac{2}{r} - \frac{1}{a}\right),$$

where V is the initial velocity, k^2 the Gaussian constant, $S + m$ the sum of the masses, r the initial distance of the bodies from each other, and a the major semi-axis. Suppose the major semi-axis has been computed by (1); it will be shown how the remaining elements can be found by the aid of very simple geometrical constructions. The initial positions of S and m, and the direction of projection of m, determine the position of the plane of the orbit, and therefore ☊ and i.

Suppose m is at the point P at the origin of time, and that it is projected in the direction PQ with the velocity V. The sun S is at one of the foci. It is known from the properties of conic sections that the lines from P to the two foci make equal angles with the tangent PQ. Draw the line PR making the same angle with the tangent that SP makes. Let r_1 represent the distance

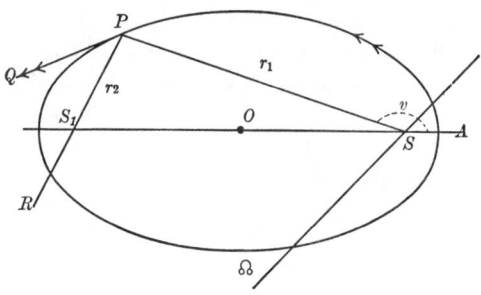

Fig. 44.

from S to P, and r_2 the distance from P to the second focus. Therefore $r_1 + r_2 = 2a$; or, $r_2 = 2a - r_1$, which defines the position of S_1. Call the mid-point of SS_1, O; then $e = \dfrac{SO}{a}$. Suppose S☊ is the line of nodes; then the angle ☊$SA = \omega$, and $\pi = \omega + $ ☊.

The only element remaining to be found is the time of perihelion passage. The angle ASP, counted in the direction of motion, is v. The eccentric anomaly is given by the equation (Art. 98)

$$(2) \qquad \tan \frac{E}{2} = \sqrt{\frac{1-e}{1+e}} \tan \frac{v}{2}.$$

After E has been found the time of perihelion passage, T, is defined by the equation (Art. 93)

$$(3) \qquad n(t - T) = E - e \sin E.$$

174. Resolution of the Disturbing Force. Whatever may be the source of the disturbing force it is convenient, in order to find its effects upon the elements, to resolve it into three rectangular components. It is possible to do this in several ways, each having advantages for particular purposes. The one will be adopted here which on the whole leads most simply to the determination of the manner in which the elements vary when the body under consideration is subject to any disturbing force. It would be possible without much difficulty to derive from geometrical considerations the expressions for the rates of change of the elements for any disturbing forces, but the object of this chapter is to explain the nature and causes of perturbations of various sorts, and the attention will not be divided by unnecessary digressions on methods of computation. This part falls naturally to the methods of analysis, which will be given in the next chapter.

The disturbing force will be resolved into three rectangular components: (a) the *orthogonal component,** S, which is perpendicular to the plane of the orbit, and which is taken positive when directed toward the north pole of the ecliptic; (b) the *tangential component, T,* which is in the line of the tangent, and which is taken positive when it acts in the direction of motion; and (c) the *normal component, N,* which is perpendicular to the tangent, and which is taken positive when directed to the interior of the orbit.

The instantaneous effects of these components upon the various elements will be discussed separately; and, unless it is otherwise stated, it always must be understood that the results refer to the way in which the elements are changing at given instants, and not to the cumulative effects of the disturbing forces. Although the effects of the different components are considered separately, yet when two or more act simultaneously it is sometimes necessary to estimate somewhat carefully the magnitude of their separate perturbations, in order to determine the character of their joint effects.

I. Effects of the Components of the Disturbing Force.

175. Disturbing Effects of the Orthogonal Component. In order to fix the ideas and abbreviate the language it will be supposed that the disturbed body is the moon moving around the earth. The perturbations arising from the disturbing action of the sun are very great and present many features of exceptional interest. Besides, this is the case which Newton treated by methods essentially the same as those employed here.† The

---

* A designation due to Sir John Herschel, *Outlines of Astronomy,* p. 420.

† *Principia,* Book I., Section 11, and Book III., Props. XXII.–XXXV.

character of the perturbations arising from positive components alone will be investigated; in every case negative components change the elements in the opposite way.

It is at once evident that the orthogonal component will not change a, e, T, and ω, if ω is counted from a fixed line in the plane of the orbit. But the ω in ordinary use is counted from the ascending node of the orbit; hence if the negative of the rate of increase of Ω be multiplied by $\cos i$ the result will be the rate of increase of ω due to the change in the origin from which it is reckoned. Consequently it is sufficient to consider the changes in Ω and i when discussing the perturbations due to the orthogonal component.

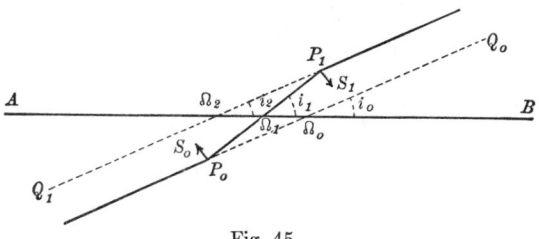

Fig. 45.

Let AB be in the plane of the ecliptic, P_0Q_0 in the plane of the undisturbed orbit, and Ω_0 and i_0 the corresponding node and inclination. Suppose there is an instantaneous impulse P_0S_0 when the moon is at P_0; it will then move in the direction P_0P_1, and the new node and inclination will be Ω_1 and i_1. It is evident at once that $i_1 > i_0$ and $\Omega_1 < \Omega_0$. Suppose a new instantaneous impulse P_1S_1 acts when the moon arrives at P_1. The new node and inclination are Ω_2 and i_2, and it is evident that $i_2 < i_1$ and $\Omega_2 < \Omega_1$. If $P_0\Omega_1 = \Omega_1P_1$, $P_0S_0 = P_1S_1$, and the velocity of the moon at P_0 equals that at P_1, then $i_0 = i_2$. The total result is a regression of the node and an unchanged inclination.

From the corresponding figure at the descending node it is seen that a negative S before node passage and a symmetrically opposite positive S after node passage will produce the same results as those which were found at the ascending node. Therefore, a positive S causes the nodes to advance if the moon is in the first or second quadrant, and to regress if it is in the third or fourth quadrant; and a positive S causes the inclination to increase if the moon is in the first or fourth quadrant, and to decrease if it is in the second or third quadrant.

The following quantitative results may be noted: The rate of change of both Ω and i is proportional to S. The rate of change of Ω is greater the smaller i; for $i = 0$ evidently Ω is not defined, but in this case in such problems as the Lunar Theory S vanishes. For a given i the rate of change of Ω is greater the nearer the point at which disturbance occurs is to midway between the two nodes. The rate at which i changes is greater the nearer the point at which the disturbance occurs is to a node.

176. Effects of the Tangential Component upon the Major Axis. Instead of deriving all the conclusions directly from geometrical constructions, it will be better to make use of some of the simple equations which have been found in chapter v. If it were desired the theorems contained in these equations could be derived from geometrical considerations, as was done by Newton in the *Principia*, but this would involve considerable labor and would add nothing to the understanding of the subject.

The major semi-axis is given in terms of the initial distance and the initial velocity by equation (1); viz.,

$$V^2 = k^2(E + m)\left(\frac{2}{r} - \frac{1}{a}\right).$$

In an elliptic orbit a is positive; hence, since a positive T increases V^2 and does not instantaneously change r, *a positive T increases the major semi-axis when the moon is in any part of its orbit.* It also follows from this equation that a given T is most effective in changing a when V has its largest value, or when the moon is at the perigee, and that the rate of change is more rapid the larger a.

Expressed in terms of partial derivatives, the dependence of a upon T is given by

$$\frac{\partial a}{\partial T} = \frac{\partial a}{\partial V}\frac{\partial V}{\partial T} = \frac{2a^2V}{k^2(E + m)}\frac{\partial V}{\partial T}.$$

177. Effects of the Tangential Component upon the Line of Apsides. The tangential component increases or decreases the speed, but does not instantaneously change the direction of motion. The focus E is of course not changed, r_1 is unchanged, and, according to the results of the last article, a is increased. Since $r_2 = 2a - r_1$ while the direction of r_2 remains the same, it follows that the focus E_1 is thrown forward to E_1', Fig. 46. The line of apsides is rotated forward from AB to $A'B'$. Hence it is easily seen that *a positive tangential component causes the line of* *apsides to rotate forward during the first half revolution, and back-ward during the second half revolution.*

The instantaneous effects are the same for points which are symmetrical with respect to the major axis. When the moon is at K or L the whole displacement of the second focus is perpendicular to the line of apsides, and at these points the rate of

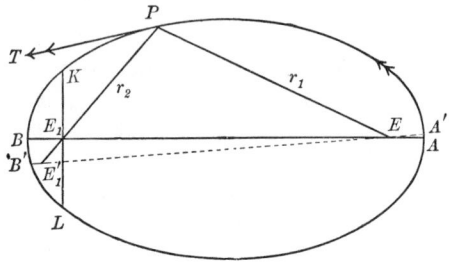

Fig. 46.

rotation of the apsides is a maximum for a given change in the major axis. But the major axis is changed most when the moon is at perigee; therefore the place at which the line of the apsides rotates most rapidly is near K and L and between these points and the perigee. The rate of rotation of the line of apsides becomes zero when the moon is at perigee or apogee. It should be remembered that the whole problem is complicated by the fact that the magnitude of T depends upon the distances of both moon and sun, and these distances continually vary.

178. Effects of the Tangential Component upon the Eccentricity.

The eccentricity is given by the equation $e = \dfrac{EE_1}{2a}$, Fig. 46.

When the moon is at the perigee EE_1 and $2a$ are increased by the same amount. Since EE_1 is less than $2a$ the eccentricity is increased at this point. When the moon is at apogee $2a$ is increased while EE_1 is decreased equally, hence the eccentricity is decreased. Consequently there is some place between perigee and apogee where the eccentricity is not changed, and it is easy to show that this place is at the end of the minor axis. Let $2\Delta a$ represent the instantaneous increase in $2a$ when the moon is at C or D, Fig. 47. Then r_2 will be increased by the quantity $2\Delta a$, and EE_1 by ΔE. If θ is the angle CE_1E, $\cos\theta = \dfrac{EE_1}{2a} = \dfrac{2ae}{2a} = e$, and, moreover, $\Delta E = 2\Delta a \cos \theta = 2e\Delta a$. Therefore

$$e' = \frac{EE_1 + \Delta E}{2a + 2\Delta a} = \frac{2ae + 2e\Delta a}{2a + 2\Delta a} = e;$$

or, the eccentricity is unchanged by the tangential component when the moon is at an end of the minor axis of its orbit.

The changes in the time of perihelion passage depend upon the changes in the period and the direction of the major axis, as well as on the direct perturbations of the longitude in the orbit. Since the period depends upon the major axis alone, whose changes

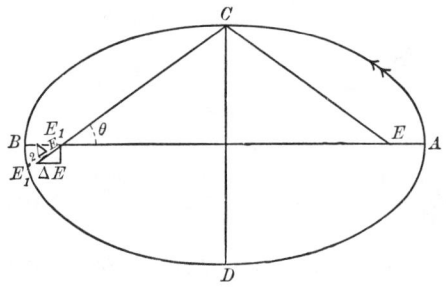

Fig. 47.

have been discussed, the foundations for an investigation of the changes in the time of perihelion passage have been laid, except in so far as they are direct perturbations in longitude; but further inquiry into this subject will be omitted because geometrical methods are not well suited to such an investigation, and because the time of perihelion passage is an element of little interest in the present connection.

179. Effects of the Normal Component upon the Major Axis.
It follows from (1) that the major axis depends upon the speed at a given point and not upon the direction of motion. Since the normal component acts at right angles to the tangent, it does not instantaneously change the speed and, therefore, leaves the major axis unchanged.

180. Effects of the Normal Component upon the Line of Apsides.
Consider the effect of an instantaneous normal component when the moon is at P, Fig. 48. Let PT represent the tangent to the orbit. The effect of the normal component will be to change it to PT'. Since the radii to the two foci make equal angles with the tangent

the radius r_2 will be changed to r_2'; and, since the normal component does not affect the length of the major axis, r_2 and r_2'
will be of equal length. Consequently, *when the moon is in the
region LAK a positive normal component will rotate the line of
apsides forward, and when it is in the region KBL, backward.* At

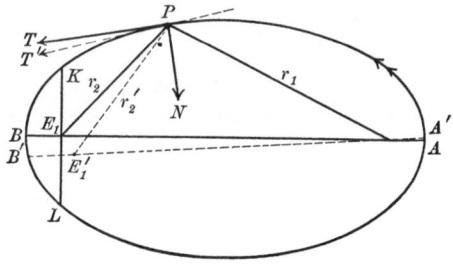

Fig. 48.

the points K and L the normal component does not change the
direction of the line of apsides.

In the applications to the perturbations of the moon it will be
important to determine the relative effectiveness of a given normal
force in changing the line of apsides when the moon is at the two
positions A and B. When the moon is at either of these two
points the second focus E_1 is displaced along the line KL. The
effectiveness of a force in changing the direction of motion of a
body is inversely proportional to the speed with which it moves;
but by the law of areas the velocities at A and B are inversely
proportional to their distances from E. Let E_A and E_B represent
the effectiveness of a given force in changing the direction of
motion at A and B respectively, and let V_A and V_B represent the
velocities at the same points. Then

$$E_A : E_B = V_B : V_A = a(1 - e) : a(1 + e).$$

The rotation of the line of apsides is directly proportional to
the displacement of E_1 along the line KL. The displacements
along KL are directly proportional to the products of the lengths
of the radii from A and B to E_1 and the angles through which they
are rotated. But the angles are proportional to E_A and E_B, and
the lengths of the radii to E_1 to $a(1 + e)$ and $a(1 - e)$. Therefore, letting R_A and R_B represent the rotation of the line of apsides
at the two points, it follows that

$$R_A : R_B = a(1 + e)E_A : a(1 - e)E_B = 1 : 1;$$

or, *equal instantaneous normal forces produce equal, but oppositely directed, rotations of the line of apsides when the moon is at apogee and at perigee.*

Suppose the forces act continuously over small arcs. Since the linear velocities are inversely as the radii, *the effectiveness, in changing the direction of the line of apsides, of a constant force acting through a small arc at A is to that of an equal force acting through an equal arc at B as* $a(1 - e)$ *is to* $a(1 + e)$. In practice the disturbing forces are not instantaneous but act continuously, their magnitudes depending upon the positions of the bodies; consequently, unless the normal component is smaller at apogee than at perigee the average rotation of the line of apsides due to a normal component always having the same sign is in the opposite direction of the rotation when the moon is at apogee.

181. Effects of the Normal Component upon the Eccentricity. If $2a$ represents the major axis, the eccentricity is given by

$$e = \frac{EE_1}{2a}.$$

After the action of the normal component the eccentricity is

$$e' = \frac{EE_1'}{2a},$$

the major axis being unchanged. It is easily seen from Fig. 48 that *a positive normal force decreases the eccentricity during the first half revolution and increases it during the second half*, EE_1' being less than EE_1 in the first case, and greater in the second. The instantaneous change in the eccentricity vanishes when the moon is at A or B.

It follows from Fig. 48 that a given change in the direction of r_2 produces a greater change in the eccentricity when the moon is in the second or third quadrant than when the moon is in a corresponding part of the first or fourth quadrant. Besides this, the moon moves slower the farther it is from the earth, and consequently a given normal component is more effective in changing the direction of motion, and therefore of r_2, when the moon is near apogee than when it is near perigee. Hence *a given normal component causes greater changes in the eccentricity if the moon is near apogee than it does if the moon is near perigee.*

182. Table of Results. The various results obtained will be of constant use in the applications which follow, and they will be most convenient when condensed into a table. The results are given for only positive values of the disturbing components; for negative components they are the opposite in every case.

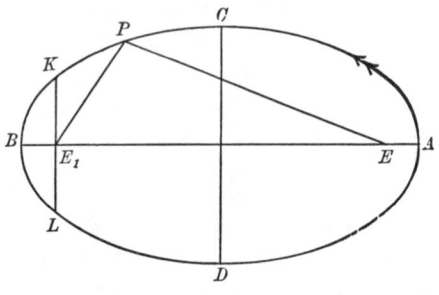

Fig. 49.

The orthogonal component, S, is positive when directed toward the north pole of the ecliptic.

The tangential component, T, is positive when directed in the direction of motion.

The normal component, N, is positive when directed to the interior of the ellipse.

Component . . .	S	T	N
Nodes	Advance in first and second quadrants; regress, in third and fourth quadrants.	0	0
Inclination. . . .	Increases in first and fourth quadrants ; decreases in second and third quadrants.	0	0
Major Axis . . .	0	Always increases	0
Line of Apsides	No effect if ω is counted from a fixed point rather than from Ω.	In interval ACB, forward; In interval BDA, backward	In interval LAK, forward; In interval KBL, backward
Eccentricity. . .	0	In interval DAC, increases; In interval CBD, decreases	In interval ACB, decreases; In interval BDA, increases

183. Disturbing Effects of a Resisting Medium. The simplest disturbance of elliptic motion is that arising from a resisting medium. The only disturbing force is a negative tangential component, which has the same magnitude for points symmetrically situated with respect to the major axis. Therefore, it is seen from the Table that: (1) Ω and i are unchanged; (2) a is continually decreased; (3) the line of apsides undergoes periodic variations, rotating backward during the first half revolution, and rotating forward equally during the second half; (4) the eccentricity decreases while the body moves through the interval DAC, and increases during the remainder of the revolution. It takes the body longer to move through the arc CBD than through DAC; but, on the other hand; if the resistance depends on a high power of the velocity, as experiment shows it does for high velocities, the change is much greater at perigee than at apogee, and the whole effect in a revolution is a decrease in the eccentricity. The application of these results to a comet, planet, or satellite resisted by meteoric matter, or possibly the ether, is evident.

184. Perturbations Arising from Oblateness of the Central Body. Consider the case of a satellite revolving around an oblate planet in the plane of its equator. It was shown in equations (30), p. 122, that the attraction under these circumstances is always greater than that of a concentric sphere of equal mass, but that

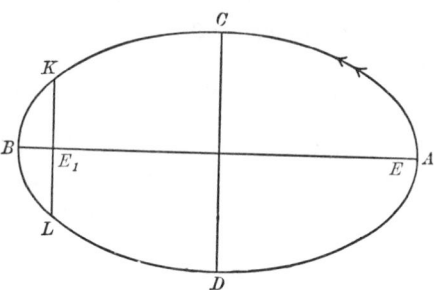

Fig. 50.

the two attractions approach equality as the satellite recedes. The excess of the attraction of the spheroid over that of an equal sphere will be considered as being the disturbing force, which, it will be observed, acts in the line of the radius vector and is always directed toward the planet. Therefore the normal com-

ponent is always positive, and is equal in value at points which
are symmetrically situated with respect to the major axis.   If the
eccentricity of the orbit is not large the tangential component is
relatively small, being negative in the interval ACB, and positive
in BDA.

(a) *Effect upon the period.*   This is most easily seen when the
orbit is a circle.   The attraction will be constant and greater
than it would be if the planet were a sphere.   This is equivalent
to increasing k^2, the acceleration per unit mass at unit distance;
therefore it is seen from the equation

$$ P = \frac{2\pi a^{\frac{3}{2}}}{k\sqrt{m_1 + m_2}} $$

that for a given orbit the period will be shorter, and for a given
period the distance greater, than it would be if the planet were a
sphere.

(b) *Effects upon the elements.*   On referring to the Table, it is
seen that: (1) Ω and i are unchanged; (2) a decreases and increases equally in a revolution; (3) the line of apsides rotates
forward during a little more than half a revolution, and that while
the disturbing force is of greatest intensity; and (4) the eccentricity
is changed equally in opposite directions in a whole revolution.
That is, *Ω and i are absolutely unchanged; a and e undergo periodic
variations which complete their period in a revolution; and the line
of apsides oscillates, but advances on the whole.*

The effects will be the greater the more oblate the planet and
the nearer the satellite.   The oblateness of the earth is so small
that it has very little effect in rotating the moon's line of apsides.
The most striking example of perturbations of this sort in the
solar system is in the orbit of the Fifth Satellite of Jupiter.   This
planet is so oblate and the satellite's orbit is so small that its
line of apsides advances about 900° in a year.

XXIII. PROBLEMS.

1. A body subject to no forces moves in a straight line with uniform speed. The elements of this orbit are the constants which define the position of the line, viz., the speed, the direction of motion in the line, and the position of the body at the time T. Show that they can be expressed in terms of six independent constants, and that it is permissible in the problem of two bodies to regard one body as always moving with respect to the other in a straight line whose position continually changes. Find the expression of these line elements in terms of the time in the case of elliptic motion.

2. Show from general considerations based on problem 1 that the methods of the variation of coördinates and the variation of parameters are essentially the same, differing only in the variables used in defining the coördinates and velocities of the bodies.

3. Suppose the sun moves through space in the line L, orthogonal to the plane Π. Take Π as the fundamental plane of reference. Let the point where the planet P_i passes through the plane Π in the direction of the motion of the sun be the ascending node, and, beginning at this point, divide the orbit into quadrants with respect to the sun as center. Suppose the ether and scattered meteoric matter slightly retard the sun and the planets, but neglect the retardation arising from the motion of the planets in their orbits around the sun.

(a) If the resistance is proportional to the masses of the respective bodies, show that the nodes and inclinations of their orbits are unchanged.

(b) Let σ and R represent the density and radius of the sun, and σ_i and R_i the corresponding quantities for the planet P_i. Then, if the resistance is proportional to the surfaces of the respective bodies, show that with respect to the plane Π the inclination and line of nodes undergo the following variations:

(1) If $\sigma_i R_i < \sigma R$.

Quadrant	1	2	3	4
Inclination.........	decreases	increases	increases	decreases
Line of nodes.......	regresses	regresses	advances	advances

(2) If $\sigma_i R_i > \sigma R$.

Quadrant	1	2	3	4
Inclination.........	increases	decreases	decreases	increases
Line of nodes.......	advances	advances	regresses	regresses

(c) If the orbits were circles the various changes in both cases would exactly balance each other in a whole revolution. How must the lines of apsides in the two cases lie with respect to the line of nodes in order that, for a few revolutions, (1) the inclination shall decrease the fastest, and (2) the line of nodes advance the fastest?

(d) Is it possible to make the relation of the line of apsides to the line of nodes such that, for a few revolutions, the inclination shall decrease and the line of nodes advance?

(e) If the line of apsides remains fixed in the plane of the orbit is it possible for the line of nodes to rotate indefinitely in one direction?

4. Suppose the orbit of a comet passes near Jupiter's orbit at one of its nodes; under what conditions will the inclination of the orbit of the comet be decreased? Show that if the major axis remains constant while the inclination is decreased the eccentricity is increased. (Use Art. 159.)

5. What is the effect of the gradual accretion of meteoric matter by a planet upon the major axis of its orbit?

6. Consider two viscous bodies revolving around their common center of mass, and rotating in the same direction with periods less than their period of revolution. They will generate tides in each other which will lag. The tidal protuberances of each body will exert a positive tangential and a positive normal component on the other, these components being greater the nearer the bodies are together. Moreover, the rotation of each body will be retarded by the action of the other on its protuberances. Suppose the bodies are initially near each other and that their orbits are slightly elliptic; follow out the evolution of all of the elements of their orbits.

II. The Lunar Theory.

185. Geometrical Resolution of the Disturbing Effects of a Third Body. The problem of the disturbance by a third body is much more difficult than those treated in Arts. 183 and 184, because the disturbing force varies in a very complicated manner.

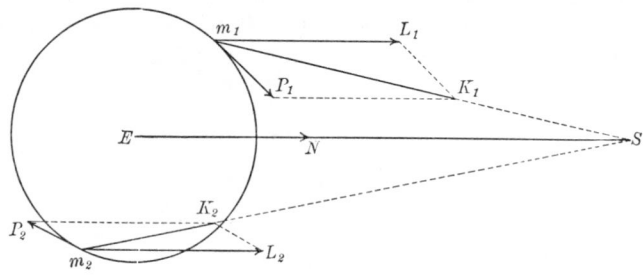

Fig. 51.

Suppose the three bodies are S, E, and m, and consider S as disturbing the motion of m around E. Two positions of m are shown at m_1 and m_2, and all the statements which are made apply for both subscripts. Let EN represent in magnitude and direction the acceleration of S on E. The order of the letters indicates the direction of the vector representing the force, and the magnitude of the vector depends upon the units employed. In the same units let mK represent in direction and amount the acceleration of S on m. The vector m_1K_1 is greater than EN because m_1S is less than ES, and m_2K_2 is less than EN because m_2S is greater than ES. By the law of gravitation they are proportional to the inverse squares of the respective distances.

Now resolve mK into two components, mL and mP, such that mL shall be equal and parallel to EN. Since mL and EN are equal and parallel these components will not disturb the relative positions of E and m. *Therefore the disturbing acceleration is mP.*

One important result is evident from Fig. 51, viz., that the disturbing acceleration is always toward the line joining E and S, or toward this line extended beyond E in the direction opposite to S when mS is greater than ES. Similar considerations applied to movable particles on the surface of the earth show why there tends to be a tide both on the side of the earth toward the moon, and also on the opposite side.

186. Analytical Resolution of the Disturbing Effects of a Third Body. Take a system of rectangular axes with the origin at the earth and with the xy-plane as the plane of the ecliptic. Let (x, y, z) and $(X, Y, 0)$ be the coördinates of the moon and sun respectively referred to this system. Let r, ρ, and R represent the distances Em, mS, and ES respectively. Let F_x, F_y, and F_z represent the components of the disturbing acceleration parallel to the x, y, and z-axes respectively. It follows from equations (24) of chapter VII., p. 272, that in the present notation

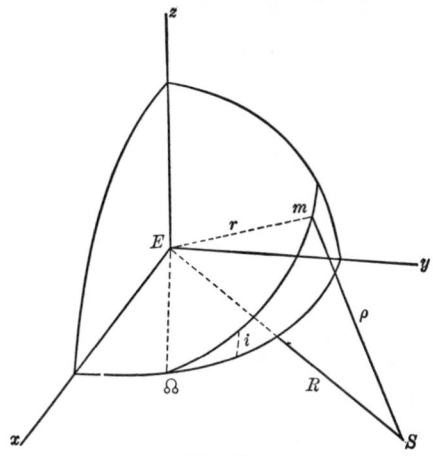

Fig. 52.

$$(4) \quad \begin{cases} F_x = k^2 S \dfrac{\partial}{\partial x}\left[\dfrac{1}{\rho} - \dfrac{xX+yY}{R^3}\right] = k^2 S\left[-\dfrac{x}{\rho^3} + X\left(\dfrac{1}{\rho^3} - \dfrac{1}{R^3}\right)\right], \\[2mm] F_y = k^2 S \dfrac{\partial}{\partial y}\left[\dfrac{1}{\rho} - \dfrac{xX+yY}{R^3}\right] = k^2 S\left[-\dfrac{y}{\rho^3} + Y\left(\dfrac{1}{\rho^3} - \dfrac{1}{R^3}\right)\right], \\[2mm] F_z = k^2 S \dfrac{\partial}{\partial z}\left[\dfrac{1}{\rho} - \dfrac{xX+yY}{R^3}\right] = k^2 S\left[-\dfrac{z}{\rho^3} + 0\left(\dfrac{1}{\rho^3} - \dfrac{1}{R^3}\right)\right]. \end{cases}$$

In order to get the components of the disturbing acceleration in any other directions it is sufficient to project these three components on lines having those directions and to take the respective sums.

Let F_r represent the component of the disturbing acceleration in the direction of the radius vector r; let F_v represent the component in a line perpendicular to r in the plane of motion of m; and let F_N represent the component which is perpendicular to both F_r and F_v. The component F_r will be taken as positive when it is directed from E; the component F_v will be taken positive when it makes with the direction of motion an angle less than 90°; and the component F_N will be taken positive when it is directed to the hemisphere which contains the positive end of the z-axis. The expression for F_r is

$$F_r = F_x \cos(xEm) + F_y \cos(yEm) + F_z \cos(zEm).$$

The expression for F_v can be obtained from this one by replacing the angle ΩEm by $\Omega Em + 90°$, because r will have the direction of the tangent at m after the body has moved forward 90° in its orbit. The expression for F_N can be conveniently obtained by first projecting F_x and F_y on a line in the xy-plane which is perpendicular to $E\Omega$, then projecting this result on the line perpendicular to the plane ΩEm, and projecting F_z directly on the same final line. Let the angle ΩEm be represented by u; then it is found from Fig. 52 by spherical trigonometry that

$$(5) \begin{cases} F_r = + F_x[\cos u \cos \Omega - \sin u \sin \Omega \cos i] \\ \qquad + F_y[\cos u \sin \Omega + \sin u \cos \Omega \cos i] \\ \qquad + F_z \sin u \sin i, \\ F_v = + F_x[- \sin u \cos \Omega - \cos u \sin \Omega \cos i] \\ \qquad + F_y[- \sin u \sin \Omega + \cos u \cos \Omega \cos i] \\ \qquad + F_z \cos u \sin i, \\ F_N = + F_x \sin \Omega \sin i - F_y \cos \Omega \sin i + F_z \cos i. \end{cases}$$

Let U represent the angle ΩES; then, since the sun moves in the xy-plane,

$$(6) \begin{cases} x = r[\cos u \cos \Omega - \sin u \sin \Omega \cos i], \\ y = r[\cos u \sin \Omega + \sin u \cos \Omega \cos i], \\ z = r \sin u \sin i; \\ X = R[\cos U \cos \Omega - \sin U \sin \Omega], \\ Y = R[\cos U \sin \Omega + \sin U \cos \Omega], \\ Z = 0. \end{cases}$$

On substituting the expressions for F_x, F_y, and F_z in (5), making use of (6), and reducing, it is found that

$$k^2 S \left\{ -\frac{r}{\rho^3} + R \left[\cos U \cos u \right. \right.$$
$$\left. \left. + \sin U \sin u \cos i \right] \left[\frac{1}{\rho^3} - \frac{1}{R^3} \right] \right\},$$

$$F_v = k^2 S \left\{ 0 + R \left[-\cos U \sin u \right. \right.$$
$$\left. \left. + \sin U \cos u \cos i \right] \left[\frac{1}{\rho^3} - \frac{1}{R^3} \right] \right\},$$

$$F_N = k^2 S \left\{ 0 - R \sin U \sin i \left[\frac{1}{\rho^3} - \frac{1}{R^3} \right] \right\}.$$

The geometry of equations (7) is important for a complete understanding of the problem. Consider a system of axes with origin at E, one axis directed toward m, another at right angles to it and 90° forward in the plane of the orbit of m, and the third perpendicular to the other two. Then it follows from the figure that the coefficients of $k^2 SR \left[\dfrac{1}{\rho^3} - \dfrac{1}{R^3} \right]$ in (7) are respectively the cosines of the angles between these axes and the line ES. Therefore F_v vanishes if the line through E parallel to the perpendicular to the radius is also perpendicular to ES, and F_N vanishes if m is in the plane of the orbit of S. They both vanish also if $R = \rho$, and then $F_r = k^2 S\left(-\dfrac{r}{\rho^3} \right)$; and in this case, if also $\rho = r$, F_r becomes simply $-\dfrac{k^2 S}{r^2}$.

Let ψ represent the angle between r and R; then

(8)
$$\begin{cases} F_r = k^2 S \left\{ -\frac{r}{\rho^3} + R \cos \psi \left[\frac{1}{\rho^3} - \frac{1}{R^3} \right] \right\}, \\ \rho^2 = R^2 + r^2 - 2Rr \cos \psi, \\ \frac{1}{\rho^3} = \frac{1}{R^3} \left[1 - \frac{2r}{R} \cos \psi + \frac{r^2}{R^2} \right]^{-\frac{3}{2}} = \frac{1}{R^3} \left[1 + 3\frac{r}{R} \cos \psi \cdots \right]. \end{cases}$$

Therefore the expression for F_r becomes

(9)
$$F_r = \frac{k^2 S}{2R^2} \left\{ \frac{r}{R} [1 + 3 \cos 2\psi] + \cdots \right\}.$$

Consequently F_r vanishes, if the terms of higher order are neglected, when

(10)
$$\begin{cases} 1 + 3 \cos 2\psi = 0, \text{ whence} \\ \psi = 54° \ 44', \quad 125° \ 16', \quad 234° \ 44', \quad 305° \ 16'. \end{cases}$$

Now consider the problem of finding the tangential and normal components of the disturbing acceleration. Let P represent a general point in the orbit, Fig. 53. Let PT be the tangent at P and PN the perpendicular to it. It follows from the elementary properties of ellipses that PN bisects the angle between r_1 and r_2.

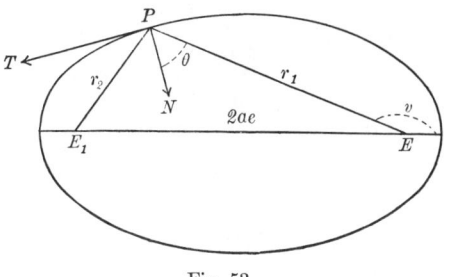

Fig. 53.

Then the tangential and normal components of the disturbing acceleration are expressed in terms of F_r and F_v by

(11)
$$\begin{cases} T = + F_r \sin \theta + F_v \cos \theta, \\ N = - F_r \cos \theta + F_v \sin \theta. \end{cases}$$

In order to complete the expressions for T and N the factors $\sin \theta$ and $\cos \theta$ must be expressed in terms of v. It follows from the geometrical properties of the ellipse and from the triangle EPE_1 that

$$\begin{cases} r_1 = \dfrac{a(1 - e^2)}{1 + e \cos v}, \\ r_1 + r_2 = 2a, \\ r_1^2 + r_2^2 - 2r_1 r_2 \cos 2\theta = 4a^2 e^2. \end{cases}$$

When r_1 and r_2 are eliminated from these three equations, it is found that

$$\begin{cases} \sin \theta = \dfrac{e \sin v}{\sqrt{1 + e^2 + 2e \cos v}}, \\ \cos \theta = \dfrac{1 + e \cos v}{\sqrt{1 + e^2 + 2e \cos v}}. \end{cases}$$

Therefore

(12)
$$\begin{cases} T = \dfrac{e \sin v}{\sqrt{1 + e^2 + 2e \cos v}} F_r + \dfrac{(1 + e \cos v)}{\sqrt{1 + e^2 + 2e \cos v}} F_v, \\ N = \dfrac{-(1 + e \cos v)}{\sqrt{1 + e^2 + 2e \cos v}} F_r + \dfrac{e \sin v}{\sqrt{1 + e^2 + 2e \cos v}} F_v. \end{cases}$$

use of (7) and the relation $u = \omega + v$, the final for the tangential and normal components of the acceleration become

(13)

$$
\begin{cases}
T = \dfrac{k^2 S}{\sqrt{1 + e^2 + 2e \cos v}} \Big\{ - e \sin v \, \dfrac{r}{\rho^3} \\
\qquad + \Big[- \cos U \, (\sin u + e \sin \omega) \\
\qquad + \sin U \cos i \, (\cos u + e \cos \omega) \Big] R \Big[\dfrac{1}{\rho^3} - \dfrac{1}{R^3} \Big] \Big\}, \\[2ex]
N = \dfrac{k^2 S}{\sqrt{1 + e^2 + 2e \cos v}} \Big\{ + \dfrac{r}{\rho^3} + \dfrac{r}{\rho^3} \, e \cos (u - \omega) \\
\qquad - \Big[\cos v \, (\cos u + e \cos \omega) \\
\qquad + \sin v \, \cos e \, (\sin u + e \sin \omega) \Big] R \Big[\dfrac{1}{\rho^3} - \dfrac{1}{R^3} \Big] \Big\}.
\end{cases}
$$

All the circumstances of the variation of T and N can be inferred from these equations.

187. Perturbations of the Node. By definition, the orthogonal component S is identical with F_N; therefore by the last of (7)

(14)   Orthog. Comp. $= S = - k^2 SR \sin U \sin i \Big[\dfrac{1}{\rho^3} - \dfrac{1}{R^3} \Big].$

The sign of the right member depends upon the signs of $\sin U$ and $\Big[\dfrac{1}{\rho^3} - \dfrac{1}{R^3} \Big]$, both of which can be either positive or negative. In order to determine which sign prevails in the long run so as to find whether on the whole there is an advance or retrogression of the line of nodes, it is necessary to expand the last factor of (14). On making use of the last equation of (8), it is found that

(15)

$$
\begin{aligned}
S &= - \dfrac{3 k^2 S r}{R^3} \sin U \sin i \cos \psi + \cdots \\[1ex]
&= - \dfrac{3 k^2 S r}{R^3} \sin U \sin i [\cos U \cos u + \sin U \sin u \cos i] + \cdots,
\end{aligned}
$$

where the S in the right member represents the mass of the sun.

The angular velocity of the sun in its orbit is slow compared to that of the moon; hence, in order to simplify the discussion, it may be supposed to stand still while the moon makes a single revolution. Since the periods of the moon and sun have no simple relation the values of sin U and cos U in the long run will be as often decreasing as increasing, and hence the assumption will cause no important error.

Suppose S is broken up into the sum of two parts, S_1 and S_2, where

$$
(16) \quad
\begin{cases}
S_1 = - \dfrac{3k^2 S r}{R^3} \sin i \, \sin U \cos U \cos u, \\[2mm]
S_2 = - \dfrac{3k^2 S r}{R^3} \sin i \, \cos i \, \sin^2 U \sin u.
\end{cases}
$$

In order to get the greatest degree of simplicity suppose the orbit of the moon is a circle so that r is a constant and $u = nt$. Suppose U has a definite value and consider the effects of S_1 during a revolution of the moon, starting with the ascending node. It follows from the table of Art. 182 that the effects of S_1 in the first and second quadrants are equal and opposite because cos u has equal numerical values and opposite signs in the two quadrants. It is the same in the third and fourth quadrants. Therefore S_1 produces only periodic perturbations in the line of nodes.

Now consider the effects of S_2. In the first half revolution, starting with the node, S_2 is negative because sin u is positive and all the other factors are positive. In the second half revolution S_2 is positive because sin u is negative. Therefore, it follows from the table of Art. 182 that S_2 *causes a continuous, but irregular, regression* (except when it is temporarily zero) *of the line of nodes.* The complete motion of the line of nodes is the resultant of the periodic oscillations due to S_1 and the periodic and continuous changes produced by S_2.

The period of revolution of the moon's line of nodes is about nineteen years. Since eclipses of the sun and moon can occur only when the sun is near a node of the moon's orbit, the times of the year at which they take place are earlier year after year, the cycle being completed in about nineteen years.

188. Perturbations of the Inclination. The expression for the orthogonal component is given in (15), which may again be broken up into the two parts S_1 and S_2. It follows from the table of Art. 182 that a positive S increases the inclination in the first and fourth quadrants and decreases it in the second and third quadrants.

Consider the effects of S_1. If sin U cos U is positive the effect in each quadrant is to decrease the inclination. But this case can be paired with that in which sin U cos U is negative and of equal numerical value. Since all possible situations can be paired in this way, S_1 produces only periodic changes in the inclination.

The case of S_2 is even simpler than that of S_1. Since sin u is positive in the first two quadrants, the effect in the second quadrant offsets that in the first. Similarly, the effects in the third and fourth quadrants mutually destroy each other. *Therefore the inclination undergoes only periodic variations.*

Some things have been neglected in this discussion to which attention should be called. No account has been taken of the eccentricities of the orbits of the moon and earth. When they are included the terms do not completely destroy one another in the simple fashion which has been described. Moreover, each perturbation has been considered independently of all other ones. As a matter of fact, each one depends on all the others. For example, if the node changes, the effects on the inclination are different from what they would otherwise have been, and conversely. It is clear that a very refined analysis is necessary in order to get accurate numerical results. But this does not mean that common-sense geometrical and physical considerations are not of the highest importance, especially in first penetrating unexplored fields.

189. Precession of the Equinoxes. Nutation. Suppose the largest sphere possible is cut out of the earth leaving an equatorial ring. Every particle in this ring may be considered as being a small satellite; then, from the principles explained in Arts. 185 and 186, the attractions of the moon and sun will exercise disturbing accelerations upon them which will tend to shift them with respect to the spherical core. But the particles of the ring are fastened to the solid earth so that it partakes of any disturbance to which they may be subject. Since their combined mass is very small compared to that of the spherical body within them, and since the disturbing forces are very slight, the changes in the motion of the earth will take place very slowly.

From the results of the last article it follows that the nodes of the orbit of every particle will have a tendency to regress on the plane of the disturbing body. The angle between the plane of the moon's orbit and that of the ecliptic may be neglected for the moment as it is small compared to the inclination of the earth's equator. They communicate this tendency to the whole earth so that the plane of the earth's equator turns in the retrograde direction on the plane of the ecliptic. On the other hand, it follows from the symmetry of the figure with respect to the nodes of the orbits of the particles of the equatorial ring that there will be no change in the inclination of the plane of the equator to that of the ecliptic or the moon's orbit. The mass moved is so great, and the forces acting are so small, that this retrograde motion, called the *precession of the equinoxes*, amounts to only about $50''.2$ annually; or, the plane of the earth's equator makes a revolution in about 26,000 years.

The moon is very near to the earth compared to the sun, and the orthogonal component arising from its attraction is greater than that coming from the sun's attraction. The main regression is, therefore, on the moon's orbit, which is inclined to the ecliptic about $5° 9'$. Since the line of the moon's nodes makes a revolution in about 19 years, the plane with respect to which the equator regresses performs a revolution in the same time. This produces a slight nodding in the motion of the pole of the equator around the pole of the ecliptic, and is called *nutation*.

The quantitative agreement between theory and observation of the rate of precession proves that the equatorial bulge is solidly attached to the remainder of the earth. If the earth were a relatively thin solid crust floating on a liquid interior, as was once supposed, it would probably slide somewhat on the interior and give a more rapid precession.

190. Resolution of the Disturbing Acceleration in the Plane of Motion. It follows from the table of Art. 182 that the orthogonal component does not produce perturbations in the major axis, longitude of perigee, and eccentricity, except indirectly as it shifts the line of nodes from which the longitude of the perigee is counted. Consequently an idea of the way these elements are perturbed can be obtained even if the inclination, with which the orthogonal component vanishes, is supposed to be zero. But it must be remembered the results obtained under these restrictions are not rigorous because T and N depend on the inclination. But the approximation is fully justified because it results in great simplifications which aid correspondingly in understanding the subject.

On taking $i = 0$ equations (13) become

$$(17) \begin{cases} T = \dfrac{k^2 S}{\sqrt{1 + e^2 + 2e \cos v}} \left\{ - e \sin v \dfrac{r}{\rho^3} \right. \\ \qquad\qquad \left. - R \left[\sin (u - U) + e \sin (\omega - U) \right] \left[\dfrac{1}{\rho^3} - \dfrac{1}{R^3} \right] \right\}, \\[2ex] N = \dfrac{k^2 S}{\sqrt{1 + e^2 + 2e \cos v}} \left\{ \dfrac{r}{\rho^3} \right. \\ \qquad\qquad \left. - R \left[\cos (u - U) + e \cos (\omega - U) \right] \left[\dfrac{1}{\rho^3} - \dfrac{1}{R^3} \right] \right\}. \end{cases}$$

Tangential Component.

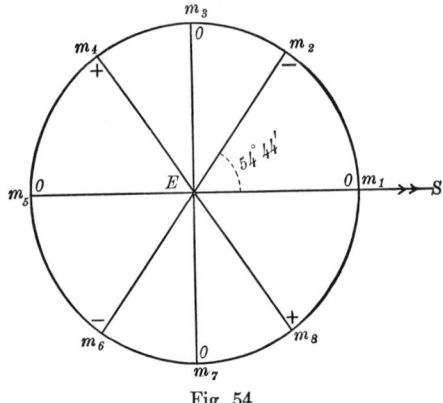

Fig. 54.

When i equals zero $\psi = u - U$, and on using the last equation of (8), it is found that

$$(18) \begin{cases} T = \dfrac{k^2 S}{\sqrt{1 + e^2 + 2e \cos v}} \left\{ - e \sin v \ \dfrac{r}{R^3} \right. \\ \qquad\qquad - 3e \dfrac{r}{R^3} \sin (\omega - U) \cos (u - U) \\ \qquad\qquad \left. - \dfrac{3}{2} \dfrac{r}{R^3} \sin 2(u - U) + \cdots \right\}, \\[2ex] N = \dfrac{k^2 S}{\sqrt{1 + e^2 + 2e \cos v}} \left\{ - 3e \dfrac{r}{R^3} \cos (\omega - U) \cos (u - U) \right. \\ \qquad\qquad \left. - \dfrac{1}{2} \dfrac{r}{R^3} [1 + 3 \cos 2(u - U)] + \cdots \right\}. \end{cases}$$

In the orbit of the moon e is approximately equal to $\frac{1}{20}$, and consequently a good idea of the numerical magnitudes of T and N and the circumstances under which they change sign can be obtained by neglecting those terms which have e as a factor. If these terms are neglected it is found that T vanishes at $u - U = 0$, $\frac{\pi}{2}$, π, and $\frac{3\pi}{2}$; it is negative in the first and third quadrants, and positive in the second and fourth quadrants. Under the same circumstances N vanishes at $54°\ 44'$, $125°\ 16'$, $234°\ 44'$, and $305°\ 16'$; it is negative from $-54°\ 44'$ to $+54°\ 44'$ and from $125°\ 16'$ to $234°\ 44'$, and is positive from $54°\ 44'$ to $125°\ 16'$ and from $234°\ 44'$ to $305°\ 16'$. If the terms depending on e and the

Normal Component.

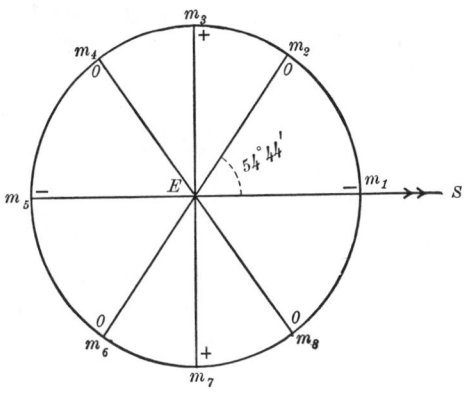

Fig. 55.

higher terms in the expansion of ρ^{-3} are retained, the points where T and N vanish are in general slightly different from those which have been found, but the differences are not important in a qualitative discussion whose aim is simply to exhibit the general characteristics of the results.

The signs of T and N for the moon in different parts of its orbit are shown in Figs. 54 and 55.

191. Perturbations of the Major Axis. If the perigee were at m_1 or m_5 the tangential component, which alone changes a, would be equal and of opposite sign at points symmetrically situated with respect to the major axis. In this case a would be unchanged at the end of a complete revolution. But this condition of affairs is only realized instantaneously, for the disturbing body S is moving in its orbit; yet, in a very large number of revolutions, when the periods are incommensurable, an equal number of equal positive and negative tangential components will have exerted a disturbing influence. The result is that in the long run a is unchanged, although it undergoes periodic variations.

192. Perturbation of the Period. The normal component is not only negative more than half a revolution, but the negative values are greater numerically than the positive ones. If the terms involving e are neglected, it is seen from the second equation of (18) that the greatest negative value of N is twice its numerically greatest positive value. One effect of the whole result is equivalent to a diminution, on the average, of the attraction of E for m; that is, to a diminution of k^2, the acceleration at unit distance. The relation of the period to the intensity of the attraction and the major axis is (Art. 89)

$$P = \frac{2\pi a^{\frac{3}{2}}}{k\sqrt{E+m}}.$$

Hence, for a given distance, P is increased if k is decreased. In this manner the sun's disturbing effect upon the orbit of the moon increases the length of the month by more than an hour. (Compare Art. 184 (a).)

193. The Annual Equation. Since the orbit of the earth is an ellipse the distance of the sun undergoes considerable variations. The farther the sun is from the earth the feebler are its disturbing effects, and in particular, the power of lengthening the month considered in the preceding article. Therefore, as the earth moves from perihelion to aphelion the disturbance which *increases* the length of the month will become less and less; that is, the length of the month will become shorter, or the moon's angular motion will be accelerated. While the earth is moving from aphelion to perihelion the moon's motion will, for the opposite reason, be retarded. This is the *Annual Equation* amounting to a little more than $11'$, and was discovered from observations by Tycho Brahe about 1590.

194. The Secular Acceleration of the Moon's Mean Motion. In the early part of the 18th century Halley found from a comparison of ancient and modern eclipses that the mean motion of the moon is gradually increasing. Nearly 100 years later (1787) Laplace gave the explanation of it, showing that it is caused by the gradual average decrease of the eccentricity of the earth's orbit, which has been going on for many thousands of years because of perturbations by the other planets, and which will continue for a long time yet before it begins to increase.

One effect of a change in the eccentricity of the earth's orbit is to change the average disturbing power of the sun on the orbit of the moon. It will now be shown that if the eccentricity decreases, the average disturbing power decreases.

The effect upon the moon's period is due almost entirely to the normal component, because it alone acts nearly along the radius of the orbit, and therefore in this discussion consideration of the tangential component may be omitted. The average value of N in a revolution of the moon, for R and U constant and e placed equal to zero, is found from the second equation of (18) to be

(19) $\qquad \begin{cases} \text{Average } N = -\frac{1}{2}k^2 S \dfrac{r}{R^3} \dfrac{n}{2\pi} \displaystyle\int_0^{\frac{2\pi}{n}} [1 + 3 \cos 2(nt - U)]dt \\[2mm] \qquad\qquad = -\frac{1}{2}k^2 S \dfrac{r}{R^3}. \end{cases}$

That is, the normal component of the disturbing acceleration on the average is very nearly proportional to the radius of the moon's orbit and the inverse third power of the radius of the earth's orbit. But if the earth's orbit is eccentric, the result for a whole year depends upon the eccentricity. When the nature of the dependence of the average N upon the eccentricity of the earth's orbit has been found, the effect of an increase or decrease in this eccentricity can be determined.

Let $\bar{N}$ represent the average N for a year. Then it follows from (19) that

(20) $\qquad\qquad \bar{N} = -\frac{1}{2}\frac{k^2 Sr}{P}\int_0^P \frac{dt}{R^3},$

where P is the earth's period of revolution. By the law of areas it follows that $h\,dt = R^2 d\theta$; hence equation (20) becomes

$$\bar{N} = -\frac{1}{2}\frac{k^2 Sr}{Ph}\int_0^{2\pi}\frac{d\theta}{R} = -\frac{1}{2}\frac{k^2 Sr}{Ph}\int_0^{2\pi}\frac{(1 + e'\cos\theta)}{a'(1 - e'^2)}\,d\theta,$$

$$= \frac{-k^2 Sr\pi}{Pha'(1 - e'^2)},$$

where a' and e' are the major semi-axis and eccentricity of the earth's orbit. But it follows from the problem of two bodies that

$$h = k\sqrt{(1 + m)a'(1 - e'^2)}, \qquad P = \frac{2\pi a'^{\frac{3}{2}}}{k\sqrt{1 + m}}.$$

Therefore the expression for $\bar{N}$ becomes

(21) $$\bar{N} = \frac{- k^2 Sr}{2a'^3(1 - e'^2)^{\frac{3}{2}}}.$$

As e' decreases $\bar{N}$ numerically decreases; therefore, as the eccentricity of the earth's orbit decreases, the efficiency of the sun in decreasing the attraction of the earth for the moon gradually decreases, and the mean motion of the moon increases correspondingly. The changes are so small that the alteration in the orbit is almost inappreciable, but in the course of centuries the longitude of the moon is sensibly increased. The theoretical amount of the acceleration is about 6″ in a century. The amount derived from a discussion of eclipses varies from 8″ to 12″. It has been suggested that tidal retardation, lengthening the day, has caused the unexplained part of the apparent change, but the subject seems to be open yet to some question.

The very long periodic variations in the eccentricity of the earth's orbit, whose effects upon the motion of the moon have just been considered, are due to the perturbations of the other planets. Although their masses are so small and they are so remote that their direct perturbations of the moon's motion are almost insensible, yet they cause this and other important variations indirectly through their disturbances of the orbit of the earth. This example of indirect action illustrates the great intricacy of the problem of the motions of the bodies of the solar system, and shows that methods of the greatest refinement must be employed in order to derive satisfactory numerical results.

195. The Variation. There is another important perturbation in the motion of the moon which does not depend upon the eccentricity of its orbit. It was discovered by Tycho Brahe, from observation, about 1590. Newton explained the cause of it in the *Principia* by a direct and elegant method which elicited the praise of Laplace.

It can be explained most readily by supposing that the undisturbed motion of the moon is in a circle. As has been shown, the normal component of the sun's disturbing acceleration is negative in the intervals $m_8 m_1 m_2$ and $m_4 m_5 m_6$, with maximum values at m_1 and m_5. Suppose the undisturbed motion at m_1 is in a circle; that is, that the acceleration due to the attraction of the earth exactly balances the centrifugal acceleration. There is no tangential component at this point but a large negative normal component. The result is that the force which tends toward E is diminished and the orbit is less curved at this point than the circle. Therefore the moon will recede to a greater distance from the earth in quadrature than in the circular orbit. At the point m_3 the tangential component is zero, the force which tends toward E is increased, and the curvature is greater than in the circle. The conditions vary continuously from those at m_1 to

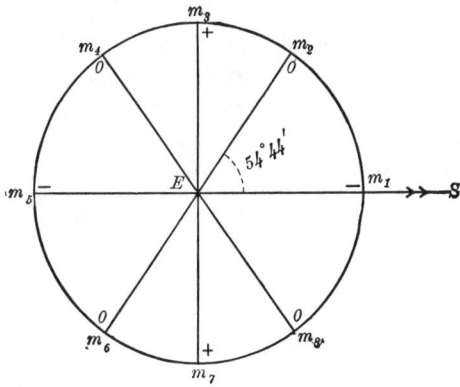

Fig. 56.

those at m_3 in the interval $m_1 m_3$. The corresponding changes in the remainder of the orbit are evident. The whole result is that the orbit is lengthened in the direction perpendicular to the line from the earth to the sun. If the sun is assumed to be so far distant that its disturbing effects in the interval $m_3 m_5 m_7$ are equal to those in the interval $m_7 m_1 m_3$, the orbit, under *proper initial conditions*, is symmetrical with respect to E as a center, and closely resembles an ellipse in form. This change of form of the orbit, and the auxiliary changes in the rate at which the radius vector sweeps over areas, give rise to an inequality in longitude between the mean position and the true position of the moon which amounts at times to about $39' 30''$, and is called the *variation*.

The variation has an interesting and important connection with the modern methods in the Lunar Theory, which were founded by G. W. Hill in his celebrated memoirs in the first volume of the *American Journal of Mathematics*, and in the *Acta Mathematica*, vol. VIII. A complete account of this method is given in Brown's *Lunar Theory* in the chapter entitled, *Method with Rect-*

angular Coördinates. Hill neglected the solar parallax; that is, he assumed that the disturbing force is equal in corresponding points in conjunction with, and opposition to, the sun. Instead of taking an ellipse as a first approximation, he took as an intermediate orbit that *variational orbit* which is closed with respect to axes rotating with the mean angular velocity of the sun, with a synodic period equal to the synodic period of the moon. The conception is not only one of great value, but the analysis was made by Hill with rare ingenuity and elegance.

196. The Parallactic Inequality. Since the sun is only a finite distance from the earth, its disturbing effects will not be exactly the same in points symmetrically situated with respect to the line $m_3 m_7$, but will be greater on the side $m_7 m_1 m_3$. For example, if the expansion of ρ^{-3} in (17) is carried one order farther so as to include the terms of the second order, that is in $\dfrac{r^2}{R^2}$, the part of N which is independent of e is found to be

$$
\begin{aligned}
N = \frac{k^2 S}{R^2} \Big\{ & -\frac{1}{2}\frac{r}{R}\left[1 + 3\cos 2(u - U)\right] \\
& -\frac{3}{8}\frac{r^2}{R^2}\left[3\cos(u - U) + 5\cos 3(u - U)\right] - \cdots \Big\}.
\end{aligned}
$$

(22)

When $u - U = 0$ the term of the second order has the same sign as the first one, and when $u - U = \pi$ it has the opposite sign. The effect of this term is relatively small because $r \div R = .0025$ nearly. The terms which are of the second order introduce a distortion in the variational orbit, which leads to an inequality of about $2' 7''$ in the longitude of the moon compared to the theoretical position in the variational orbit. Since it is due to the parallax of the sun it has been called the *parallactic inequality.* Laplace remarked that, when it has been determined with very great accuracy from a long series of observations, it will furnish a satisfactory method of obtaining the distance of the sun. The chief practical difficulty is that the troublesome problem of finding the relative masses of the earth and moon must be solved before the method can be applied.*

197. The Motion of the Line of Apsides. On account of the more complicated manner in which the different components affect the motion of the line of apsides, the perturbations of this

---

* See Brown's *Lunar Theory*, p. 127.

element present greater difficulties than those heretofore considered. Suppose first that the line of apsides coincides with the line ES, and that the perigee is at m_1. The normal component at m_1 is negative, and therefore (Table, Art. 182) produces a retrogression of the line of apsides. On the other hand, when the moon is at m_5 the negative normal component causes the line of apsides to advance. It was shown in Art. 180 that the effectiveness of a normal component acting while the moon describes a short arc at apogee is to that of an equal normal component acting while an equal arc is described at perigee as $a(1 + e)$ is to $a(1 - e)$. Moreover, the second equation of (18) shows that the normal component varies directly as the distance of the moon from the earth. Therefore the normal component is greater at apogee, and is more effective in proportion to its magnitude, than the corresponding acceleration at perigee. The

Normal Component.

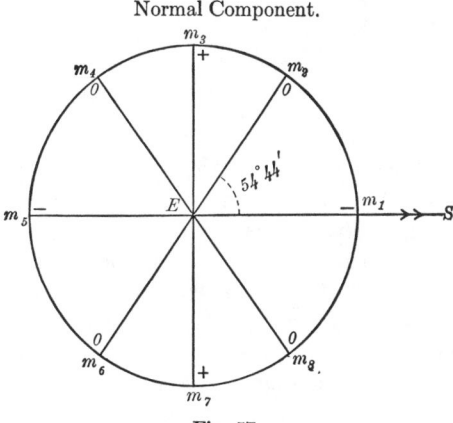

Fig. 57.

normal component is positive, though comparatively small, in the intervals $m_2 m_3 m_4$ and $m_6 m_7 m_8$. These intervals are almost equally divided by K and L (Fig. 48) where the effect of the normal component on the line of apsides vanishes. Therefore it follows from the Table that the total effect in these intervals is very small. Hence when the perigee is at m_1 the result in a whole revolution is to rotate the line of apsides forward through a considerable angle. Similar reasoning leads to precisely the same results when the perigee is at m_5.

When the perigee is at m_1 the tangential component is equal in numerical value and opposite in sign on opposite sides of the major axis. Hence it follows from the Table that the effects are in the same direction and equal in magnitude for points symmetrically situated on opposite sides of the major axis. But the effects in

Tangential Component.

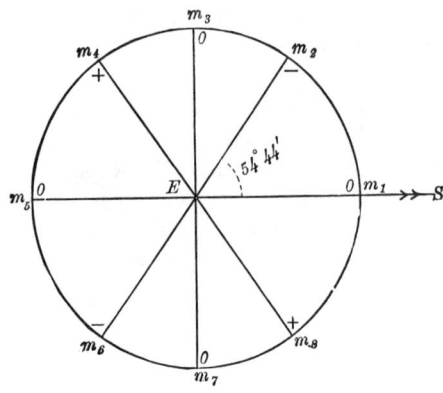

Fig. 58.

the second and third quadrants are opposite in sign to those in the first and fourth quadrants; moreover, they are a little greater in the second and third quadrants because then r is greatest and the tangential component, by (18), is proportional to r. Hence when the perigee is at m_1 the total effect of the tangential component in a whole revolution is to rotate the apsides forward. Now pair this with the case where the perigee is at m_5, a condition which will arise because of the motion of the sun even if the apsides were stationary. Under these circumstances the apsides are rotated backward, and the rotations in the two cases offset each other.

Suppose now that the line of apsides is perpendicular to the line ES. It is immaterial in this discussion at which end of the line the perigee is, but, to fix the ideas, it will be taken at m_3. The normal component is positive in the interval $m_2m_3m_4$, and, according to the Table, rotates the line of apsides forward. It is also positive in the interval $m_6m_7m_8$ and there rotates the line of apsides backward. In the latter case the disturbing acceleration is greater, and more effective for its magnitude, so that the whole result is a retrogression. The intervals $m_8m_1m_2$ and $m_4m_5m_6$, in which the normal components are negative, are divided nearly equally by L and K; hence it is seen from the Table that their results almost exactly balance each other in a whole revolution. Therefore, when the perigee is at m_3, the result of the normal component on the line of apsides for a whole revolution is a considerable retrogression.

When the perigee is at m_3 the tangential component is positive in the interval $m_3 m_5$ and negative in $m_5 m_7$. From the Table it is seen that a positive T in the interval $m_3 m_5 m_7$ causes the line of apsides to rotate forward, and a negative, backward. Since the sign of T is opposite in the two nearly equal parts of the interval the whole result upon the line of apsides is very small. The result is the same in the half revolution $m_7 m_1 m_3$. Thus it is seen that the combined effects of the normal and tangential components in a whole revolution is to rotate the line of apsides *backward* when it is perpendicular to the line from the earth to the sun.

It was found that the line of apsides rotates *forward* when it coincides with the line from the earth to the sun. The next question to be answered is whether the advance or the retrogression is the greater. It is noticed that the total changes arising from the action of the tangential components are the differences of nearly equal tendencies, and therefore small. The same may be said of the normal components which act in the vicinity of the ends of the minor axis of the ellipse. Moreover, in the two positions considered they act in opposite directions so that their whole result is still smaller. The most important changes arise from the normal components which act in the vicinity of the ends of the major axis. It follows from the second equation of (18) that in the first case, in which the line of apsides advances, they are about twice as great as in the second, in which the line of apsides regresses. Therefore, the whole change for the two positions of the line of apsides is an advance. The results for the positions near the two considered will be similar, but less in amount up to some intermediate points, where the rotation of the line of apsides in a whole revolution of the moon will be zero. From the way in which the tangential components change sign (Fig. 58) it is evident that these points will be nearer to m_3 and m_7 than to m_1 and m_5; therefore *the average results for all possible positions of the perigee is an advance in the line of apsides.*

198. Secondary Effects. The results thus far have been derived as though the sun were stationary. It moves, however, in the same direction as the moon. It has been shown that when the moon is near apogee and the sun near the line of apsides, the normal component makes the apsides advance. This advance tends to preserve the relation of the orbit with reference to the position of the sun, and the advance of the apsides is prolonged and increased. On the other hand, when the moon is at perigee and the sun near the line of apsides the line of apsides moves backward; the sun moving one way and the line of apsides the other, this particular relation of the sun and the moon's orbit is quickly destroyed, and the retrogression is less than it would have been if the sun had remained stationary. In a similar manner, for every relative position of the line of apsides, the advance is increased and the retrogression is decreased.

There is another important secondary effect which depends upon the tangential component and is independent of the motion of the sun. As an example, take the case in which the line of apsides passes through the sun with the perigee at m_1. The tangential component in $m_3 m_5$ is positive, and, according to the Table, rotates the line of apsides forward until the moon arrives at apogee. But, as the line of apsides advances, the moon will arrive at apogee later, and the effect of this component will be increased. When the motion of the sun is also included this *secondary effect* becomes of still greater importance. In this manner, perturbation exaggerates perturbation, and it is clear what astronomers mean when they say that nearly half the motion of the lunar perigee is due to the square of the disturbing force, or that it is obtained in a second approximation.

The theoretical determination of the motion of the moon's line of apsides has been one of the most troublesome problems of Celestial Mechanics; the secondary effects escaped Newton when he wrote the *Principia*,* and were not explained until Clairaut developed his Lunar Theory in 1749. The most successful and masterful analysis of the subject yet made is undoubtedly that of G. W. Hill, in the *Acta Mathematica*, vol. VIII., which, for the terms treated, leaves nothing to be desired. The line of apsides of the moon's orbit makes a complete revolution in about $9\frac{1}{2}$ years.

199. Perturbations of the Eccentricity. Suppose the line of apsides passes through the sun and that the perigee is at m_1.

---

* In the manuscripts which Newton left, and which are now known as the Portsmouth Collection, having been published but recently, a correct explanation of the motion of the line of apsides is given, and nearly correct numerical results are obtained.

From the symmetry of the normal components with respect to the line ES and the results given in the Table, it follows that the increase and the decrease in the eccentricity in a complete revolution due to this component, are exactly equal under these circumstances. From the way in which the tangential component changes sign, and from the results given in the Table, it follows that the changes in the eccentricity, due to this component, also exactly balance. Therefore there is no change in the eccentricity in a complete revolution of the moon under the conditions postulated. In a similar manner the same results are reached when the perigee is at m_5.

Suppose the line of apsides has the direction $m_3 m_7$. It can be shown as before that neither the normal nor the tangential component makes any permanent change in the eccentricity.

Now consider the case in which the line of apsides is in some intermediate position; for simplicity suppose it is in the line $m_2 m_6$ with the perigee at m_2. Consider simultaneously with this case that in which the perigee is at m_6. First consider only the effects of the normal component. It follows from Fig. 57 and the Table of Art. 182 that when the perigee is at m_2 and the moon is in the region $m_2 m_4$, the normal component decreases the eccentricity; and when the perigee is at m_6, increases the eccentricity. The two effects largely destroy each other. But it was shown in Art. 181 that a given normal component is more effective in changing the eccentricity when the moon is near apogee than it is when the moon is correspondingly near perigee. Besides this, since N is proportional to r, as follows from the second equation of (18), the normal component is larger the greater the moon's distance. For both of these reasons, while the moon is in the arc $m_2 m_4$ the increase of the eccentricity with the perigee at m_6 is greater than the decrease with the perigee at m_2. The two cases combined give a small second order residual increase in the eccentricity which may be represented by $+ \Delta_1 e$. Similarly, while the moon is in the region $m_4 m_6$ the effects of the normal component on the eccentricity with the perigee at m_2 and m_6 are respectively an increase and a decrease. On paying heed to the relative positions of the apogee, it is seen that the combined effect on the eccentricity is a second order residual increase $+ \Delta_2 e$. By analogous discussions, the combined effects for the moon in the arcs $m_6 m_8$ and $m_8 m_2$ are the positive second order residuals $+ \Delta_3 e$ and $+ \Delta_4 e$.

The question arises whether the second order residuals are not in some way destroyed.   In order to show that they also vanish consider the case in which the line of apsides has a symmetrically opposite position with respect to the line ES, that is, the case in which the perigee is at m_8 or m_4.   When the perigee is at m_4 and the moon in the region $m_2 m_4$ the eccentricity is increased by the normal component; when the perigee is at m_8, the eccentricity is decreased.   The decrease in the latter case is greater than the increase in the former because when the perigee is at m_8 the region $m_2 m_4$ is near the apogee.   Therefore the combined effect is a second order decrease in the eccentricity; and, since the arc $m_2 m_4$ is not only situated the same relatively with respect to the earth and sun but also with respect to the moon's orbit as when the line of apsides was the line $m_2 m_6$, it follows that the second order decrease in the eccentricity is $- \Delta_1 e$.   It is found similarly that when the moon is in the arcs $m_4 m_6$, $m_6 m_8$, and $m_8 m_2$ the sums of the changes of the eccentricity when the perigee is at m_4 and m_8 are respectively $- \Delta_2 e$, $- \Delta_3 e$, and $- \Delta_4 e$.   When these second order residuals are added to those obtained when the line of apsides was the line $m_2 m_6$ the result is zero.   A corresponding discussion leads to the same results for any other position of the line of apsides, viz., it can be paired with another which is symmetrically opposite with respect to the line ES so that when the perigee is taken in both directions on each line the total effect of the normal component on the eccentricity is zero.   *Therefore the normal component in the long run makes no permanent change in the eccentricity of the moon's orbit;* and a somewhat similar discussion establishes the same result for the tangential component.

The sun does not, however, stand still while the moon makes its revolution, and the conditions which have been assumed are never exactly fulfilled.   Nevertheless, it is useful to show how the different configurations, even though changing from instant to instant, may be paired.   In a very great number of revolutions the supplementary configurations will have occurred an equal number of times, and the eccentricity will have returned to its original value.   The period required for this cycle of change depends in the first place upon the periods of the sun and the moon; in the second place, upon the eccentricity of the sun's orbit (the earth's orbit); and lastly, upon the manner in which the lines of apsides of the sun's and moon's orbits rotate.

The present system, with abundant geological and biological evidence of a very long existence for the earth in at least approxi-

mately its present condition, shows with reasonable certainty that
the system is nearly stable, if not quite. It is an interesting fact,
though, that those two elements, the line of nodes and the line of
apsides, which may change continually in one direction without
threatening the stability of the system do, on the average, respectively retrograde and advance forever.

200. The Evection. It has just been shown that the eccentricity
does not change in the long run; yet it undergoes periodic variations of considerable magnitude which give rise to the largest lunar
perturbation, known as the *evection*. At its maximum effect it
displaces the moon in geocentric longitude through an angle of
about 1° 15′ compared to its position in the undisturbed elliptic
orbit. This variation was discovered by Hipparchus and was
carefully observed by Ptolemy.

The perturbations of the elements, and of the eccentricity in
particular, depend upon two things, the position of the moon in
its orbit, and the position of the moon with respect to the earth
and sun. Suppose the moon and sun start in conjunction with
the perigee at m_1. Consider the motion throughout one synodic
revolution. It follows from the Table of Art. 182 and Figs. 57
and 58 that the eccentricity is not changing when the moon is at
m_1; that it is decreasing, or zero, when the moon is at m_2, m_3,
and m_4; that it is not changing when the moon is at m_5; that it is
increasing, or zero, when the moon is at m_6, m_7, and m_8; and that
it ceases to change when the moon has returned to m_1 again.
This is true only under the hypothesis that the perigee has remained at m_1 throughout the whole revolution; or, in other words,
that the line of apsides advances as fast as the sun moves in its
orbit. Now, the actual case is that the sun moves about 8.5
times as fast as the line of apsides rotates. Since the synodic
period of the moon is about 29.5 days while the sun moves about
one degree daily, the moon will be about 26° past its perigee when
it arrives at m_1. What modification in the conclusions does this
introduce? The normal component is negative and, in this part
of the orbit, causes an increase in the eccentricity, while the
tangential makes no change, since it is zero. As the moon proceeds past m_1 the normal component becomes less in numerical
value, while the tangential component becomes negative and tends
to decrease the eccentricity. The tendencies of the two components to change the eccentricity in opposite directions balance
when the moon is at some point between m_1 and m_2, instead of

at m_1, after which the eccentricity decreases. There is a corresponding advance of the point near m_5 at which the eccentricity ceases to decrease and begins to increase. Similar conclusions are reached starting from any other initial configuration.

The results may be summarized thus: The perturbations of the sun decrease the eccentricity of the moon's orbit somewhat more than half of a synodical revolution, and then increase it for an equal time. These changes in the eccentricity cause deviations in the geocentric longitude from the ones given by the elliptic theory, which constitute the *evection*. The appropriate methods show that the period of this inequality is about 31.8 days.

201. Gauss' Method of Computing Secular Variations. It has been shown in the preceding articles that some of the elements, such as the line of nodes and the line of apsides, vary in one direction without limit. This change is not at a uniform rate, for in addition to the general variations, there are many short period oscillations which are of such magnitude that the element frequently varies in the opposite direction. When the results are put into the symbols of analysis, the general average advance is represented by a term proportional to the time, called the *secular variation*, while the deviations from this uniform change are represented by a sum of periodic terms having various periods and phases. Thus it is seen that the secular variations are caused by a sort of average of the disturbing forces when the disturbing and disturbed bodies occupy every possible position with respect to each other.

There are other elements, such as the inclination and the eccentricity which, though periodic in the long run, vary continuously in one direction on the average for many thousands of years. These changes may be regarded as secular variations also, and they likewise result from a sort of average of perturbations.

In 1818 Gauss published a memoir upon the theory of secular variations based upon the conceptions just outlined. His method has been applied especially in the computation of the secular variations of the elements of the planetary orbits. Instead of considering the motions of the bodies, Gauss supposed that the mass of each planet is spread out in an elliptical ring coinciding with its orbit in such a manner that the density at each point is inversely as the velocity with which the body moves at that point. He then showed how to compute the attraction of one ring upon the other, and the rate at which their positions and shapes change under the influence of these forces.

The method of Gauss has been the subject of quite a number of memoirs. Probably the most useful for practical purposes is by G. W. Hill in vol. I. of the *Astronomical Papers of the American Ephemeris and Nautical Almanac.* Hill's formulas have been applied by Professor Eric Doolittle with great success, the results which he obtained agreeing very closely with those found by Leverrier and Newcomb by entirely different methods.

202. The Long Period Inequalities. In the theories of the mutual perturbations of the planets very large terms of long periods occur. They arise only when the periods of the two bodies considered are nearly commensurable, and it is easy to discover their cause from geometrical considerations.

Since the most important variation occurs in the mutual perturbations of Jupiter and Saturn the explanation will be adapted to that case. Five times the period of Jupiter is a little more than twice the period of Saturn. Suppose that the two planets are in conjunction at the origin of time on the line l_0. After five

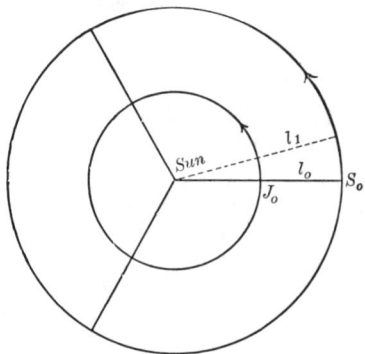

Fig. 59.

revolutions of Jupiter and two of Saturn they will be in conjunction again on a line l_1 very near l_0, but having a little greater longitude. This continues indefinitely, each conjunction occurring at a little greater longitude than the preceding. Conjunctions occurring frequently at about the same points in the orbits cause very large perturbations, and the *Long Period* is the time which it takes the point of conjunction to make a complete revolution. In the case of Jupiter and Saturn it is about 918 years. This inequality, which is the greatest in the longitudes of the planets, displacing Jupiter 21' and Saturn 49', long baffled astronomers in their attempts to explain it as a necessary consequence of the law of gravitation. Laplace finally made one of his many important contributions to Celestial Mechanics by pointing out its true cause, and showing that theory and observation agree.

XXIV. PROBLEMS.

1. Prove that the locus of the point at which the attractions of the sun and earth are equal is a sphere whose radius is $\dfrac{R\sqrt{SE}}{S-E}$, and whose center is on the line joining the sun and earth, at the distance $\dfrac{ER}{S-E}$ from the center of the earth opposite to the sun, where S and E represent the mass of the sun and earth respectively, and R the distance from the sun to the earth.

If $R = 93,000,000$ miles, and $\dfrac{S}{E} = 330,000$, then

$$\frac{R\sqrt{SE}}{S-E} = 161,550 \text{ miles,}$$

$$\frac{RE}{S-E} = 281 \text{ miles.}$$

Since the moon's orbit has a radius of about 240,000 miles, it is always attracted more by the sun than by the earth.

2. The moon may be regarded as revolving around the earth and disturbed by the sun, or as revolving around the sun and disturbed by the earth. Assume that the moon's orbit is a circle, and find the position at which the disturbing effects of the sun will be a maximum; show that the disturbing effects due to the earth, regarding the moon as revolving around the sun, are a minimum for the same position.

3. Find the ratio of the greatest disturbing effect of the sun to the least disturbing effect of the earth.

Ans. Let R equal the distance from the sun to the earth, ρ the distance from the sun to the moon, and r the distance from the earth to the moon; then

$$\frac{D_S}{D_E} = \frac{S}{E} \cdot \frac{r^2}{\rho^2} \cdot \frac{R^2 - \rho^2}{R^2 - r^2} = \frac{S}{E} \cdot \frac{r^3}{\rho^3} \cdot \frac{R + \rho}{R + r} = .0114.$$

4. Find the ratio of the sun's disturbing force at its maximum value to the attraction of the sun, and to the attraction of the earth.

Ans. $\dfrac{D_S}{A_S} = \dfrac{r(R + \rho)}{R^2} = .005,$ $\dfrac{D_S}{A_E} = \dfrac{S}{M} \dfrac{r^2(R + \rho)}{R^2\rho^2} = \mathbf{.011.}$

5. Prove in detail the conclusion of Art. 199 that the tangential component produces no secular changes in the eccentricity of the moon's orbit.

6. Suppose a planet disturbs the motion of another planet which is near to the sun. Find the way in which all the elements of the orbit of the inner planet are changed for all relative positions of the bodies in their orbits.

7. Show that, if the rates of change of the elements are known when the planet is in a particular position in its orbit, the intensity and direction of the disturbing force can be found. Show that, if it is assumed that the distance of the disturbing body from the sun is known, its direction and mass can be found. (This is part of the problem solved by Adams and Leverrier when they predicted the apparent position of Neptune from the knowledge of its perturbations of the motion of Uranus. There are troublesome practical difficulties which arise on account of the minuteness of the quantities involved which do not appear in the simple statement given here.)

HISTORICAL SKETCH AND BIBLIOGRAPHY.

The first treatment of the Problem of Three Bodies, as well as of Two Bodies, was due to Newton. It was given in Book I., Section XI., of the *Principia*, and it was said by Airy to be " the most valuable chapter that was ever written on physical science." It contained a somewhat complete explanation of the variation, the parallactic inequality, the annual equation, the motion of the perigee, the perturbations of the eccentricity, the revolution of the nodes, and the perturbations of the inclination. The value of the motion of the lunar perigee found by Newton from theory was only half that given by observations. In 1872, in certain of Newton's unpublished manuscripts, known as the Portsmouth Collection, it was found that Newton had accounted for the entire motion of the perigee by including perturbations of the second order. (See Art. 198.) This work being unknown to astronomers, the motion of the lunar perigee was not otherwise derived from theory until the year 1749, when Clairaut found the true explanation, after being on the point of sub-

stituting for Newton's law of attraction one of the form $\alpha = \frac{\mu}{r^2} + \frac{\nu}{r^3}$. Newton

regarded the Lunar Theory as being very difficult, and he is said to have told his friend Halley in despair that it " made his head ache and kept him awake so often that he would think of it no more."

Since the days of Newton the methods of Analysis have succeeded those of Geometry, except in elementary explanations of the causes of different sorts of perturbations. In the eighteenth century the development of the Lunar Theory, and of Celestial Mechanics in general, was almost entirely the work of five men: Euler (1707–1783), a Swiss, born at Basle, living at St. Petersburg from 1727 to 1747, at Berlin from 1747 to 1766, and at St. Petersburg from 1766 to 1783; Clairaut (1713–1765), born at Paris, and spending nearly all his life in his native city; D'Alembert (1717–1783), also a native and an inhabitant of Paris; Lagrange (1736–1813), born at Turin, Italy, but of French descent, Professor of Mathematics in a military school in Turin from 1753 to 1766, succeeding Euler at Berlin and spending twenty years there, going to Paris and spending the remainder of his life in the French capital; and Laplace (1749–1827), son of a French peasant of Beaumont, in Normandy, Professor in l'École Militaire and in l'École Normale in Paris, where he spent most of his life after he was eighteen years of age. The only part of their work which will be mentioned here will be that relating to the Lunar Theory. The account of investigations in the general planetary theories comes more properly in the next chapter.

There was a general demand for accurate lunar tables in the eighteenth century for the use of navigators in determining their positions at sea. This, together with the fact that the motions of the moon presented the best test of the Newtonian Theory, induced the English Government and a number of scientific societies to offer very substantial prizes for lunar tables agreeing with observations within certain narrow limits. Euler published some rather imperfect lunar tables in 1746. In 1747, Clairaut and d'Alembert presented to the Paris Academy on the same day memoirs on the Lunar Theory. Each had trouble in explaining the motion of the perigee. As has been stated, Clairaut found the source of the difficulty in 1749, and it was also discovered by both Euler and d'Alembert a little later. Clairaut won the prize offered by the St. Petersburg Academy in 1752 for his *Théorie de la Lune*. Both he and d'Alembert published theories and numerical tables in 1754. They were revised and extended later. Euler published a Lunar Theory in 1753, in the appendix of which the analytical method of the variation of the elements was partially worked out. Tobias Mayer (1723–1762), of Göttingen, compared Euler's tables with observations and corrected them so successfully that he and Euler were each granted a reward of £3000 by the English Government. In 1772 Euler published a second Lunar Theory which possessed many new features of great importance.

Lagrange did little in the Lunar Theory except to point out general methods. On the other hand, Laplace gave much attention to this subject, and made one of his important contributions to Celestial Mechanics in 1787, when he explained the cause of the secular acceleration of the moon's mean motion. He also proposed to determine the distance of the sun from the parallactic inequality. Laplace's theory is contained in the third volume of his *Mécanique Céleste*.

Damoiseau (1768–1846) carried out Laplace's method to a high degree of approximation in 1824–28, and the tables which he constructed were used quite generally until Hansen's tables were constructed in 1857. Plana (1781–1869) published a theory in 1832, similar in most respects to that of Laplace. An incomplete theory was worked out by Lubbock (1803–1865) in 1830–4. A great advance along new lines was made by Hansen (1795–1874) in 1838, and again in 1862–4. His tables published in 1857 were very generally adopted for Nautical Almanacs. De Pontécoulant (1795–1874) published his *Théorie Analytique du Système du Monde* in 1846. The fourth volume contains his Lunar Theory worked out in detail. It is in its essentials similar to that of Lubbock. A new theory of great mathematical elegance, and carried out to a very high degree of approximation, was published by Delaunay (1816–1872) in 1860 and 1867.

A most remarkable new theory based on new conceptions, and developed by new mathematical methods, was published by G. W. Hill in 1878 in the *American Journal of Mathematics*. The first fundamental idea was to take the variational orbit as an approximate solution instead of the ellipse. Expressions for the coordinates of the variational orbit were developed with rare mathematical skill, and are noteworthy for the rapidity of their convergence. A second approximation giving part of the motion of the perigee was published in volume VIII. of *Acta Mathematica*. This memoir contained the first solution of a linear differential equation having periodic coefficients, and introduced into mathematics the infinite determinant. Hill's researches have been extended to higher approximations, and completed, by a series of papers published by E. W. Brown in the *American Journal of Mathematics*, vols. XIV., XV., and XVII., and in the *Monthly Notices of the R.A.S.*, LII., LIV., and LV. As it now stands the work of Brown is numerically the most perfect Lunar Theory in existence, and from this point of view leaves little to be desired. The motion of the moon's nodes was found by Adams (1819–1892) by methods similar to those used by Hill in determining the motion of the perigee.

For the treatment of perturbations from geometrical considerations consult the *Principia*, Airy's (1801–1892) *Gravitation*, and Sir John Herschel's (1792–1871) *Outlines of Astronomy*. For the analytical treatment, aside from the original memoirs quoted, one cannot do better than to consult Tisserand's *Mécanique Céleste*, vol. III., and Brown's *Lunar Theory*. Both volumes are most excellent ones in both their contents and clearness of exposition. Brown's *Lunar Theory* especially is complete in those points, such as the meaning of the constants employed, which are apt to be somewhat obscure to one just entering this field.

CHAPTER X.

203. Introductory Remarks. The subject of the mutual perturbations of the motions of the heavenly bodies has been one to which many of the great mathematicians, from Newton's time on, have devoted a great deal of attention. It is needless to say that the problem is very difficult and that many methods of attacking it have been devised. Since the general solutions of the problem have not been obtained it has been necessary to treat special classes of perturbations by special methods. It has been found convenient to divide the cases which arise in the solar system into three general classes, (a) the Lunar Theory and satellite theories; (b) the mutual perturbations of the planets; and (c) the perturbations of comets by planets. The method which will be given in this chapter is applicable to the planetary theories, and it will be shown in the proper places why it is not applicable to the other cases. References were given in the last chapter to treatises on the Lunar Theory, especially to those of Tisserand and Brown. Some hints will be given in this chapter on the method of computing the perturbations of comets.

The chief difficulties which arise in getting an understanding of the theories of perturbations come from the large number of variables which it is necessary to use, and the very long transformations which must be made, in order to put the equations in a form suitable for numerical computations. It is not possible, because of the lack of space, to develop here in detail the explicit expressions adapted to computation; and, indeed, it is not desired to emphasize this part, for it is much more important to get an accurate understanding of the nature of the problem, the mathematical features of the methods employed, the limitations which are necessary, the exact places where approximations are introduced, if at all, and their character, the origin of the various sorts of terms, and the foundations upon which the celebrated theorems regarding the stability of the solar system rest.

There are two general methods of considering perturbations, (a) as the variations of the coördinates of the various bodies, and (b) as the variations of the elements of their orbits. These two conceptions were explained in the beginning of the preceding chapter. Their analytical development was begun by Euler and Clairaut and was carried to a high degree of perfection by Lagrange and Laplace. Yet there were points at which pure assumptions were made, it having become possible to establish completely the legitimacy of the proceedings, under the proper restrictions, only during the latter half of the nineteenth century by the aid of the work in pure Mathematics of Cauchy, Weierstrass, and Poincaré.

204. Illustrative Example. The mathematical basis for the theory of perturbations is often obscured by the large number of variables and the complicated formulas which must be used. Many of the essential features of the method of computing perturbations can be illustrated by simpler examples which are not subject to the complexities of many variables and involved formulas. One will be selected in which the physical relations are also simple.

Consider the solution of

$$(1) \qquad \frac{d^2x}{dt^2} + k^2x = -\mu \left(\frac{dx}{dt} \right)^3 + \nu \cos lt,$$

where k^2, μ, ν, and l are positive constants. If μ and ν were zero the differential equation would be that which defines simple harmonic motion. It arises in many physical problems, such as that of the simple pendulum, and of all classes of musical instruments. In order to make the interpretation definite, suppose it belongs to the problem of the vibrations of a tuning fork. The first term in the right member may be interpreted as being due to the resistance of the medium in which the tuning fork vibrates. It is not asserted, of course, that the resistance to the vibrations of a tuning fork varies as the third power of the velocity. An odd power is taken so that the differential equation will have the same form whether the motion is in the positive or negative direction, and the first power is not taken because then the differential equation would be linear and could be completely integrated in finite terms without any difficulty.

The left member of equation (1) will be considered as defining the undisturbed motion of the tuning fork. The first term on the right introduces a perturbation which depends upon the velocity of the tuning fork; the second term on the right introduces a perturbation which is independent of the position and velocity of the tuning fork. The first is analogous to the mutual perturbations of the planets, which depend upon their relative positions; the second is more of the nature of the forces which produce the tides, for they are exterior to the earth. The tides are defined by equations analogous to (1).

In order to have equation (1) in the form of the equations which arise in the theory of perturbations, let

$$(2) \qquad x = x_1, \qquad \frac{dx}{dt} = x_2.$$

Then (1) becomes

$$(3) \qquad \begin{cases} \dfrac{dx_1}{dt} - x_2 = 0, \\[2mm] \dfrac{dx_2}{dt} + k^2 x_1 = -\mu x_2{}^3 + \nu \cos lt. \end{cases}$$

The corresponding differential equations for undisturbed motion are

$$(4) \qquad \begin{cases} \dfrac{dx_1}{dt} - x_2 = 0, \\[2mm] \dfrac{dx_2}{dt} + k^2 x_1 = 0. \end{cases}$$

Equations (4) are easily integrated, and their general solution is

$$(5) \qquad \begin{cases} x_1 = +\alpha \cos kt + \beta \sin kt, \\ x_2 = -k\alpha \sin kt + k\beta \cos kt, \end{cases}$$

where α and β are the arbitrary constants of integration. In the terminology of Celestial Mechanics, α and β are the elements of the orbit of the tuning fork.

Now consider the problem of finding the solutions of equations (3). Physically speaking, the elements α and β must be so varied that the equations shall be satisfied for all values of t. Mathematically considered, equations (5) are relations between the original dependent variables x_1 and x_2, and the new dependent variables α and β which make it possible to transform the differential equations (3) from one set of variables to the other. This would be true whether (5) were solutions of (4) or not, but since (5) are solutions of (4) and (4) are a part of (3), a number of terms drop out after the transformation has been made. On regarding

(5) as a set of equations relating the variables x_1 and x_2 to α and β, and making direct substitution in (3), it is found that

$$(6) \quad \begin{cases} + \cos kt \dfrac{d\alpha}{dt} + \sin kt \dfrac{d\beta}{dt} = 0, \\[2mm] - \sin kt \dfrac{d\alpha}{dt} + \cos kt \dfrac{d\beta}{dt} = \mu k^2 [\alpha \sin kt - \beta \cos kt]^3 + \dfrac{\nu}{k} \cos lt. \end{cases}$$

These equations are linear in $\dfrac{d\alpha}{dt}$ and $\dfrac{d\beta}{dt}$ and can be solved for these derivatives because the determinant of their coefficients is unity. The solution is

$$(7) \quad \begin{cases} \dfrac{d\alpha}{dt} = - \mu k^2 [\alpha \sin kt - \beta \cos kt]^3 \sin kt - \dfrac{\nu}{k} \cos lt \sin kt, \\[2mm] \dfrac{d\beta}{dt} = + \mu k^2 [\alpha \sin kt - \beta \cos kt]^3 \cos kt + \dfrac{\nu}{k} \cos lt \cos kt. \end{cases}$$

The problem of solving (7) is as difficult as that of solving (3) because their right members involve the unknown quantities α and β in as complicated manner as x_1 and x_2 enter the right members of (3). But suppose μ and ν are very small; then, since they enter as factors in the right members of equations (7), the dependent variables α and β change very slowly. Consequently, for a considerable time they will be given with sufficient approximation if equations (7) are integrated regarding them as constants in the right members. To assist in seeing this mathematically consider the simpler equation

$$(8) \qquad \frac{d\alpha}{dt} = \mu\alpha(1 + k \cos kt).$$

The solution of this equation is

$$\alpha = C e^{\mu(t + \sin kt)},$$

where C is the constant of integration. If the right member of this equation is expanded, the expression for α becomes

$$(9) \quad \alpha = C \left[1 + \mu(t + \sin kt) + \frac{\mu^2}{2} (t + \sin kt)^2 + \cdots \right].$$

If μ is very small and if t is not too great the right member of this equation is nearly equal to its first two terms. If it were not for the term t which is not in the trigonometric function no limitations on t would be necessary. But in general such limitations are necessary; and in most cases, though not in the present one, they are necessary in order to secure convergence of the series.

It is observed that the solution (9) is in reality a power series in the parameter μ, and the coefficients involve t. If it is desired equation (8) can be integrated directly as a power series in μ. The process is, in fact, a general one which can be used in solving (7), and equations (10), which follow, are the first terms of the solution. The conditions of validity of this method of integration are given in Art. 207.

The fact that when μ is very small α and β may be regarded as constants in the right members of (7) for not too great values of t can be seen from a physical illustration. Consider the perturbation theory. The changes in the elements of an orbit depend upon the elements of the orbits of the mutually disturbing bodies and upon the relative positions of the bodies in their orbits. It is intuitionally clear that only a slight error in the computation of the mutual disturbances of two planets would be committed if constant elements were used which differ a little, say a degree in the case of angular elements, from the true slowly changing ones.

If equations (7) are integrated regarding α and β as constants in the right members, it is found that

$$
(10)
\begin{cases}
\begin{aligned}
\alpha = \alpha_0 - \mu k^2 &\left\{ \frac{3\alpha}{8}(\alpha^2 + \beta^2)t + \frac{\beta}{8k}(3\alpha^2 + \beta^2)[\cos 2kt - 1] \right. \\
&\quad - \frac{\beta}{32k}(3\alpha^2 - \beta^2)[\cos 4kt - 1] \\
&\quad \left. - \frac{\alpha^3}{4k}\sin 2kt + \frac{\alpha}{32k}(\alpha^2 - 3\beta^2)\sin 4kt \right\} \\
&\quad + \frac{\nu}{2k(l+k)}[\cos (l+k)t - 1] \\
&\quad - \frac{\nu}{2k(l-k)}[\cos (l-k)t - 1], \\
\beta = \beta_0 + \mu k^2 &\left\{ -\frac{3\beta}{8}(\alpha^2 + \beta^2)t - \frac{\alpha}{8k}(\alpha^2 + 3\beta^2)[\cos 2kt - 1] \right. \\
&\quad + \frac{\alpha}{32k}(\alpha^2 - 3\beta^2)[\cos 4kt - 1] \\
&\quad \left. - \frac{\beta^3}{4k}\sin 2kt + \frac{\beta}{32k}(3\alpha^2 - \beta^2)\sin 4kt \right\} \\
&\quad + \frac{\nu}{2k(l+k)}\sin (l+k)t \\
&\quad + \frac{\nu}{2k(l-k)}\sin (l-k)t,
\end{aligned}
\end{cases}
$$

where α_0 and β_0 are the values of α and β respectively at $t = 0$. When these values of α and β are substituted in (5) the values of x_1 and x_2 are determined approximately for all values of t which are not too remote from the initial time.

Consider equations (10). The right member of each of them has a term which contains t only as a simple factor, while elsewhere t appears only in the sine and cosine terms. The terms which are proportional to t seem to indicate that α and β increase or decrease indefinitely with the time; but it must be remembered that equations (10) are only approximate expressions for α and β, which are useful only for a limited time. It might be that the rigorous expressions would contain higher powers of t, and that the sums would have bounded values, just as

$$\sin t = t - \frac{t^3}{3!} + \frac{t^5}{5!} - \cdots$$

is an expression whose numerical value does not exceed unity, though a consideration of the first term alone would lead to the conclusion that it becomes indefinitely great with t. On the other hand the presence of terms which increase proportionally to the time may indicate an actual indefinite increase of the elements α and β. For example, it was found in the preceding chapter that the line of nodes and the apsides of the moon's orbit respectively regress and advance continually. The terms which change proportionally to t are called *secular terms*.

The right members of equations (10) also contain periodic terms having the periods $\frac{\pi}{k}$, $\frac{\pi}{2k}$, $\frac{2\pi}{l+k}$, and $\frac{2\pi}{l-k}$. These are known as *periodic terms*. If l and k are nearly equal the terms which involve sines or cosines of $(l-k)t$ have very long periods, and are called *long period terms*. Sometimes terms arise which are the products of t and periodic terms. These mixed terms are called *Poisson terms* because they were encountered by Poisson in the discussion of the variations of the major axes of the planetary orbits. If (10) are substituted in (5) the resulting expressions for x_1 and x_2 contain Poisson terms but no secular terms.

The physical interpretation of equations (10) is simple. The elements α and β continually decrease because of the secular terms; that is, the amplitudes of the oscillations indicated in (5) continually diminish. This reduction is entirely due to the resistance to the motion as is shown by the fact that these terms contain the coefficient μ as a factor. There are terms in x_1 and x_2 of period three times and five times the undisturbed period which are also due to the resistance. And the periodic disturbing force introduces in α and β terms whose periods depend both on the period of the disturbing force and also on the natural period of the tuning fork. But it is noticed that the periods of the terms which they introduce into the expressions for x_1 and x_2 are the period of the disturbing force and the natural period of the tuning fork.

205. Equations in the Problem of Three Bodies. Consider the motion of two planets, m_1 and m_2, with respect to the sun, S. Take the center of the sun as origin and let the coördinates of m_1 be (x_1, y_1, z_1), and of m_2, (x_2, y_2, z_2). Let the distances of m_1 and m_2 from the sun be r_1 and r_2 respectively, and let $r_{1,2}$ represent the distance from m_1 to m_2. Then the differential equations of motion, as derived in Art. 148, are

$$(11) \begin{cases} \dfrac{d^2x_1}{dt^2} + k^2(S + m_1)\dfrac{x_1}{r_1^3} = m_2\dfrac{\partial R_{1,2}}{\partial x_1}, \\[2mm] \dfrac{d^2y_1}{dt^2} + k^2(S + m_1)\dfrac{y_1}{r_1^3} = m_2\dfrac{\partial R_{1,2}}{\partial y_1}, \\[2mm] \dfrac{d^2z_1}{dt^2} + k^2(S + m_1)\dfrac{z_1}{r_1^3} = m_2\dfrac{\partial R_{1,2}}{\partial z_1}; \\[2mm] \dfrac{d^2x_2}{dt^2} + k^2(S + m_2)\dfrac{x_2}{r_2^3} = m_1\dfrac{\partial R_{2,1}}{\partial x_2}, \\[2mm] \dfrac{d^2y_2}{dt^2} + k^2(S + m_2)\dfrac{y_2}{r_2^3} = m_1\dfrac{\partial R_{2,1}}{\partial y_2}, \\[2mm] \dfrac{d^2z_2}{dt^2} + k^2(S + m_2)\dfrac{z_2}{r_2^3} = m_1\dfrac{\partial R_{2,1}}{\partial z_2}, \\[2mm] R_{1,2} = k^2\left[\dfrac{1}{r_{1,2}} - \dfrac{x_1x_2 + y_1y_2 + z_1z_2}{r_2^3}\right], \\[2mm] R_{2,1} = k^2\left[\dfrac{1}{r_{1,2}} - \dfrac{x_2x_1 + y_2y_1 + z_2z_1}{r_1^3}\right]. \end{cases}$$

The right members of equations (11) are multiplied by the factors m_1 and m_2 which are very small compared to S; therefore they will be of slight importance in comparison with the terms on the left which come from the attraction of the sun, at least for a considerable time. If m_1 and m_2 are put equal to zero in the right members, the first three equations and the second three form two sets which are independent of each other, and the problem for each set of three equations reduces to that of two bodies, and can be completely solved.

It will be advantageous to reduce the six equations (11) of the second order to twelve of the first order. Let

$$x' = \frac{dx}{dt}, \qquad y' = \frac{dy}{dt}, \qquad z' = \frac{dz}{dt};$$

then equations (11) become

$$(12) \quad \begin{cases} \dfrac{dx_1}{dt} - x_1' = 0, & \dfrac{dx_1'}{dt} + k^2(S + m_1)\dfrac{x_1}{r_1{}^3} = m_2\dfrac{\partial R_{1,\,2}}{\partial x_1}, \\[2ex] \dfrac{dy_1}{dt} - y_1' = 0, & \dfrac{dy_1'}{dt} + k^2(S + m_1)\dfrac{y_1}{r_1{}^3} = m_2\dfrac{\partial R_{1,\,2}}{\partial y_1}, \\[2ex] \dfrac{dz_1}{dt} - z_1' = 0, & \dfrac{dz_1'}{dt} + k^2(S + m_1)\dfrac{z_1}{r_1{}^3} = m_2\dfrac{\partial R_{1,\,2}}{\partial z_1}, \end{cases}$$

and similar equations in which the subscript is 2.

If the motions of m_1 and m_2 were not disturbed by each other equations (12) would become

$$(13) \quad \begin{cases} \dfrac{dx_1}{dt} - x_1' = 0, & \dfrac{dx_1'}{dt} + k^2(S + m_1)\dfrac{x_1}{r_1{}^3} = 0, \\[2ex] \dfrac{dy_1}{dt} - y_1' = 0, & \dfrac{dy_1'}{dt} + k^2(S + m_1)\dfrac{y_1}{r_1{}^3} = 0, \\[2ex] \dfrac{dz_1}{dt} - z_1' = 0, & \dfrac{dz_1'}{dt} + k^2(S + m_1)\dfrac{z_1}{r_1{}^3} = 0, \end{cases}$$

and an independent system of similar equations in which the subscript is 2. Let $\Omega_1 = \frac{1}{2}(x_1'^2 + y_1'^2 + z_1'^2) - k^2\dfrac{(S + m_1)}{r_1}$; then equations (13) take the form

$$(14) \quad \begin{cases} \dfrac{dx_1}{dt} = \dfrac{\partial \Omega_1}{\partial x_1'}, & \dfrac{dx_1'}{dt} = -\dfrac{\partial \Omega_1}{\partial x_1}, \\[2ex] \dfrac{dy_1}{dt} = \dfrac{\partial \Omega_1}{\partial y_1'}, & \dfrac{dy_1'}{dt} = -\dfrac{\partial \Omega_1}{\partial y_1}, \\[2ex] \dfrac{dz_1}{dt} = \dfrac{\partial \Omega_1}{\partial z_1'}, & \dfrac{dz_1'}{dt} = -\dfrac{\partial \Omega_1}{\partial z_1}. \end{cases}$$

This form of the differential equations is convenient in connection with the problem of transforming equations so that the elliptic elements become the dependent variables whose values in terms of t are required.

206. Transformation of Variables. In order to avoid confusion in the analysis, and to be able to say where and how the approximations are introduced, the method of the variation of parameters must be regarded in the first instance as simply a transformation of variables, which is perfectly legitimate for all values of the time for which the equations of transformation are valid. From this point of view the whole process is mathematically simple and lucid, the only trouble arising from the number of variables involved and the complicated relations among them.

In chapter v. it was shown how to express the coördinates in the Problem of Two Bodies in terms of the elements and the time. Let $\alpha_1, \cdots, \alpha_6$ represent the elements of the orbit m_1, and $\beta_1, \cdots, \beta_6$ those of m_2. Then the equations for the coördinates in the Problem of Two Bodies may be written

$$(15) \quad \begin{cases} x_1 = f(\alpha_1, \cdots, \alpha_6, t), & x_1' = \theta(\alpha_1, \cdots, \alpha_6, t), \\ y_1 = g(\alpha_1, \cdots, \alpha_6, t), & y_1' = \phi(\alpha_1, \cdots, \alpha_6, t), \\ z_1 = h(\alpha_1, \cdots, \alpha_6, t), & z_1' = \psi(\alpha_1, \cdots, \alpha_6, t), \\ x_2 = f(\beta_1, \cdots, \beta_6, t), & x_2' = \theta(\beta_1, \cdots, \beta_6, t), \\ y_2 = g(\beta_1, \cdots, \beta_6, t), & y_2' = \phi(\beta_1, \cdots, \beta_6, t), \\ z_2 = h(\beta_1, \cdots, \beta_6, t), & z_2' = \psi(\beta_1, \cdots, \beta_6, t). \end{cases}$$

A transformation of variables in equations (12) will now be made. Let it be forgotten for the moment that equations (15) are the solutions of the Problem of Two Bodies, and that the α_i and β_i are the elements of the two orbits; but let (15) be considered as being the equations which transform equations (12) in the old variables, $x_1, y_1, z_1, x_1', y_1', z_1', x_2, y_2, z_2, x_2', y_2', z_2'$, into an equivalent system in the new variables, $\alpha_1, \cdots, \alpha_6, \beta_1, \cdots, \beta_6$. The transformations are effected by computing the derivatives occurring in (12) and making direct substitutions. The derivatives of equations (15) with respect to t are

$$(16) \quad \begin{cases} \dfrac{dx_1}{dt} = \dfrac{\partial x_1}{\partial t} + \sum_{i=1}^{6} \dfrac{\partial x_1}{\partial \alpha_i} \dfrac{d\alpha_i}{dt}, \\ \cdot \quad \cdot \quad \cdot \quad \cdot \quad \cdot \quad \cdot \quad \cdot \\ \dfrac{dx_1'}{dt.} = \dfrac{\partial x_1'}{\partial t} + \sum_{i=1}^{6} \dfrac{\partial x_1'}{\partial \alpha_i} \dfrac{d\alpha_i}{dt}, \\ \cdot \quad \cdot \quad \cdot \quad \cdot \quad \cdot \quad \cdot \quad \cdot \end{cases}$$

The direct substitution of (16) in (12) gives

$$
(17) \begin{cases}
\dfrac{\partial x_1}{\partial t} - x_1' + \sum_{i=1}^{6} \dfrac{\partial x_1}{\partial \alpha_i} \dfrac{d\alpha_i}{dt} = 0, \\[2mm]
\dfrac{\partial y_1}{\partial t} - y_1' + \sum_{i=1}^{6} \dfrac{\partial y_1}{\partial \alpha_i} \dfrac{d\alpha_i}{dt} = 0, \\[2mm]
\dfrac{\partial z_1}{\partial t} - z_1' + \sum_{i=1}^{6} \dfrac{\partial z_1}{\partial \alpha_i} \dfrac{d\alpha_i}{dt} = 0, \\[2mm]
\dfrac{\partial x_1'}{\partial t} + k^2(S + m_1)\dfrac{x_1}{r_1^3} + \sum_{i=1}^{6} \dfrac{\partial x_1'}{\partial \alpha_i} \dfrac{d\alpha_i}{dt} = m_2 \dfrac{\partial R_{1,2}}{\partial x_1}, \\[2mm]
\dfrac{\partial y_1'}{\partial t} + k^2(S + m_1)\dfrac{y_1}{r_1^3} + \sum_{i=1}^{6} \dfrac{\partial y_1'}{\partial \alpha_i} \dfrac{d\alpha_i}{dt} = m_2 \dfrac{\partial R_{1,2}}{\partial y_1}, \\[2mm]
\dfrac{\partial z_1'}{\partial t} + k^2(S + m_1)\dfrac{z_1}{r_1^3} + \sum_{i=1}^{6} \dfrac{\partial z_1'}{\partial \alpha_i} \dfrac{d\alpha_i}{dt} = m_2 \dfrac{\partial R_{1,2}}{\partial z_1},
\end{cases}
$$

and similar equations in $x_2, \cdots, z_2'$, and $\beta_1, \cdots, \beta_6$. These equations are linear in the derivatives $\dfrac{d\alpha_i}{dt}$ and can be solved for them, expressing them in terms of $\alpha_1, \cdots, \alpha_6, \beta_1, \cdots, \beta_6$, and t, provided the determinant of their coefficients is distinct from zero.

But if equations (15) are the solution of the problem of undisturbed elliptic motion equations (17) are greatly simplified, for it is seen from (13) that, when $\alpha_1, \cdots, \alpha_6$ are constant, $\dfrac{dx_1}{dt} - x_1' = 0$ for all values of t. The partial derivative $\dfrac{\partial x_1}{\partial t}$, when $\alpha_1, \cdots, \alpha_6$ are regarded as variables, is identical with $\dfrac{dx_1}{dt}$ when they are regarded as constants. Therefore $\dfrac{\partial x_1}{\partial t} - x_1' \equiv 0$; and similarly $\dfrac{\partial x_1'}{\partial t} + k^2(S + m_1)\dfrac{x_1}{r_1^3} \equiv 0$, and similar equations in y and z. As a consequence of these relations equations (17) reduce to

$$
(18) \begin{cases}
\displaystyle\sum_{i=1}^{6} \dfrac{\partial x_1}{\partial \alpha_i} \dfrac{d\alpha_i}{dt} = 0, \qquad \sum_{i=1}^{6} \dfrac{\partial x_1'}{\partial \alpha_i} \dfrac{d\alpha_i}{dt} = m_2 \dfrac{\partial R_{1,2}}{\partial x_1}, \\[2mm]
\displaystyle\sum_{i=1}^{6} \dfrac{\partial y_1}{\partial \alpha_i} \dfrac{d\alpha_i}{dt} = 0, \qquad \sum_{i=1}^{6} \dfrac{\partial y_1'}{\partial \alpha_i} \dfrac{d\alpha}{dt} = m_2 \dfrac{\partial R_{1,2}}{\partial y_1}, \\[2mm]
\displaystyle\sum_{i=1}^{6} \dfrac{\partial z_1}{\partial \alpha_i} \dfrac{d\alpha_i}{dt} = 0, \qquad \sum_{i=1}^{6} \dfrac{\partial z_1'}{\partial \alpha_i} \dfrac{d\alpha_i}{dt} = m_2 \dfrac{\partial R_{1,2}}{\partial z_1},
\end{cases}
$$

and similar equations in the β_i. These equations are linear in the derivatives $\dfrac{d\alpha_i}{dt}$ and can be solved for them unless the determinant of their coefficients is zero. But the determinant of the linear system (18) is the Jacobian of the first set of equations (15) with respect to $\alpha_1, \cdots, \alpha_6$, and cannot vanish if these functions are independent and give a simple and unique determination of the elements.* These functions are independent, and in general they give simple and unique values for the elements since they are the expressions for the coördinates in the Problem of Two Bodies. The problem of determining the elements from the values of the coördinates and components of velocity was solved in chap. v.

If $m_2 = 0$ equations (18) are linear and homogeneous, and since the determinant is not zero they can be satisfied only by $\dfrac{d\alpha_i}{dt} = 0$, $(i = 1, \cdots, 6)$. That is, the elements are constants, which, of course, is nothing new.

On solving equations (18), it is found that

$$(19) \quad \begin{cases} \dfrac{d\alpha_i}{dt} = m_2\,\phi_i(\alpha_1, \cdots, \alpha_6;\ \beta_1, \cdots, \beta_6;\ t), & (i = 1, \cdots, 6), \\[2ex] \dfrac{d\beta_i}{dt} = m_1\,\psi_i(\alpha_1, \cdots, \alpha_6;\ \beta_1, \cdots, \beta_6;\ t), & (i = 1, \cdots, 6). \end{cases}$$

It will be remembered that in determining the coördinates in the Problem of Two Bodies the first step, viz., the computation of the mean anomaly, involved the mean motion, defined by the equation

$$n_j = \frac{k\sqrt{S + m_j}}{a_j^{\frac{3}{2}}}, \qquad (j = 1, 2).$$

Since the n_j involve the masses of the planets the right members of (15), and consequently of (19), involve m_1 and m_2 implicitly.

In order to justify mathematically the precise method of integrating equations (19) which is employed by astronomers, some remarks are necessary upon m_1 and m_2. In those places where they occur implicitly in the functions φ_i and ψ_i they will be regarded as fixed numbers; as they appear as factors of the ψ_i and φ_i respectively they will be regarded as parameters in powers of which the solutions may be expanded. Such a generalization of parameters is clearly permissible because, if a function involves a parameter in two different ways, there is no reason why it may

---

* See Baltzer's *Determinanten*, p. 141.

not be expanded with respect to the parameter so far as it is involved in one way and not with respect to it as it is involved in the other. If the function, instead of being given explicitly, is defined by a set of differential equations the same things regarding the expansions in terms of parameters are true. If the attractions of bodies depended on something besides their masses (measured by their inertias) and their distances, as for example, on their rates of rotation or temperatures, then m_1 and m_2 so far as they enter in the φ_i and ψ_i implicitly through n_1 and n_2, where they would be defined numerically by their individual mutual attractions for the sun, would be different from their values where they occur as factors of the φ_i and ψ_i, for in the latter places they would be defined by their attractions for each other.

Hence, the values of the masses m_1 and m_2 entering implicitly in equations (15) and (19) are treated as fixed numbers, given in advance, and do not need to be retained explicitly; on the other hand, the m_1 and m_2 which are factors of the perturbing terms of the equations are retained explicitly, being supposed capable of taking any values not exceeding certain limits.

207. Method of Solution. Equations (11) are the general differential equations of motion for the Problem of Three Bodies. Equations (12) are equally general. No approximations were introduced in making the transformation of variables by (15); therefore equations (19) are general and rigorous. The difference is that if (19) were integrated the elements would be found instead of the coordinates as in (11), but as the latter can always be found from the former this must be regarded as the solution of the problem.

Instead of interrupting the course of mathematical reasoning by working out the explicit forms of (19), it will be preferable to show first by what methods they are solved. Explicit mention will be made at the appropriate times of all points at which assumptions or approximations are made.

When m_1 and m_2 are very small compared to S, as they are in the solar system, the orbits are very nearly fixed ellipses, and therefore α_i and β_i change very slowly. Consequently if they were regarded as constants in the right members of (19) and the equations integrated, approximate values of the α_i and the β_i would be obtained for values of t not too remote from the initial time. This is the method adopted in the illustrative example of the preceding article, and has been the point of view often taken by astronomers, especially in the pioneer days of Celestial Mechanics. But any theory which is only approximate, even though it is numerically adequate, does not measure up to the ideals of science.

Equations (19) are of the type which Cauchy and Poincaré have shown can be integrated as power series in m_1 and m_2. Cauchy proved that m_1, m_2, and t can all be taken so small that the series converge. Poincaré proved the more general theorem* that if the orbits in which the bodies are instantaneously moving at the initial time do not intersect, then for any finite range of values of t, the m_1 and m_2 can be taken so small that the solutions converge for every value of t in the interval. However, the masses cannot be chosen arbitrarily small but are given by Nature. Hence the practical importance of the additional theorem that, whatever the values of m_1 and m_2, there exists a range for t so restricted that the solutions of equations (19) as power series in the parameters m_1 and m_2 converge for every value of t in the range. In general, the larger the values of the parameters the more restricted the range. This is, of course, a special case of a general theorem respecting the expansion of solutions of differential equations of the type to which (19) belong as power series in parameters.†

It follows from the last theorem quoted that, if the range of t is not too great, the solutions of equations (19) can be expressed in convergent power series in m_1 and m_2, of the form

$$(20) \quad \begin{cases} \alpha_i = \sum_{j=0}^{\infty} \sum_{k=0}^{\infty} \alpha_i^{(j,\,k)} m_1{}^j m_2{}^k, \\ \beta_i = \sum_{j=0}^{\infty} \sum_{k=0}^{\infty} \beta_i^{(j,\,k)} m_1{}^j m_2{}^k, \end{cases}$$

where the superfixes on the α_i and β_i simply indicate the order of the coefficient. The $\alpha_i^{(j,\,k)}$ and $\beta_i^{(j,\,k)}$ are functions of the time which are to be determined. It has been customary in the theory of perturbations to assume without proof that this expansion is valid for any desired length of time. As has been stated, it can be proved that it is valid for a sufficiently small interval of time; but as the method of demonstration gives only a limit within which the series certainly converge, and not the longest time

---

* *Les Méthodes Nouvelles de la Mécanique Céleste*, vol. I., p. 58.

† See Picard's *Traité d'Analyse*, vol. II., chap. XI., and vol. III.

during which they converge, and as the limit is almost certainly far too small, it has never been computed. It is to be understood, therefore, that the method which is just to be explained, is valid for a certain interval of time, which in the planetary theories is doubtless several hundreds of years.

On substituting (20) in (19) and developing with respect to m_1 and m_2, it is found that

(21)
$$
\begin{aligned}
&\frac{d\alpha_i^{(0,\,0)}}{dt} + \frac{d\alpha_i^{(0,\,1)}}{dt}m_2 + \frac{d\alpha_i^{(1,\,0)}}{dt}m_1 + \frac{d\alpha_i^{(1,\,1)}}{dt}m_1 m_2 \\
&\qquad + \frac{d\alpha_i^{(0,\,2)}}{dt}m_2{}^2 + \frac{d\alpha_i^{(2,\,0)}}{dt}m_1{}^2 + \cdots \\[2mm]
&\quad = m_2 \phi_i(\alpha_1^{(0,\,0)}, \cdots, \alpha_6^{(0,\,0)}; \beta_1^{(0,\,0)}, \cdots, \beta_6^{(0,\,0)}; t) \\[2mm]
&\qquad + m_2 \sum_{j=1}^{6} \frac{\partial \phi_i}{\partial \alpha_j}(\alpha_j^{(0,\,1)}m_2 + \alpha_j^{(1,\,0)}m_1) \\[2mm]
&\qquad + m_2 \sum_{j=1}^{6} \frac{\partial \phi_i}{\partial \beta_j}(\beta_j^{(0,\,1)}m_2 + \beta_j^{(1,\,0)}m_1) \\[2mm]
&\qquad + \text{higher powers in } m_1 \text{ and } m_2, \\[3mm]
&\frac{d\beta_i^{(0,\,0)}}{dt} + \frac{d\beta_i^{(0,\,1)}}{dt}m_2 + \frac{d\beta_i^{(1,\,0)}}{dt}m_1 + \frac{d\beta_i^{(1,\,1)}}{dt}m_1 m_2 \\
&\qquad + \frac{d\beta_i^{(0,\,2)}}{dt}m_2{}^2 + \frac{d\beta_i^{(2,\,0)}}{dt}m_1{}^2 + \cdots \\[2mm]
&\quad = m_1 \psi_i(\alpha_1^{(0,\,0)}, \cdots, \alpha_6^{(0,\,0)}; \beta_1^{(0,\,0)}, \cdots, \beta_6^{(0,\,0)}; t) \\[2mm]
&\qquad + m_1 \sum_{j=1}^{6} \frac{\partial \psi_i}{\partial \alpha_j}(\alpha_j^{(0,\,1)}m_2 + \alpha_j^{(1,\,0)}m_1) \\[2mm]
&\qquad + m_1 \sum_{j=1}^{6} \frac{\partial \psi_i}{\partial \beta_j}(\beta_j^{(0,\,1)}m_2 + \beta_j^{(1,\,0)}m_1) \\[2mm]
&\qquad + \text{higher powers in } m_1 \text{ and } m_2, \quad (i = 1, \cdots, 6).
\end{aligned}
$$

In the partial derivatives it is to be understood that α_i and β_i are replaced by $\alpha_i^{(0,\,0)}$ and $\beta_i^{(0,\,0)}$ respectively. If m_1 and m_2 were not regarded as fixed numbers in the left members of equations (11), ϕ_i, ψ_i, $\dfrac{\partial \phi_i}{\partial \alpha_j}$, $\dfrac{\partial \phi_i}{\partial \beta_j}$, etc., would have to be developed as power series in m_1 and m_2, thus adding greatly to the complexity of the work.

Within the limits of convergence the coefficients of like powers of m_1 and m_2 on the two sides of the equations are equal. Hence, on equating them, it follows that

$$(22) \quad \begin{cases} \dfrac{d\alpha_i^{(0,\,0)}}{dt} = 0, \qquad (i = 1,\, \cdots,\, 6), \\[2mm] \dfrac{d\beta_i^{(0,\,0)}}{dt} = 0. \end{cases}$$

$$(23) \quad \begin{cases} \dfrac{d\alpha_i^{(0,\,1)}}{dt} = \phi_i(\alpha_1^{(0,\,0)},\, \cdots,\, \alpha_6^{(0,\,0)};\, \beta_1^{(0,\,0)},\, \cdots,\, \beta_6^{(0,\,0)};\, t), \\[2mm] \dfrac{d\alpha_i^{(1,\,0)}}{dt} = 0, \\[2mm] \dfrac{d\beta_i^{(0,\,1)}}{dt} = 0, \\[2mm] \dfrac{d\beta_i^{(1,\,0)}}{dt} = \psi_i(\alpha_1^{(0,\,0)},\, \cdots,\, \alpha_6^{(0,\,0)};\, \beta_1^{(0,\,0)},\, \cdots,\, \beta_6^{(0,\,0)};\, t). \end{cases}$$

$$(24) \quad \begin{cases} \dfrac{d\alpha_i^{(1,\,1)}}{dt} = \sum_{j=1}^{6} \dfrac{\partial \phi_i}{\partial \alpha_j} \alpha_j^{(1,\,0)} + \sum_{j=1}^{6} \dfrac{\partial \phi_i}{\partial \beta_j} \beta_j^{(1,\,0)}, \\[2mm] \dfrac{d\alpha_i^{(0,\,2)}}{dt} = \sum_{j=1}^{6} \dfrac{\partial \phi_i}{\partial \alpha_j} \alpha_j^{(0,\,1)} + \sum_{j=1}^{6} \dfrac{\partial \phi_i}{\partial \beta_j} \beta_j^{(0,\,1)}, \\[2mm] \dfrac{d\alpha_i^{(2,\,0)}}{dt} = 0, \\[2mm] \dfrac{d\beta_i^{(1,\,1)}}{dt} = \sum_{j=1}^{6} \dfrac{\partial \psi_i}{\partial \alpha_j} \alpha_j^{(0,\,1)} + \sum_{j=1}^{6} \dfrac{\partial \psi_i}{\partial \beta_j} \beta_j^{(0,\,1)}, \\[2mm] \dfrac{d\beta_i^{(2,\,0)}}{dt} = \sum_{j=1}^{6} \dfrac{\partial \psi_i}{\partial \alpha_j} \alpha_j^{(1,\,0)} + \sum_{j=1}^{6} \dfrac{\partial \psi_i}{\partial \beta_j} \beta_j^{(1,\,0)}, \\[2mm] \dfrac{d\beta_i^{(0,\,2)}}{dt} = 0, \end{cases}$$

$$\cdot \quad \cdot \quad \cdot \quad \cdot \quad \cdot \quad \cdot \quad \cdot \quad \cdot \quad \cdot \quad \cdot \quad \cdot \quad \cdot \quad \cdot \quad \cdot \quad \cdot$$

On integrating equations (22) and substituting the values of $\alpha_i^{(0,\,0)}$ and $\beta_i^{(0,\,0)}$ thus obtained in (23), the latter are reduced to quadratures and can be integrated; on integrating (23) and substituting the expressions for $\alpha_i^{(0,\,1)}$, $\alpha_i^{(1,\,0)}$, $\beta_i^{(0,\,1)}$, $\beta_i^{(1,\,0)}$ in (24), the latter are reduced to quadratures and can be integrated; and this process can be continued indefinitely. In this manner the coefficients of the series (20) can be determined, and the values of α_i and β_i can be found to any desired degree of precision for values of the time for which the series converge.

208. Determination of the Constants of Integration. A new constant of integration is introduced when equations (22), (23), $\cdots$ are integrated for each $\alpha_i{}^{(j,\,k)}$, $\beta_i{}^{(j,\,k)}$. These constants will now be determined.

Let the constant which is introduced with the $\alpha_i{}^{(j,\,k)}$ be denoted by $-a_i{}^{(j,\,k)}$ and with the $\beta_i{}^{(j,\,k)}$, by $-b_i{}^{(j,\,k)}$. Since the first set of differential equations have m_2 as a factor in their right members, while the second set have m_1 as a factor, it follows that

$$\alpha_i{}^{(j,\,0)} = a_i{}^{(j,\,0)}, \qquad (j = 0, \,\cdots\infty),$$
$$\beta_i{}^{(0,\,k)} = b_i{}^{(0,\,k)}, \qquad (k = 0, \,\cdots\infty).$$

Since the $\alpha_i{}^{(j,\,k)}$ and $\beta_i{}^{(j,\,k)}$ are defined by quadratures all the constants of integration are simply added to functions of t. That is, the $\alpha_i{}^{(j,\,k)}$ and $\beta_i{}^{(j,\,k)}$ have the form

$$\begin{cases} \alpha_i{}^{(j,\,k)} = f_i{}^{(j,\,k)}(t) - a_i{}^{(j,\,k)}, \\ \beta_i{}^{(j,\,k)} = g_i{}^{(j,\,k)}(t) - b_i{}^{(j,\,k)}. \end{cases}$$

Therefore equations (20) become

$$(25) \quad \begin{cases} \alpha_i = \displaystyle\sum_{j=0}^{\infty} a_i{}^{(j,\,0)} m_1{}^j + \sum_{j=0}^{\infty} \sum_{k=1}^{\infty} (f_i{}^{(j,\,k)} - a_i{}^{(j,\,k)}) m_1{}^j m_2{}^k, \\ \beta_i = \displaystyle\sum_{k=0}^{\infty} b_i{}^{(0,\,k)} m_2{}^k + \sum_{j=1}^{\infty} \sum_{k=0}^{\infty} (g_i{}^{(j,\,k)} - b_i{}^{(j,\,k)}) m_1{}^j m_2{}^k. \end{cases}$$

Let the values of α_i and β_i at $t = t_0$ be $\alpha_i{}^{(0)}$ and $\beta_i{}^{(0)}$ respectively. Then, at $t = t_0$, equations (25) become

$$\begin{cases} \alpha_i{}^{(0)} = \displaystyle\sum_{j=0}^{\infty} a_i{}^{(j,\,0)} m_1{}^j + \sum_{j=0}^{\infty} \sum_{k=1}^{\infty} (f_i{}^{(j,\,k)} - a_i{}^{(j,\,k)})_0\, m_1{}^j m_2{}^k, \\ \beta_i{}^{(0)} = \displaystyle\sum_{k=0}^{\infty} b_i{}^{(0,\,k)} m_2{}^k + \sum_{j=1}^{\infty} \sum_{k=0}^{\infty} (g_i{}^{(j,\,k)} - b_i{}^{(j,\,k)})_0\, m_1{}^j m_2{}^k. \end{cases}$$

Since these equations must be true for all values of m_1 and m_2 below certain limits, the coefficients of corresponding powers of m_1 and m_2 in the right and left members are equal; whence

$$(26) \quad \begin{cases} \alpha_i{}^{(0,\,0)} = \alpha_i{}^{(0)}, \qquad \alpha_i{}^{(j,\,0)} = 0, \qquad (j = 1, \,\cdots\infty), \\ \beta_i{}^{(0,\,0)} = \beta_i{}^{(0)}, \qquad \beta_i{}^{(0,\,k)} = 0, \qquad (k = 1, \,\cdots\infty), \\ f_i{}^{(j,\,k)}(t_0) - a_i{}^{(j,\,k)} = 0, \qquad (j = 1, \,\cdots\infty\,;\, k = 1, \,\cdots\infty), \\ g_i{}^{(j,\,k)}(t_0) - b_i{}^{(j,\,k)} = 0, \qquad (j = 1, \,\cdots\infty\,;\, k = 1, \,\cdots\infty). \end{cases}$$

Since all the terms of the right members of (25) except the first vanish at $t = t_0$, it follows that $\alpha_i^{(0,\,0)}$ and $\beta_i^{(0,\,0)}$ are the osculating elements [Art. 172] of the orbits of m_1 and m_2 respectively at the time $t = t_0$, and that the other coefficients of (20) are the definite integrals of the differential equations which define them taken between the limits $t = t_0$ and $t = t$.

209. The Terms of the First Order. The terms of the first order with respect to the masses are defined by equations (23). Since the terms of order zero are the osculating elements at t_0, the differential equations become

$$(27) \quad \begin{cases} \dfrac{d\alpha_i^{(0,\,1)}}{dt} = \phi_i(\alpha_1^{(0)}, \, \cdots, \, \alpha_6^{(0)}; \; \beta_1^{(0)}, \, \cdots, \, \beta_6^{(0)}; \; t), \\[3mm] \dfrac{d\beta_i^{(1,\,0)}}{dt} = \psi_i(\alpha_1^{(0)}, \, \cdots, \, \alpha_6^{(0)}; \; \beta_1^{(0)}, \, \cdots, \, \beta_6^{(0)}; \; t). \end{cases}$$

The right members of these equations are proportional to the rates at which the several elements of the orbits of the two planets would vary at any time t, if the two planets were moving at that instant strictly in the original ellipses. The integrals of (27) are, therefore, the sums of the instantaneous effects; or, in other words, they are the sums of the changes which would be produced if the forces and their instantaneous results were always exactly equal to those in the undisturbed orbits. Of course the perturbations modify these conditions and produce secondary, tertiary, and higher order effects. They are included in the coefficients of higher powers of m_1 and m_2 in (20).

The quantities $\alpha_i^{(0,\,1)}$ and $\beta_i^{(1,\,0)}$ are usually called perturbations of the first order with respect to the masses. The reason is clearly because they are the coefficients of the first powers of the masses in the series (20). In the planetary theories it is not necessary to go to perturbations of higher orders except in the case of the larger planets which are near each other, and then comparatively few terms are great enough to be sensible. It is not necessary in the present state of the planetary theories to include terms of the third order except in the mutual perturbations of Jupiter and Saturn.

Instead of there being but two planets and the sun there are eight planets and the sun, so that the actual theory is not quite so simple as that which has been outlined. Yet, as will be shown, the increased complexity comes chiefly in the perturbations of higher orders. If there were a third planet m_3 whose orbit had the elements $\gamma_1, \, \cdots, \, \gamma_6$, equations (23) would become

$$(28) \begin{cases} \dfrac{d\alpha_i{}^{(1,0,0)}}{dt} = 0, \\[2ex] \dfrac{d\alpha_i{}^{(0,1,0)}}{dt} = \phi_i(\alpha_1{}^{(0)}, \cdots, \alpha_6{}^{(0)};\ \beta_1{}^{(0)}, \cdots, \beta_6{}^{(0)};\ t), \\[2ex] \dfrac{d\alpha_i{}^{(0,0,1)}}{dt} = \phi_i(\alpha_1{}^{(0)}, \cdots, \alpha_6{}^{(0)};\ \gamma_1{}^{(0)}, \cdots, \gamma_6{}^{(0)};\ t), \\[2ex] \dfrac{d\beta_i{}^{(1,0,0)}}{dt} = \psi_i(\alpha_1{}^{(0)}, \cdots, \alpha_6{}^{(0)};\ \beta_1{}^{(0)}, \cdots, \beta_6{}^{(0)};\ t), \\[2ex] \dfrac{d\beta_i{}^{(0,1,0)}}{dt} = 0, \\[2ex] \dfrac{d\beta_i{}^{(0,0,1)}}{dt} = \psi_i(\beta_1{}^{(0)}, \cdots, \beta_6{}^{(0)};\ \gamma_1{}^{(0)}, \cdots, \gamma_6{}^{(0)};\ t), \\[2ex] \dfrac{d\gamma_i{}^{(1,0,0)}}{dt} = \chi_i(\alpha_1{}^{(0)}, \cdots, \alpha_6{}^{(0)};\ \gamma_1{}^{(0)}, \cdots, \gamma_6{}^{(0)};\ t), \\[2ex] \dfrac{d\gamma_i{}^{(0,1,0)}}{dt} = \chi_i(\beta_1{}^{(0)}, \cdots, \beta_6{}^{(0)};\ \gamma_1{}^{(0)}, \cdots, \gamma_6{}^{(0)};\ t), \\[2ex] \dfrac{d\gamma_i{}^{(0,0,1)}}{dt} = 0. \end{cases}$$

If there were more planets more equations of the same type would be added. Consider the perturbations of the first order of the elements of the orbits m_1; they are composed of two distinct parts given by the second and third equations of (28), one coming from the attraction of m_2, and the other from the attraction of m_3. Therefore, the statement of astronomers that the perturbing effects of the various planets may be considered separately, is true for the perturbations of the first order with respect to the masses.

210. The Terms of the Second Order. It has been shown that $\alpha_i{}^{(1,0)} = \alpha_i{}^{(2,0)} = \beta_i{}^{(0,1)} = \beta_i{}^{(0,2)} = 0$; therefore it follows from (24) that the terms of the second order with respect to the masses are determined by the equations

$$(29) \begin{cases} \dfrac{d\alpha_i{}^{(1,1)}}{dt} = \sum_{j=1}^{6} \dfrac{\partial\phi_i(\alpha_1{}^{(0)}, \cdots, \alpha_6{}^{(0)};\ \beta_1{}^{(0)}, \cdots, \beta_6{}^{(0)};\ t)}{\partial\beta_j} \beta_j{}^{(1,0)}, \\[3ex] \dfrac{d\alpha_i{}^{(0,2)}}{dt} = \sum_{j=1}^{6} \dfrac{\partial\phi_i(\alpha_1{}^{(0)}, \cdots, \alpha_6{}^{(0)};\ \beta_1{}^{(0)}, \cdots, \beta_6{}^{(0)};\ t)}{\partial\alpha_j} \alpha_j{}^{(0,1)}, \\[3ex] \dfrac{d\beta_i{}^{(1,1)}}{dt} = \sum_{j=1}^{6} \dfrac{\partial\psi_i(\alpha_1{}^{(0)}, \cdots, \alpha_6{}^{(0)};\ \beta_1{}^{(0)}, \cdots, \beta_6{}^{(0)};\ t)}{\partial\alpha_j} \alpha_j{}^{(0,1)}, \\[3ex] \dfrac{d\beta_i{}^{(2,0)}}{dt} = \sum_{j=1}^{6} \dfrac{\partial\psi_i(\alpha_1{}^{(0)}, \cdots, \alpha_6{}^{(0)};\ \beta_1{}^{(0)}, \cdots, \beta_6{}^{(0)};\ t)}{\partial\beta_j} \beta_j{}^{(1,0)}. \end{cases}$$

The perturbations of the first order are those which would result if the disturbing forces at every instant were the same as they would be if the bodies were moving in the original ellipses. If the bodies m_1 and m_2 move in curves differing from the original ellipses the rates at which the elements change at every instant are different from the values given by equations (27). The perturbations of the elements of the orbit of m_1 due to the fact that m_2 departs from its original ellipse by perturbations of the first order are given by the equations of the type of the first of (29), for, if $\beta_j^{(1, 0)} = 0$, it follows that $\alpha_i^{(1, 1)} = 0$ also. The perturbations of the elements of the orbit of m_1 due to the fact that m_1 departs from its original ellipse by perturbations of the first order are given by the equations of the type of the second of (29), for, if $\alpha_j^{(0, 1)} = 0$, it follows that $\alpha_i^{(0, 2)} = 0$ also. The terms $\beta_i^{(1, 1)}$ and $\beta_i^{(2, 0)}$ in the elements of the orbit of m_2 arise from similar causes. Thus the perturbations of the second order correct the errors in the terms of the first order, and those of the third order correct the errors in the second, and so on.

As has been said, the solutions expressed as power series in the masses converge if the interval of time is taken not too great. In a general way, the smaller the masses of the planets the longer the time during which the series converge. In the Lunar Theory the sun plays the rôle of the disturbing planet. Since its mass is very great compared to that of the central body, the earth, the series in powers of the masses as given above would converge for only a very short time, probably only a few months instead of years. Such a Lunar Theory would be entirely unsatisfactory. On this account the perturbations in the Lunar Theory are developed in powers of the ratio of the distances of the moon and the sun from the earth, and special artifices are employed to avoid secular terms in all the elements except the nodes and perigee.

If there is a third planet the perturbations of the second order are considerably more complicated. Let the planets be m_1, m_2, and m_3, and consider the perturbations of the second order of the elements of the orbit of m_1. From purely physical considerations it is seen that the following sorts of terms will arise: (a) terms arising from the disturbing action of m_2 and m_3, due respectively to the perturbations of the first order of the elements of m_2 and m_3 by m_1; (b) terms arising from the disturbing action of m_2 and m_3, due to the perturbations of the first order of the elements of the orbit of m_1 by m_2 and m_3; (c) terms arising from the disturbing action of m_2, due to the perturbations of the first order of the elements of the orbit of m_1 by m_3; (d) terms arising from the disturbing action of m_2, due to the perturbations of the first order of the elements of the orbit of m_2 by m_3; (e) terms arising from the disturbing action of m_3, due to the perturbations of the first order of the elements of the orbit of m_1 by m_2; and (f) terms arising from the disturbing action of m_3, due to the perturbations of the first order of the elements of m_3 by m_2.

Under the supposition that there are three planets, the terms of the second order with respect to the masses are found from equations (19) and (20) to be

$$(30)\begin{cases} \dfrac{d\alpha_i^{(1,1,0)}}{dt} = \sum_{j=1}^{6} \dfrac{\partial\phi_i(\alpha_1^{(0)}, \cdots, \alpha_6^{(0)}; \beta_1^{(0)}, \cdots, \beta_6^{(0)}; t)}{\partial\beta_j} \beta_j^{(1,0,0)}, \\[2ex] \dfrac{d\alpha_i^{(1,0,1)}}{dt} = \sum_{j=1}^{6} \dfrac{\partial\phi_i(\alpha_1^{(0)}, \cdots, \alpha_6^{(0)}; \gamma_1^{(0)}, \cdots, \gamma_6^{(0)}; t)}{\partial\gamma_j} \gamma_j^{(1,0,0)}, \\[2ex] \dfrac{d\alpha_i^{(0,2,0)}}{dt} = \sum_{j=1}^{6} \dfrac{\partial\phi_i(\alpha_1^{(0)}, \cdots, \alpha_6^{(0)}; \beta_1^{(0)}, \cdots, \beta_6^{(0)}; t)}{\partial\alpha_j} \alpha_j^{(0,1,0)}, \\[2ex] \dfrac{d\alpha_i^{(0,0,2)}}{dt} = \sum_{j=1}^{6} \dfrac{\partial\phi_i(\alpha_1^{(0)}, \cdots, \alpha_6^{(0)}; \gamma_1^{(0)}, \cdots, \gamma_6^{(0)}; t)}{\partial\alpha_j} \alpha_j^{(0,0,1)}, \\[2ex] \dfrac{d\alpha_i^{(0,1,1)}}{dt} = \sum_{j=1}^{6} \dfrac{\partial\phi_i(\alpha_1^{(0)}, \cdots, \alpha_6^{(0)}; \beta_1^{(0)}, \cdots, \beta_6^{(0)}; t)}{\partial\alpha_j} \alpha_j^{(0,0,1)} \\[2ex] \quad + \sum_{j=1}^{6} \dfrac{\partial\phi_i(\alpha_1^{(0)}, \cdots, \alpha_6^{(0)}; \beta_1^{(0)}, \cdots, \beta_6^{(0)}; t)}{\partial\beta_j} \beta_j^{(0,0,1)} \\[2ex] \quad + \sum_{j=1}^{6} \dfrac{\partial\phi_i(\alpha_1^{(0)}, \cdots, \alpha_6^{(0)}; \gamma_1^{(0)}, \cdots, \gamma_6^{(0)}; t)}{\partial\alpha_j} \alpha_j^{(0,1,0)} \\[2ex] \quad + \sum_{j=1}^{6} \dfrac{\partial\phi_i(\alpha_1^{(0)}, \cdots, \alpha_6^{(0)}; \gamma_1^{(0)}, \cdots, \gamma_6^{(0)}; t)}{\partial\gamma_j} \gamma_j^{(0,1,0)}, \end{cases}$$

and similar equations for $\dfrac{d\beta_i}{dt}$ and $\dfrac{d\gamma_i}{dt}$.

The first two equations give the perturbations of the class (a), for, $\phi_i(\alpha, \beta)$ and $\phi_i(\alpha, \gamma)$ are the portions of the perturbative function given by m_2 and m_3 respectively, while $\beta_j^{(1,0,0)}$ and $\gamma_j^{(1,0,0)}$ are the perturbations of the first order of the elements of the orbits of m_2 and m_3 by m_1. Similarly, the third and fourth equations give the perturbations of the class (b); the first term of the fifth equation, those of class (c); the second term, of class (d); the third term, of class (e); and the fourth term, of the class (f).

It appears from this that the terms of the second order cannot be computed separately for each of the disturbing planets.

The types of terms which will arise in the perturbations of the third order can be similarly predicted from physical considerations, and the predictions can be verified by a detailed discussion of the equations.

XXV. PROBLEMS.

1. In equations (3) take the term $\nu \cos lt$ to the left member before starting the integration, and include it in equations (4). Carry out the whole process of integration with this variation in the procedure.

2. If equations (7) are integrated as power series in μ and ν, what types of functions of t will arise in the terms of the second order?

3. Write the equations defining the terms of order zero, one, and two in the masses when equations (11) are integrated as series in m_1 and m_2. Show that the terms of order zero are the coördinates that m_1 and m_2 would have if they were particles moving around the sun in ellipses defined by their initial conditions. Show that the equations defining the terms of the first and higher orders are linear and non-homogeneous, instead of being reduced to quadratures as they are after the method of the variation of parameters has been used.

4. Suppose there are four planets, m_1, m_2, m_3, m_4; write all the terms of the second order with respect to the masses according to (30) and interpret each.

5. Suppose there are two planets m_1 and m_2; write all of the terms of the third order with respect to the masses and interpret each.

6. Suppose $m_1 = m_2 = m_3$ and that the planets are arranged in the order m_1, m_2, m_3 with respect to their distance from the sun. Show that of the perturbations defined by equations (30) the most important are those given by the first and third equations and the second term of the fifth; that the perturbations next in importance are given by the first, third, and fourth terms of the fifth equation; and that the least important are given by the second and fourth equations.

211. Choice of Elements. In order to exhibit the manner in which the various sorts of terms enter in the perturbations of the first order, it will be necessary to develop equations (19) explicitly. This was deferred, on account of the length of the transformations which are necessary, until a general view of the mathematical principles involved could be given.

If terms of the first order alone are considered the functions $\phi_i(\alpha, \beta)$ can be considered independently of $\psi_i(\alpha, \beta)$. Any independent functions of the elements may be used in place of the ordinary elements. In fact, one of the elements already employed, $\pi = \omega + \Omega$, is the sum of two geometrically simpler elements. Now the form of $\phi_i(\alpha, \beta)$ will depend upon the elements chosen; with certain elements they are rather simple, and with others very complicated. They will be taken in the first example which follows so that those functions shall become as simple as possible.

212. Lagrange's Brackets. Lagrange has made the following transformation which greatly facilitates the computation of (19). Multiply (18) by $-\dfrac{\partial x_1'}{\partial \alpha_1},\ -\dfrac{\partial y_1'}{\partial \alpha_1},\ -\dfrac{\partial z_1'}{\partial \alpha_1},\ \dfrac{\partial x_1}{\partial \alpha_1},\ \dfrac{\partial y_1}{\partial \alpha_1},\ \dfrac{\partial z_1}{\partial \alpha_1}$ respectively and add. The result is

$$(31)\quad
\left\{
\begin{aligned}
&\frac{d\alpha_2}{dt}\left\{\frac{\partial x_1}{\partial \alpha_1}\frac{\partial x_1'}{\partial \alpha_2}-\frac{\partial x_1'}{\partial \alpha_1}\frac{\partial x_1}{\partial \alpha_2}+\frac{\partial y_1}{\partial \alpha_1}\frac{\partial y_1'}{\partial \alpha_2}-\frac{\partial y_1'}{\partial \alpha_1}\frac{\partial y_1}{\partial \alpha_2}\right.\\
&\hspace{5cm}\left.+\frac{\partial z_1}{\partial \alpha_1}\frac{\partial z_1'}{\partial \alpha_2}-\frac{\partial z_1'}{\partial \alpha_1}\frac{\partial z_1}{\partial \alpha_2}\right\}\\
&+\frac{d\alpha_3}{dt}\left\{\frac{\partial x_1}{\partial \alpha_1}\frac{\partial x_1'}{\partial \alpha_3}-\frac{\partial x_1'}{\partial \alpha_1}\frac{\partial x_1}{\partial \alpha_3}+\cdots\right\}\\
&\cdot\quad\cdot\quad\cdot\quad\cdot\quad\cdot\quad\cdot\quad\cdot\quad\cdot\quad\cdot\quad\cdot\quad\cdot\quad\cdot\\
&+\frac{d\alpha_6}{dt}\left\{\frac{\partial x_1}{\partial \alpha_1}\frac{\partial x_1'}{\partial \alpha_6}-\frac{\partial x_1'}{\partial \alpha_1}\frac{\partial x_1}{\partial \alpha_6}+\cdots\right\}\\
&\quad=m_2\frac{\partial R_{1,\,2}}{\partial x_1}\frac{\partial x_1}{\partial \alpha_1}+m_2\frac{\partial R_{1,\,2}}{\partial y_1}\frac{\partial y_1}{\partial \alpha_1}+m_2\frac{\partial R_{1,\,2}}{\partial z_1}\frac{\partial z_1}{\partial \alpha_1}\\
&\quad=m_2\frac{\partial R_{1,\,2}}{\partial \alpha_1}.
\end{aligned}
\right.$$

Lagrange's brackets $[\alpha_i,\ \alpha_j]$ are defined by

$$(32)\quad
\begin{aligned}
[\alpha_i,\ \alpha_j]\equiv &\frac{\partial x_1}{\partial \alpha_i}\frac{\partial x_1'}{\partial \alpha_j}-\frac{\partial x_1'}{\partial \alpha_i}\frac{\partial x_1}{\partial \alpha_j}+\frac{\partial y_1}{\partial \alpha_i}\frac{\partial y_1'}{\partial \alpha_j}-\frac{\partial y_1'}{\partial \alpha_i}\frac{\partial y_1}{\partial \alpha_j}\\
&+\frac{\partial z_1}{\partial \alpha_i}\frac{\partial z_1'}{\partial \alpha_j}-\frac{\partial z_1'}{\partial \alpha_i}\frac{\partial z_1}{\partial \alpha_j}.
\end{aligned}$$

Form the equations corresponding to (31) in $\alpha_2, \cdots, \alpha_6$; the resulting system of equations is

$$(33) \quad \begin{cases} \sum_{i=1}^{6} [\alpha_1, \alpha_i] \dfrac{d\alpha_i}{dt} = m_2 \dfrac{\partial R_{1,2}}{\partial \alpha_1}, \\[2mm] \sum_{i=1}^{6} [\alpha_2, \alpha_i] \dfrac{d\alpha_i}{dt} = m_2 \dfrac{\partial R_{1,2}}{\partial \alpha_2}, \\ \cdot \quad \cdot \quad \cdot \quad \cdot \quad \cdot \quad \cdot \quad \cdot \quad \cdot \\ \sum_{i=1}^{6} [\alpha_6, \alpha_i] \dfrac{d\alpha_i}{dt} = m_2 \dfrac{\partial R_{1,2}}{\partial \alpha_6}. \end{cases}$$

These equations are equivalent to the system (18) and will be used in place of them.

213. Properties of Lagrange's Brackets. It follows at once from the definitions of Lagrange's brackets that

$$(34) \quad \begin{cases} [\alpha_i, \alpha_i] = 0, \\ [\alpha_i, \alpha_j] = - [\alpha_j, \alpha_i]. \end{cases}$$

A more important property is that they do not contain the time explicitly; that is,

$$(35) \quad \frac{\partial [\alpha_i, \alpha_j]}{\partial t} = 0, \qquad (i = 1, \cdots, 6; j = 1, \cdots, 6),$$

as will be proved immediately.

Many complicated expressions will arise in the following discussion which are symmetrical in x, y, and z. In order to abbreviate the writing let S, standing before a function of x, indicate that the same functions of y and z are to be added. Thus, for example,

$$S(x_1 x_2' - x_2 x_1') \equiv (x_1 x_2' - x_2 x_1') + (y_1 y_2' - y_2 y_1') + (z_1 z_2' - z_2 z_1').$$

In starting from the definitions of the brackets and omitting the subscripts of $x, \cdots, z'$, which will not be of use in what follows, it is found that

$$\begin{aligned} \frac{\partial [\alpha_i, \alpha_j]}{\partial t} &= S \left\{ \frac{\partial^2 x}{\partial \alpha_i \partial t} \frac{\partial x'}{\partial \alpha_j} + \frac{\partial x}{\partial \alpha_i} \frac{\partial^2 x'}{\partial \alpha_j \partial t} - \frac{\partial^2 x'}{\partial \alpha_i \partial t} \frac{\partial x}{\partial \alpha_j} - \frac{\partial x'}{\partial \alpha_i} \frac{\partial^2 x}{\partial \alpha_j \partial t} \right\} \\ &= \frac{\partial}{\partial \alpha_i} S \left\{ \frac{\partial x}{\partial t} \frac{\partial x'}{\partial \alpha_j} - \frac{\partial x'}{\partial t} \frac{\partial x}{\partial \alpha_j} \right\} + S \left\{ - \frac{\partial x}{\partial t} \frac{\partial^2 x'}{\partial \alpha_i \partial \alpha_j} + \frac{\partial x'}{\partial t} \frac{\partial^2 x}{\partial \alpha_i \partial \alpha_j} \right\} \\ &\qquad\qquad + S \left\{ \frac{\partial x}{\partial \alpha_i} \frac{\partial^2 x'}{\partial \alpha_j \partial t} - \frac{\partial x'}{\partial \alpha_i} \frac{\partial^2 x}{\partial \alpha_j \partial t} \right\} \\ &= \frac{\partial}{\partial \alpha_i} S \left\{ \frac{\partial x}{\partial t} \frac{\partial x'}{\partial \alpha_j} - \frac{\partial x'}{\partial t} \frac{\partial x}{\partial \alpha_j} \right\} - \frac{\partial}{\partial \alpha_j} S \left\{ \frac{\partial x}{\partial t} \frac{\partial x'}{\partial \alpha_i} - \frac{\partial x'}{\partial t} \frac{\partial x}{\partial \alpha_i} \right\}. \end{aligned}$$

The partial derivatives of the coördinates with respect to the time are the same in disturbed motion as the total derivatives in undisturbed motion. Therefore this equation becomes as a consequence of (14)

$$\frac{\partial[\alpha_i, \alpha_j]}{\partial t} = \frac{\partial}{\partial \alpha_i} S \left\{ \frac{\partial \Omega}{\partial x} \frac{\partial x}{\partial \alpha_j} + \frac{\partial \Omega}{\partial x'} \frac{\partial x'}{\partial \alpha_j} \right\} - \frac{\partial}{\partial \alpha_j} S \left\{ \frac{\partial \Omega}{\partial x} \frac{\partial x}{\partial \alpha_i} + \frac{\partial \Omega}{\partial x'} \frac{\partial x'}{\partial \alpha_i} \right\}$$

$$= \frac{\partial}{\partial \alpha_i} \left(\frac{\partial \Omega}{\partial \alpha_j} \right) - \frac{\partial}{\partial \alpha_j} \left(\frac{\partial \Omega}{\partial \alpha_i} \right) = \frac{\partial^2 \Omega}{\partial \alpha_i \partial \alpha_j} - \frac{\partial^2 \Omega}{\partial \alpha_j \partial \alpha_i} = 0,$$

which proves the theorem that the brackets do not contain t explicitly. This would hardly be anticipated since each of the quantities which appears in the brackets is an explicit function of t.

Since the brackets do not contain the time explicitly they may be computed for any epoch whatever, and in particular for $t = t_0$. The equations become very simple if the coördinates at the time $t = t_0$ are taken for the elements $\alpha_1, \cdots, \alpha_6$. This is permissible since the ordinary elements are defined by these quantities, and conversely. It must not be supposed that they are constants; they are such quantities that if the elements are computed from them, and then if the coördinates at any time t are computed using these elements, the correct results will be obtained. Since in disturbed motion the elements vary with the time, the values of the coördinates at $t = t_0$ also vary. Otherwise considered, if the osculating elements at t are used and if the coördinates at the time $t = t_0$ are computed, it will be found in the case of disturbed motion that the coördinates at $t = t_0$ vary, and these values of the coördinates are the ones in question.

Let the coördinates at the time $t = t_0$ be $x_0, \cdots, z_0'$; then

$$[x_0, y_0] = S \left\{ \frac{\partial x_0}{\partial x_0} \frac{\partial x_0'}{\partial y_0} - \frac{\partial x_0'}{\partial x_0} \frac{\partial x_0}{\partial y_0} \right\},$$

which equals zero because x_0' is independent of y_0 and x_0. Similarly,

(36) $\begin{cases} [y_0, \ z_0] = [z_0, \ x_0] = [x_0', \ y_0'] = [y_0', \ z_0'] = [z_0', \ x_0'] = [x_0, \ y_0] = 0, \\ [x_0, \ y_0'] = [x_0, \ z_0'] = [y_0, \ x_0'] = [y_0, \ z_0'] = [z_0, \ x_0'] = [z_0, \ y_0'] = 0. \end{cases}$

But

(37) $[x_0, \ x_0'] = [y_0, \ y_0'] = [z_0, \ z_0'] = 1.$

Therefore equations (33) become in this case

$$(38) \quad \begin{cases} \dfrac{dx_0}{dt} = m_2 \dfrac{\partial R_{1,2}}{\partial x_0'}, & \dfrac{dx_0'}{dt} = -m_2 \dfrac{\partial R_{1,2}}{\partial x_0}, \\[2.5ex] \dfrac{dy_0}{dt} = m_2 \dfrac{\partial R_{1,2}}{\partial y_0'}, & \dfrac{dy_0'}{dt} = -m_2 \dfrac{\partial R_{1,2}}{\partial y_0}, \\[2.5ex] \dfrac{dz_0}{dt} = m_2 \dfrac{\partial R_{1,2}}{\partial z_0'}, & \dfrac{dz_0'}{dt} = -m_2 \dfrac{\partial R_{1,2}}{\partial z_0}. \end{cases}$$

Any system of differential equations of the form (38) is known as a *canonical system*, and they possess properties which make them particularly valuable in theoretical investigations. There is a theorem that any dynamical problem in which the forces can be represented as partial derivatives of a potential function can be expressed in this form; and if it is possible to put a problem in the canonical form it is possible to do so in infinitely many systems of dependent variables.

If equations (38) were solved they would give the values of the coördinates at t_0 which would have to be used to obtain the true coördinates at the time t, under the supposition that the planet moved in an undisturbed ellipse during $t - t_0$. If the variables were the elliptic elements the solutions of the equations would give the elements which would have to be used to compute the coördinates at the time t, when they are supposed to have been constant during the interval $t - t_0$. Thus, when the elements have been found the remainder of the computation is that of undisturbed motion.

214. Transformation to the Ordinary Elements. The elements used in Astronomy are not the coördinates at $t = t_0$, but Ω, i, a, e, π, and T (or $\epsilon = \pi - nT$), which were expressed in terms of the initial conditions in Arts. 86, 87, and 88. It will be necessary, therefore, to transform equations (38) to the corresponding ones which involve only the elements which are actually in use by astronomers.

Let s represent any one of the elements Ω, i, a, e, π, ϵ. It may be expressed symbolically in terms of the initial conditions by

$$(39) \qquad s = f(x_0, y_0, z_0, x_0', y_0', z_0').$$

Hence it follows that

$$\frac{ds}{dt} = S\left\{ \frac{\partial f}{\partial x_0}\frac{dx_0}{dt} + \frac{\partial f}{\partial x_0'}\frac{dx_0'}{dt} \right\};$$

or, because of (38),

$$(40) \qquad \frac{ds}{dt} = m_2 S \left\{ \frac{\partial f}{\partial x_0} \frac{\partial R_{1,\,2}}{\partial x_0'} - \frac{\partial f}{\partial x_0'} \frac{\partial R_{1,\,2}}{\partial x_0} \right\}.$$

The partial derivatives of $R_{1,\,2}$ are expressed in terms of the partial derivatives with respect to the new variables by the equations

$$(41) \begin{cases} \dfrac{\partial R_{1,\,2}}{\partial x_0} = \dfrac{\partial R_{1,\,2}}{\partial \Omega} \dfrac{\partial \Omega}{\partial x_0} + \dfrac{\partial R_{1,\,2}}{\partial i} \dfrac{\partial i}{\partial x_0} + \dfrac{\partial R_{1,\,2}}{\partial a} \dfrac{\partial a}{\partial x_0} + \dfrac{\partial R_{1,\,2}}{\partial e} \dfrac{\partial e}{\partial x_0} \\[2ex] \qquad\qquad\qquad + \dfrac{\partial R_{1,\,2}}{\partial \pi} \dfrac{\partial \pi}{\partial x_0} + \dfrac{\partial R_{1,\,2}}{\partial \epsilon} \dfrac{\partial \epsilon}{\partial x_0}, \\[2ex] \cdot\quad\cdot\quad\cdot\quad\cdot\quad\cdot\quad\cdot\quad\cdot\quad\cdot\quad\cdot\quad\cdot\quad\cdot\quad\cdot\quad\cdot \\[2ex] \dfrac{\partial R_{1,\,2}}{\partial z_0'} = \dfrac{\partial R_{1,\,2}}{\partial \Omega} \dfrac{\partial \Omega}{\partial z_0'} + \dfrac{\partial R_{1,\,2}}{\partial i} \dfrac{\partial i}{\partial z_0'} + \dfrac{\partial R_{1,\,2}}{\partial a} \dfrac{\partial a}{\partial z_0'} + \dfrac{\partial R_{1,\,2}}{\partial e} \dfrac{\partial e}{\partial z_0'} \\[2ex] \qquad\qquad\qquad + \dfrac{\partial R_{1,\,2}}{\partial \pi} \dfrac{\partial \pi}{\partial z_0'} + \dfrac{\partial R_{1,\,2}}{\partial \epsilon} \dfrac{\partial \epsilon}{\partial z_0'}. \end{cases}$$

On carrying out the complicated computations of $\dfrac{\partial s}{\partial x_0}, \cdots, \dfrac{\partial s}{\partial z_0'}$ by means of the equations given in Arts. 86, 87, and 88, and expressing all the partial derivatives in terms of the new variables, the partial derivatives $\dfrac{\partial R_{1,\,2}}{\partial x_0}, \cdots, \dfrac{\partial R_{1,\,2}}{\partial z_0'}$ are found in terms of the elements and $\dfrac{\partial R_{1,\,2}}{\partial \Omega}, \cdots, \dfrac{\partial R_{1,\,2}}{\partial \epsilon}$. On substituting in (40) and expressing $\dfrac{\partial f}{\partial x_0}, \cdots, \dfrac{\partial f}{\partial z_0'}$ in terms of the elements, $\dfrac{ds}{dt}$ is found in terms of the elements and the derivatives of the perturbative function, $R_{1,\,2}$, with respect to the elements.

215. Method of Direct Computation of Lagrange's Brackets.
The transformations required in the method of the preceding article are very laborious, and the direct computation of the brackets, though considerably involved, is to be preferred from a practical point of view. *All of the computation in the transformations of this sort might be avoided by using canonical variables;* but, in order to employ them, a lengthy digression upon the properties of canonical systems would be necessary, and such a discussion is outside the limits of this work. Still, the labor may be notably reduced by first taking elements somewhat different from those defined in chapter v., and then transforming to those in more ordinary use. The following is based on Tisserand's exposition of Lagrange's method.*

* Tisserand's *Mécanique Céleste*, vol. I., p. 179.

Let the xy-plane be the plane of the ecliptic, ΩP the projection of the orbit upon the celestial sphere, Π the projection of the perihelion point, and P the projection of the position of the planet at the time t. In place of π and ϵ, adopt the new elements ω and σ defined by the equations

(42)
$$\left\{ \begin{array}{l} \omega = \pi - \Omega, \\ \sigma = - nT. \end{array} \right.$$

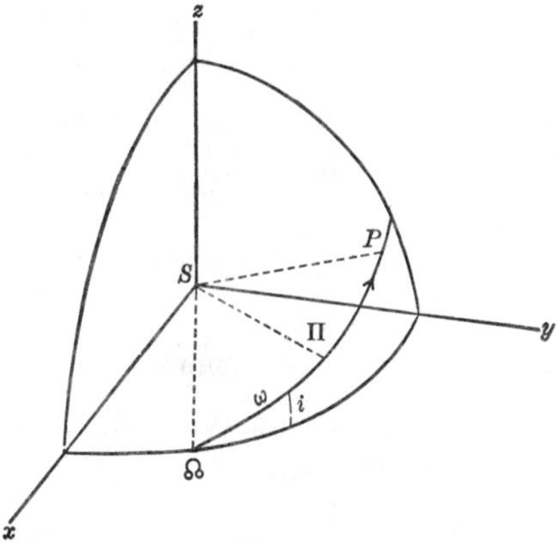

Fig. 60.

The following equations are either given in Art. 98, or are obtained from Fig. 60 by the fundamental formulas of Trigonometry:

(43)
$$\left\{ \begin{array}{l} n = \dfrac{k \sqrt{S + m_1}}{a^{\frac{3}{2}}}, \\[2mm] E - e \sin E = nt + \sigma, \\[2mm] r = a(1 - e \cos E), \\[2mm] \tan \dfrac{v}{2} = \sqrt{\dfrac{1 + e}{1 - e}} \tan \dfrac{E}{2}, \\[2mm] \cos v = \dfrac{\cos E - e}{1 - e \cos E}, \\[2mm] \sin v = \dfrac{\sqrt{1 - e^2} \sin E}{1 - e \cos E}, \\[2mm] x = r\{\cos (v + \omega) \cos \Omega - \sin (v + \omega) \sin \Omega \cos i\}, \\[2mm] y = r\{\cos (v + \omega) \sin \Omega + \sin (v + \omega) \cos \Omega \cos i\}, \\[2mm] z = r \sin (v + \omega) \sin i. \end{array} \right.$$

From these equations and their derivatives with respect to the time the partial derivatives of the coördinates with respect to the elements can be computed. The elements have been chosen in such a manner that they are divided into two groups having distinct properties; Ω, i, and ω define the position of the plane of motion and the orientation of the orbit in the plane, and a, e, and σ define the dimensions and shape of the orbit and the position of the planet in its orbit. Therefore the coördinates in the orbit can be expressed in terms of the elements of the second group alone, and from them, the coördinates in space can be found by means of the first group alone.

Take a new system of axes with the origin at the sun, the positive end of the ξ-axis directed to the perihelion point, the η-axis $90°$ forward in the plane of the orbit, and the ζ-axis perpendicular to the plane of the orbit. Let the direction cosines between the x-axis and the ξ, η, and ζ-axes be α, α', α''; between the y-axis and the ξ, η, and ζ-axes be β, β', β''; and between the z-axis and the ξ, η, and ζ-axes be γ, γ', γ''. Then it follows from Fig. 60 that

$$(44) \begin{cases} \alpha = \cos\omega\cos\Omega - \sin\omega\sin\Omega\cos i, \\ \beta = \cos\omega\sin\Omega + \sin\omega\cos\Omega\cos i, \\ \gamma = \sin\omega\sin i, \\ \alpha' = -\sin\omega\cos\Omega - \cos\omega\sin\Omega\cos i, \\ \beta' = -\sin\omega\sin\Omega + \cos\omega\cos\Omega\cos i, \\ \gamma' = \cos\omega\sin i, \\ \alpha'' = \sin\Omega\sin i, \\ \beta'' = -\cos\Omega\sin i, \\ \gamma'' = \cos i. \end{cases}$$

There exist among these nine direction cosines, as can easily be verified, the relations

$$(45) \begin{cases} \alpha^2 + \beta^2 + \gamma^2 = 1, \qquad \alpha\alpha' + \beta\beta' + \gamma\gamma' = 0, \\ \alpha'^2 + \beta'^2 + \gamma'^2 = 1, \qquad \alpha'\alpha'' + \beta'\beta'' + \gamma'\gamma'' = 0, \\ \alpha''^2 + \beta''^2 + \gamma''^2 = 1, \qquad \alpha''\alpha + \beta''\beta + \gamma''\gamma = 0, \\ \alpha = \beta'\gamma'' - \gamma'\beta'', \quad \alpha' = \beta''\gamma - \gamma''\beta, \quad \alpha'' = \beta\gamma' - \gamma\beta', \\ \beta = \gamma'\alpha'' - \alpha'\gamma'', \quad \beta' = \gamma''\alpha - \alpha''\gamma, \quad \beta'' = \gamma\alpha' - \alpha\gamma', \\ \gamma = \alpha'\beta'' - \beta'\alpha'', \quad \gamma' = \alpha''\beta - \beta''\alpha, \quad \gamma'' = \alpha\beta' - \beta\alpha'. \end{cases}$$

It follows from (43) and (44) and the definition of the new system of axes that

$$(46) \begin{cases} \xi = r \cos v = a(\cos E - e), \quad \eta = a\sqrt{1-e^2}\sin E, \\[1mm] \dfrac{\partial E}{\partial t} = \dfrac{n}{1 - e\cos E}, \\[2mm] \xi' = \dfrac{-na\sin E}{1 - e\cos E} = \dfrac{-k\sqrt{S+m_1}\sin E}{\sqrt{a}(1 - e\cos E)}, \\[2mm] \eta' = \dfrac{na\sqrt{1-e^2}\cos E}{1 - e\cos E} = \dfrac{k\sqrt{S+m_1}\sqrt{1-e^2}\cos E}{\sqrt{a}(1 - e\cos E)}, \\[2mm] x = \alpha\xi + \alpha'\eta, \quad y = \beta\xi + \beta'\eta, \quad z = \gamma\xi + \gamma'\eta, \\[1mm] x' = \alpha\xi' + \alpha'\eta', \quad y' = \beta\xi' + \beta'\eta', \quad z' = \gamma\xi' + \gamma'\eta', \end{cases}$$

where the accents on x, y, z, ξ, η, and ζ indicate first derivatives with respect to t.

The partial derivatives of $\alpha, \cdots, \gamma''$ with respect to the elements may be computed once for all; they are found from (44) to be

$$(47) \begin{cases} \dfrac{\partial \alpha}{\partial \omega} = \alpha', & \dfrac{\partial \alpha'}{\partial \omega} = -\alpha, & \dfrac{\partial \alpha''}{\partial \omega} = 0, \\[2mm] \dfrac{\partial \beta}{\partial \omega} = \beta', & \dfrac{\partial \beta'}{\partial \omega} = -\beta, & \dfrac{\partial \beta''}{\partial \omega} = 0, \\[2mm] \dfrac{\partial \gamma}{\partial \omega} = \gamma', & \dfrac{\partial \gamma'}{\partial \omega} = -\gamma, & \dfrac{\partial \gamma''}{\partial \omega} = 0; \end{cases}$$

$$(48) \begin{cases} \dfrac{\partial \alpha}{\partial \Omega} = -\beta, & \dfrac{\partial \alpha'}{\partial \Omega} = -\beta', & \dfrac{\partial \alpha''}{\partial \Omega} = -\beta'', \\[2mm] \dfrac{\partial \beta}{\partial \Omega} = \alpha, & \dfrac{\partial \beta'}{\partial \Omega} = \alpha', & \dfrac{\partial \beta''}{\partial \Omega} = \alpha'', \\[2mm] \dfrac{\partial \gamma}{\partial \Omega} = 0, & \dfrac{\partial \gamma'}{\partial \Omega} = 0, & \dfrac{\partial \gamma''}{\partial \Omega} = 0; \end{cases}$$

$$(49) \begin{cases} \dfrac{\partial \alpha}{\partial i} = \alpha''\sin\omega, & \dfrac{\partial \alpha'}{\partial i} = \alpha''\cos\omega, & \dfrac{\partial \alpha''}{\partial i} = +\sin\Omega\cos i, \\[2mm] \dfrac{\partial \beta}{\partial i} = \beta''\sin\omega, & \dfrac{\partial \beta'}{\partial i} = \beta''\cos\omega, & \dfrac{\partial \beta''}{\partial i} = -\cos\Omega\cos i, \\[2mm] \dfrac{\partial \gamma}{\partial i} = \gamma''\sin\omega, & \dfrac{\partial \gamma'}{\partial i} = \gamma''\cos\omega, & \dfrac{\partial \gamma''}{\partial i} = -\sin i. \end{cases}$$

There are as many brackets to be computed as there are combinations of the six elements taken two at a time, or $\dfrac{6!}{2!\,4!} = 15$. Three of them involve elements of only the first group; nine, one element of the first group and one of the second; and three, elements of only the second group. Let K and L represent any of the elements of the first group, Ω, i, ω; and P and Q any of the elements of the second group, a, e, σ. Then the Lagrangian brackets to be computed are

$$(50) \begin{cases} (a) & [K, L] = S\left\{ \dfrac{\partial x}{\partial K}\dfrac{\partial x'}{\partial L} - \dfrac{\partial x'}{\partial K}\dfrac{\partial x}{\partial L} \right\}, \quad \text{(3 equations)}, \\[2ex] (b) & [K, P] = S\left\{ \dfrac{\partial x}{\partial K}\dfrac{\partial x'}{\partial P} - \dfrac{\partial x'}{\partial K}\dfrac{\partial x}{\partial P} \right\}, \quad \text{(9 equations)}, \\[2ex] (c) & [P, Q] = S\left\{ \dfrac{\partial x}{\partial P}\dfrac{\partial x'}{\partial Q} - \dfrac{\partial x'}{\partial P}\dfrac{\partial x}{\partial Q} \right\}, \quad \text{(3 equations)}. \end{cases}$$

It is found from (46) that

$$(51) \begin{cases} \dfrac{\partial x}{\partial K} = \xi\dfrac{\partial \alpha}{\partial K} + \eta\dfrac{\partial \alpha'}{\partial K}, & \dfrac{\partial x'}{\partial K} = \xi'\dfrac{\partial \alpha}{\partial K} + \eta'\dfrac{\partial \alpha'}{\partial K}, \\[2ex] \dfrac{\partial x}{\partial P} = \alpha\dfrac{\partial \xi}{\partial P} + \alpha'\dfrac{\partial \eta}{\partial P}, & \dfrac{\partial x'}{\partial P} = \alpha\dfrac{\partial \xi'}{\partial P} + \alpha'\dfrac{\partial \eta'}{\partial P}, \end{cases}$$

and similar equations in y and z.

216. Computation of $[\omega, \Omega]$, $[\Omega, i]$, $[i, \omega]$. Let S indicate that the sum of the functions, symmetrical in α, β, and γ, is to be taken. Then the first equation of (50) becomes as a consequence of (51)

$$[K, L] = (\xi\eta' - \eta\xi')S\left\{ \dfrac{\partial \alpha}{\partial K}\dfrac{\partial \alpha'}{\partial L} - \dfrac{\partial \alpha'}{\partial K}\dfrac{\partial \alpha}{\partial L} \right\}.$$

But the law of areas [Art. 89] gives

$$\xi\eta' - \eta\xi' = \xi\dfrac{d\eta}{dt} - \eta\dfrac{d\xi}{dt} = k\sqrt{(S + m_1)a(1 - e^2)} = na^2\sqrt{1 - e^2}.$$

Therefore

$$(52) \qquad [K, L] = na^2\sqrt{1 - e^2}\,S\left\{ \dfrac{\partial \alpha}{\partial K}\dfrac{\partial \alpha'}{\partial L} - \dfrac{\partial \alpha'}{\partial K}\dfrac{\partial \alpha}{\partial L} \right\}.$$

On computing the right member of this equation by means of (47),

(48), and (49), and reducing by means of (45), the brackets involving elements of only the first group are found to be

$$(53) \begin{cases} [\omega, \Omega] = na^2 \sqrt{1-e^2} \, (-\alpha\beta - \alpha'\beta' + \alpha\beta + \alpha'\beta') = 0, \\ [\Omega, i] = na^2 \sqrt{1-e^2} \, \{(\alpha\beta'' - \beta\alpha'') \cos \omega \\ \qquad\qquad + (\beta'\alpha'' - \alpha'\beta'') \sin \omega\} \\ \qquad = na^2 \sqrt{1-e^2} \, (-\gamma' \cos \omega - \gamma \sin \omega) \\ \qquad = -na^2 \sqrt{1-e^2} \sin i, \\ [i, \ \omega] = -na^2 \sqrt{1-e^2} \, \{(\alpha'\alpha'' + \beta'\beta'' + \gamma'\gamma'') \cos \omega \\ \qquad\qquad + (\alpha''\alpha + \beta''\beta + \gamma''\gamma) \sin \omega\} = 0. \end{cases}$$

217. Computation of $[K, P]$. The second equations of (50) become, as a consequence of (51),

$$[K, P] = S\left\{ \left[\xi \frac{\partial \alpha}{\partial K} + \eta \frac{\partial \alpha'}{\partial K} \right] \left[\alpha \frac{\partial \xi'}{\partial P} + \alpha' \frac{\partial \eta'}{\partial P} \right] \right.$$

$$\left. - \left[\xi' \frac{\partial \alpha}{\partial K} + \eta' \frac{\partial \alpha'}{\partial K} \right] \left[\alpha \frac{\partial \xi}{\partial P} + \alpha' \frac{\partial \eta}{\partial P} \right] \right\}$$

$$= + \left[\alpha \frac{\partial \alpha}{\partial K} + \beta \frac{\partial \beta}{\partial K} + \gamma \frac{\partial \gamma}{\partial K} \right] \left[\xi \frac{\partial \xi'}{\partial P} - \xi' \frac{\partial \xi}{\partial P} \right]$$

$$+ \left[\alpha' \frac{\partial \alpha'}{\partial K} + \beta' \frac{\partial \beta'}{\partial K} + \gamma' \frac{\partial \gamma'}{\partial K} \right] \left[\eta \frac{\partial \eta'}{\partial P} - \eta' \frac{\partial \eta}{\partial P} \right]$$

$$+ \left[\alpha \frac{\partial \alpha'}{\partial K} + \beta \frac{\partial \beta'}{\partial K} + \gamma \frac{\partial \gamma'}{\partial K} \right] \left[\eta \frac{\partial \xi'}{\partial P} - \eta' \frac{\partial \xi}{\partial P} \right]$$

$$+ \left[\alpha' \frac{\partial \alpha}{\partial K} + \beta' \frac{\partial \beta}{\partial K} + \gamma' \frac{\partial \gamma}{\partial K} \right] \left[\xi \frac{\partial \eta'}{\partial P} - \xi' \frac{\partial \eta}{\partial P} \right].$$

It follows from equations (45), (47), (48), and (49) that

$$\alpha \frac{\partial \alpha}{\partial K} + \beta \frac{\partial \beta}{\partial K} + \gamma \frac{\partial \gamma}{\partial K} = 0,$$

$$\alpha' \frac{\partial \alpha'}{\partial K} + \beta' \frac{\partial \beta'}{\partial K} + \gamma' \frac{\partial \gamma'}{\partial K} = 0,$$

$$\alpha \frac{\partial \alpha'}{\partial K} + \beta \frac{\partial \beta'}{\partial K} + \gamma \frac{\partial \gamma'}{\partial K} = - \left[\alpha' \frac{\partial \alpha}{\partial K} + \beta' \frac{\partial \beta}{\partial K} + \gamma' \frac{\partial \gamma}{\partial K} \right].$$

Therefore

$$(54) \begin{cases} [K, P] = \left[\alpha' \dfrac{\partial \alpha}{\partial K} + \beta' \dfrac{\partial \beta}{\partial K} + \gamma' \dfrac{\partial \gamma}{\partial K} \right] \\ \qquad\qquad \times \left[\xi \dfrac{\partial \eta'}{\partial P} + \eta' \dfrac{\partial \xi}{\partial P} - \xi' \dfrac{\partial \eta}{\partial P} - \eta \dfrac{\partial \xi'}{\partial P} \right] \\ = \left[\alpha' \dfrac{\partial \alpha}{\partial K} + \beta' \dfrac{\partial \beta}{\partial K} + \gamma' \dfrac{\partial \gamma}{\partial K} \right] \dfrac{\partial (\xi \eta' - \eta \xi')}{\partial P} \\ = k \sqrt{S + m_1} \left[\alpha' \dfrac{\partial \alpha}{\partial K} + \beta' \dfrac{\partial \beta}{\partial K} + \gamma' \dfrac{\partial \gamma}{\partial K} \right] \dfrac{\partial \sqrt{p}}{\partial P}. \end{cases}$$

Let $P = a, e, \sigma$ in succession. Then it is found that

$$(55) \begin{cases} k \sqrt{S + m_1} \dfrac{\partial \sqrt{a(1 - e^2)}}{\partial a} = \dfrac{na}{2} \sqrt{1 - e^2}, \\ k \sqrt{S + m_1} \dfrac{\partial \sqrt{a(1 - e^2)}}{\partial e} = - \dfrac{na^2 e}{\sqrt{1 - e^2}}, \\ k \sqrt{S + m_1} \dfrac{\partial \sqrt{a(1 - e^2)}}{\partial \sigma} = 0. \end{cases}$$

Let $K = \omega, \Omega, i$ in turn in (54), and make use of (55); then it is found that

$$(56) \begin{cases} [\omega, a] = \dfrac{na}{2} \sqrt{1 - e^2}, & [\omega, e] = \dfrac{- na^2 e}{\sqrt{1 - e^2}}, & [\omega, \sigma] = 0, \\ [\Omega, a] = \dfrac{na}{2} \sqrt{1 - e^2} \cos i, & [i, a] = 0, & [i, e] = 0, \\ [\Omega, e] = \dfrac{- na^2 e}{\sqrt{1 - e^2}} \cos i, & [\Omega, \sigma] = 0, & [i, \sigma] = 0. \end{cases}$$

218. Computation of $[a, e]$, $[e, \sigma]$, $[\sigma, a]$. The third equation of (50) becomes, as a consequence of (51),

$$[P, Q] = S \left\{ \left[\alpha \dfrac{\partial \xi}{\partial P} + \alpha' \dfrac{\partial \eta}{\partial P} \right] \left[\alpha \dfrac{\partial \xi'}{\partial Q} + \alpha' \dfrac{\partial \eta'}{\partial Q} \right] \right.$$

$$\left. - \left[\alpha \dfrac{\partial \xi'}{\partial P} + \alpha' \dfrac{\partial \eta'}{\partial P} \right] \left[\alpha \dfrac{\partial \xi}{\partial Q} + \alpha' \dfrac{\partial \eta}{\partial Q} \right] \right\}$$

$$= + (\alpha^2 + \beta^2 + \gamma^2) \left[\dfrac{\partial \xi}{\partial P} \dfrac{\partial \xi'}{\partial Q} - \dfrac{\partial \xi}{\partial Q} \dfrac{\partial \xi'}{\partial P} \right]$$

$$+ (\alpha'^2 + \beta'^2 + \gamma'^2) \left[\dfrac{\partial \eta}{\partial P} \dfrac{\partial \eta'}{\partial Q} - \dfrac{\partial \eta}{\partial Q} \dfrac{\partial \eta'}{\partial P} \right]$$

$$+ (\alpha \alpha' + \beta \beta' + \gamma \gamma') \left[\dfrac{\partial \xi}{\partial P} \dfrac{\partial \eta'}{\partial Q} - \dfrac{\partial \xi}{\partial Q} \dfrac{\partial \eta'}{\partial P} + \dfrac{\partial \xi'}{\partial Q} \dfrac{\partial \eta}{\partial P} - \dfrac{\partial \xi'}{\partial P} \dfrac{\partial \eta}{\partial Q} \right].$$

As a consequence of equations (45), the right member of this equation reduces to

$$(57) \qquad [P, Q] = \frac{\partial \xi}{\partial P} \frac{\partial \xi'}{\partial Q} - \frac{\partial \xi}{\partial Q} \frac{\partial \xi'}{\partial P} + \frac{\partial \eta}{\partial P} \frac{\partial \eta'}{\partial Q} - \frac{\partial \eta}{\partial Q} \frac{\partial \eta'}{\partial P}.$$

Since the brackets do not contain the time explicitly t may be given any value after the partial derivatives have been formed. The partial derivatives become the simplest when $t = T$, the time of perihelion passage. For this value of t, $E = 0$, $r = a(1 - e)$, and it is found from equations (46) that*

$$(58) \quad \begin{cases} \dfrac{\partial \xi}{\partial a} = 1 - e, & \dfrac{\partial \eta}{\partial a} = 0, & \dfrac{\partial \xi'}{\partial a} = 0, & \dfrac{\partial \eta'}{\partial a} = - \dfrac{n}{2} \sqrt{\dfrac{1 + e}{1 - e}}, \\[3mm] \dfrac{\partial \xi}{\partial e} = - a, & \dfrac{\partial \eta}{\partial e} = 0, & \dfrac{\partial \xi'}{\partial e} = 0, & \dfrac{\partial \eta'}{\partial e} = \dfrac{1}{1 - e} \cdot \dfrac{na}{\sqrt{1 - e^2}}, \\[3mm] \dfrac{\partial \xi}{\partial \sigma} = 0, & \dfrac{\partial \eta}{\partial \sigma} = a \sqrt{\dfrac{1 + e}{1 - e}}, & \dfrac{\partial \xi'}{\partial \sigma} = \dfrac{- na}{(1 - e)^2}, & \dfrac{\partial \eta'}{\partial \sigma} = 0. \end{cases}$$

Then equation (57) gives

$$(59) \qquad [a, e] = 0, \qquad [e, \sigma] = 0, \qquad [\sigma, a] = \frac{na}{2}.$$

On making use of the fact that $[\alpha_i, \alpha_j] = - [\alpha_j, \alpha_i]$ and equations (53), (56), and (59), equations (33) become

$$(60) \quad \begin{cases} \dfrac{na}{2} \sqrt{1 - e^2} \dfrac{da}{dt} - \dfrac{na^2 e}{\sqrt{1 - e^2}} \dfrac{de}{dt} = m_2 \dfrac{\partial R_{1,2}}{\partial \omega}, \\[3mm] - na^2 \sqrt{1 - e^2} \sin i \dfrac{di}{dt} + \dfrac{na}{2} \sqrt{1 - e^2} \cos i \dfrac{da}{dt} \\[3mm] \qquad\qquad - \dfrac{na^2 e}{\sqrt{1 - e^2}} \cos i \dfrac{de}{dt} = m_2 \dfrac{\partial R_{1,}}{\partial \delta\delta}, \\[3mm] na^2 \sqrt{1 - e^2} \sin i \dfrac{d\delta\delta}{dt} = m_2 \dfrac{\partial R_{1,2}}{\partial i}, \end{cases}$$

---

* It should be remembered that a and e enter explicitly and also implicitly through E and n, for E is defined by the equation

$$E - e \sin E = n(t - T) = \frac{k \sqrt{S + m_1}}{a^{\frac{3}{2}}} (t - T).$$

Then, e. g., $\dfrac{\partial \xi}{\partial a} = \cos E - e - a \sin E \dfrac{\partial E}{\partial a} = 1 - e$ when $t = T$, etc.

$$(60) \begin{cases} -\dfrac{na}{2}\sqrt{1-e^2}\,\dfrac{d\omega}{dt} - \dfrac{na}{2}\sqrt{1-e^2}\cos i\,\dfrac{d\Omega}{\partial t} - \dfrac{na}{2}\dfrac{d\sigma}{dt} = m_2\dfrac{\partial R_{1,\,2}}{\partial a}, \\[2.5ex] \dfrac{na^2 e}{\sqrt{1-e^2}}\dfrac{d\omega}{dt} + \dfrac{na^2 e\cos i}{\sqrt{1-e^2}}\dfrac{d\Omega}{dt} = m_2\dfrac{\partial R_{1,\,2}}{\partial e}, \\[2.5ex] \dfrac{na}{2}\dfrac{da}{dt} = m_2\dfrac{\partial R_{1,\,2}}{\partial \sigma}. \end{cases}$$

These equations are easily solved for the derivatives, and give

$$(61) \begin{cases} \dfrac{d\Omega}{dt} = \dfrac{m_2}{na^2\sqrt{1-e^2}\sin i}\dfrac{\partial R_{1,\,2}}{\partial i}, \\[2.5ex] \dfrac{di}{dt} = \dfrac{m_2\cos i}{na^2\sqrt{1-e^2}\sin i}\dfrac{\partial R_{1,\,2}}{\partial \omega} - \dfrac{m_2}{na^2\sqrt{1-e^2}\sin i}\dfrac{\partial R_{1,\,2}}{\partial \Omega}, \\[2.5ex] \dfrac{d\omega}{dt} = \dfrac{-m_2\cos i}{na^2\sqrt{1-e^2}\sin i}\dfrac{\partial R_{1,\,2}}{\partial i} + \dfrac{m_2\sqrt{1-e^2}}{na^2 e}\dfrac{\partial R_{1,\,2}}{\partial e}, \\[2.5ex] \dfrac{da}{dt} = \dfrac{2m_2}{na}\dfrac{\partial R_{1,\,2}}{\partial \sigma}, \\[2.5ex] \dfrac{de}{dt} = \dfrac{m_2(1-e^2)}{na^2 e}\dfrac{\partial R_{1,\,2}}{\partial \sigma} - \dfrac{m_2\sqrt{1-e^2}}{na^2 e}\dfrac{\partial R_{1,\,2}}{\partial \omega}, \\[2.5ex] \dfrac{\partial \sigma}{dt} = -\dfrac{m_2(1-e^2)}{na^2 e}\dfrac{\partial R_{1,\,2}}{\partial e} - \dfrac{2m_2}{na}\dfrac{\partial R_{1,\,2}}{\partial a}. \end{cases}$$

The perturbative function $R_{1,\,2}$ involves the element a explicitly, and also implicitly through n which enters only in the combination $nt + \sigma$. Consequently the last equation of (61) becomes

$$(62)\quad \dfrac{d\sigma}{dt} = -\dfrac{m_2(1-e^2)}{na^2 e}\dfrac{\partial R_{1,\,2}}{\partial e} - \dfrac{2m_2}{na}\left(\dfrac{\partial R_{1,\,2}}{\partial a}\right) - \dfrac{2m_2}{na}\dfrac{\partial R_{1,\,2}}{\partial n}\dfrac{\partial n}{\partial a},$$

where the partial derivative in parenthesis indicates the derivative is taken only so far as the parameter appears explicitly.

It follows from the combination $nt + \sigma$ that

$$(63)\quad \dfrac{2m_2}{na}\dfrac{\partial R_{1,\,2}}{\partial n} = \dfrac{2m_2 t}{na}\dfrac{\partial R_{1,\,2}}{\partial \sigma} = t\dfrac{da}{dt}.$$

It will be shown [Arts. 225–227] that $\dfrac{\partial R_{1,\,2}}{\partial \sigma}$ is a sum of periodic terms; therefore σ, as defined by (62), contains terms which are the products of t and trigonometric terms. It is obvious that such an element is inconvenient when large values of t are to be used.

In order to avoid this difficulty Leverrier used* in place of σ the mean longitude from the perihelion as an element. It is defined by

$$(64) \qquad\qquad l = \int n\,dt + \sigma,$$

whence

$$(65) \qquad\qquad \frac{dl}{d} = n + t\frac{dn}{dt} + \frac{d\sigma}{dt}.$$

Since $n = \dfrac{k\sqrt{S + m_1}}{a^{\frac{3}{2}}}$, it follows that

$$(66) \qquad \frac{\partial n}{\partial a} = -\frac{3}{2}\frac{n}{a}, \qquad\qquad \frac{dn}{dt} = -\frac{3n}{2a}\frac{da}{dt}.$$

Therefore equation (65) becomes, on making use of (62),

$$(67) \qquad \frac{dl}{dt} = n - \frac{m_2(1 - e^2)}{na^2 e}\frac{\partial R_{1,\,2}}{\partial e} - \frac{2m_2}{na}\left(\frac{\partial R_{1,\,2}}{\partial a}\right).$$

Since $\dfrac{\partial R_{1,\,2}}{\partial \sigma} = \dfrac{\partial R_{1,\,2}}{\partial l}$, the fourth and fifth equations, where alone the partial derivative of $R_{1,\,2}$ with respect to σ occurs, will not be changed in form. Hence, if l is used in place of σ throughout (61), the equations will be unchanged in form, and the partial derivative of $R_{1,\,2}$ with respect to a is to be taken only so far as a occurs explicitly.

219. Change from Ω, ω, and σ to Ω, π, and ϵ. The transformation from the elements Ω, ω, and σ to Ω, π, and ϵ is readily made because the relations between the ω and σ and the π and ϵ are very simple. It follows from the definitions of Arts. 214 and 215 that

$$(68) \qquad \begin{cases} \Omega = \Omega, \\ \omega = \pi - \Omega, \\ \sigma = \epsilon - \pi; \end{cases}$$

whence

$$(69) \qquad \begin{cases} \dfrac{d\Omega}{dt} = \dfrac{d\Omega}{dt}, \\[2mm] \dfrac{d\omega}{dt} = \dfrac{d\pi}{dt} - \dfrac{d\Omega}{dt}, \\[2mm] \dfrac{d\sigma}{dt} = \dfrac{d\epsilon}{dt} - \dfrac{d\pi}{dt}. \end{cases}$$

On solving (68) for Ω, π, and ϵ in terms of Ω, ω, and σ, it is found that

* *Annales de l'Observatoire de Paris*, vol. I., p. 255.

$$(70) \quad \begin{cases} \Omega = \Omega, \\ \pi = \omega + \Omega, \\ \epsilon = \sigma + \pi = \sigma + \omega + \Omega. \end{cases}$$

Hence the transformations in the partial derivatives are given by the equations

$$(71) \quad \begin{cases} \dfrac{\partial R_{1,2}}{\partial \Omega} = \left(\dfrac{\partial R_{1,2}}{\partial \Omega} \right) \dfrac{\partial \Omega}{\partial \Omega} + \left(\dfrac{\partial R_{1,2}}{\partial \pi} \right) \dfrac{\partial \pi}{\partial \Omega} + \left(\dfrac{\partial R_{1,2}}{\partial \epsilon} \right) \dfrac{\partial \epsilon}{\partial \Omega} \\[2mm] \qquad = \left(\dfrac{\partial R_{1,2}}{\partial \Omega} \right) + \left(\dfrac{\partial R_{1,2}}{\partial \pi} \right) + \left(\dfrac{\partial R_{1,2}}{\partial \epsilon} \right), \\[4mm] \dfrac{\partial R_{1,2}}{\partial \omega} = \left(\dfrac{\partial R_{1,2}}{\partial \Omega} \right) \dfrac{\partial \Omega}{\partial \omega} + \left(\dfrac{\partial R_{1,2}}{\partial \pi} \right) \dfrac{\partial \pi}{\partial \omega} + \left(\dfrac{\partial R_{1,2}}{\partial \epsilon} \right) \dfrac{\partial \epsilon}{\partial \omega} \\[2mm] \qquad = \left(\dfrac{\partial R_{1,2}}{\partial \pi} \right) + \left(\dfrac{\partial R_{1,2}}{\partial \epsilon} \right), \\[4mm] \dfrac{\partial R_{1,2}}{\partial \sigma} = \left(\dfrac{\partial R_{1,2}}{\partial \Omega} \right) \dfrac{\partial \Omega}{\partial \sigma} + \left(\dfrac{\partial R_{1,2}}{\partial \pi} \right) \dfrac{\partial \pi}{\partial \sigma} + \left(\dfrac{\partial R_{1,2}}{\partial \epsilon} \right) \dfrac{\partial \epsilon}{\partial \sigma} \\[2mm] \qquad = \left(\dfrac{\partial R_{1,2}}{\partial \epsilon} \right). \end{cases}$$

On substituting (69) and (71) in (61) and omitting the parentheses around the partial derivatives, and on solving for the derivatives of the elements with respect to t, it is found that

$$(72) \quad \begin{cases} \dfrac{d\Omega}{dt} = \dfrac{m_2}{na^2 \sqrt{1 - e^2} \sin i} \dfrac{\partial R_{1,2}}{\partial i}, \\[4mm] \dfrac{di}{dt} = \dfrac{- m_2}{na^2 \sqrt{1-e^2} \sin i} \dfrac{\partial R_{1,2}}{\partial \Omega} - \dfrac{m_2 \tan \frac{i}{2}}{na^2 \sqrt{1-e^2}} \left[\dfrac{\partial R_{1,2}}{\partial \pi} + \dfrac{\partial R_{1,2}}{\partial \epsilon} \right], \\[4mm] \dfrac{d\pi}{dt} = \dfrac{m_2 \tan \frac{i}{2}}{na^2 \sqrt{1-e^2}} \dfrac{\partial R_{1,2}}{\partial i} + \dfrac{m_2 \sqrt{1 - e^2}}{na^2 e} \dfrac{\partial R_{1,2}}{\partial e}, \\[4mm] \dfrac{da}{dt} = \dfrac{2m_2}{na} \dfrac{\partial R_{1,2}}{\partial \epsilon}, \\[4mm] \dfrac{de}{dt} = - m_2 \sqrt{1-e^2} \dfrac{1 - \sqrt{1 - e^2}}{na^2 e} \dfrac{\partial R_{1,2}}{\partial \epsilon} - \dfrac{m_2 \sqrt{1 - e^2}}{na^2 e} \dfrac{\partial R_{1,2}}{\partial \pi}, \\[4mm] \dfrac{d\epsilon}{dt} = \dfrac{m_2 \tan \frac{i}{2}}{na^2 \sqrt{1 - e^2}} \dfrac{\partial R_{1,2}}{\partial i} + m_2 \sqrt{1 - e^2} \dfrac{1 - \sqrt{1 - e^2}}{na^2 e} \dfrac{\partial R_{1,2}}{\partial e} \\[4mm] \qquad\qquad - \dfrac{2m_2}{na} \dfrac{\partial R_{1,2}}{\partial a}. \end{cases}$$

These equations,* together with the corresponding ones for the elements of the planet m_2, constitute a rigorous system of differential equations for the determination of the motion of the planets m_1 and m_2 with respect to the sun when there are no other forces than the mutual attractions of the three bodies.

If $R_{1, 2}$ is expressed in terms of the time and the osculating elements at the epoch t_0, equations (72) become the explicit expressions for the first half of the system (27), and define the perturbations of the elements which are of the first order with respect to the masses.

220. Introduction of Rectangular Components of the Disturbing Acceleration.   Equations (72) require for their application that $R_{1, 2}$ shall be expressed first in terms of the elements, after which the partial derivatives must be formed.   In some cases, especially in the orbits of comets, it is advantageous to have the rates of variation of the elements expressed in terms of three rectangular components of the disturbing acceleration.

The disturbing acceleration will be resolved into three rectangular components W, S, R, where W is the component of acceleration perpendicular to the plane of the orbit with the positive direction toward the north pole; S is the component in the plane of the orbit which acts at right angles to the radius vector with the positive direction making an angle less than 90° with the direction of motion; R is the component acting along the radius vector with the positive direction away from the sun. The components used in the preceding chapter evidently might be employed here instead of these, but the resulting equations would be less simple.

In order to obtain the desired equations it is only necessary to express the partial derivatives of $R_{1, 2}$ with respect to the elements in terms of W, S, and R, and to substitute them in (61) or (72), depending upon the set of elements used.   The transformation will be made for the elements used in equations (61).

The quantities $m_2 \dfrac{\partial R_{1, 2}}{\partial x}$, $m_2 \dfrac{\partial R_{1, 2}}{\partial y}$, $m_2 \dfrac{\partial R_{1, 2}}{\partial z}$ are the components of the disturbing acceleration parallel to the fixed axes of reference.   It follows from the elementary properties of the

---

* The subscript 1, which was omitted from the coördinates and elements in Art. 213, should be replaced when the equations for more than one planet are written.

resolution and composition of accelerations that $m_2 \dfrac{\partial R_{1,2}}{\partial x}$ is equal to the sum of the projections of W, S, and R upon the x-axis, and similarly for the others.

Let u represent the argument of the latitude, or the distance from the ascending node to the planet P, Fig. 61. Then it follows

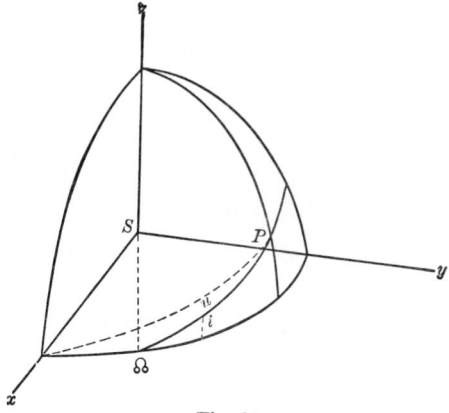

Fig. 61.

from the fundamental formulas of Trigonometry that

$$(73) \begin{cases} m_2 \dfrac{\partial R_{1,2}}{\partial x} = + R(\cos u \cos \Omega - \sin u \sin \Omega \cos i) \\ \qquad - S(\sin u \cos \Omega + \cos u \sin \Omega \cos i) \\ \qquad + W \sin \Omega \sin i, \\ m_2 \dfrac{\partial R_{1,2}}{\partial y} = + R(\cos u \sin \Omega + \sin u \cos \Omega \cos i) \\ \qquad - S(\sin u \sin \Omega - \cos u \cos \Omega \cos i) \\ \qquad - W \cos \Omega \sin i, \\ m_2 \dfrac{\partial R_{1,2}}{\partial z} = + R \sin u \sin i + S \cos u \sin i + W \cos i. \end{cases}$$

Let s represent any of the elements $\Omega, \cdots, \sigma$; then

$$(74) \qquad \frac{\partial R_{1,2}}{\partial s} = \frac{\partial R_{1,2}}{\partial x} \frac{\partial x}{\partial s} + \frac{\partial R_{1,2}}{\partial y} \frac{\partial y}{\partial s} + \frac{\partial R_{1,2}}{\partial z} \frac{\partial z}{\partial s}.$$

The derivatives $\dfrac{\partial R_{1,2}}{\partial x}, \dfrac{\partial R_{1,2}}{\partial y}, \dfrac{\partial R_{1,2}}{\partial z}$ are given in (73) and when $\dfrac{\partial x}{\partial s}, \dfrac{\partial y}{\partial s},$ and $\dfrac{\partial z}{\partial s}$ have been found, the transformation can be completed at once.

It follows from equations (51) that

$$(75) \begin{cases} \dfrac{\partial x}{\partial K} = \xi \dfrac{\partial \alpha}{\partial K} + \eta \dfrac{\partial \alpha'}{\partial K}, & \dfrac{\partial x}{\partial P} = \alpha \dfrac{\partial \xi}{\partial P} + \alpha' \dfrac{\partial \eta}{\partial P}, \\[2mm] \dfrac{\partial y}{\partial K} = \xi \dfrac{\partial \beta}{\partial K} + \eta \dfrac{\partial \beta'}{\partial K}, & \dfrac{\partial y}{\partial P} = \beta \dfrac{\partial \xi}{\partial P} + \beta' \dfrac{\partial \eta}{\partial P}, \\[2mm] \dfrac{\partial z}{\partial K} = \xi \dfrac{\partial \gamma}{\partial K} + \eta \dfrac{\partial \gamma'}{\partial K}, & \dfrac{\partial z}{\partial P} = \gamma \dfrac{\partial \xi}{\partial P} + \gamma' \dfrac{\partial \eta}{\partial P}, \end{cases}$$

where K is any of the elements Ω, i, ω, and P any of the elements a, e, σ. The quantities $\alpha, \cdots, \gamma'$ are defined in (44), and their derivatives are given in (47), (48), and (49); the derivatives $\dfrac{\partial \xi}{\partial P}$ and $\dfrac{\partial \eta}{\partial P}$ are to be computed from (46).

It is found after some rather long but simple reductions that

$$(76) \begin{cases} m_2 \dfrac{\partial R_{1,2}}{\partial \Omega} = Sr \cos i - Wr \cos u \sin i, \\[3mm] m_2 \dfrac{\partial R_{1,2}}{\partial i} = Wr \sin u, \\[3mm] m_2 \dfrac{\partial R_{1,2}}{\partial \omega} = Sr, \\[3mm] m_2 \dfrac{\partial R_{1,2}}{\partial a} = R \dfrac{r}{a}, \\[3mm] m_2 \dfrac{\partial R_{1,2}}{\partial e} = - Ra \cos v + S \left[1 + \dfrac{r}{p} \right] a \sin v, \\[3mm] m_2 \dfrac{\partial R_{1,2}}{\partial \sigma} = \dfrac{Rae}{\sqrt{1 - e^2}} \sin v + S \dfrac{a^2}{r} \sqrt{1 - e^2}. \end{cases}$$

Therefore equations (61) become

$$(77) \begin{cases} \dfrac{d\Omega}{dt} = \dfrac{r \sin u}{na^2 \sqrt{1 - e^2} \sin i} W, \\[3mm] \dfrac{di}{dt} = \dfrac{r \cos u}{na^2 \sqrt{1 - e^2}} W, \\[3mm] \dfrac{d\omega}{dt} = \dfrac{- \sqrt{1 - e^2} \cos v}{nae} R + \dfrac{\sqrt{1 - e^2}}{nae} \left[1 + \dfrac{r}{p} \right] \sin v \, S \\[3mm] \qquad\qquad - \dfrac{r \sin u \cot i}{na^2 \sqrt{1 - e^2}} W, \end{cases}$$

$$(77) \quad \begin{cases} \dfrac{da}{dt} = \dfrac{2e \sin v}{n \sqrt{1 - e^2}} R + \dfrac{2a \sqrt{1 - e^2}}{nr} S, \\[2ex] \dfrac{de}{dt} = \dfrac{\sqrt{1 - e^2} \sin v}{na} R + \dfrac{\sqrt{1 - e^2}}{na^2 e} \left[\dfrac{a^2(1 - e^2)}{r} - r \right] S, \\[2ex] \dfrac{d\sigma}{dt} = -\dfrac{1}{na} \left[\dfrac{2r}{a} - \dfrac{1 - e^2}{e} \cos v \right] R \\[2ex] \qquad\qquad\qquad - \dfrac{(1 - e^2)}{nae} \left[1 + \dfrac{r}{p} \right] \sin v \, S. \end{cases}$$

XXVI. PROBLEMS.

1. Find the components S and R of this chapter in terms of T and N, which were used in chapter IX., Art. 174.

$$\textit{Ans.} \quad \begin{cases} S = \dfrac{(1 + e \cos v)}{\sqrt{1 + e^2 + 2e \cos v}} T + \dfrac{e \sin v}{\sqrt{1 + e^2 + 2e \cos v}} N, \\[2ex] R = \dfrac{e \sin v}{\sqrt{1 + e^2 + 2e \cos v}} T - \dfrac{1 + e \cos v}{\sqrt{1 + e^2 + 2e \cos v}} N. \end{cases}$$

2. By means of the equations of problem 1 express the variations of the elements $\Omega, \cdots, \sigma$ in terms of T and N, and verify all the results contained in the Table of Art. 182.

3. Explain why $\dfrac{d\omega}{dt}$ contains a term depending upon W.

4. Suppose the disturbed body moves in a resisting medium; find the equations for the variations of the elements.

$$\textit{Ans.} \quad \begin{cases} \dfrac{d\Omega}{dt} = 0, \\[2ex] \dfrac{di}{dt} = 0, \\[2ex] \dfrac{d\omega}{dt} = \dfrac{2 \sqrt{1 - e^2}}{nae} \dfrac{\sin v}{\sqrt{1 + e^2 + 2e \cos v}} T, \\[2ex] \dfrac{da}{dt} = \dfrac{2 \sqrt{1 + e^2 + 2e \cos v}}{n \sqrt{1 - e^2}} T, \\[2ex] \dfrac{de}{dt} = \dfrac{2 \sqrt{1 - e^2}\,(\cos v + e)}{na \sqrt{1 + e^2 + 2e \cos v}} T, \\[2ex] \dfrac{d\sigma}{dt} = -\dfrac{2(1 - e^2)(1 + e^2 + e \cos v) \sin v}{nae(1 + e \cos v) \sqrt{1 + e^2 + 2e \cos v}} T. \end{cases}$$

5. Discuss the way in which the elements vary in the last problem, including the values of v for which the maxima and minima in their rates of change occur, when T is a constant, and when it varies as the square of the velocity.

6. Derive the equations corresponding to (77) for the elements Ω, i, π, a, e, and ϵ.

Ans.

$$
\begin{cases}
\dfrac{d\Omega}{dt} = \dfrac{r\,\sin u}{na\,\sqrt{1-e^2}\,\sin i}\,W, \\[2mm]
\dfrac{di}{dt} = \dfrac{r\,\cos u}{na^2\,\sqrt{1-e^2}}W, \\[2mm]
\dfrac{d\pi}{dt} = 2\,\sin^2\dfrac{i}{2}\dfrac{d\Omega}{dt} + \dfrac{\sqrt{1-e^2}}{nae}\left\{ -R\cos v + S\left(1+\dfrac{r}{p}\right)\sin v\right\}, \\[2mm]
\dfrac{da}{dt} = \dfrac{2}{n\sqrt{1-e^2}}\left(Re\sin v + S\dfrac{p}{r}\right), \\[2mm]
\dfrac{de}{dt} = \dfrac{\sqrt{1-e^2}}{na}\left\{ R\sin v + S\left(\dfrac{e+\cos v}{1+e\cos v}+\cos v\right)\right\}, \\[2mm]
\dfrac{d\epsilon}{dt} = -\dfrac{2rR}{na^2} + \dfrac{e^2}{1+\sqrt{1-e^2}}\dfrac{d\pi}{dt} + 2\,\sqrt{1-e^2}\,\sin^2\dfrac{i}{2}\dfrac{d\Omega}{dt}.
\end{cases}
$$

221. Development of the Perturbative Function. In order to apply equations (72) the perturbative function $R_{1,2}$ must be developed explicitly in terms of the elements and the time. From this point on only perturbations of the first order will be considered; therefore, in accordance with the results of Art. 208, the elements which appear in $R_{1,2}$ are the osculating elements at the time t_0.

In the notation of Art. 205 the perturbative function is

$$
(78)\quad
\begin{cases}
R_{1,2} = k^2\left[\dfrac{1}{r_{1,2}} - \dfrac{x_1 x_2 + y_1 y_2 + z_1 z_2}{r_2^{3}}\right], \\[2mm]
r_{1,2} = \sqrt{(x_2-x_1)^2 + (y_2-y_1)^2 + (z_2-z_1)^2}, \\[2mm]
r_2 = \sqrt{x_2^{2} + y_2^{2} + z_2^{2}}.
\end{cases}
$$

The perturbing forces evidently depend upon the mutual inclinations of the orbits, rather than upon their inclinations independently to the fixed plane of reference. It will be convenient, therefore, to develop $R_{1,2}$ in terms of the mutual inclination. Since this angle is expressible in terms of i_1, i_2, Ω_1, and Ω_2, the partial derivatives of $R_{1,2}$ with respect to these elements will depend in part on their occurring implicitly in this angle.

The development of the perturbative function consists of three steps:*

---

* There are many more or less important variations of the method outlined here, which is based on the work of Leverrier in the *Annales de l'Observatoire de Paris*, vol I.

(a) Development of $R_{1,2}$ as a power series in the square of the sine of half the mutual inclination of the orbits.

(b) Development of the coefficients of the series obtained in (a) into power series in e_1 and e_2.

(c) Development of the coefficients of the preceding series into Fourier series in the mean longitudes of the two planets and the angular variables π_1, π_2, Ω_1, and Ω_2.

In the little space available here it will not be possible to give more than a general outline of the operations which are necessary to effect the complete development. A detailed discussion is given in Tisserand's *Mécanique Céleste*, vol. I., chapters XII. to XVIII. inclusive.

222. (a) **Development of** $R_{1,2}$ **in the Mutual Inclination.** Let S represent the angle between the radii r_1 and r_2; then

$$(79) \qquad \frac{1}{r_{1,2}} = (r_1{}^2 + r_2{}^2 - 2r_1r_2 \cos S)^{-\frac{1}{2}}.$$

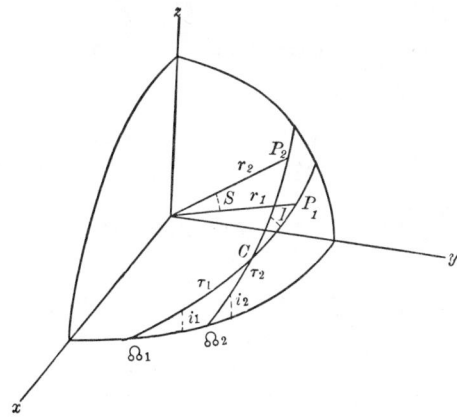

Fig. 62.

Let the angles between r_1 and the x, y, and z-axes be α_1, β_1, γ_1 respectively, and in the case of r_2, α_2, β_2, and γ_2. Then it follows that

$$(80) \qquad x_1 = r_1 \cos \alpha_1, \qquad y_1 = r_1 \cos \beta_1, \qquad z_1 = r_1 \cos \gamma_1, \ \text{etc.,}$$

and

$$(81) \qquad \begin{aligned} x_1x_2 + y_1y_2 + z_1z_2 &= r_1r_2(\cos \alpha_1 \cos \alpha_2 + \cos \beta_1 \cos \beta_2 \\ &\quad + \cos \gamma_1 \cos \gamma_2) = r_1r_2 \cos S. \end{aligned}$$

Let I represent the angle between the two orbits, and τ_1 and τ_2

the distances from their ascending nodes to their point of inter-section. From the spherical triangle $P_1 P_2 C$ the value of $\cos S$ is found to be

$$(82) \begin{cases} \cos S = \cos (u_1 - \tau_1) \cos (u_2 - \tau_2) \\ \qquad\qquad + \sin (u_1 - \tau_1) \sin (u_2 - \tau_2) \cos I, \quad \text{or} \\ \cos S = \cos (u_1 - u_2 + \tau_2 - \tau_1) \\ \qquad\qquad - 2 \sin (u_1 - \tau_1) \sin (u_2 - \tau_2) \sin^2 \dfrac{I}{2}, \\ u_1 - \tau_1 = v_1 + \pi_1 - \Omega_1 - \tau_1, \\ u_2 - \tau_2 = v_2 + \pi_2 - \Omega_2 - \tau_2. \end{cases}$$

The quantities I, τ_1, and τ_2 are determined by the formulas of Gauss applied to the triangle $\Omega_1 \Omega_2 C$:

$$(83) \begin{cases} \sin I \sin \tau_1 = \sin i_2 \sin (\Omega_1 - \Omega_2), \\ \sin I \sin \tau_2 = \sin i_1 \sin (\Omega_1 - \Omega_2), \\ \sin I \cos \tau_1 = \sin i_1 \cos i_2 - \cos i_1 \sin i_2 \cos (\Omega_1 - \Omega_2), \\ \sin I \cos \tau_2 = - \cos i_1 \sin i_2 + \sin i_1 \cos i_2 \cos (\Omega_1 - \Omega_2), \\ \cos I = \cos i_1 \cos i_2 + \sin i_1 \sin i_2 \cos (\Omega_1 - \Omega_2). \end{cases}$$

For simplicity I, τ_1, and τ_2 will be retained, but it must be remembered when the partial derivatives of $R_{1,2}$ are taken that they are functions of i_1, i_2, Ω_1, and Ω_2.

As a consequence of (79), (81), and (82), the perturbative function can be written in the form

$$(84) \begin{cases} R_{1,2} = [r_1^2 + r_2^2 - 2 r_1 r_2 \cos (u_1 - u_2 + \tau_2 - \tau_1)]^{-\frac{1}{2}} \\ \quad \times \left[1 + \dfrac{4 r_1 r_2 \sin (u_1 - \tau_1) \sin (u_2 - \tau_2) \sin^2 \dfrac{I}{2}}{r_1^2 + r_2^2 - 2 r_1 r_2 \cos (u_1 - u_2 + \tau_2 - \tau_1)} \right]^{-\frac{1}{2}} \\ \quad - \dfrac{r_1}{r_2^2} \left[\cos (u_1 - u_2 + \tau_2 - \tau_1) \right. \\ \qquad\qquad \left. - 2 \sin (u_1 - \tau_1) \sin (u_2 - \tau_2) \sin^2 \dfrac{I}{2} \right]. \end{cases}$$

The radii r_1 and r_2 are independent of I. The second factor of the first term of the right member of this equation can be expanded by the binomial theorem into an absolutely converging power series in $\sin^2 \dfrac{I}{2}$ so long as the numerical value of

$$(85) \qquad \frac{4r_1r_2 \sin (u_1 - \tau_1) \sin (u_2 - \tau_2) \sin^2 \dfrac{I}{2}}{r_1^2 + r_2^2 - 2r_1r_2 \cos (u_1 - u_2 + \tau_2 - \tau_1)}$$

is less than unity. This fraction is less than, or at most equal to,

$$(86) \qquad \frac{4r_1r_2 \sin^2 \dfrac{I}{2}}{(r_1 - r_1)^2}.$$

If this expression is less than unity for all the values which r_1 and r_2 can take in the given ellipses the expansion of (84) is valid for all values of the time. In the case of the major planets it is always very small, the greatest value of $\sin^2 \dfrac{I}{2}$ being for Mercury and Mars, 0.0118. In the perturbations of the planetoids by Jupiter it often fails, for I is sometimes of considerable magnitude while $r_2 - r_1$ may become very small. In the case of Mars and Eros $r_2 - r_1$ may actually vanish and this mode of development consequently fails. It is needless to say that it is not generally applicable in the cometary orbits.

In those cases in which the expansion of (84) does not fail, the expression for $R_{1,2}$ becomes

$$(87) \quad \left\{ \begin{aligned} R_{1,2} = &+ [r_1^2 + r_2^2 - 2r_1r_2 \cos (u_1 - u_2 + \tau_2 - \tau_1)]^{\frac{1}{2}} \\ &- r_1r_2[r_1^2 + r_2^2 - 2r_1r_2 \cos (u_1 - u_2 + \tau_2 - \tau_1)]^{-\frac{3}{2}} \\ &\qquad \times 2 \sin (u_1 - \tau_1) \sin (u_2 - \tau_2) \sin^2 \frac{I}{2} \\ &+ r_1^2r_2^2[r_1^2 + r_2^2 - 2r_1r_2 \cos (u_1 - u_2 + \tau_2 - \tau_1)]^{\frac{1}{2}} \\ &\qquad \times 6 \sin^2 (u_1 - \tau_1) \sin^2 (u_2 - \tau_2) \sin^4 \frac{I}{2} \\ &+ \quad \cdot \quad \cdot \quad \cdot \quad \cdot \quad \cdot \quad \cdot \quad \cdot \quad \cdot \quad \cdot \quad \cdot \\ &- \frac{r_1}{r_2^2} \cos (u_1 - u_2 + \tau_2 - \tau_1) \\ &+ \frac{2r_1}{r_2^2} \sin (u_1 - \tau_1) \sin (u_2 - \tau_2) \sin^2 \frac{I}{2}. \end{aligned} \right.$$

223. (b) **Development of the Coefficients in powers of e_1 and e_2.** The radii r_1 and r_2 vary from $a_1(1 - e_1)$ and $a_2(1 - e_2)$ to $a_1(1 + e_1)$ and $a_2(1 + e_2)$ respectively. Let

$$(88) \qquad \left\{ \begin{aligned} r_1 &= a_1(1 + \rho_1), \\ r_2 &= a_2(1 + \rho_2). \end{aligned} \right.$$

The angles u_1 and u_2 are expressed in terms of the true anomalies, v_1 and v_2, and the elements by (82). The true anomalies are equal to the mean anomalies plus the equations of the center, which may be denoted by w_1 and w_2. Let l_1 and l_2 represent the mean longitudes counted from the x-axis [Fig. (62)]; then

$$(89) \qquad \begin{cases} u_1 - \tau_1 = l_1 - \Omega_1 - \tau_1 + w_1, \\ u_2 - \tau_2 = l_2 - \Omega_2 - \tau_2 + w_2. \end{cases}$$

It follows from (81) that $R_{1,2}$ can be written in the form

$$(89a) \qquad R_{1,2} = F[a_1(1 + \rho_1), a_2(1 + \rho_2)],$$

where F is a homogeneous function of a_1 and a_2 of degree -1. Therefore

$$(90) \qquad R_{1,2} = \frac{1}{1 + \rho_2} F\left[a_1 + a_1 \frac{\rho_1 - \rho_2}{1 + \rho_2}, a_2 \right].$$

The right member of this equation can be developed by Taylor's formula, giving

$$(91) \qquad \begin{aligned} R_{1,2} = \frac{1}{1 + \rho_2} \Bigg\{ & F(a_1, a_2) + \frac{\rho_1 - \rho_2}{1 + \rho_2} \frac{a_1}{1} \frac{\partial F(a_1, a_2)}{\partial a_1} \\ & + \left(\frac{\rho_1 - \rho_2}{1 + \rho_2} \right)^2 \frac{a_1^2}{1 \cdot 2} \frac{\partial^2 F(a_1, a_2)}{\partial a_1^2} + \cdots \Bigg\}. \end{aligned}$$

The expressions $\left(\dfrac{\rho_1 - \rho_2}{1 + \rho_2} \right)^i$ can be developed as power series in ρ_1 and ρ_2. But in Art. 100, equation (62), ρ is given as a power series in e whose coefficients are cosines of multiples of the mean anomaly. On making these expansions and substitutions in (91), $R_{1,2}$ can be arranged as a power series in e_1 and e_2. These operations are to be actually performed upon the separate terms of the series (87), so the resulting series is arranged according to powers of e_1, e_2, and $\sin^2 \dfrac{I}{2}$. The angles w_1 and w_2 also depend upon e_1 and e_2 respectively, but their developments will not be introduced until after the next step.

224. (c) **Developments in Fourier Series.** The first term within the bracket of (91) is obtained by replacing r_1 and r_2 by a_1 and a_2 respectively in (87). The higher terms involve the derivatives of the first with respect to a_1. On referring to the explicit series in (87), it is seen that the development of the expressions of the type

$$(a_1 a_2)^{\frac{\nu-1}{2}} [a_1^2 + a_2^2 - 2a_1 a_2 \cos (u_1 - u_2 + \tau_2 - \tau_1)]^{-\frac{\nu}{2}},$$

where ν is an odd integer, must be considered.

Let $u_1 - u_2 + \tau_2 - \tau_1 = \psi$. It is known from the theory of Fourier series when a_1 and a_2 are unequal, as is assumed, that $[a_1^2 + a_2^2 - 2a_1 a_2 \cos \psi]^{-\frac{\nu}{2}}$ can be developed into a series of cosines of multiples of ψ, which is convergent for all values of ψ. That is,

$$(92) \quad (a_1 a_2)^{\frac{\nu-1}{2}} [a_1^2 + a_2^2 - 2a_1 a_2 \cos \psi]^{-\frac{\nu}{2}} = \frac{1}{2} \sum_{i=-\infty}^{+\infty} B_\nu^{(i)} \cos i\psi,$$

where $B_\nu^{(i)} = B_\nu^{(-i)}$.

The coefficients $B_\nu^{(i)}$ are of course given by Fourier's integral

$$B_\nu^{(i)} = \frac{1}{\pi} \int_0^{2\pi} (a_1 a_2)^{\frac{\nu-1}{2}} [a_1^2 + a_2^2 - 2a_1 a_2 \cos \psi]^{-\frac{\nu}{2}} \cos i\psi \, d\psi,$$

but the difficulty of finding the integral makes it advisable in this particular problem to proceed otherwise.

Let $z = e^{\sqrt{-1}\psi}$, where e represents the Napierian base. Then

$$2 \cos \psi = z + z^{-1}, \qquad 2 \cos i\psi = z^i + z^{-i}.$$

Suppose $a_2 > a_1$ and let $\dfrac{a_1}{a_2} = \alpha$; then (92) becomes

$$(93) \quad \frac{\alpha^{\frac{\nu-1}{2}}}{a_2} (1 + \alpha^2 - 2\alpha \cos \psi)^{-\frac{\nu}{2}} = \frac{1}{2} \sum_{i=-\infty}^{+\infty} B_\nu^{(i)} \cos i\psi.$$

Let

$$(1 + \alpha^2 - 2\alpha \cos \psi)^{-\frac{\nu}{2}} = (1 - \alpha z)^{-\frac{\nu}{2}} (1 - \alpha z^{-1})^{-\frac{\nu}{2}} = \frac{1}{2} \sum_{i=-\infty}^{+\infty} b_\nu^{(i)} z^i;$$

therefore

$$(94) \qquad B_\nu^{(i)} = \frac{\alpha^{\frac{\nu-1}{2}}}{a_2} b_\nu^{(i)}.$$

Since the absolute values of αz and αz^{-1} are less than unity for all real values of ψ, the factors $(1 - \alpha z)^{-\frac{\nu}{2}}$ and $(1 - \alpha z^{-1})^{-\frac{\nu}{2}}$ can be expanded by the binomial theorem into convergent power series in αz and αz^{-1}. The coefficient of z^i in the product of these series is $\frac{1}{2} b_\nu^{(i)}$, after which $B_\nu^{(i)}$ is obtained from (94). The general term of the product of the expansions is easily found to be

$$(95) \quad \tfrac{1}{2}b_\nu{}^{(i)} = \frac{\dfrac{\nu}{2}\left(\dfrac{\nu}{2}+1\right)\cdots\left(\dfrac{\nu}{2}+i-1\right)}{i!}\,\alpha^i\left[1+\frac{\dfrac{\nu}{2}}{1}\cdot\frac{\dfrac{\nu}{2}+i}{i+1}\alpha^2\right.$$

$$\left.+\frac{\dfrac{\nu}{2}\left(\dfrac{\nu}{2}+1\right)}{1\cdot2}\cdot\frac{\left(\dfrac{\nu}{2}+i\right)\left(\dfrac{\nu}{2}+i+1\right)}{(i+1)(i+2)}\alpha^4+\cdots\right].$$

In this manner the coefficients of $\rho_1{}^{j_1}\rho_2{}^{j_2}\left(\sin^2\dfrac{I}{2}\right)^k$ are developed in Fourier series in $\cos i(u_1 - u_2 + \tau_2 - \tau_1)$. But these functions are multiplied by the factors $\sin(u_1 - \tau_1)\sin(u_2 - \tau_2)$ raised to different powers [equation (87)]. These powers of sines are to be reduced to sines and cosines of multiples of the arguments, and the products formed with $\cos i(u_1 - u_2 + \tau_2 - \tau_1)$, and the reduction again made to sines and cosines of multiples of arcs. The final trigonometrical terms will have the form $\cos(j_1u_1 + j_2u_2 + k_1\tau_1 + k_2\tau_2)$, where j_1, j_2, k_1, and k_2 are integers. As a consequence of (89) this expression can be developed into

$$(96)\begin{cases}\cos(j_1l_1 + j_2l_2 - j_1\Omega_1 - j_2\Omega_2 + k_1\tau_1 + k_2\tau_2 + j_1w_1 + j_2w_2)\\[4pt] = \cos(j_1l_1 + j_2l_2 - j_1\Omega_1 - j_2\Omega_2 + k_1\tau_1 + k_2\tau_2)\\[4pt] \qquad\times\{\cos(j_1w_1)\cos(j_2w_2) - \sin(j_1w_1)\sin(j_2w_2)\}\\[4pt] \quad - \sin(j_1l_1 + j_2l_2 - j_1\Omega_1 - j_2\Omega_2 + k_1\tau_1 + k_2\tau_2)\\[4pt] \qquad\times\{\sin(j_1w_1)\cos(j_2w_2) + \cos(j_1w_1)\sin(j_2w_2)\}.\end{cases}$$

Since

$$\begin{cases}l_1 = \Omega_1 + \omega_1 + n_1(t_0 - T_1) + n_1(t - t_0) = n_1t + \epsilon_1,\\[4pt] l_2 = \Omega_2 + \omega_2 + n_2(t_0 - T_2) + n_2(t - t_0) = n_2t + \epsilon_2,\end{cases}$$

the first factors of the terms in the right member of this equation are independent of e_1 and e_2. $\cos(j_1w_1)$, etc., are to be expanded into power series in w_1 and w_2 by the usual methods. Now $w_1 = v_1 - M_1$, $w_2 = v_2 - M_2$, and these quantities were developed into power series in e_1 and e_2 [Art. 100, eq. (64)] whose coefficients were Fourier series with multiples of the mean anomaly as arguments. On substituting these series for w_1 and w_2 in the expansions of the second factors of the terms of the right member of (96), and reducing the powers of sines and cosines of the mean anomaly to sines and cosines of multiples of the mean anomaly, and multiplying by the factors

$$\cos(j_1l_1 + j_2l_2 - j_1\Omega_1 - j_2\Omega_2 + k_1\tau_1 + k_2\tau_2)$$

and

$$\sin (j_1 l_1 + j_2 l_2 - j_1 \Omega_1 - j_2 \Omega_2 + k_1 \tau_1 + k_2 \tau_2),$$

and again reducing to sines and cosines of multiples of the arguments, the expression (96) is developed as a power series in e_1 and e_2 whose coefficients are series in sines and cosines of sums of multiples of l_1, l_2, Ω_1, Ω_2, τ_1, τ_2, M_1, M_2. But $M_1 = l_1 - \pi_1$, $M_2 = l_2 - \pi_2$; therefore the arguments will be l_1, l_2, Ω_1, Ω_2, τ_1, τ_2, π_1, π_2, where τ_1 and τ_2 are functions of Ω_1, Ω_2, i_1, and i_2 defined by (83).

When the several expansions and reductions which have been described have all been made, $R_{1,\,2}$ will be developed in a power series in e_1, e_2, and $\sin^2 \dfrac{I}{2}$, the coefficients of which are series of sines and cosines of multiples of l_1, l_2, Ω_1, Ω_2, τ_1, τ_2, π_1, π_2, the coefficient of each trigonometric term depending upon the ratio of the major semi-axes. If the signs of Ω_1, Ω_2, π_1, π_2, τ_1, τ_2, ϵ_1, ϵ_2, and t are changed the value of $R_{1,\,2}$, as defined in (84), obviously is unchanged; therefore the expansion in question contains only cosines of the argument. Hence

$$(97) \quad \left\{ \begin{aligned} R_{1,\,2} &= \Sigma C \cos D, \\ D &= j_1(n_1 t + \epsilon_1) + j_2(n_2 t + \epsilon_2) - j_1' \Omega_1 - j_2' \Omega_2 \\ &\qquad + k_1 \tau_1 + k_2 \tau_2 + k_1' \pi_1 + k_2' \pi_2, \\ C &= f\left(a_1,\ a_2,\ e_1,\ e_2,\ \sin^2 \frac{I}{2} \right), \end{aligned} \right.$$

in which j_1, $\cdots$, k_2' take all integral values, positive, negative, and zero, the summation being extended over all of these terms.

It is clear from the foregoing that the series for $R_{1,\,2}$ is very complicated and that much labor is required to expand it in any particular case. Leverrier has carried out the literal development of all terms up to the seventh order inclusive in e_1, e_2, $\sin^2 \dfrac{I}{2}$, and the length of the work is such that fifty-three quarto pages of the first volume of the *Annales de l'Observatoire de Paris* are required in order to write out the result.

225. Periodic Variations. It follows from equations (72) and (97) that the rates of change of the elements of m_1 are given by

$$(98) \begin{cases} \dfrac{d\Omega_1}{dt} = \dfrac{m_2}{n_1 a_1{}^2 \sqrt{1 - e_1{}^2}\sin i_1} \sum \left\{ \dfrac{\partial C}{\partial i_1}\cos D \right. \\ \left. \qquad\qquad\qquad - \left[k_1 \dfrac{\partial \tau_1}{\partial i_1} + k_2 \dfrac{\partial \tau_2}{\partial i_1} \right] C \sin D \right\}, \\[2ex] \dfrac{di_1}{dt} = \dfrac{-m_2}{n_1 a_1{}^2 \sqrt{1 - e_1{}^2}\sin i_1} \sum \left\{ j_1' - k_1 \dfrac{\partial \tau_1}{\partial \Omega_1} - k_2 \dfrac{\partial \tau_2}{\partial \Omega_1} \right\} \\ \qquad\qquad\qquad\qquad\qquad\qquad\qquad \times C \sin D \\[1ex] \qquad + \dfrac{m_2 \tan \dfrac{i_1}{2}}{n_1 a_1{}^2 \sqrt{1 - e_1{}^2}} \sum \left\{ k_1' + j_1 + k_1 \dfrac{\partial \tau_1}{\partial \pi_1} + k_2 \dfrac{\partial \tau_2}{\partial \pi_1} \right\} \\ \qquad\qquad\qquad\qquad\qquad\qquad\qquad \times C \sin D, \\[2ex] \dfrac{d\pi_1}{dt} = \dfrac{m_2 \tan \dfrac{i_1}{2}}{n_1 a_1{}^2 \sqrt{1 - e_1{}^2}} \sum \left\{ \dfrac{\partial C}{\partial i_1}\cos D - \left[k_1 \dfrac{\partial \tau_1}{\partial i_1} + k_2 \dfrac{\partial \tau_2}{\partial i_1} \right] \right. \\ \left. \qquad\qquad \times C \sin D \right\} + \dfrac{m_2 \sqrt{1 - e_1{}^2}}{n_1 a_1{}^2 e_1} \sum \dfrac{\partial C}{\partial e_1}\cos D, \\[2ex] \dfrac{da_1}{dt} = \dfrac{-2m_2}{n_1 a_1} \sum j_1 C \sin D, \\[2ex] \dfrac{de_1}{dt} = m_2 \sqrt{1 - e_1{}^2} \dfrac{1 - \sqrt{1 - e_1{}^2}}{n_1 a_1{}^2 e_1} \sum j_1 C \sin D \\ \qquad + \dfrac{m_2 \sqrt{1 - e_1{}^2}}{n_1 a_1{}^2 e_1} \sum \left\{ k_1' + k_1 \dfrac{\partial \tau_1}{\partial \pi_1} + k_2 \dfrac{\partial \tau_2}{\partial \pi_1} \right\} C \sin D, \\[2ex] \dfrac{d\epsilon_1}{dt} = \dfrac{m_2 \tan \dfrac{i_1}{2}}{n_1 a_1{}^2 \sqrt{1 - e_1{}^2}} \sum \left\{ \dfrac{\partial C}{\partial i_1}\cos D - \left[k_1 \dfrac{\partial \tau_1}{\partial i_1} + k_2 \dfrac{\partial \tau_2}{\partial i_1} \right] \right. \\ \left. \qquad \times C \sin D \right\} + m_2 \sqrt{1 - e_1{}^2} \dfrac{1 - \sqrt{1 - e_1{}^2}}{n_1 a_1{}^2 e_1} \sum \dfrac{\partial C}{\partial e_1}\cos D \\ \qquad\qquad\qquad\qquad - \dfrac{2m_2}{n_1 a_1} \sum \dfrac{\partial C}{\partial a_1}\cos D. \end{cases}$$

The perturbations of the elements of the orbit of m_1 of the first order with respect to the mass m_2 are the integrals of these equations regarding the elements as constants in the right members. Similar terms must be added for each disturbing planet.

There are terms in $R_{1, 2}$ of three classes: (a) those in which $j_1 n_1 + j_2 n_2$ is distinct from zero and not small; (b) those in which $j_1 n_1 + j_2 n_2$ is very small, but distinct from zero; and (c) those in which $j_1 n_1 + j_2 n_2$ equals zero. Denote the fact that $R_{1, 2}$ contains these three sorts of terms by writing

$$R_{1,\,2} = \Sigma C_0 \cos D_0 + \Sigma C_1 \cos D_1 + \Sigma C_2 \cos D_2,$$

where the three sums in the right member include these three classes of terms respectively. Hence the perturbations of the elements of m_1 by m_2 of the first order and of the first class are

$$(99)\begin{cases}
(\Omega_1{}^{(0,\,1)}) - (\Omega_1{}^{(0,\,1)})_{t_0} = \dfrac{m_2}{n_1 a_1{}^2 \sqrt{1-e_1{}^2}\,\sin i_1} \\[2ex]
\qquad \times \sum \left\{ \dfrac{\partial C_0}{\partial i_1} \dfrac{\sin D_0}{j_1 n_1 + j_2 n_2} + \left[k_1 \dfrac{\partial \tau_1}{\partial i_1} + k_2 \dfrac{\partial \tau_2}{\partial i_1} \right] \dfrac{C_0 \cos D_0}{j_1 n_1 + j_2 n_2} \right\}, \\[2ex]
(i_1{}^{(0,\,1)}) - (i_1{}^{(0,\,1)})_{t_0} = \dfrac{m_2}{n_1 a_1{}^2 \sqrt{1-e_1{}^2}\,\sin i_1} \\[2ex]
\qquad \times \sum \left\{ j_1' - k_1 \dfrac{\partial \tau_1}{\partial \Omega_1} - k_2 \dfrac{\partial \tau_2}{\partial \Omega_1} \right\} \dfrac{C_0 \cos D_0}{j_1 n_1 + j_2 n_2} \\[2ex]
\qquad - \dfrac{m_2 \tan \dfrac{i_1}{2}}{n_1 a_1{}^2 \sqrt{1-e_1{}^2}} \sum \left\{ k_1' + j_1 + k_1 \dfrac{\partial \tau_1}{\partial \pi_1} + k_2 \dfrac{\partial \tau_2}{\partial \pi_1} \right\} \dfrac{C_0 \cos D_0}{j_1 n_1 + j_2 n_2}, \\[2ex]
(\pi_1{}^{(0,\,1)}) - (\pi_1{}^{(0,\,1)})_{t_0} = \dfrac{m_2 \tan \dfrac{i_1}{2}}{n_1 a_1{}^2 \sqrt{1-e_1{}^2}} \sum \left\{ \dfrac{\partial C_0}{\partial i_1} \dfrac{\sin D_0}{j_1 n_1 + j_2 n_2} \right. \\[2ex]
\qquad \left. + \left[k_1 \dfrac{\partial \tau_1}{\partial i_1} + k_2 \dfrac{\partial \tau_2}{\partial i_1} \right] \dfrac{C_0 \cos D_0}{j_1 n_1 + j_2 n_2} \right\} \\[2ex]
\qquad + \dfrac{m_2 \sqrt{1-e_1{}^2}}{n_1 a_1{}^2 e_1} \sum \dfrac{\partial C_0}{\partial e_1} \dfrac{\sin D_0}{j_1 n_1 + j_2 n_2}, \\[2ex]
(a_1{}^{(0,\,1)}) - (a_1{}^{(0,\,1)})_{t_0} = \dfrac{2 m_2}{n_1 a_1} \sum j_1 \dfrac{C_0 \cos D_0}{j_1 n_1 + j_2 n_2}, \\[2ex]
(e_1{}^{(0,\,1)}) - (e_1{}^{(0,\,1)})_{t_0} = - m_2 \sqrt{1-e_1{}^2} \dfrac{1-\sqrt{1-e_1{}^2}}{n_1 a_1{}^2 e_1} \sum j_1 \dfrac{C_0 \cos D_0}{j_1 n_1 + j_2 n_2} \\[2ex]
\qquad - \dfrac{m_2 \sqrt{1-e_1{}^2}}{n_1 a_1{}^2 e_1} \sum \left\{ k_1' + k_1 \dfrac{\partial \tau_1}{\partial \pi_1} + k_2 \dfrac{\partial \tau_2}{\partial \pi_1} \right\} \dfrac{C_0 \cos D_0}{j_1 n_1 + j_2 n_2}, \\[2ex]
(\epsilon_1{}^{(0,\,1)}) - (\epsilon_1{}^{(0,\,1)})_{t_0} = \dfrac{m_2 \tan \dfrac{i_1}{2}}{n_1 a_1{}^2 \sqrt{1-e_1{}^2}} \sum \left\{ \dfrac{\partial C_0}{\partial i_1} \dfrac{\sin D_0}{j_1 n_1 + j_2 n_2} \right. \\[2ex]
\qquad \left. + \left[k_1 \dfrac{\partial \tau_1}{\partial i_1} + k_2 \dfrac{\partial \tau_2}{\partial i_1} \right] \dfrac{C_0 \cos D_0}{j_1 n_1 + j_2 n_2} \right\} \\[2ex]
\qquad + m_2 \sqrt{1-e_1{}^2} \dfrac{1-\sqrt{1-e_1{}^2}}{n_1 a_1{}^2 e_1} \sum \dfrac{\partial C_0}{\partial e_1} \dfrac{\sin D_0}{j_1 n_1 + j_2 n_2} \\[2ex]
\qquad - \dfrac{2 m_2}{n_1 a_1} \sum \dfrac{\partial C_0}{\partial a_1} \dfrac{\sin D_0}{j_1 n_1 + j_2 n_2}.
\end{cases}$$

These terms are purely periodic with periods $\dfrac{2\pi}{j_1 n_1 + j_2 n_2}$, and constitute the *periodic variations*. Every element is subject to them, depending upon an infinity of such terms whose periods are different. The larger $j_1 n_1 + j_2 n_2$ is, the shorter is the period of the term and in general the smaller is its coefficient.

The method of representing the motion of the planets by a series of periodic terms is somewhat analogous to the epicycloid theory of Ptolemy, for each term alone is equivalent to the adding of a small circular motion to that previously existing. This theory is more complex than that of Ptolemy in that it adds epicycloid upon epicycloid without limit; it is simpler than that of Ptolemy in that it flows from one simple principle, the law of gravitation.

226. Long Period Variations. The letters j_1 and j_2 represent all positive and negative integers and zero. Therefore, unless n_1 and n_2 are incommensurable j_1 and j_2 exist such that $j_1 n_1 + j_2 n_2 = 0$, where j_1 and j_2 are not zero. But then D is a constant and the integral is not formed this way. However, whether n_1 and n_2 are incommensurable or not, such a pair of numbers can be found that $j_1 n_1 + j_2 n_2$ is very small. The corresponding term will be large unless its C is very small. It is shown in a complete discussion of the development of $R_{1, 2}$ that the order of C in e_1, e_2, $\sin^2 \dfrac{I}{2}$ is at the least equal to the numerical value of $j_1 + j_2$ (see Tisserand's *Méc. Cél.*, vol I., p. 308). Since n_1 and n_2 are both positive, one of the numbers j_1, j_2 must be positive and the other negative in order that the sum $j_1 n_1 + j_2 n_2$ shall be small. The more nearly equal j_1 and j_2 are numerically the smaller the numerical value of $j_1 + j_2$ is, and consequently, the larger C will be. When the mean motions of the two planets are such that they are nearly commensurable with the ratio of n_1 to n_2 expressible in small integers, then large terms in the perturbations will arise from the presence of these small divisors. The period of such a term is $\dfrac{2\pi}{j_1 n_1 + j_2 n_2}$, which is very great, whence the appellation *long period*. These terms are given by equations of the same form as (99), but with the restriction that $j_1 n_1 + j_2 n_2$ shall be very small.

Geometrically considered, the condition that the periods shall be nearly commensurable with the ratio expressible in small integers means that the points of conjunction occur at nearly the same part of the orbits with only a few other conjunctions intervening. The extreme case is that in which there are no conjunctions intervening, i. e., when j_1 and j_2 differ in numerical value by unity.

The mean motions of Jupiter and Saturn are nearly in the ratio of five to two. Consequently $j_1 = 2$, $j_2 = -5$ gives a long period term, and the order of the coefficient C is the absolute value of $2 - 5$, or 3. The cause of the long period inequality of Jupiter and Saturn was discovered by Laplace in 1784 in computing the perturbations of the third order in e_1 and e_2. The length of the period in the case of these two planets is about 850 years.

227. Secular Variations. The expression D is independent of the time for all of those terms in which $j_1 = j_2 = 0$. The partial derivatives of D with respect to the elements are also independent of the time; hence, on taking these terms of (98) and integrating, it is found that

$$
(100)
\begin{cases}
[\Omega_1^{(0,\,1)}] = \dfrac{m_2}{n_1 a_1{}^2 \sqrt{1 - e_1{}^2}\,\sin i_1} \sum \left\{ \dfrac{\partial C_2}{\partial i_1} \cos D_2 \right. \\
\qquad\qquad \left. - \left[k_1 \dfrac{\partial \tau_1}{\partial i_1} + k_2 \dfrac{\partial \tau_2}{\partial i_1} \right] C_2 \sin D_2 \right\} (t - t_0), \\[4pt]
[i_1^{(0,\,1)}] = \dfrac{-m_2}{n_1 a_1{}^2 \sqrt{1 - e_1{}^2}\,\sin i_1} \sum \left\{ j_1' - k_1 \dfrac{\partial \tau_1}{\partial \Omega_1} \right. \\
\qquad\qquad\qquad \left. - k_2 \dfrac{\partial \tau_2}{\partial \Omega_1} \right\} C_2 \sin D_2 \cdot (t - t_0) \\[4pt]
\qquad\quad + \dfrac{m_2 \tan \dfrac{i_1}{2}}{n_1 a_1{}^2 \sqrt{1 - e_1{}^2}} \sum \left\{ k_1' + k_1 \dfrac{\partial \tau_1}{\partial \pi_1} \right. \\
\qquad\qquad\qquad \left. + k_2 \dfrac{\partial \tau_2}{\partial \pi_1} \right\} C_2 \sin D_2 \cdot (t - t_0), \\[4pt]
[\pi_1^{(0,\,1)}] = \dfrac{m_2 \tan \dfrac{i_1}{2}}{n_1 a_1{}^2 \sqrt{1 - e_1{}^2}} \sum \left\{ \dfrac{\partial C_2}{\partial i_1} \cos D_2 \right. \\
\qquad\qquad \left. - \left[k_1 \dfrac{\partial \tau_1}{\partial i_1} + k_2 \dfrac{\partial \tau_2}{\partial i_1} \right] C_2 \sin D_2 \right\} (t - t_0) \\[4pt]
\qquad\quad + \dfrac{m_2 \sqrt{1 - e_1{}^2}}{n_1 a_1{}^2 e_1} \sum \dfrac{\partial C_2}{\partial e_1} \cos D_2 \cdot (t - t_0),
\end{cases}
$$

28

$$(100) \begin{cases} [a_1^{(0,\,1)}] = 0, \\[2mm] [e_1^{(0,\,1)}] = \dfrac{m_2\sqrt{1 - e_1^2}}{n_1 a_1^2 e_1} \sum \left\{ k_1' + k_1 \dfrac{\partial \tau_1}{\partial \pi_1} \right. \\[3mm] \qquad\qquad \left. + k_2 \dfrac{\partial \tau_2}{\partial \pi_1} \right\} C_2 \sin D_2 \cdot (t - t_0), \\[4mm] [\epsilon_1^{(0,\,1)}] = \dfrac{m_2 \tan \dfrac{i_1}{2}}{n_1 a_1^2 \sqrt{1 - e_1^2}} \sum \left\{ \dfrac{\partial C_2}{\partial i_1} \cos D_2 \right. \\[3mm] \qquad\qquad \left. - \left[k_1 \dfrac{\partial \tau_1}{\partial i_1} + k_2 \dfrac{\partial \tau_2}{\partial i_1} \right] C_2 \sin D_2 \right\} (t - t_0) \\[3mm] \qquad\qquad + m_2 \sqrt{1 - e_1^2}\, \dfrac{1 - \sqrt{1 - e_1^2}}{n_1 a_1^2 e_1} \\[3mm] \qquad\qquad\qquad \times \sum \dfrac{\partial C_2}{\partial e_1} \cos \dot{D}_2 \cdot (t - t_0) \\[3mm] \qquad\qquad - \dfrac{2 m_2}{n_1 a_1} \sum \dfrac{\partial C_2}{\partial a_1} \cos D_2 \cdot (t - t_0). \end{cases}$$

It follows that there are no secular terms of this type of the first order with respect to the masses in the perturbations of a. This constitutes the first theorem on the stability of the solar system. It was proved up to the second powers of the eccentricities by Laplace in 1773,* when he was but twenty-four years of age, in a memoir upon the mutual perturbations of Jupiter and Saturn; it was shown by Lagrange in 1776 that it is true for all powers of the eccentricities.† It was proved by Poisson in 1809 that there are no secular terms in a in the perturbations of the second order with respect to the masses, but that there are terms of the type $t \cos D$, where D contains the time.‡ Terms of this type are commonly called *Poisson terms*.

All of the elements except a have secular terms. It appears to have been supposed that the secular terms, which apparently cause the elements to change without limit, alone prevent the use of equations (72) for computing the perturbations for any time however great. Many methods of computing perturbations have been devised in order to avoid the appearance of secular terms; yet it is clear that, whether or not terms proportional to the time

---

* Memoir presented to the *Paris Academy of Sciences.*
† *Memoirs of the Berlin Academy,* 1776.
‡ *Journal de l'École Polytechnique,* vol. xv.

appear, the method is strictly valid for only those values of the time for which the series (20) of Art. 207 are convergent.

Secular terms may enter in another way, usually not considered. If $j_1n_1 + j_2n_2 = 0$ with $j_1 \neq 0$, $j_2 \neq 0$, D is independent of the time and the corresponding terms are secular. In this case D is not independent of ϵ_1 and there will be secular terms in the perturbations of a. As has been remarked, this condition will always be fulfilled by an infinity of values of j_1 and j_2 if n_1 and n_2 are not incommensurable. But it is impossible to determine from observations whether or not n_1 and n_2 are incommensurable, for there is always a limit to the accuracy with which observations can be made, and within this limit there exist infinitely many commensurable and incommensurable numbers. There is as much reason, therefore, to say that secular terms in a of this type exist as that they do not. However, they are of no practical importance because the ratio of n_1 to n_2 cannot be expressed in small integers, and the coefficients of these terms, if they do exist, are so small that they are not sensible for such values of the time as are ordinarily used.

228. Terms of the Second Order with Respect to the Masses. The terms of the second order are defined by equations (29), Art. 210. The right members of these equations are the products of the partial derivatives, with respect to the elements, of the right members which occur in the terms of the first order, and the perturbations of the first order of the corresponding elements. Thus, the second order perturbations of the node are determined by the equations

$$(101) \quad \begin{cases} \dfrac{d\Omega_1^{(0,\,2)}}{dt} = \dfrac{m_2}{n_1 a_1{}^2\sqrt{1-e_1{}^2}\sin i_1} \sum_{s_1} \dfrac{\partial^2 R_{1,\,2}}{\partial i_1 \partial s_1} s_1^{(0,\,1)}, \\[4mm] \dfrac{d\Omega_1^{(1,\,1)}}{dt} = \dfrac{m_2}{n_1 a_1{}^2\sqrt{1-e_1{}^2}\sin i_1} \sum_{s_2} \dfrac{\partial^2 R_{1,\,2}}{\partial i_1 \partial s_2} s_2^{(1,\,0)}, \end{cases}$$

where s_1 and s_2 represent the elements of the orbits of m_1 and m_2 respectively. The partial derivative $\dfrac{\partial^2 R_{1,\,2}}{\partial i_1 \partial s_1}$ is a sum of periodic and constant terms; $s_1^{(0,\,1)}$ and $s_2^{(1,\,0)}$ are sums of periodic terms and terms containing the time to the first degree as a factor. The products $\dfrac{\partial^2 R_{1,\,2}}{\partial i_1 \partial s_1} s_1^{(0,\,1)}$ and $\dfrac{\partial^2 R_{1,\,2}}{\partial i_1 \partial s_2} s_2^{(1,\,0)}$ therefore contain terms of four types: $(a)\ \frac{\sin}{\cos}\ D$, where D contains the time; $(b)\ t\frac{\sin}{\cos}\ D$; $(c)\ \frac{\sin}{\cos}\ D_2$, where D_2 is independent of the time; and $(d)\ t\frac{\sin}{\cos}\ D_2$. The integrals of these four types are respectively:

$$(a)\quad \frac{-\frac{\cos}{\sin}D}{j_1n_1 + j_2n_2};\qquad (b)\quad t\frac{-\frac{\cos}{\sin}D}{j_1n_1 + j_2n_2} + \frac{\frac{\sin}{\cos}D}{(j_1n_1 + j_2n_2)^2};$$

$$(c)\quad t\frac{\sin}{\cos}D_2;\qquad (d)\quad \frac{t^2}{2}\frac{\sin}{\cos}D_2.$$

Therefore, the perturbations of the second order with respect to the masses have purely periodic terms; Poisson terms, or terms in which the trigonometric terms are multiplied by the time; secular terms where the time occurs to the first degree; and secular terms where the time occurs to the second degree. This is true for all of the elements except the major semi-axis, in the case of which the coefficients of the terms of the third and fourth types are zero, as Poisson first proved.

In the terms of the third order with respect to the masses there are secular terms in the perturbations of all the elements except a_1, which are proportional to the third power of the time, and so on.

229. Lagrange's Treatment of the Secular Variations. The presence of the secular terms in the expressions for the elements seems to indicate that, if it is assumed that the series represent the elements for all values of the time, then the elements change without limit with the time. But this conclusion is by no means necessarily true. For example, consider the function

$$(102)\qquad \sin{(cmt)} = cmt - \frac{c^3m^3t^3}{3!} + \cdots,$$

where c is a constant and m a very small factor which may take the place of a mass. The series in the right member converges for all values of t. This function is never greater than unity for any value of the time; yet if its expansion in powers of m were given, and if the first few terms were considered without the law of the coefficients being known, it might seem that the series represents a function which increases indefinitely in numerical value with the time.

On following out the idea that the secular terms may be ex-

pansions of functions which are always finite, Lagrange has shown
(see *Collected Works*, vols. v. and vi.), under certain assumptions
which have not been logically justified, that the secular terms are
in reality the expansions of periodic terms of very long period.
These terms differ from the long period variations (Art. 226) in
that they come from the small uncompensated parts of the periodic
variations, instead of directly from special conditions of conjunctions. As a rule these terms are very small, and their periods
are much longer than those of the sensible long period terms. It
will not be possible to give here more than a very general idea of
the method of Lagrange.

The first step in the method of Lagrange is a transformation of
variables by the equations

(103)
$$\begin{cases} h_j = e_j \sin \pi_j, \\ l_j = e_j \cos \pi_j, \end{cases}$$

and

(104)
$$\begin{cases} p_j = \tan i_j \sin \Omega_j, \\ q_j = \tan i_j \cos \Omega_j, \end{cases}$$

where e_j, π_j, etc., are the elements of the orbit of m_j, and l_j is a
new variable not to be confused with the mean longitude. These
transformations are to be made simultaneously in the elements of
the orbits of all of the planets. The elements a_j and ϵ_j remain
without transformation. On omitting the subscripts, it is found
from (103) and (104) that

(105)
$$\begin{cases} \dfrac{dh}{dt} = + e \cos \pi \dfrac{d\pi}{dt} + \sin \pi \dfrac{de}{dt}, \\[2mm] \dfrac{dl}{dt} = - e \sin \pi \dfrac{d\pi}{dt} + \cos \pi \dfrac{de}{dt}, \\[2mm] \dfrac{\partial R}{\partial e} = \dfrac{\partial R}{\partial h} \dfrac{\partial h}{\partial e} + \dfrac{\partial R}{\partial l} \dfrac{\partial l}{\partial e} = \sin \pi \dfrac{\partial R}{\partial h} + \cos \pi \dfrac{\partial R}{\partial l}, \\[2mm] \dfrac{\partial R}{\partial \pi} = \dfrac{\partial R}{\partial h} \dfrac{\partial h}{\partial \pi} + \dfrac{\partial R}{\partial l} \dfrac{\partial l}{\partial \pi} = e \cos \pi \dfrac{\partial R}{\partial h} - e \sin \pi \dfrac{\partial R}{\partial l}, \\[2mm] \dfrac{dp}{dt} = + \tan i \cos \Omega \dfrac{d\Omega}{dt} + \sec^2 i \sin \Omega \dfrac{di}{dt}, \\[2mm] \dfrac{dq}{dt} = - \tan i \sin \Omega \dfrac{d\Omega}{dt} + \sec^2 i \cos \Omega \dfrac{di}{dt}, \end{cases}$$

$$(105) \begin{cases} \dfrac{\partial R}{\partial \Omega} = \dfrac{\partial R}{\partial p}\dfrac{\partial p}{\partial \Omega} + \dfrac{\partial R}{\partial q}\dfrac{\partial q}{\partial \Omega} \\[2mm] \qquad = \tan i \cos \Omega \, \dfrac{\partial R}{\partial p} - \tan i \sin \Omega \, \dfrac{\partial R}{\partial q}, \\[3mm] \dfrac{\partial R}{\partial i} = \dfrac{\partial R}{\partial p}\dfrac{\partial p}{\partial i} + \dfrac{\partial R}{\partial q}\dfrac{\partial q}{\partial i} \\[2mm] \qquad = \sec^2 i \sin \Omega \, \dfrac{\partial R}{\partial p} + \sec^2 i \cos \Omega \, \dfrac{\partial R}{\partial q}. \end{cases}$$

Then it follows from (72) that

$$(106) \begin{cases} \dfrac{dh}{dt} = \dfrac{m_2\sqrt{1-h^2-l^2}}{na^2}\dfrac{\partial R}{\partial l} \\[3mm] \qquad - \dfrac{m_2\sqrt{1-h^2-l^2}}{na^2}\dfrac{h}{1+\sqrt{1-h^2-l^2}}\dfrac{\partial R}{\partial \epsilon} \\[3mm] \qquad + \dfrac{m_2 l \tan\frac{i}{2}}{na^2\sqrt{1-h^2-l^2}}\dfrac{\partial R}{\partial i}, \\[3mm] \dfrac{dl}{dt} = \dfrac{-m_2\sqrt{1-h^2-l^2}}{na^2}\dfrac{\partial R}{\partial h} \\[3mm] \qquad - \dfrac{m_2\sqrt{1-h^2-l^2}}{na^2}\dfrac{l}{1+\sqrt{1-h^2-l^2}}\dfrac{\partial R}{\partial \epsilon} \\[3mm] \qquad - \dfrac{m_2 h \tan\frac{i}{2}}{na^2\sqrt{1-h^2-l^2}}\dfrac{\partial R}{\partial i}, \\[3mm] \dfrac{dp}{dt} = \dfrac{m_2}{na^2\sqrt{1-h^2-l^2}\,\cos^3 i}\dfrac{\partial R}{\partial q} \\[3mm] \qquad - \dfrac{m_2 p}{2na^2\sqrt{1-h^2-l^2}\,\cos i \,\cos^2\frac{i}{2}}\left[\dfrac{\partial R}{\partial \pi}+\dfrac{\partial R}{\partial \epsilon}\right], \\[3mm] \dfrac{dq}{dt} = \dfrac{-m_2}{na^2\sqrt{1-h^2-l^2}\,\cos^3 i}\dfrac{\partial R}{\partial p} \\[3mm] \qquad - \dfrac{m_2 q}{2na^2\sqrt{1-h^2-l^2}\,\cos i \,\cos^2\frac{i}{2}}\left[\dfrac{\partial R}{\partial \pi}+\dfrac{\partial R}{\partial \epsilon}\right]. \end{cases}$$

On developing the right members of these equations and neglecting all terms of degree higher than the first* in h, l, p, and q, these

* The terms of order higher than the first are neglected throughout in a later step in the method.

equations reduce to

$$
(107) \quad
\begin{cases}
\dfrac{dh}{dt} = + \dfrac{m_2}{na^2}\dfrac{\partial R}{\partial l}, \\[2ex]
\dfrac{dl}{dt} = - \dfrac{m_2}{na^2}\dfrac{\partial R}{\partial h}, \\[2ex]
\dfrac{dp}{dt} = + \dfrac{m_2}{na^2}\dfrac{\partial R}{\partial q}, \\[2ex]
\dfrac{dq}{dt} = - \dfrac{m_2}{na^2}\dfrac{\partial R}{\partial p}.
\end{cases}
$$

The terms which involve the derivative of R with respect to ϵ, i, and π do not appear in these equations because they involve h, l, p, or q as a factor. This fact follows from the properties of C given in Art. 226 and the form of equations (103) and (104).

Each perturbing planet contributes terms in the right members of equations (107) similar to the ones written which come from m_2. These differential equations are not strictly correct, since the first approximation has already been made in neglecting the higher powers of the variables.

The second step is in the method of treating the differential equations. The expansions of the $R_{i,\,j}$ contain certain terms which are independent of the time, which in the ordinary method give rise to the secular terms. Let $R^{(0)}{}_{i,\,j}$ represent these terms. Lagrange then treated the differential equations by neglecting the periodic terms in $R_{i,\,j}$, and writing

$$
(108) \quad
\begin{cases}
\dfrac{dh_i}{dt} = + \displaystyle\sum_{j=1}^{n} m_j \dfrac{\partial R^{(0)}{}_{i,\,j}}{\partial l_i}, \qquad (i = 1,\ \cdots,\ n;\ j \neq i), \\[3ex]
\dfrac{dl_i}{dt} = - \displaystyle\sum_{j=1}^{n} m_j \dfrac{\partial R^{(0)}{}_{i,\,j}}{\partial h_i}, \\[3ex]
\dfrac{dp_i}{dt} = + \displaystyle\sum_{j=1}^{n} m_j \dfrac{\partial R^{(0)}{}_{i,\,j}}{\partial q_i}, \\[3ex]
\dfrac{dq_i}{dt} = - \displaystyle\sum_{j=1}^{n} m_j \dfrac{\partial R^{(0)}{}_{i,\,j}}{\partial p_i}.
\end{cases}
$$

The values of h_i, l_i, p_i, and q_i determined from equations (108) are used instead of the secular terms obtained by the method of Art. 227. The process of breaking up a differential equation in this manner is not permissible except as a first approximation, and any conclusions based on it are open to suspicion.

In spite of the logical defects of the method and the fact that it cannot be generally applied, there is little doubt that in the present case it gives an accurate idea of the actual manner in which the elements vary.

The right members of equations (108) are expanded in powers of h_i, l_i, p_i, and q_i, and all of the terms except those of the first degree are neglected; consequently the terms omitted in (107) would have disappeared here if they had been retained up to this point. The system becomes linear, and the detailed discussion of the $R_{i,j}$ shows that it is homogeneous, giving equations of the form

$$
(109)\quad
\begin{cases}
\dfrac{dh_1}{dt} - \displaystyle\sum_{j=1}^{n} c_{1j}l_j = 0, \\[2mm]
\dfrac{dl_1}{dt} + \displaystyle\sum_{j=1}^{n} c_{1j}h_j = 0, \\[2mm]
\dfrac{dh_2}{dt} - \displaystyle\sum_{j=1}^{n} c_{2j}l_j = 0, \\[2mm]
\dfrac{dl_2}{dt} + \displaystyle\sum_{j=1}^{n} c_{2j}h_j = 0, \\[2mm]
\quad\cdot\quad\cdot\quad\cdot\quad\cdot\quad\cdot\quad\cdot \\[2mm]
\dfrac{dh_n}{dt} - \displaystyle\sum_{j=1}^{n} c_{nj}l_j = 0, \\[2mm]
\dfrac{dl_n}{dt} + \displaystyle\sum_{j=1}^{n} c_{nj}h_j = 0,
\end{cases}
$$

and a similar system of equations in the p_j and the q_j.

The coefficients c_{ij} depend only on the major axes (the ϵ_j not appearing in the secular terms) which are considered as being constants, since the major axes have no secular terms in the perturbations of the first and second orders with respect to the masses. It is to be noted here that the assumption that the c_{ij} are constants is not strictly true because the major axes have periodic perturbations which may be of considerable magnitude.

When these linear equations are solved by the method used in Art. 160, the values of the variables are found in the form

$$
(110)\quad
\begin{aligned}
&h_i = \sum_{j=1}^{n} H_{ij}e^{\lambda_j t}, &\qquad& l_i = \sum_{j=1}^{n} L_{ij}e^{\lambda_j t}, \\[2mm]
&p_i = \sum_{j=1}^{n} P_{ij}e^{\mu_j t}, &\qquad& q_i = \sum_{j=1}^{n} Q_{ij}e^{\mu_j t},
\end{aligned}
$$

where the H_{ij}, L_{ij}, P_{ij}, and Q_{ij}, are constants depending upon the initial conditions. A detailed discussion shows that the λ_j and μ_j are all pure imaginaries with very small absolute values; therefore the h_i, l_i, p_i, and q_i oscillate around mean values with very long periods. Or, since the e_j and $\tan i_j$ are expressible as the sums of squares of the h_j, l_j, p_j, and q_j, it follows that they also perform small oscillations with long periods; for example, the eccentricity of the earth's orbit is now decreasing and will continue to decrease for about 24,000 years.

Equations (109) admit integrals first found by Laplace in 1784, which lead practically to the same theorem. They are

$$(111) \quad \begin{cases} \displaystyle\sum_{j=1}^{n} m_j n_j a_j{}^2 (h_j{}^2 + l_j{}^2) = \text{Constant} = C, \\[2mm] \displaystyle\sum_{j=1}^{n} m_j n_j a_j{}^2 (p_j{}^2 + q_j{}^2) = C'; \end{cases}$$

or, because of (103) and (104),

$$(112) \quad \begin{cases} \displaystyle\sum_{j=1}^{n} m_j n_j a_j{}^2 e_j{}^2 = C, \\[2mm] \displaystyle\sum_{j=1}^{n} m_j n_j a_j{}^2 \tan^2 i_j = C', \end{cases}$$

where n_j is the mean motion of m_j. The constants C and C' as determined by the initial conditions are very small, and since the left members of (112) are made up of positive terms alone, no e_j or i_j can ever become very great. There might be an exception if the corresponding m_j were very small compared to the others.

Equations (112) give the celebrated theorems of Laplace that the eccentricities and inclinations cannot vary except within very narrow limits. Although the demonstration lacks complete rigor, yet the results must be considered as remarkable and significant. Equations (112) do not give the periods and amplitudes of the oscillations as do equations (110).

230. Computation of Perturbations by Mechanical Quadratures. If the second term of the second factor of (84) in absolute value is greater than unity, the series (87) does not converge and cannot be used in computing perturbations. The expansions may fail because r_1 and r_2 are very nearly equal; or, sometimes when they are not nearly equal, because I is large. In the latter case another mode of expansion sometimes can be employed,* but there are cases in which neither method leads to valid results. They both fail if the two orbits placed in the same plane would intersect, for in this case

$$r^2_{1,\,2} = r_1{}^2 + r_2{}^2 - 2r_1r_2 \cos (u_1 - u_2 + \tau_2 - \tau_1),$$

would vanish when the two bodies arrive at a point of intersection of their orbits at the same time. Unless the periods are commensurable in a special way this would always happen. Of course, it is not necessary that $r_{1,\,2}$ should actually vanish in order that the expansion of (84) should fail to converge.

Perturbations can be computed by the method of mechanical quadratures without expanding the perturbative function explicitly in terms of the time. Consequently, this method can be used in computing the disturbing effects of planets on comets and in other cases where the expansion of $R_{1,\,2}$ fails altogether or converges slowly. Let s represent an element of the orbit of m_1; then equations (77) can be written in the form

$$\frac{ds}{dt} = f_s(t),$$

and the perturbations of the first order in the interval $t_n - t_0$ are

$$(113) \qquad\qquad s = s_0 + \int_{t_0}^{t_n} f_s(t)dt,$$

where s_0 is the value of s at $t = t_0$.

The only difficulty in computing perturbations is in forming the integrals indicated in (113). When the perturbative function cannot be expanded explicitly in terms of t the primitive of the function $f_s(t)$ cannot be found. But in any case the values of $f_s(t)$ can be found for any values of t, and from the values of $f_s(t)$ for special values of t an approximation to the integral can be obtained. Geometrically considered, the integral (113) is the area comprised between the t-axis and the curve $f = f_s(t)$ and the ordinates t_0 and t_n. An approximate value of the integral is

$$s \doteq s_0 + f_s(t_0)(t_1 - t_0) + f_s(t_1)(t_2 - t_1) + \cdots + f_s(t_{n-1})(t_n - t_{n-1}).$$

The intervals $t_1 - t_0,\ t_2 - t_1,\ \cdots,\ t_n - t_{n-1}$ can be taken so small that the approximation will be as close as may be desired.

Another method of obtaining an approximate value of the inte-

* Tisserand, *Mécanique Céleste*, vol. I., chap. XXVIII.

gral is to replace the curve $f_s(t)$, whose explicit value in convenient form may not be obtainable, by a polynomial curve of the nth degree which agrees in value with $f_s(t)$ at $t = t_0, t_1, \cdots, t_n$. The equation of this polynomial is

$$
\begin{aligned}
f_s = {} &+ \frac{(t - t_1)(t - t_2) \cdots (t - t_n)}{(t_0 - t_1)(t_0 - t_2) \cdots (t_0 - t_n)} f_s(t_0) \\
&+ \frac{(t - t_0)(t - t_2) \cdots (t - t_n)}{(t_1 - t_0)(t_1 - t_2) \cdots (t_1 - t_n)} f_s(t_1) \\
&\qquad \cdot \quad \cdot \quad \cdot \quad \cdot \quad \cdot \quad \cdot \quad \cdot \quad \cdot \quad \cdot \quad \cdot \\
&+ \frac{(t - t_0)(t - t_1) \cdots (t - t_{n-1})}{(t_n - t_0)(t_n - t_1) \cdots (t_n - t_{n-1})} f_s(t_n).
\end{aligned}
$$

Since there is no trouble in forming the integral of a polynomial there is no trouble in computing the perturbation of s for the interval $t_n - t_0$. If the value of the function $f_s(t)$ is not changing very rapidly or irregularly, its representation by a polynomial is very exact provided the intervals $t_1 - t_0, \cdots, t_n - t_{n-1}$ are not too great.

However, the area between the polynomial, the t-axis, and the limiting ordinates is not the best approximation to the value of the integral that can be obtained from the values of $f_s(t)$ at t_0, $\cdots, t_n$. The values of the function give information respecting the nature of the curvature of the curve between the ordinates (this being true, of course, only because the function $f_s(t)$ is a regular function of t), and corrections of the area due to these curvatures can easily be made. Ordinarily they would involve the derivatives of $f_s(t)$ at $t_0, \cdots, t_n$, which would require a vast amount of labor to compute; but the derivatives can be expressed with sufficient approximation in terms of the successive differences of the function, and the differences are obtained directly from the tabular values by simple subtraction. The derivation of the most convenient explicit formulas is a lengthy matter and must be omitted.*

Suppose the computation of the integrals from the values of $f_s(t)$ at $t = t_0, \cdots, t_n$ has not given results which are sufficiently exact. More exact ones can be obtained by dividing the interval $t_n - t_0$ into a greater number of sub-intervals. A little experience usually makes it unnecessary to subdivide the intervals first chosen.

---

* See Tisserand's *Mécanique Céleste*, vol. IV., chaps. X. and XI.; and Charlier's *Mechanik des Himmels*, vol. II., chap. 1.

There is a second reason why the results obtained by mechanical quadratures may not be sufficiently exact. It has so far been assumed that $f_s(t)$ is a function of t alone; or, in other words, that the elements of the orbits on which it depends are constants. This is the assumption in computing perturbations of the first order. If it is not exact enough, new values of $f_s(t_1)$, $\cdots$, $f_s(t_n)$ can be computed, on using in them the respective values of the elements s which were found by the first integration. From the new values of $f_s(t_1)$, $\cdots$, $f_s(t_n)$ a more approximate value of the integral can be obtained. Unless the interval $t_n - t_0$ is too great this process converges and the integral can be found with any desired degree of approximation, because this method is simply Picard's method of successive approximations whose validity has been established.* In practice it is always advisable to choose the interval $t_n - t_0$ so short that no repetition of the computation with improved values of the function at the ends of the sub-intervals will be required. At each new stage of the integration the values of the elements at the end of the preceding step are employed. It follows that the method, as just explained, enables one to compute not only the perturbations of the first order, but perturbations of all orders except for the limitations that the intervals cannot be taken indefinitely small and the computation cannot be made with indefinitely many places.

The process of computing perturbations by the method of mechanical quadratures, as compared with that of using the expanded form of the perturbative function, has its advantages and its disadvantages. It is an advantage that in employing mechanical quadratures it is not necessary to express the perturbing forces explicitly in terms of the elements and the time. This is sometimes of great importance, for, in cases where the eccentricities and inclinations are large, as in some of the asteroid orbits, these expressions, which are series, are very slowly convergent; and in the case of orbits whose eccentricities exceed 0.6627, or of orbits which have any radius of one equal to any radius of the other the series are divergent and cannot be used. The method of mechanical quadratures is equally applicable to all kinds of orbits, the only restriction being that the intervals shall be taken sufficiently short. It is the method actually employed, in one of its many forms, in computing the perturbations of the orbits of comets.

* Picard's *Traité d'Analyse*, vol. II., chap. XI., section 2.

The disadvantages are that, in order to find by mechanical quadratures the values of the elements at any particular time, it is necessary to compute them at all of the intermediate epochs. Being purely numerical, it throws no light whatever on the general character of perturbations, and leads to no general theorems regarding the stability of a system. These are questions of great interest, and some of the most brilliant discoveries in Celestial Mechanics have been made respecting them.

231. General Reflections. Astronomy is the oldest science and in a certain sense the parent of all the others. The relatively simple and regularly recurring celestial phenomena first taught men, in the days of the ancient Greeks, that Nature is systematic and orderly. The importance of this lesson can be inferred from the fact that it is the foundation on which all science is based. For a long time progress was painfully slow. Centuries of observations and attempts at theories for explaining them were necessary before it was finally possible for Kepler to derive the laws which are a first approximation to the description of the way in which the planets move. The wonder is that, in spite of the distractions of the constant struggles incident to an unstable social order, there should have been so many men who found their greatest pleasure in patiently making the laborious observations which were necessary to establish the laws of the celestial motions.

The work of Kepler closed the preliminary epoch of two thousand years, or more, and the brilliant discoveries of Newton opened another. The invention of the Calculus by Newton and Leibnitz furnished for the first time mathematical machinery which was at all suitable for grappling with such difficult problems as the disturbing effects of the sun on the motion of the moon, or the mutual perturbations of the planets. It was fortunate that the telescope was invented about the same time; for, without its use, it would not have been possible to have made the accurate observations which furnished the numerical data for the mathematical theories and by which they were tested. The history of Celestial Mechanics during the eighteenth century is one of a continuous series of triumphs. The analytical foundations laid by Clairaut, d'Alembert, and Euler formed the basis for the splendid achievements of Lagrange and Laplace. Their successors in the nineteenth century pushed forward, by the same methods on the whole, the theories of the motions of the moon and planets to higher orders of approximation and compared them with more and better observations. In this connection the names of Leverrier, Delaunay, Hansen, and Newcomb will be especially remembered. Near the close of the nineteenth century a third epoch was entered. It is distinguished by new points of view and new methods which, in power and mathematical rigor, enormously surpass all those used before. It was inaugurated by Hill in his *Researches on the Lunar Theory*, but owes most to the brilliant contributions of Poincaré to the Problem of Three Bodies.

At the present time Celestial Mechanics is entitled to be regarded as the most perfect science and one of the most splendid achievements of the human mind. No other science is based on so many observations extending over so long a time. In no other science is it possible to test so critically its conclusions, and in no other are theory and experience in so perfect accord. There are thousands of small deviations from conic section motion in the orbits of the planets, satellites, and comets where theory and the observations exactly agree, while the only unexplained irregularities (probably due to unknown forces) are a very few small ones in the motion of the moon and the motion of the perihelion of the orbit of Mercury. Over and over again theory has outrun practise and indicated the existence of peculiarities of motion which had not yet been derived from observations. Its perfection during the time covered by experience inspires confidence in following it back into the past to a time before observations began, and into the future to a time when perhaps they shall have ceased. As the telescope has brought within the range of the eye of man the wonders of an enormous space, so Celestial Mechanics has brought within reach of his reason the no lesser wonders of a correspondingly enormous time. It is not to be marveled at that he finds profound satisfaction in a domain where he is largely freed from the restrictions of both space and time.

XXVII. PROBLEMS.

1. Suppose (a) that $R_{1,2}$ is large and nearly constant; (b) that $R_{1,2}$ is large and changing rapidly; (c) that $R_{1,2}$ is small and nearly constant. If the perturbations are computed by mechanical quadratures how should the $t_n - t_0$ be chosen relatively in the three cases, and how should the numbers of subdivisions of $t_n - t_0$ compare?

2. The perturbative function involves the reciprocal of the distance from the disturbing to the disturbed planets. This is called the *principal part* and gives the most difficulty in the development. How many separate reciprocal

distances must be developed in order to compute, in a system of one sun and n planets, (*a*) the perturbations of the first order of one planet; (*b*) the perturbations of the first order of two planets; (*c*) the perturbations of the second order of one planet; and (*d*) the perturbations of the third order of one planet?

3. What simplifications would there be in the development of the perturbative function if the mutual inclinations of the orbits were zero, and if the orbits were circles?

4. What sorts of terms will in general appear in perturbations of the third order with respect to the masses?

HISTORICAL SKETCH AND BIBLIOGRAPHY.

The theory of perturbations, as applied to the Lunar Theory, was developed from the geometrical standpoint by Newton. The memoirs of Clairaut and D'Alembert in 1747 contained important advances, making the solutions depend upon the integration of the differential equations in series. Clairaut soon had occasion to apply his processes of integration to the perturbations of Halley's comet by the planets Jupiter and Saturn. This comet had been observed in 1531, 1607, and 1682. If its period were constant it would pass the perihelion again about the middle of 1759. Clairaut computed the perturbations due to the attractions of Jupiter and Saturn, and predicted that the perihelion passage would be April 13, 1759. He remarked that the time was uncertain to the extent of a month because of the uncertainties in the masses of Jupiter and Saturn and the possibility of perturbations from unknown planets beyond these two. The comet passed the perihelion March 13, giving a striking proof of the value of Clairaut's methods.

The theory of the perturbations of the planets was begun by Euler, whose memoirs on the mutual perturbations of Jupiter and Saturn gained the prizes of the French Academy in 1748 and 1752. In these memoirs was given the first analytical development of the method of the variation of parameters. The equations were not entirely general as he had not considered the elements as being all simultaneously variables. The first steps in the development of the perturbative function were also given by Euler.

Lagrange, whose contributions to Celestial Mechanics were of the most brilliant character, wrote his first memoir in 1766 on the perturbations of Jupiter and Saturn. In this work he developed still further the method of the variation of parameters, leaving his final equations, however, still incorrect by regarding the major axes and the epochs of the perihelion passages as constants in deriving the equations for the variations. The equations for the inclination, node, and longitude of the perihelion from the node were perfectly correct. In the expressions for the mean longitudes of the planets there were terms proportional to the first and second powers of the time. These were entirely due to the imperfections of the method, their true form being that of the long period terms, as was shown by Laplace in 1784 by considering terms of the third order in the eccentricities. The method of the variation of parameters was completely developed for the first time in 1782 by Lagrange in a prize memoir on the perturbations of comets moving in elliptical orbits. By far the most extensive use of the method of variation of parameters is due to Delaunay, whose Lunar Theory is essentially a long succession of the applications of the process, each step of it removing a term from the perturbative function.

In 1773 Laplace presented his first memoir to the French Academy of Sciences. In it he proved his celebrated theorem that, up to the second powers of the eccentricities, the major axes, and consequently the mean motions of the planets, have no secular terms. This theorem was extended by Lagrange in 1774 and 1776 to all powers of the eccentricities and of the sine of the angle of the mutual inclination, for perturbations of the first order with respect to the masses. Poisson proved in 1809 that the major axes have no purely secular terms in the perturbations of the second order with respect to the masses. Haretu proved in his Dissertation at Sorbonne in 1878 that there *are* secular variations in the expressions for the major axes in the terms of the third order with respect to the masses. In vol. XIX. of *Annales de l'Observatoire de Paris*, Eginitis considered terms of still higher order with respect to the masses.

Lagrange began the study of the secular terms in 1774, introducing the variables h, l, p, and q. The investigations were carried on by Lagrange and Laplace, each supplementing and extending the work of the other, until 1784 when their work became complete by Laplace's discovery of his celebrated equations

$$\begin{cases} \sum_{j=1}^{n} m_j n_j a_j^2 e_j^2 = C, \\[2ex] \sum_{j=1}^{n} m_j n_j a_j^2 \tan^2 i_j = C'. \end{cases}$$

These equations were derived by using only the linear terms in the differential equations. Leverrier, Hill, and others have extended the work by methods of successive approximations to terms of higher degree. Newcomb (*Smithsonian Contributions to Science*, vol. XXI., 1876) has established the more far-reaching results that it is possible, in the case of the planetary perturbations, to represent the elements by purely periodic functions of the time which formally satisfy the differential equations of motion. If these series were convergent the stability of the solar system would be assured; but Poincaré has shown that they are in general divergent (*Les Méthodes Nouvelles de la Mécanique Céleste*, chap. IX.). Lindstedt and Gyldèn have also succeeded in integrating the equations of the motion of n bodies in periodic series, which, however, are in general divergent.

Gauss, Airy, Adams, Leverrier, Hansen, and many others have made important contributions to the planetary theory in some of its many aspects. Adams and Leverrier are noteworthy for having predicted the existence and apparent position of Neptune from the unexplained irregularities in the motion of Uranus. More recently Poincaré turned his attention to Celestial Mechanics, publishing a prize memoir in the *Acta Mathematica*, vol. XIII. This memoir was enlarged and published in book form with the title *Les Méthodes Nouvelles de la Mécanique Céleste*. Poincaré applied to the problem all the resources of modern mathematics with unrivaled genius; he brought into the investigation such a wealth of ideas, and he devised methods of such immense power that the subject in its theoretical aspects has been entirely revolutionized in his hands. It cannot be doubted that much of the work of the next fifty years will be in amplifying and applying the processes which he explained.

The following works should be consulted:

Laplace's *Mécanique Céleste*, containing practically all that was known of Celestial Mechanics at the time it was written (1799–1805).

On the variation of parameters—*Annales de l'Observatoire de Paris*, vol. I.; Tisserand's *Mécanique Céleste*, vol. I.; Brown's *Lunar Theory;* Dziobek's *Planeten-Bewegungen.*

On the development of the perturbative function—*Annales de l'Observatoire de Paris*, vol. I.; Tisserand's *Mécanique Céleste*, vol. I.; Hansen's *Entwickelung des Products einer Potenz des Radius-Vectors mit dem Sinus oder Cosinus eines Vielfachen der wahren Anomalie*, etc., *Abh. d. K. Sächs. Ges. zu Leipzig*, vol. II.; Newcomb's memoir on the General Integrals of Planetary Motion; Poincaré, *Les Méthodes Nouvelles*, vol. I., chap. VI.

On the stability of the solar system—Tisserand's *Mécanique Céleste*, vol. I., chaps. XI., XXV., XXVI., and vol. IV., chap. XXVI.; Gylden, *Traité Analytique des Orbites absolues*, vol. I.; Newcomb, *Smithsonian Cont.*, vol. XXI.; Poincaré, *Les Méthodes Nouvelles de la Mécanique Céleste*, vol. II., chap. X.

On the subject of Celestial Mechanics as a whole there is no better work available than that of Tisserand, which should be in the possession of every one giving special attention to this subject. Another noteworthy work is Charlier's *Mechanik des Himmels*, which, besides maintaining a high order of general excellence, is unequaled by other treatises in its discussion of periodic solutions of the Problem of Three Bodies.

INDEX.

A CATALOG OF SELECTED

DOVER BOOKS
IN SCIENCE AND MATHEMATICS

A CATALOG OF SELECTED
DOVER BOOKS
IN SCIENCE AND MATHEMATICS

QUALITATIVE THEORY OF DIFFERENTIAL EQUATIONS, V.V. Nemytskii and V.V. Stepanov. Classic graduate-level text by two prominent Soviet mathematicians covers classical differential equations as well as topological dynamics and ergodic theory. Bibliographies. 523pp. 5⅜ x 8½. 65954-2 Pa. $14.95

MATRICES AND LINEAR ALGEBRA, Hans Schneider and George Phillip Barker. Basic textbook covers theory of matrices and its applications to systems of linear equations and related topics such as determinants, eigenvalues and differential equations. Numerous exercises. 432pp. 5⅜ x 8½. 66014-1 Pa. $10.95

QUANTUM THEORY, David Bohm. This advanced undergraduate-level text presents the quantum theory in terms of qualitative and imaginative concepts, followed by specific applications worked out in mathematical detail. Preface. Index. 655pp. 5⅜ x 8½. 65969-0 Pa. $14.95

ATOMIC PHYSICS (8th edition), Max Born. Nobel laureate's lucid treatment of kinetic theory of gases, elementary particles, nuclear atom, wave-corpuscles, atomic structure and spectral lines, much more. Over 40 appendices, bibliography. 495pp. 5⅜ x 8½. 65984-4 Pa. $13.95

ELECTRONIC STRUCTURE AND THE PROPERTIES OF SOLIDS: The Physics of the Chemical Bond, Walter A. Harrison. Innovative text offers basic understanding of the electronic structure of covalent and ionic solids, simple metals, transition metals and their compounds. Problems. 1980 edition. 582pp. 6⅛ x 9¼. 66021-4 Pa. $16.95

BOUNDARY VALUE PROBLEMS OF HEAT CONDUCTION, M. Necati Özisik. Systematic, comprehensive treatment of modern mathematical methods of solving problems in heat conduction and diffusion. Numerous examples and problems. Selected references. Appendices. 505pp. 5⅜ x 8½. 65990-9 Pa. $12.95

A SHORT HISTORY OF CHEMISTRY (3rd edition), J.R. Partington. Classic exposition explores origins of chemistry, alchemy, early medical chemistry, nature of atmosphere, theory of valency, laws and structure of atomic theory, much more. 428pp. 5⅜ x 8½. (Available in U.S. only) 65977-1 Pa. $11.95

A HISTORY OF ASTRONOMY, A. Pannekoek. Well-balanced, carefully reasoned study covers such topics as Ptolemaic theory, work of Copernicus, Kepler, Newton, Eddington's work on stars, much more. Illustrated. References. 521pp. 5⅜ x 8½. 65994-1 Pa. $12.95

PRINCIPLES OF METEOROLOGICAL ANALYSIS, Walter J. Saucier. Highly respected, abundantly illustrated classic reviews atmospheric variables, hydrostatics, static stability, various analyses (scalar, cross-section, isobaric, isentropic, more). For intermediate meteorology students. 454pp. 6⅛ x 9¼. 65979-8 Pa. $14.95

RELATIVITY, THERMODYNAMICS AND COSMOLOGY, Richard C. Tolman. Landmark study extends thermodynamics to special, general relativity; also applications of relativistic mechanics, thermodynamics to cosmological models. 501pp. 5⅜ x 8½. 65383-8 Pa. $13.95

APPLIED ANALYSIS, Cornelius Lanczos. Classic work on analysis and design of finite processes for approximating solution of analytical problems. Algebraic equations, matrices, harmonic analysis, quadrature methods, much more. 559pp. 5⅜ x 8½. 65656-X Pa. $13.95

INTRODUCTION TO ANALYSIS, Maxwell Rosenlicht. Unusually clear, accessible coverage of set theory, real number system, metric spaces, continuous functions, Riemann integration, multiple integrals, more. Wide range of problems. Undergraduate level. Bibliography. 254pp. 5⅜ x 8½. 65038-3 Pa. $8.95

INTRODUCTION TO QUANTUM MECHANICS With Applications to Chemistry, Linus Pauling & E. Bright Wilson, Jr. Classic undergraduate text by Nobel Prize winner applies quantum mechanics to chemical and physical problems. Numerous tables and figures enhance the text. Chapter bibliographies. Appendices. Index. 468pp. 5⅜ x 8½. 64871-0 Pa. $12.95

ASYMPTOTIC EXPANSIONS OF INTEGRALS, Norman Bleistein & Richard A. Handelsman. Best introduction to important field with applications in a variety of scientific disciplines. New preface. Problems. Diagrams. Tables. Bibliography. Index. 448pp. 5⅜ x 8½. 65082-0 Pa. $12.95

MATHEMATICS APPLIED TO CONTINUUM MECHANICS, Lee A. Segel. Analyzes models of fluid flow and solid deformation. For upper-level math, science and engineering students. 608pp. 5⅜ x 8½. 65369-2 Pa. $14.95

ELEMENTS OF REAL ANALYSIS, David A. Sprecher. Classic text covers fundamental concepts, real number system, point sets, functions of a real variable, Fourier series, much more. Over 500 exercises. 352pp. 5⅜ x 8½. 65385-4 Pa. $11.95

PHYSICAL PRINCIPLES OF THE QUANTUM THEORY, Werner Heisenberg. Nobel Laureate discusses quantum theory, uncertainty, wave mechanics, work of Dirac, Schroedinger, Compton, Wilson, Einstein, etc. 184pp. 5⅜ x 8½. 60113-7 Pa. $6.95

INTRODUCTORY REAL ANALYSIS, A.N. Kolmogorov, S.V. Fomin. Translated by Richard A. Silverman. Self-contained, evenly paced introduction to real and functional analysis. Some 350 problems. 403pp. 5⅜ x 8½. 61226-0 Pa. $10.95

PROBLEMS AND SOLUTIONS IN QUANTUM CHEMISTRY AND PHYSICS, Charles S. Johnson, Jr. and Lee G. Pedersen. Unusually varied problems, detailed solutions in coverage of quantum mechanics, wave mechanics, angular momentum, molecular spectroscopy, scattering theory, more. 280 problems plus 139 supplementary exercises. 430pp. 6½ x 9¼. 65236-X Pa. $13.95

ASYMPTOTIC METHODS IN ANALYSIS, N.G. de Bruijn. An inexpensive, comprehensive guide to asymptotic methods–the pioneering work that teaches by explaining worked examples in detail. Index. 224pp. 5$\frac{3}{8}$ x 8$\frac{1}{2}$. 64221-6 Pa. $7.95

OPTICAL RESONANCE AND TWO-LEVEL ATOMS, L. Allen and J. H. Eberly. Clear, comprehensive introduction to basic principles behind all quantum optical resonance phenomena. 53 illustrations. Preface. Index. 256pp. 5$\frac{3}{8}$ x 8$\frac{1}{2}$. 65533-4 Pa. $8.95

COMPLEX VARIABLES, Francis J. Flanigan. Unusual approach, delaying complex algebra till harmonic functions have been analyzed from real variable viewpoint. Includes problems with answers. 364pp. 5$\frac{3}{8}$ x 8$\frac{1}{2}$. 61388-7 Pa. $9.95

ATOMIC SPECTRA AND ATOMIC STRUCTURE, Gerhard Herzberg. One of best introductions; especially for specialist in other fields. Treatment is physical rather than mathematical. 80 illustrations. 257pp. 5$\frac{3}{8}$ x 8$\frac{1}{2}$. 60115-3 Pa. $7.95

APPLIED COMPLEX VARIABLES, John W. Dettman. Step-by-step coverage of fundamentals of analytic function theory–plus lucid exposition of five important applications: Potential Theory; Ordinary Differential Equations; Fourier Transforms; Laplace Transforms; Asymptotic Expansions. 66 figures. Exercises at chapter ends. 512pp. 5$\frac{3}{8}$ x 8$\frac{1}{2}$. 64670-X Pa. $12.95

ULTRASONIC ABSORPTION: An Introduction to the Theory of Sound Absorption and Dispersion in Gases, Liquids and Solids, A.B. Bhatia. Standard reference in the field provides a clear, systematically organized introductory review of fundamental concepts for advanced graduate students, research workers. Numerous diagrams. Bibliography. 440pp. 5$\frac{3}{8}$ x 8$\frac{1}{2}$. 64917-2 Pa. $11.95

UNBOUNDED LINEAR OPERATORS: Theory and Applications, Seymour Goldberg. Classic presents systematic treatment of the theory of unbounded linear operators in normed linear spaces with applications to differential equations. Bibliography. 199pp. 5$\frac{3}{8}$ x 8$\frac{1}{2}$. 64830-3 Pa. $7.95

LIGHT SCATTERING BY SMALL PARTICLES, H.C. van de Hulst. Comprehensive treatment including full range of useful approximation methods for researchers in chemistry, meteorology and astronomy. 44 illustrations. 470pp. 5$\frac{3}{8}$ x 8$\frac{1}{2}$. 64228-3 Pa. $12.95

CONFORMAL MAPPING ON RIEMANN SURFACES, Harvey Cohn. Lucid, insightful book presents ideal coverage of subject. 334 exercises make book perfect for self-study. 55 figures. 352pp. 5$\frac{5}{8}$ x 8$\frac{1}{4}$. 64025-6 Pa. $11.95

OPTICKS, Sir Isaac Newton. Newton's own experiments with spectroscopy, colors, lenses, reflection, refraction, etc., in language the layman can follow. Foreword by Albert Einstein. 532pp. 5$\frac{3}{8}$ x 8$\frac{1}{2}$. 60205-2 Pa. $12.95

GENERALIZED INTEGRAL TRANSFORMATIONS, A.H. Zemanian. Graduate-level study of recent generalizations of the Laplace, Mellin, Hankel, K. Weierstrass, convolution and other simple transformations. Bibliography. 320pp. 5$\frac{3}{8}$ x 8$\frac{1}{2}$. 65375-7 Pa. $8.95

THE ELECTROMAGNETIC FIELD, Albert Shadowitz. Comprehensive undergraduate text covers basics of electric and magnetic fields, builds up to electromagnetic theory. Also related topics, including relativity. Over 900 problems. 768pp. 5⅝ x 8¼. 65660-8 Pa. $18.95

FOURIER SERIES, Georgi P. Tolstov. Translated by Richard A. Silverman. A valuable addition to the literature on the subject, moving clearly from subject to subject and theorem to theorem. 107 problems, answers. 336pp. 5⅜ x 8½. 63317-9 Pa. $9.95

THEORY OF ELECTROMAGNETIC WAVE PROPAGATION, Charles Herach Papas. Graduate-level study discusses the Maxwell field equations, radiation from wire antennas, the Doppler effect and more. xiii + 244pp. 5⅜ x 8½. 65678-0 Pa. $6.95

DISTRIBUTION THEORY AND TRANSFORM ANALYSIS: An Introduction to Generalized Functions, with Applications, A.H. Zemanian. Provides basics of distribution theory, describes generalized Fourier and Laplace transformations. Numerous problems. 384pp. 5⅜ x 8½. 65479-6 Pa. $11.95

THE PHYSICS OF WAVES, William C. Elmore and Mark A. Heald. Unique overview of classical wave theory. Acoustics, optics, electromagnetic radiation, more. Ideal as classroom text or for self-study. Problems. 477pp. 5⅜ x 8½. 64926-1 Pa. $13.95

CALCULUS OF VARIATIONS WITH APPLICATIONS, George M. Ewing. Applications-oriented introduction to variational theory develops insight and promotes understanding of specialized books, research papers. Suitable for advanced undergraduate/graduate students as primary, supplementary text. 352pp. 5⅜ x 8½. 64856-7 Pa. $9.95

A TREATISE ON ELECTRICITY AND MAGNETISM, James Clerk Maxwell. Important foundation work of modern physics. Brings to final form Maxwell's theory of electromagnetism and rigorously derives his general equations of field theory. 1,084pp. 5⅜ x 8½. 60636-8, 60637-6 Pa., Two-vol. set $25.90

AN INTRODUCTION TO THE CALCULUS OF VARIATIONS, Charles Fox. Graduate-level text covers variations of an integral, isoperimetrical problems, least action, special relativity, approximations, more. References. 279pp. 5⅜ x 8½. 65499-0 Pa. $8.95

HYDRODYNAMIC AND HYDROMAGNETIC STABILITY, S. Chandrasekhar. Lucid examination of the Rayleigh-Benard problem; clear coverage of the theory of instabilities causing convection. 704pp. 5⅜ x 8¼. 64071-X Pa. $14.95

CALCULUS OF VARIATIONS, Robert Weinstock. Basic introduction covering isoperimetric problems, theory of elasticity, quantum mechanics, electrostatics, etc. Exercises throughout. 326pp. 5⅜ x 8½. 63069-2 Pa. $9.95

DYNAMICS OF FLUIDS IN POROUS MEDIA, Jacob Bear. For advanced students of ground water hydrology, soil mechanics and physics, drainage and irrigation engineering and more. 335 illustrations. Exercises, with answers. 784pp. 6⅛ x 9¼. 65675-6 Pa. $19.95

NUMERICAL METHODS FOR SCIENTISTS AND ENGINEERS, Richard Hamming. Classic text stresses frequency approach in coverage of algorithms, polynomial approximation, Fourier approximation, exponential approximation, other topics. Revised and enlarged 2nd edition. 721pp. 5⅜ x 8½.   65241-6 Pa. $15.95

THEORETICAL SOLID STATE PHYSICS, Vol. 1: Perfect Lattices in Equilibrium; Vol. II: Non-Equilibrium and Disorder, William Jones and Norman H. March. Monumental reference work covers fundamental theory of equilibrium properties of perfect crystalline solids, non-equilibrium properties, defects and disordered systems. Appendices. Problems. Preface. Diagrams. Index. Bibliography. Total of 1,301pp. 5⅜ x 8½. Two volumes.   Vol. I: 65015-4 Pa. $16.95
Vol. II: 65016-2 Pa. $16.95

OPTIMIZATION THEORY WITH APPLICATIONS, Donald A. Pierre. Broad spectrum approach to important topic. Classical theory of minima and maxima, calculus of variations, simplex technique and linear programming, more. Many problems, examples. 640pp. 5⅜ x 8½.   65205-X Pa. $16.95

THE CONTINUUM: A Critical Examination of the Foundation of Analysis, Hermann Weyl. Classic of 20th-century foundational research deals with the conceptual problem posed by the continuum. 156pp. 5⅜ x 8½.   67982-9 Pa. $6.95

ESSAYS ON THE THEORY OF NUMBERS, Richard Dedekind. Two classic essays by great German mathematician: on the theory of irrational numbers; and on transfinite numbers and properties of natural numbers. 115pp. 5⅜ x 8½.
21010-3 Pa. $5.95

THE FUNCTIONS OF MATHEMATICAL PHYSICS, Harry Hochstadt. Comprehensive treatment of orthogonal polynomials, hypergeometric functions, Hill's equation, much more. Bibliography. Index. 322pp. 5⅜ x 8½. 65214-9 Pa. $9.95

NUMBER THEORY AND ITS HISTORY, Oystein Ore. Unusually clear, accessible introduction covers counting, properties of numbers, prime numbers, much more. Bibliography. 380pp. 5⅜ x 8½.   65620-9 Pa. $10.95

THE VARIATIONAL PRINCIPLES OF MECHANICS, Cornelius Lanczos. Graduate level coverage of calculus of variations, equations of motion, relativistic mechanics, more. First inexpensive paperbound edition of classic treatise. Index. Bibliography. 418pp. 5⅜ x 8½.   65067-7 Pa. $12.95

MATHEMATICAL TABLES AND FORMULAS, Robert D. Carmichael and Edwin R. Smith. Logarithms, sines, tangents, trig functions, powers, roots, reciprocals, exponential and hyperbolic functions, formulas and theorems. 269pp. 5⅜ x 8½.
60111-0 Pa. $6.95

THEORETICAL PHYSICS, Georg Joos, with Ira M. Freeman. Classic overview covers essential math, mechanics, electromagnetic theory, thermodynamics, quantum mechanics, nuclear physics, other topics. First paperback edition. xxiii + 885pp. 5⅜ x 8½.   65227-0 Pa. $21.95

HANDBOOK OF MATHEMATICAL FUNCTIONS WITH FORMULAS, GRAPHS, AND MATHEMATICAL TABLES, edited by Milton Abramowitz and Irene A. Stegun. Vast compendium: 29 sets of tables, some to as high as 20 places. 1,046pp. 8 x 10½. 61272-4 Pa. $26.95

MATHEMATICAL METHODS IN PHYSICS AND ENGINEERING, John W. Dettman. Algebraically based approach to vectors, mapping, diffraction, other topics in applied math. Also generalized functions, analytic function theory, more. Exercises. 448pp. 5⅜ x 8¼. 65649-7 Pa. $10.95

A SURVEY OF NUMERICAL MATHEMATICS, David M. Young and Robert Todd Gregory. Broad self-contained coverage of computer-oriented numerical algorithms for solving various types of mathematical problems in linear algebra, ordinary and partial, differential equations, much more. Exercises. Total of 1,248pp. 5⅜ x 8½.
Two volumes. Vol. I: 65691-8 Pa. $16.95
Vol. II: 65692-6 Pa. $16.95

TENSOR ANALYSIS FOR PHYSICISTS, J.A. Schouten. Concise exposition of the mathematical basis of tensor analysis, integrated with well-chosen physical examples of the theory. Exercises. Index. Bibliography. 289pp. 5⅜ x 8½. 65582-2 Pa. $8.95

INTRODUCTION TO NUMERICAL ANALYSIS (2nd Edition), F.B. Hildebrand. Classic, fundamental treatment covers computation, approximation, interpolation, numerical differentiation and integration, other topics. 150 new problems. 669pp. 5⅜ x 8½. 65363-3 Pa. $16.95

INVESTIGATIONS ON THE THEORY OF THE BROWNIAN MOVEMENT, Albert Einstein. Five papers (1905–8) investigating dynamics of Brownian motion and evolving elementary theory. Notes by R. Fürth. 122pp. 5⅜ x 8½.
60304-0 Pa. $5.95

CATASTROPHE THEORY FOR SCIENTISTS AND ENGINEERS, Robert Gilmore. Advanced-level treatment describes mathematics of theory grounded in the work of Poincaré, R. Thom, other mathematicians. Also important applications to problems in mathematics, physics, chemistry and engineering. 1981 edition. References. 28 tables. 397 black-and-white illustrations. xvii + 666pp. 6⅛ x 9¼.
67539-4 Pa. $17.95

AN INTRODUCTION TO STATISTICAL THERMODYNAMICS, Terrell L. Hill. Excellent basic text offers wide-ranging coverage of quantum statistical mechanics, systems of interacting molecules, quantum statistics, more. 523pp. 5⅜ x 8½.
65242-4 Pa. $12.95

STATISTICAL PHYSICS, Gregory H. Wannier. Classic text combines thermodynamics, statistical mechanics and kinetic theory in one unified presentation of thermal physics. Problems with solutions. Bibliography. 532pp. 5⅜ x 8½.
65401-X Pa. $12.95

CATALOG OF DOVER BOOKS

ORDINARY DIFFERENTIAL EQUATIONS, Morris Tenenbaum and Harry Pollard. Exhaustive survey of ordinary differential equations for undergraduates in mathematics, engineering, science. Thorough analysis of theorems. Diagrams. Bibliography. Index. 818pp. 5⅜ x 8½. 64940-7 Pa. $18.95

STATISTICAL MECHANICS: Principles and Applications, Terrell L. Hill. Standard text covers fundamentals of statistical mechanics, applications to fluctuation theory, imperfect gases, distribution functions, more. 448pp. 5⅜ x 8½. 65390-0 Pa. $11.95

ORDINARY DIFFERENTIAL EQUATIONS AND STABILITY THEORY: An Introduction, David A. Sánchez. Brief, modern treatment. Linear equation, stability theory for autonomous and nonautonomous systems, etc. 164pp. 5⅜ x 8¼. 63828-6 Pa. $6.95

THIRTY YEARS THAT SHOOK PHYSICS: The Story of Quantum Theory, George Gamow. Lucid, accessible introduction to influential theory of energy and matter. Careful explanations of Dirac's anti-particles, Bohr's model of the atom, much more. 12 plates. Numerous drawings. 240pp. 5⅜ x 8½. 24895-X Pa. $7.95

THEORY OF MATRICES, Sam Perlis. Outstanding text covering rank, nonsingularity and inverses in connection with the development of canonical matrices under the relation of equivalence, and without the intervention of determinants. Includes exercises. 237pp. 5⅜ x 8½. 66810-X Pa. $8.95

GREAT EXPERIMENTS IN PHYSICS: Firsthand Accounts from Galileo to Einstein, edited by Morris H. Shamos. 25 crucial discoveries: Newton's laws of motion, Chadwick's study of the neutron, Hertz on electromagnetic waves, more. Original accounts clearly annotated. 370pp. 5⅜ x 8½. 25346-5 Pa. $10.95

INTRODUCTION TO PARTIAL DIFFERENTIAL EQUATIONS WITH APPLICATIONS, E.C. Zachmanoglou and Dale W. Thoe. Essentials of partial differential equations applied to common problems in engineering and the physical sciences. Problems and answers. 416pp. 5⅜ x 8½. 65251-3 Pa. $11.95

BURNHAM'S CELESTIAL HANDBOOK, Robert Burnham, Jr. Thorough guide to the stars beyond our solar system. Exhaustive treatment. Alphabetical by constellation: Andromeda to Cetus in Vol. 1; Chamaeleon to Orion in Vol. 2; and Pavo to Vulpecula in Vol. 3. Hundreds of illustrations. Index in Vol. 3. 2,000pp. 6⅛ x 9¼. 23567-X, 23568-8, 23673-0 Pa., Three-vol. set $44.85

CHEMICAL MAGIC, Leonard A. Ford. Second Edition, Revised by E. Winston Grundmeier. Over 100 unusual stunts demonstrating cold fire, dust explosions, much more. Text explains scientific principles and stresses safety precautions. 128pp. 5⅜ x 8½. 67628-5 Pa. $5.95

AMATEUR ASTRONOMER'S HANDBOOK, J.B. Sidgwick. Timeless, comprehensive coverage of telescopes, mirrors, lenses, mountings, telescope drives, micrometers, spectroscopes, more. 189 illustrations. 576pp. 5⅜ x 8¼. (Available in U.S. only) 24034-7 Pa. $11.95

SPECIAL FUNCTIONS, N.N. Lebedev. Translated by Richard Silverman. Famous Russian work treating more important special functions, with applications to specific problems of physics and engineering. 38 figures. 308pp. 5⅜ x 8½. 60624-4 Pa. $9.95

OBSERVATIONAL ASTRONOMY FOR AMATEURS, J.B. Sidgwick. Mine of useful data for observation of sun, moon, planets, asteroids, aurorae, meteors, comets, variables, binaries, etc. 39 illustrations. 384pp. 5⅜ x 8¼. (Available in U.S. only)
24033-9 Pa. $8.95

INTEGRAL EQUATIONS, F.G. Tricomi. Authoritative, well-written treatment of extremely useful mathematical tool with wide applications. Volterra Equations, Fredholm Equations, much more. Advanced undergraduate to graduate level. Exercises. Bibliography. 238pp. 5⅜ x 8½. 64828-1 Pa. $8.95

POPULAR LECTURES ON MATHEMATICAL LOGIC, Hao Wang. Noted logician's lucid treatment of historical developments, set theory, model theory, recursion theory and constructivism, proof theory, more. 3 appendixes. Bibliography. 1981 edition. ix + 283pp. 5⅜ x 8½. 67632-3 Pa. $8.95

MODERN NONLINEAR EQUATIONS, Thomas L. Saaty. Emphasizes practical solution of problems; covers seven types of equations. ". . . a welcome contribution to the existing literature...."–*Math Reviews*. 490pp. 5⅜ x 8½. 64232-1 Pa. $13.95

FUNDAMENTALS OF ASTRODYNAMICS, Roger Bate et al. Modern approach developed by U.S. Air Force Academy. Designed as a first course. Problems, exercises. Numerous illustrations. 455pp. 5⅜ x 8½. 60061-0 Pa. $10.95

INTRODUCTION TO LINEAR ALGEBRA AND DIFFERENTIAL EQUATIONS, John W. Dettman. Excellent text covers complex numbers, determinants, orthonormal bases, Laplace transforms, much more. Exercises with solutions. Undergraduate level. 416pp. 5⅜ x 8½. 65191-6 Pa. $11.95

INCOMPRESSIBLE AERODYNAMICS, edited by Bryan Thwaites. Covers theoretical and experimental treatment of the uniform flow of air and viscous fluids past two-dimensional aerofoils and three-dimensional wings; many other topics. 654pp. 5⅜ x 8½. 65465-6 Pa. $16.95

INTRODUCTION TO DIFFERENCE EQUATIONS, Samuel Goldberg. Exceptionally clear exposition of important discipline with applications to sociology, psychology, economics. Many illustrative examples; over 250 problems. 260pp. 5⅜ x 8½. 65084-7 Pa. $8.95

LAMINAR BOUNDARY LAYERS, edited by L. Rosenhead. Engineering classic covers steady boundary layers in two- and three- dimensional flow, unsteady boundary layers, stability, observational techniques, much more. 708pp. 5⅜ x 8½. 65646-2 Pa. $18 95

LECTURES ON CLASSICAL DIFFERENTIAL GEOMETRY, Second Edition, Dirk J. Struik. Excellent brief introduction covers curves, theory of surfaces, fundamental equations, geometry on a surface, conformal mapping, other topics. Problems. 240pp. 5⅜ x 8½. 65609-8 Pa. $8.95

ROTARY-WING AERODYNAMICS, W.Z. Stepniewski. Clear, concise text covers aerodynamic phenomena of the rotor and offers guidelines for helicopter performance evaluation. Originally prepared for NASA. 537 figures. 640pp. 6⅛ x 9¼.
64647-5 Pa. $16.95

DIFFERENTIAL GEOMETRY, Heinrich W. Guggenheimer. Local differential geometry as an application of advanced calculus and linear algebra. Curvature, transformation groups, surfaces, more. Exercises. 62 figures. 378pp. 5⅜ x 8½.
63433-7 Pa. $9.95

INTRODUCTION TO SPACE DYNAMICS, William Tyrrell Thomson. Comprehensive, classic introduction to space-flight engineering for advanced undergraduate and graduate students. Includes vector algebra, kinematics, transformation of coordinates. Bibliography. Index. 352pp. 5⅜ x 8½.
65113-4 Pa. $9.95

A SURVEY OF MINIMAL SURFACES, Robert Osserman. Up-to-date, in-depth discussion of the field for advanced students. Corrected and enlarged edition covers new developments. Includes numerous problems. 192pp. 5⅜ x 8½.
64998-9 Pa. $8.95

ANALYTICAL MECHANICS OF GEARS, Earle Buckingham. Indispensable reference for modern gear manufacture covers conjugate gear-tooth action, gear-tooth profiles of various gears, many other topics. 263 figures. 102 tables. 546pp. 5⅜ x 8½.
65712-4 Pa. $14.95

SET THEORY AND LOGIC, Robert R. Stoll. Lucid introduction to unified theory of mathematical concepts. Set theory and logic seen as tools for conceptual understanding of real number system. 496pp. 5⅜ x 8¼.
63829-4 Pa. $12.95

A HISTORY OF MECHANICS, René Dugas. Monumental study of mechanical principles from antiquity to quantum mechanics. Contributions of ancient Greeks, Galileo, Leonardo, Kepler, Lagrange, many others. 671pp. 5⅜ x 8½.
65632-2 Pa. $14.95

FAMOUS PROBLEMS OF GEOMETRY AND HOW TO SOLVE THEM, Benjamin Bold. Squaring the circle, trisecting the angle, duplicating the cube: learn their history, why they are impossible to solve, then solve them yourself. 128pp. 5⅜ x 8½.
24297-8 Pa. $4.95

MECHANICAL VIBRATIONS, J.P. Den Hartog. Classic textbook offers lucid explanations and illustrative models, applying theories of vibrations to a variety of practical industrial engineering problems. Numerous figures. 233 problems, solutions. Appendix. Index. Preface. 436pp. 5⅜ x 8½.
64785-4 Pa. $11.95

CURVATURE AND HOMOLOGY, Samuel I. Goldberg. Thorough treatment of specialized branch of differential geometry. Covers Riemannian manifolds, topology of differentiable manifolds, compact Lie groups, other topics. Exercises. 315pp. 5⅜ x 8½.
64314-X Pa. $9.95

HISTORY OF STRENGTH OF MATERIALS, Stephen P. Timoshenko. Excellent historical survey of the strength of materials with many references to the theories of elasticity and structure. 245 figures. 452pp. 5⅜ x 8½.
61187-6 Pa. $12.95

GEOMETRY OF COMPLEX NUMBERS, Hans Schwerdtfeger. Illuminating, widely praised book on analytic geometry of circles, the Moebius transformation, and two-dimensional non-Euclidean geometries. 200pp. 5⅝ x 8¼. 63830-8 Pa. $8.95

MECHANICS, J.P. Den Hartog. A classic introductory text or refresher. Hundreds of applications and design problems illuminate fundamentals of trusses, loaded beams and cables, etc. 334 answered problems. 462pp. 5⅜ x 8½. 60754-2 Pa. $11.95

TOPOLOGY, John G. Hocking and Gail S. Young. Superb one-year course in classical topology. Topological spaces and functions, point-set topology, much more. Examples and problems. Bibliography. Index. 384pp. 5⅝ x 8¼. 65676-4 Pa. $10.95

STRENGTH OF MATERIALS, J.P. Den Hartog. Full, clear treatment of basic material (tension, torsion, bending, etc.) plus advanced material on engineering methods, applications. 350 answered problems. 323pp. 5⅜ x 8½. 60755-0 Pa. $9.95

ELEMENTARY CONCEPTS OF TOPOLOGY, Paul Alexandroff. Elegant, intuitive approach to topology from set-theoretic topology to Betti groups; how concepts of topology are useful in math and physics. 25 figures. 57pp. 5⅜ x 8½. 60747-X Pa. $3.95

ADVANCED STRENGTH OF MATERIALS, J.P. Den Hartog. Superbly written advanced text covers torsion, rotating disks, membrane stresses in shells, much more. Many problems and answers. 388pp. 5⅜ x 8½. 65407-9 Pa. $10.95

COMPUTABILITY AND UNSOLVABILITY, Martin Davis. Classic graduate-level introduction to theory of computability, usually referred to as theory of recurrent functions. New preface and appendix. 288pp. 5⅜ x 8½. 61471-9 Pa. $8.95

GENERAL CHEMISTRY, Linus Pauling. Revised 3rd edition of classic first-year text by Nobel laureate. Atomic and molecular structure, quantum mechanics, statistical mechanics, thermodynamics correlated with descriptive chemistry. Problems. 992pp. 5⅜ x 8½. 65622-5 Pa. $19.95

AN INTRODUCTION TO MATRICES, SETS AND GROUPS FOR SCIENCE STUDENTS, G. Stephenson. Concise, readable text introduces sets, groups, and most importantly, matrices to undergraduate students of physics, chemistry, and engineering. Problems. 164pp. 5⅜ x 8½. 65077-4 Pa. $7.95

THE HISTORICAL BACKGROUND OF CHEMISTRY, Henry M. Leicester. Evolution of ideas, not individual biography. Concentrates on formulation of a coherent set of chemical laws. 260pp. 5⅜ x 8½. 61053-5 Pa. $8.95

THE PHILOSOPHY OF MATHEMATICS: An Introductory Essay, Stephan Körner. Surveys the views of Plato, Aristotle, Leibniz & Kant concerning propositions and theories of applied and pure mathematics. Introduction. Two appendices. Index. 198pp. 5⅜ x 8½. 25048-2 Pa. $8.95

THE DEVELOPMENT OF MODERN CHEMISTRY, Aaron J. Ihde. Authoritative history of chemistry from ancient Greek theory to 20th-century innovation. Covers major chemists and their discoveries. 209 illustrations. 14 tables. Bibliographies. Indices. Appendices. 851pp. 5⅜ x 8½. 64235-6 Pa. $18.95

DE RE METALLICA, Georgius Agricola. The famous Hoover translation of greatest treatise on technological chemistry, engineering, geology, mining of early modern times (1556). All 289 original woodcuts. 638pp. 6¾ x 11.   60006-8 Pa. $21.95

SOME THEORY OF SAMPLING, William Edwards Deming. Analysis of the problems, theory and design of sampling techniques for social scientists, industrial managers and others who find statistics increasingly important in their work. 61 tables. 90 figures. xvii + 602pp. 5⅜ x 8½.   64684-X Pa. $16.95

THE VARIOUS AND INGENIOUS MACHINES OF AGOSTINO RAMELLI: A Classic Sixteenth-Century Illustrated Treatise on Technology, Agostino Ramelli. One of the most widely known and copied works on machinery in the 16th century. 194 detailed plates of water pumps, grain mills, cranes, more. 608pp. 9 x 12.
28180-9 Pa. $24.95

LINEAR PROGRAMMING AND ECONOMIC ANALYSIS, Robert Dorfman, Paul A. Samuelson and Robert M. Solow. First comprehensive treatment of linear programming in standard economic analysis. Game theory, modern welfare economics, Leontief input-output, more. 525pp. 5⅜ x 8½.   65491-5 Pa. $14.95

ELEMENTARY DECISION THEORY, Herman Chernoff and Lincoln E. Moses. Clear introduction to statistics and statistical theory covers data processing, probability and random variables, testing hypotheses, much more. Exercises. 364pp. 5⅜ x 8½.   65218-1 Pa. $10.95

THE COMPLEAT STRATEGYST: Being a Primer on the Theory of Games of Strategy, J.D. Williams. Highly entertaining classic describes, with many illustrated examples, how to select best strategies in conflict situations. Prefaces. Appendices. 268pp. 5⅜ x 8½.   25101-2 Pa. $7.95

CONSTRUCTIONS AND COMBINATORIAL PROBLEMS IN DESIGN OF EXPERIMENTS, Damaraju Raghavarao. In-depth reference work examines orthogonal Latin squares, incomplete block designs, tactical configuration, partial geometry, much more. Abundant explanations, examples. 416pp. 5⅜ x 8¼.
65685-3 Pa. $10.95

THE ABSOLUTE DIFFERENTIAL CALCULUS (CALCULUS OF TENSORS), Tullio Levi-Civita. Great 20th-century mathematician's classic work on material necessary for mathematical grasp of theory of relativity. 452pp. 5⅜ x 8½.
63401-9 Pa. $11.95

VECTOR AND TENSOR ANALYSIS WITH APPLICATIONS, A.I. Borisenko and I.E. Tarapov. Concise introduction. Worked-out problems, solutions, exercises. 257pp. 5⅝ x 8¼.   63833-2 Pa. $8.95

THE FOUR-COLOR PROBLEM: Assaults and Conquest, Thomas L. Saaty and Paul G. Kainen. Engrossing, comprehensive account of the century-old combinatorial topological problem, its history and solution. Bibliographies. Index. 110 figures. 228pp. 5⅜ x 8½.   65092-8 Pa. $7.95

CATALYSIS IN CHEMISTRY AND ENZYMOLOGY, William P. Jencks. Exceptionally clear coverage of mechanisms for catalysis, forces in aqueous solution, carbonyl- and acyl-group reactions, practical kinetics, more. 864pp. 5⅜ x 8½.
65460-5 Pa. $19.95

PROBABILITY: An Introduction, Samuel Goldberg. Excellent basic text covers set theory, probability theory for finite sample spaces, binomial theorem, much more. 360 problems. Bibliographies. 322pp. 5⅜ x 8½.
65252-1 Pa. $10.95

LIGHTNING, Martin A. Uman. Revised, updated edition of classic work on the physics of lightning. Phenomena, terminology, measurement, photography, spectroscopy, thunder, more. Reviews recent research. Bibliography. Indices. 320pp. 5⅜ x 8¼.
64575-4 Pa. $8.95

PROBABILITY THEORY: A Concise Course, Y.A. Rozanov. Highly readable, self-contained introduction covers combination of events, dependent events, Bernoulli trials, etc. Translation by Richard Silverman. 148pp. 5⅜ x 8¼.
63544-9 Pa. $7.95

AN INTRODUCTION TO HAMILTONIAN OPTICS, H. A. Buchdahl. Detailed account of the Hamiltonian treatment of aberration theory in geometrical optics. Many classes of optical systems defined in terms of the symmetries they possess. Problems with detailed solutions. 1970 edition. xv + 360pp. 5⅜ x 8½.
67597-1 Pa. $10.95

STATISTICS MANUAL, Edwin L. Crow, et al. Comprehensive, practical collection of classical and modern methods prepared by U.S. Naval Ordnance Test Station. Stress on use. Basics of statistics assumed. 288pp. 5⅜ x 8½.
60599-X Pa. $7.95

DICTIONARY/OUTLINE OF BASIC STATISTICS, John E. Freund and Frank J. Williams. A clear concise dictionary of over 1,000 statistical terms and an outline of statistical formulas covering probability, nonparametric tests, much more. 208pp. 5⅜ x 8½.
66796-0 Pa. $7.95

STATISTICAL METHOD FROM THE VIEWPOINT OF QUALITY CONTROL, Walter A. Shewhart. Important text explains regulation of variables, uses of statistical control to achieve quality control in industry, agriculture, other areas. 192pp. 5⅜ x 8½.
65232-7 Pa. $7.95

METHODS OF THERMODYNAMICS, Howard Reiss. Outstanding text focuses on physical technique of thermodynamics, typical problem areas of understanding, and significance and use of thermodynamic potential. 1965 edition. 238pp. 5⅜ x 8½.
69445-3 Pa. $8.95

STATISTICAL ADJUSTMENT OF DATA, W. Edwards Deming. Introduction to basic concepts of statistics, curve fitting, least squares solution, conditions without parameter, conditions containing parameters. 26 exercises worked out. 271pp. 5⅜ x 8½.
64685-8 Pa. $9.95

TENSOR CALCULUS, J.L. Synge and A. Schild. Widely used introductory text covers spaces and tensors, basic operations in Riemannian space, non-Riemannian spaces, etc. 324pp. 5⅜ x 8¼.
63612-7 Pa. $9.95

CATALOG OF DOVER BOOKS

A CONCISE HISTORY OF MATHEMATICS, Dirk J. Struik. The best brief history of mathematics. Stresses origins and covers every major figure from ancient Near East to 19th century. 41 illustrations. 195pp. 5⅜ x 8½. 60255-9 Pa. $8.95

A SHORT ACCOUNT OF THE HISTORY OF MATHEMATICS, W.W. Rouse Ball. One of clearest, most authoritative surveys from the Egyptians and Phoenicians through 19th-century figures such as Grassman, Galois, Riemann. Fourth edition. 522pp. 5⅜ x 8½. 20630-0 Pa. $11.95

HISTORY OF MATHEMATICS, David E. Smith. Nontechnical survey from ancient Greece and Orient to late 19th century; evolution of arithmetic, geometry, trigonometry, calculating devices, algebra, the calculus. 362 illustrations. 1,355pp. 5⅜ x 8½. 20429-4, 20430-8 Pa., Two-vol. set $26.90

THE GEOMETRY OF RENÉ DESCARTES, René Descartes. The great work founded analytical geometry. Original French text, Descartes' own diagrams, together with definitive Smith-Latham translation. 244pp. 5⅜ x 8½. 60068-8 Pa. $8.95

THE ORIGINS OF THE INFINITESIMAL CALCULUS, Margaret E. Baron. Only fully detailed and documented account of crucial discipline: origins; development by Galileo, Kepler, Cavalieri; contributions of Newton, Leibniz, more. 304pp. 5⅜ x 8½. (Available in U.S. and Canada only) 65371-4 Pa. $9.95

THE HISTORY OF THE CALCULUS AND ITS CONCEPTUAL DEVELOPMENT, Carl B. Boyer. Origins in antiquity, medieval contributions, work of Newton, Leibniz, rigorous formulation. Treatment is verbal. 346pp. 5⅜ x 8½. 60509-4 Pa. $9.95

THE THIRTEEN BOOKS OF EUCLID'S ELEMENTS, translated with introduction and commentary by Sir Thomas L. Heath. Definitive edition. Textual and linguistic notes, mathematical analysis. 2,500 years of critical commentary. Not abridged. 1,414pp. 5⅜ x 8½. 60088-2, 60089-0, 60090-4 Pa., Three-vol. set $32.85

GAMES AND DECISIONS: Introduction and Critical Survey, R. Duncan Luce and Howard Raiffa. Superb nontechnical introduction to game theory, primarily applied to social sciences. Utility theory, zero-sum games, n-person games, decision-making, much more. Bibliography. 509pp. 5⅜ x 8½. 65943-7 Pa. $13.95

THE HISTORICAL ROOTS OF ELEMENTARY MATHEMATICS, Lucas N.H. Bunt, Phillip S. Jones, and Jack D. Bedient. Fundamental underpinnings of modern arithmetic, algebra, geometry and number systems derived from ancient civilizations. 320pp. 5⅜ x 8½. 25563-8 Pa. $8.95

CALCULUS REFRESHER FOR TECHNICAL PEOPLE, A. Albert Klaf. Covers important aspects of integral and differential calculus via 756 questions. 566 problems, most answered. 431pp. 5⅜ x 8½. 20370-0 Pa. $8.95

CATALOG OF DOVER BOOKS

CHALLENGING MATHEMATICAL PROBLEMS WITH ELEMENTARY
SOLUTIONS, A.M. Yaglom and I.M. Yaglom. Over 170 challenging problems on
probability theory, combinatorial analysis, points and lines, topology, convex poly-
gons, many other topics. Solutions. Total of 445pp. 5⅜ x 8½. Two-vol. set.

Vol. I: 65536-9 Pa. $7.95
Vol. II: 65537-7 Pa. $7.95

FIFTY CHALLENGING PROBLEMS IN PROBABILITY WITH SOLUTIONS,
Frederick Mosteller. Remarkable puzzlers, graded in difficulty, illustrate elementary
and advanced aspects of probability. Detailed solutions. 88pp. 5⅜ x 8½.
65355-2 Pa. $4.95

EXPERIMENTS IN TOPOLOGY, Stephen Barr. Classic, lively explanation of one
of the byways of mathematics. Klein bottles, Moebius strips, projective planes, map
coloring, problem of the Koenigsberg bridges, much more, described with clarity
and wit. 43 figures. 210pp. 5⅜ x 8½. 25933-1 Pa. $6.95

RELATIVITY IN ILLUSTRATIONS, Jacob T. Schwartz. Clear nontechnical treatment makes relativity more accessible than ever before. Over 60 drawings illustrate
concepts more clearly than text alone. Only high school geometry needed.
Bibliography. 128pp. 6⅛ x 9¼. 25965-X Pa. $7.95

AN INTRODUCTION TO ORDINARY DIFFERENTIAL EQUATIONS, Earl
A. Coddington. A thorough and systematic first course in elementary differential
equations for undergraduates in mathematics and science, with many exercises and
problems (with answers). Index. 304pp. 5⅜ x 8½. 65942-9 Pa. $8.95

FOURIER SERIES AND ORTHOGONAL FUNCTIONS, Harry F. Davis. An
incisive text combining theory and practical example to introduce Fourier series,
orthogonal functions and applications of the Fourier method to boundary-value
problems. 570 exercises. Answers and notes. 416pp. 5⅜ x 8½. 65973-9 Pa. $11.95

AN INTRODUCTION TO ALGEBRAIC STRUCTURES, Joseph Landin. Superb
self-contained text covers "abstract algebra": sets and numbers, theory of groups, theory of rings, much more. Numerous well-chosen examples, exercises. 247pp. 5⅜ x 8½.
65940-2 Pa. $8.95

STARS AND RELATIVITY, Ya. B. Zel'dovich and I. D. Novikov. Vol. 1 of
Relativistic Astrophysics by famed Russian scientists. General relativity, properties of
matter under astrophysical conditions, stars and stellar systems. Deep physical
insights, clear presentation. 1971 edition. References. 544pp. 5⅜ x 8½.
69424-0 Pa. $14.95

Prices subject to change without notice.

Available at your book dealer or write for free Mathematics and Science Catalog to Dept. GI,
Dover Publications, Inc., 31 East 2nd St., Mineola, N.Y. 11501. Dover publishes more than 250
books each year on science, elementary and advanced mathematics, biology, music, art, literature, history, social sciences and other areas.